ASTRONOMICAL CONSTANTS

Sun:
mass $\qquad M_\odot = 1.99 \times 10^{33}$ g
radius $\qquad R_\odot = 6.96 \times 10^{10}$ cm
surface gravity $\qquad g_\odot = 2.74 \times 10^4$ cm/sec^2
luminosity $\qquad \mathscr{L}_\odot = 3.9 \times 10^{33}$ erg/sec

Earth:
mass $\qquad M_\oplus = 5.98 \times 10^{27}$ g
equatorial radius $\qquad R_\oplus = 6.38 \times 10^8$ cm
polar radius $\qquad R'_\oplus = R_\oplus - 2.15 \times 10^6$ cm
surface gravity $\qquad g = 9.81 \times 10^2$ cm/sec^2
moment of inertia:
  about polar axis $\qquad I^{33} = 0.331\ M_\oplus R_\oplus{}^2$
  about equatorial axis $\qquad I^{22} = I^{11} = 0.329\ M_\oplus R_\oplus{}^2$
period of rotation $\qquad$ 1 sidereal day = 8.62×10^4 sec
mean distance to sun $\qquad$ 1 A.U. = 1.50×10^{13} cm
orbital period $\qquad$ 1 sidereal year = 3.16×10^7 sec
orbital velocity $\qquad$ 29.8 km/sec

Moon:
mass $\qquad M_\math022 = 7.35 \times 10^{25}$ g
radius $\qquad R_\math022 = 1.74 \times 10^8$ cm
mean distance from earth $\qquad 3.84 \times 10^{10}$ cm
orbital period $\qquad$ 1 sidereal month = 27.3 days

10-17-96

Gravitation and Spacetime
SECOND EDITION

Gravitation and Spacetime

SECOND EDITION

Hans Ohanian
Rensselaer Polytechnic Institute

Remo Ruffini
University of Rome

W. W. Norton & Company
New York • London

Library of Congress Cataloging-in-Publication Data

Ohanian, Hans.
   Gravitation and Spacetime / by Hans C. Ohanian. —2nd ed.
       p. cm.
    1. Gravitation.   2. Space and time.   I. Title.
QC178.035   1994
530.1'4—dc20   93-34408

ISBN 0-393-96501-5

W. W. Norton & Company, Inc. 500 Fifth Avenue, New York, N. Y. 10110

W. W. Norton & Company Ltd., 10 Coptic Street, London WC1A 1PU

       2   3   4   5   6   7   8   9   0

To John A. Wheeler,
who showed us the way

Contents

Preface

Einstein discovered his theory of gravitation in 1916. By rights, this theory should not have been discovered until twenty years later, when physicists acquired a clear understanding of relativistic field theory and of gauge invariance. Einstein's profound and premature insights into the nature of gravitation had more to do with intuition than with logic. In contrast to the admirably precise and clear operational foundations on which he based his theory of special relativity, the foundations on which he based general relativity were vague and obscure. As has been emphasized by Synge and by Fock, even the very name of the theory indicates a misconception: there is no such thing as a relativity more general than special relativity. But, whatever murky roads he may have taken, in the end Einstein's intuition led him to create a theory of dazzling beauty. If, using Arthur Koestler's image, we regard Copernicus, Kepler, and Newton as sleepwalkers, who knew where they wanted to go and managed to get there without quite knowing how, then Einstein was the greatest sleepwalker of them all.

It is the objective of this book to develop gravitational theory in the most logical and straightforward way--in the way it probably would have developed without Einstein's intervention. This means that we will begin with the linear approximation and regard gravitation as a field theory, entirely analogous to electrodynamics. The geometrical interpretation and the nonlinear Einstein equations gradually emerge as we attempt to understand and improve the equations of the linear approximation. This approach is not new: Gupta, Feynman, Thirring, and Weinberg have presented it from somewhat different points of view and with varying amounts of detail. One advantage of this approach is that it gives a clearer insight into how and why gravitation is geometry. Another advantage is that the linear theory permits us to delve immediately into the physics: light deflection, retardation, redshift, gravitational lensing, and gravitational radiation can be directly treated in the context of the linear approximation, without any lengthy preliminary digressions on the mathematics of Riemannian spacetime geometry.

In this second edition of the book, as in the first, we place considerable emphasis on the description of experimental results. The last thirty years have seen a blossoming of experimental and observational data, and we have tried to make the lists of such results as complete and up-to-date as possible.

All the chapters of the first edition have been drastically revised. On a small scale, this involved corrections of some mistakes and improvements of explanations. On a larger scale, it involved reorganization and additions. For instance, the linear approximation is now presented in two chapters (Chapter 3 on the theoretical basis, and Chapter 4 on applications), and the early universe has expanded to fill all of the new Chapter 10, which now includes details on the thermal equilibrium of particles in the early universe, on nucleosynthesis, and on the inflationary scenario. Other additions are a broad discussion of the hypothetical deviations from the inverse-square law (including the ill-fated speculations on the "fifth force") in Chapter 1, a thorough treatment of the theoretical and observational aspects of gravitational lenses in Chapter 4, discussion of the Hulse-Taylor pulsar and of laser interferometric gravitational wave detectors (LIGOs) in Chapter 5, details on the geodetic precession of a gyroscope and the Stanford Gravity Probe B experiment in Chapter 7, a semiquantitative treatment of particle creation by black holes (the Hawking process) in Chapter 8, and discussion of the latest determinations of the Hubble constant, dark matter, and the COBE results on the cosmic microwave radiation in Chapter 9.

The language of differential forms and the exterior calculus of Cartan is now widely used in general relativity. New sections added to Chapters 2 and 6 provide self-contained introductions to this language. However, the bulk of the book relies on ordinary tensor calculus, since this is easier to grasp for students in their first encounter with general relativity.

The exercises that are scattered throughout the chapters are an integral part of the text; they amplify discussions, supply proofs, and are intended to be done while the book is being read. Only a fanatic will find the time to do them all; the reader is invited to look upon these exercises as challenges which should not always be refused. The collection of problems at the ends of the chapters that appeared in the first edition has been much expanded, mostly by the addition of problems from examinations that were given to students at Rensselaer Polytechnic Institute.

We thank Charles J. Goebel (University of Wisconsin, Madison), Stuart L. Shapiro (Cornell University), and Lawrence C. Shepley (University of Texas at Austin) for their careful reviews of this second edition, and for their many suggestions for improvements. We also thank our students, upon whom a trial edition of this book was inflicted, for their patience and their comments. One of the authors (H.C.O.) would like to thank the Università di Roma "La Sapienza" and the Specola Vaticana, Castel Gandolfo, for their hospitality during work on this book.

H. C. O.
R. R.

April 1994

Notation

The components of a three-dimensional vector **A** with respect to the three-dimensional rectangular coodinates will be indicated by *superscripts* with the values 1, 2, 3:

$$A^1 = A_x , \quad A^2 = A_y , \quad A^3 = A_z$$

For the position vector **x**, the components are

$$x^1 = x , \quad x^2 = y , \quad x^3 = z$$

The symbol A^k, where the Latin superscript k takes on the values $k = 1, 2, 3$, then stands for the kth component of the vector. If no particular value of k is specified, the symbol A^k will also stand for the set (A^1, A^2, A^3) of all the components taken together; in the latter case, A^k represents the entire vector **A**.

The Einstein summation convention applies: when a repeated Latin index appears in a term in an equation, a summation is to be carried out over the values 1, 2, 3 of that index, for example,

$$A^n B^n \equiv \sum_{n=1}^{3} A^n B^n$$

The Kronecker delta will be written as

$$\delta_m^n = \begin{cases} 1 \text{ if } m = n \\ 0 \text{ if } m \neq n \end{cases}$$

Integration over a three-dimensional volume will be written as

$$\int f(\mathbf{x}) \, d^3x \equiv \iiint f(\mathbf{x}) \, dxdydz$$

The components of a four-dimensional vector will be indicated by superscripts with the values 0, 1, 2, 3. In flat four-dimensional space-time, with rectangular coordinates *ct, x, y, z,*

$$A^0 = A_t \, , \qquad A^1 = A_x \, , \qquad A^2 = A_y \, , \qquad A^3 = A_z$$

and

$$x^0 = ct \, , \qquad x^1 = x \, , \qquad x^2 = y \, , \qquad x^3 = z$$

The symbol A^μ, where the Greek superscript takes on the values $\mu = 0, 1, 2, 3$, stands for the μth component of the vector; it also stands for the set (A^0, A^1, A^2, A^3), and in the latter case represents the entire four-dimensional vector.

The definition of the four-dimensional Kronecker delta is the same as in the three-dimensional case,

$$\delta^\nu_\mu = \begin{cases} 1 \text{ if } \mu = \nu \\ 0 \text{ if } \mu \neq \nu \end{cases}$$

When a repeated Greek index appears in a term of an equation, a summation is to be carried out over the values 0, 1, 2, 3 of that index, for example,

$$\eta_{\mu\nu} A^\nu \equiv \sum_{\nu=0}^{3} \eta_{\mu\nu} A^\nu$$

The Minkowski metric tensor of flat spacetime is taken as

$$\eta_{\mu\nu} = \begin{pmatrix} \eta_{00} & \eta_{01} & \eta_{02} & \eta_{03} \\ \eta_{10} & \eta_{11} & \eta_{12} & \eta_{13} \\ \eta_{20} & \eta_{21} & \eta_{22} & \eta_{23} \\ \eta_{30} & \eta_{31} & \eta_{32} & \eta_{33} \end{pmatrix} = \begin{pmatrix} 1 & 0 & 0 & 0 \\ 0 & -1 & 0 & 0 \\ 0 & 0 & -1 & 0 \\ 0 & 0 & 0 & -1 \end{pmatrix}$$

The spacetime interval of flat spacetime is

$$ds^2 = \eta_{\mu\nu}\, dx^\mu\, dx^\nu \;=\; \sum_{\mu=0}^{3} \sum_{\nu=0}^{3} \eta_{\mu\nu}\, dx^\mu\, dx^\nu$$

$$= (cdt)^2 - (dx)^2 - (dy)^2 - (dz)^2$$

and the spacetime interval of curved spacetime is

$$ds^2 = g_{\mu\nu}\, dx^\mu\, dx^\nu \;=\; \sum_{\mu=0}^{3} \sum_{\nu=0}^{3} g_{\mu\nu}\, dx^\mu\, dx^\nu$$

If x^0 is the time coordinate, then $g_{00} > 0$ (timelike sign convention).

In general, indices are raised and lowered with the metric tensor of curved spacetime, for instance,

$$A_\mu = g_{\mu\nu} A^\nu$$

However, in all equations that are written in the linear approximation, the indices are raised and lowered with the Minkowski metric $\eta_{\mu\nu}$.

Partial derivatives are indicated by a comma or by the differential operator ∂:

$$\frac{\partial f}{\partial x^\mu} = f_{,\mu} = \partial_\mu f$$

A dot over a variable indicates a derivative with respect to time (for example, $\dot{z} = dz/dt$ in Chapter 5 and in Appendix 1), or a derivative with respect to proper time (for example, $\dot{r} = dr/d\tau$ in Chapter 8), or a derivative with respect to a "time parameter" (for example, $\dot{a} = da/d\eta$ in Chapter 10).

Gravitation and Spacetime
SECOND EDITION

1. Newton's Gravitational Theory

IT WAS OCCASIONED BY THE FALL OF AN APPLE, AS HE
SAT IN A CONTEMPLATIVE MOOD...
William Stukeley, *Memoirs of Sir Isaac Newton's Life*

Few theories can compare in the accuracy of their predictions with
Newton's theory of universal gravitation. The predictions of celestial
mechanics for the positions of the major planets agree with observation to within a few seconds of arc over time intervals of many years.
The discovery of Neptune and the rediscovery of Ceres are among the
spectacular successes that testify to the accuracy of the theory. But
Newton's theory is not perfect: the predicted motions of the perihelia
for the inner planets deviate somewhat from the observed values. In
the case of Mercury the excess perihelion precession amounts to 43
arcseconds per century. This small deviation was discovered through
calculations by LeVerrier in 1845, and it was recalculated by
Newcomb in 1882. The explanation of this perihelion precession was
one of the early successes of Einstein's relativistic theory of gravitation.

Optical observations of planetary angular positions stretching over
hundreds of years are needed to detect the excess perihelion precession. However, with the recent development of radar astronomy it has
become possible to measure the distances to the inner planets directly
and very accurately by means of the travel time of a radio signal sent
from the Earth to the planet and reflected back. With such radar
observations of distances, the small deviations from Newton's theory

can be detected after just a few years of observation (Reasenberg and Shapiro, 1976).

Although Newton's theory is not perfect, it is in excellent agreement with observation in the limiting case of motion at low velocity in a weak gravitational field. Any relativistic theory of gravitation ought to agree with Newton's theory in this limiting case. We therefore begin with a brief exposition of some aspects of Newton's theory.

1.1 THE LAW OF UNIVERSAL GRAVITATION

According to Newton, the law governing gravitational interactions is "that there is a power of gravity pertaining to all bodies, proportional to the several quantities of matter which they contain . . . The force of gravity towards the several equal parts of any body is inversely as the square of the distance of places from the particles" (Newton, 1686a). If one particle is at the origin and the other at a radial distance r, then the equation for the force takes the form

$$F = - \frac{Gmm'}{r^2} \qquad [1]$$

where the negative sign indicates that the force is attractive. The value of the gravitational constant in Eq. [1] is $G = 6.673 \times 10^{-8}$ dyne·cm²/g².

Strictly speaking, the mass that enters the force law [1] is the *gravitational mass,* which is the source of gravitation, in the same way that the electric charge is the source of electromagnetism. We will worry about the distinction between gravitational mass and inertial mass later on. In the following discussion of gravitational fields and potentials (Sections 1.1-1.4), the masses are always gravitational.

If we adopt a naive interpretation of the above law, gravitation is action-at-distance: a mass at one point acts directly and instantaneously on another mass even though the other mass is not in contact with it. Newton had serious misgivings about such a ghostly tug-of-war of distant masses and suggested that the interaction should be conveyed by some material medium. The modern view is that gravitation, like electromagnetism and all other fundamental interactions, acts locally through fields: a mass at one point produces a field, and this field acts on whatever masses it comes in contact with. The gravitational field may be regarded as the material medium sought by Newton; the field is material because it possesses an energy density. The description of interactions by means of local fields has the further advantage that it leads to a relativistic theory in which gravitational effects propagate at finite velocity. Instantaneous action-at-distance makes no sense as a relativistic theory because of the lack of an absolute time; what is instantaneous propagation in one reference frame need not be instantaneous in another. Of course, in the case of static or quasistatic mass distributions, retardation effects are insignificant, and there is no practical distinction between local interaction and action-at-distance.

In the Solar System, Newton's theory is an excellent approximation. The condition for the validity of Newton's theory can be conveniently stated in terms of the potential energy $V(r)$, which for the inverse-square force [1] is

$$V(r) = - \frac{Gmm'}{r} \qquad [2]$$

In general, we can say that relativistic effects will be small provided that the potential energy is much less than the rest-mass energy and that the speeds are much less than the speed of light. For a mass m orbiting with speed v around a central mass m' we can express these conditions as

$$|V| \ll mc^2 \quad \text{and} \quad v \ll c$$

where c is the speed of light. Note that the condition on the potential is equivalent to $r \gg Gm'/c^2$. Hence the deviations from Newton's theory are expected to be very small if the distance from the central mass is sufficiently large and the speed sufficiently low. For the Sun, with a mass $m' = M_\odot \simeq 2.0 \times 10^{33}$ g, we have $Gm'/c^2 \simeq 1.5$ km, and the condition $r \gg 1.5$ km is obviously very well satisfied, even for comets with a perihelion close to the surface of the Sun.

The gravitational constant G that appears in Eq. [1] is not known very precisely. Whereas the values of e and $\hbar$ are known to seven or eight significant figures, the value of G is known to only four significant figures. Measurements of G are difficult because of the extreme weakness of the gravitational force between masses of laboratory size. The gravitational force between masses of planetary size is not so weak, but this is of no help in determining G because only the combination GM (where M is the mass of the attracting body) appears in the equations of motion of bodies with purely gravitational interactions; hence, planetary observations cannot determine the separate values of G and M. Table 1.1 summarizes results of laboratory measurements of G. The value of G adopted for the most recent (1986) list of best values of the physical constants is $G = (6.6726 \pm 0.0008) \times 10^{-8}$ dyne·cm²/g².

Fig. 1.1 shows the torsion balance used by Cavendish in his pioneering measurements of G. The balance consists of a beam with two small masses (B, B) suspended from a thin fiber. The small masses are gravitationally attracted by the two large lead spheres (W, W), and this results in a measurable deflection of the beam of the balance through some angle around the vertical. From the known torsional constant of the fiber and the geometry of the apparatus, the gravitational constant can then be calculated.

The measurement by Luther and Towler (1982), listed last in Table 1.1, has given the most precise value for *G*. The method of periods, or the "dynamical" method, used in this and several other recent measurements was first devised by Eötvös. It uses an arrangement similar to that of a Cavendish balance, but instead of a measurement of the deflection produced by the presence of the large masses, the experiment involves a measurement of the period of oscillation, first when the large masses are placed near the equilibrium position of the small

TABLE 1.1 LABORATORY MEASUREMENTS OF THE GRAVITATIONAL CONSTANT

Experimenter(s)*	Year	Method	G (10^{-8} dyne·cm^2/g^2)
Cavendish	1798	torsion-balance deflection	6.75(±5)**
Reich	1838	torsion-balance deflection	6.64(±6)
Baily	1843	torsion-balance deflection	6.63(±7)
Cornu and Baille	1872	torsion-balance deflection	6.64(±2)
Jolly	1873	beam balance	6.4(±1)
Richarz and Krigar-Menzel	1888	beam balance	6.68(±1)
Wilsing	1889	metronome balance	6.6(±1)
Poynting	1891	beam balance	6.70(±4)
Boys	1895	torsion-balance deflection	6.658(±7)
Eotvos	1896	torsion-balance period	6.66(±1)
Braun	1897	torsion-balance deflection	6.649(±2)
Burgess	1902	torsion-balance deflection	6.64(±?)
Cremieu	1909	torsion-balance deflection	6.67(±?)
Heyl	1930	torsion-balance period	6.673(±3)
Zahradnicek	1933	torsion-balance resonance	6.66(±4)
Heyl and Chrzanowski	1942	torsion-balance period	6.673(±3)
Rose et al.	1969	rotating torsion balance	6.670(±1)
Facy and Pontikis	1970	torsion-balance resonance	6.6714(±6)
Renner	1973	torsion-balance period	6.670(±8)
Karagioz et al.	1976	torsion-balance period	6.668(±2)
Sagitov	1977	torsion-balance period	6.6745(±8)
Luther and Towler	1982	torsion-balance period	6.6726(±5)

* References for experiments before 1909 are given by Poynting (1911) and by de Boer (1984).

** The number in parentheses is the experimental uncertainty (rms error) in the last decimal listed.

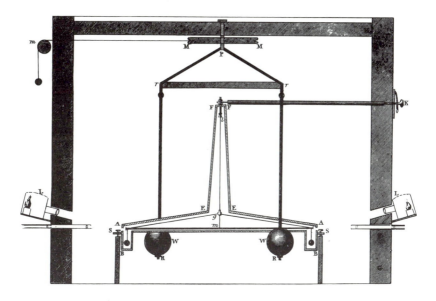

Fig. 1.1 The apparatus used by Cavendish. The large lead spheres (W, W) attract the small spheres (B, B) which are attached to the beam of the torsion balance. (From Cavendish, 1798.)

Fig. 1.2 The apparatus of Luther and Towler. The small masses are two tungsten disks mounted on a small rod (a dumbbell) suspended by a quartz fiber. The large masses are two tungsten spheres, of about 10 kg each. In the diagram, the large masses are aligned with the equilibrium positions of the small masses. If the beam with the small masses is rotated away from this alignment, the gravitational force of the large masses contributes a restoring force.

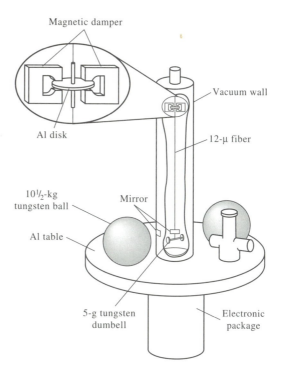

masses and again when the large masses are removed from the apparatus. In the former configuration (see Fig. 1.2), the period is shortened since the large masses produce an extra restoring force, which adds to the restoring force of the suspension fiber. The change in the period determines *G*. The advantage of the method is that the period can be measured more precisely than the deflection.

1.2 TESTS OF THE INVERSE-SQUARE LAW

Is it possible that there are deviations from the inverse-square law at large distances or at small distances? By "large distances" we mean distances of up to 10^4 or 10^5 light-years; such distances are large compared with the dimensions of the Solar System, but small compared with the typical dimensions of the universe.* There are some general properties of relativistic field theory that place tight restrictions on a possible alternative to the inverse-square law. It is easiest to express these restrictions in terms of the potential. The inverse-square law has the special potential given by Eq. [2]. The general potential consistent with field theory turns out to be

$$V(r) = - Gmm' \; \frac{e^{-r/\lambda}}{r} \tag{3}$$

where λ is a constant. This is called a *Yukawa potential;* obviously, the $1/r$ potential [2] is a special Yukawa potential with $\lambda = \infty$. The constant λ is called the *range* of the potential--if the distance r appreciably exceeds λ, the potential, and the force it produces, becomes negligible. Besides [3], the only other possibility is some combination of several Yukawa potentials, which would mean that we are dealing with several gravitational fields. In this case, the large-distance behavior of the net potential is dominated by the Yukawa potential of the longest range, since this potential will dominate over the others.

   If the gravitational potential is as given by [3], then what can we say about λ? We know that the range of the gravitational force is long. We know that our Galaxy is held together by gravitation, and this implies that the gravitational potential does not deviate much from $1/r$ out to distances of $r \simeq$ [galactic radius] $\simeq 10^{22}$ cm. Hence we can conclude that

$$\lambda > 10^{22} \text{ cm} \tag{4}$$

---

* At very large distances (more than 10^7 light-years), there may be cosmological deviations from the $1/r^2$ force (see Section 7.3). These deviations are not our concern in the present context.

Incidentally: The value of λ is related to the mass of the graviton, a (hypothetical) particle of spin 2, which is to gravitation what the photon is to electromagnetism. According to relativistic quantum theory, the mass of the graviton is inversely proportional to the range of the Yukawa potential:

$$m_\Gamma = \hbar/\lambda c \qquad [5]$$

If the gravitational force is inverse-square, the range of the force is infinite, and the mass of the graviton is zero. If we rely on the observational limit given by the inequality [4], we obtain

$$m_\Gamma < 10^{-59} \text{ g}$$

It is interesting to note that by setting analogous observational limits on possible deviations from Maxwell's equations, we find that the limit on the mass of the photon is of the same order of magnitude,

$$m_\gamma < 10^{-59} \text{ g}$$

This limit is obtained from examination of the galactic magnetic fields. The limits on the masses of gravitons and photons are the same because both are obtained from examination of phenomena extending over the same range of distances.

Since λ is certainly very large, and since a value $\lambda = \infty$ is consistent with all our observational data, we will hereafter assume that $\lambda = \infty$. This means there are no deviations from the inverse-square law at large distances.

There remains the question of possible deviations from the inverse-square law at *short* distances. Such deviations have been suggested in connection with a hypothetical "fifth force" coupled to baryon number that some experimenters claimed to have detected (see Section 1.7); however, the question of short-distance deviations from the inverse-square law is worth exploring, whether the speculations about the fifth force have merit or not. To explore such deviations, we have to contemplate a superposition of a $1/r$ potential (which accounts for the long-distance behavior) and a Yukawa potential with a finite value of λ (which represents the short-distance deviation):

$$V(r) = -\frac{Gmm'}{r} - \alpha \frac{Gmm'}{r} e^{-r/\lambda} \qquad [6]$$

Here α is a constant that characterizes the strength of the Yukawa potential relative to the $1/r$ potential. Eq. [6] gives an inverse-square force at large distances, but a complicated behavior for distances smaller than $r \simeq \lambda$. Note that for $r \ll \lambda$, the potential reduces to

$$V(r) \simeq - \frac{Gmm'}{r} (1 + \alpha) \qquad\qquad [7]$$

and the force is, again, an inverse-square force, but with a new, modified value $(1 + \alpha)G$ of the gravitational constant. If the range λ of the Yukawa potential in Eq. [6] is of the order of a few hundred meters, then the gravitational constant measured in laboratory experiments is $(1 + \alpha)G$, whereas the gravitational constant for interplanetary forces is G.

Limits on α and on λ can be extracted from a variety of orbital, laboratory, and geophysical observations and experiments.

Orbital Observations High-precision measurements of the distances to Mercury, Venus, Mars, and Jupiter have been obtained by radar ranging, either with radar signals directly reflected by the surface of the planet, or with signals returned by a transponder on a spacecraft during a flyby of the planet. In combination with data on the orbital period, obtained by traditional astronomical observations, the distance data permit a rigorous test of Kepler's third law, and therefore a test of the inverse-square law. A recent analysis of all the available data (Talmadge et al., 1988) sets a tight limit on the strength of the extra Yukawa potential:

$$|\alpha| < 10^{-9} \text{ for } \lambda \simeq 10^{13} \text{ cm}$$

Somewhat less tight limits apply for larger or smaller choices of λ.

An analogous test can be performed for the orbits of the Moon or of artificial satellites around the Earth. The distance to the Moon has been measured with high precision by laser ranging, by means of laser pulses reflected from the reflectors placed on the Moon during the *Apollo 11* mission. Such precise measurements have also been performed on the LAGEOS artificial satellite. In combination with a determination of the orbital period, the data show no detectable deviations from the inverse-square law at distances larger than 10^9 cm (Smith et al., 1985). To set a quantitative limit on α and λ, we can compare the centripetal acceleration of the Moon with the acceleration of gravity at the surface of the Earth, and check whether they are in the ratio of the inverse squares of the distances (this is a repetition of the calculation that led Newton to the discovery of the inverse-square law). The calculation is far from trivial--the acceleration of gravity at the surface of the Earth varies from place to place, and we must extrapolate to an average, smooth ellipsoidal surface (geoid) for an ideal Earth, without irregularities on its surface. With this extrapolation, we find that the acceleration of the Moon agrees with the expected value to within the errors, about 1 part in 10^6 (Rapp, 1987).

Accordingly, the strength of the Yukawa potential is restricted to

$$|\alpha| < 10^{-6} \text{ for } \lambda \simeq 10^9 \text{ cm}$$

Laboratory Measurements A simple way to test the inverse-square law is to compare the results of determinations of G by different experimenters. Most of these determinations were made with torsion balances. If the force between the masses deviates from the inverse-square law, then the result of a determination of G will depend on the size of the torsion balance. Cavendish used a rather large balance, with a beam of about 2 m; modern versions of the experiment used beams as small as 2 cm. The agreement between such determinations of G suggests that there are no substantial deviations from inverse-square. However, in view of the rather large experimental uncertainties in the determinations of G, the comparison does not yield any stringent limits (see de Boer, 1984).

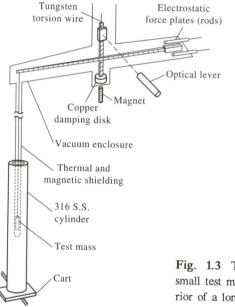

Tungsten torsion wire

Electrostatic force plates (rods)

Optical lever

Copper damping disk

Magnet

Vacuum enclosure

Thermal and magnetic shielding

316 S.S. cylinder

Test mass

Cart

Fig. 1.3 Torsion balance with a small test mass suspended in the interior of a long cylindrical shell.

Better limits on deviations from the inverse-square law have been obtained by experiments specifically designed for this purpose. An elegant experiment by Spero et al. (1980) used a torsion balance to explore the force field inside a long cylindrical shell (see Fig. 1.3). If, and only if, the inverse-square law is valid, the force that such a cylindrical shell exerts on a small test mass in its interior is exactly zero. In the experiment, the cylinder was moved back and forth, to see whether the small test mass experiences any force when near the wall of the cylinder. The absence of any detectable force set a limit of $|\alpha| < 0.0002$ for $\lambda = 2$ cm. The limit obtained for α actually depends on the assumed value of λ, since this value determines the sensitivity of the experiment. Throughout this section we will always report the value of λ that makes the experiment most sensitive and yields the most stringent limit on α.

Other torsion-balance experiments compared the force exerted by a small mass placed near a torsion balance with the force exerted by a larger mass placed farther from the torsion balance. The best of these experiments set limits of $-\alpha < 0.00006$ for $\lambda = 4$ cm and $|\alpha| < 0.001$ for $\lambda = 10$ cm (Chen et al., 1984; Hoskins et al., 1985).

Less stringent limits on α, but for larger hypothetical values of λ, were obtained by experiments performed with hydroelectric pumped-storage reservoirs. The water level of such reservoirs often rises or falls by tens of meters in just a few hours, and the change of gravity that this produces in the region above the water depends on α and λ. The change of gravity can be measured with a beam balance that has one of its pans above the water level, and the other pan below water level, in a long tube (Stacey et al., 1987). Alternatively, the change of gravity can be measured with a high-precision gravimeter, that is, a delicate spring balance (Müller et al., 1990).

Geophysical Measurements   Geophysical investigations of the inverse-square law are based on a method for the determination of G first proposed by Airy in 1856. This method hinges on the variation of the acceleration of gravity with depth below the surface of the Earth (or height above the surface). If we descend into a deep mine shaft, we find that g varies with depth. For a uniform-density sphere, g would decrease linearly with depth. However, the Earth is not of uniform density, and g at first increases with depth, then decreases. For illustrative purposes, assume that the mass distribution of the Earth is spherical, with a density $\rho(r)$ and a mass $M(r)$ enclosed within the radius r. Then

$$g = G\frac{M(r)}{r^2} \tag{8}$$

and

$$\frac{dg}{dr} = -2G\frac{M(r)}{r^3} + \frac{G}{r^2}\frac{dM}{dr}$$

$$= -\frac{2g}{r} + \frac{G}{r^2}4\pi r^2 \rho(r) \tag{9}$$

According to this equation, the value of G can be calculated from the measured values of g and dg/dr, provided we know the density ρ. Eq. [9] is only a crude approximation; for an accurate determination of G

TABLE 1.2 TESTS FOR SHORT-RANGE DEVIATIONS FROM INVERSE-SQUARE

Experimenter(s)	Year	Method	Result
Smith et al.	1985	Moon and satellite orbits	not detected for $r > 1.2 \times 10^9$ cm
Rapp	1987	acceleration of Moon vs. g	$\|\alpha\lambda\| < 1.4 \times 10^3$ cm
Yu et al.	1979	gravity near oil tank	$\|\alpha\| < 0.2$ for $\lambda = 1$ to 5 m
Panov and Frontov	1979	torsion balance	$\|\alpha\| < 0.01$ for $\lambda = 1$ m
Spero et al.	1980	torsion balance	$\|\alpha\| < 0.0002$ for $\lambda = 2$ cm
Ogawa et al.	1982	quadrupole interaction	$\|\alpha\| < 0.01$ for $\lambda = 1$ m
Chan et al.	1982	gradiometer	$\|\alpha\| < 0.06$ for $\lambda = 1$ m
Chen et al.	1984	torsion balance	$-\alpha < 0.00006$ for $\lambda = 4$ cm
Hoskins et al.	1985	torsion balance	$\|\alpha\| < 0.001$ for $\lambda = 10$ cm
Stacey and Tuck	1981	gravity in Gulf of Mexico	$\alpha = -0.02$
Stacey et al.	1981-86	gravity in mine shafts	$\alpha = -0.008$ or -0.01 for $\lambda = 200$ to 1000 m
Stacey et al.	1986-89	gravity near pumped reservoir	$\|\alpha\| < 0.01$ for $\lambda = 1$ to 7 m
Ander et al.	1989	gravity in icecap	$\alpha \simeq -0.01$ for $\lambda = 1$ km
Eckhardt et al.	1988	gravity on tower	$\|\alpha\| < 0.001$ for $\lambda > 100$ m
Thomas et al.	1989	gravity on tower	$-0.005 < \alpha < 0.0005$ for $\lambda = 100$ m to 10 km
Thomas and Vogel	1990	gravity in borehole	$\|\alpha\| < 0.04$
Muller et al.	1990	gravity near pumped reservoir	$\|\alpha\| < 0.004$ for $\lambda = 1$ to 20 m
Zumberge et al.	1991	gravity in Pacific Ocean	$\|\alpha\| < 0.002$ for $\lambda = 1$ m to 5 km

via this method, we must also take into account the rotation of the Earth and its ellipsoidal shape.

The practical application of the Airy method involves the following: Find some region where the density ρ is known, and measure g as a function of depth in the ground; then calculate G from g and dg/dr, by means of Eq. [9] or, rather, by means of the accurate version of this equation. If the result of this determination of G agrees with the laboratory value $G = 6.673 \times 10^{-8}$ dyne·cm²/gm², then the result verifies the inverse-square law; if not, then it disproves the inverse-square law.

Attempting to apply the Airy method, experimenters have measured gravity as a function of depth in mine shafts (Stacey et al., 1987), in boreholes in the ground (Thomas and Vogel, 1990) and in the Greenland icecap (Ander et al., 1989), and underwater in the ocean (Stacey and Tuck, 1981; Zumberge et al., 1991). In a variant of the Airy method, experimenters have also measured gravity as a function of height on TV transmitter towers several hundred meters high (Eckhardt et al., 1988; Thomas et al., 1989).

Table 1.2 lists the results of tests for deviations from the inverse-square law obtained by orbital, laboratory, and geophysical observations and experiments. Almost all these tests agree that there is no deviation from the inverse-square law, to within experimental errors. The two exceptions are the results of Stacey and Tuck (1981) and of Stacey et al. (1987), who found definite, nonzero values of α; but these results are contradicted by the other results, and they are also inconsistent with each other. Thus, we conclude that there is no credible evidence for deviations from the inverse-square law for gravitation.

1.3 THE GRAVITATIONAL POTENTIAL

The Newtonian gravitational force obeys the principle of linear superposition: the gravitational force exerted by a system of particles is the vector sum of the individual forces of the particles. If N particles are located at positions $\mathbf{x}_1$, $\mathbf{x}_2$, $\mathbf{x}_3$,..., $\mathbf{x}_N$, then the force these exert on a particle of mass m located at $\mathbf{x}$ is given by*

$$\mathbf{F}(\mathbf{x}) = -Gm \sum_{i=1}^{N} \frac{m_i}{\left|\mathbf{x} - \mathbf{x}_i\right|^3} (\mathbf{x} - \mathbf{x}_i) \qquad [10]$$

---

*$\mathbf{x}$ is the position vector, $\mathbf{x} = x\hat{\mathbf{x}} + y\hat{\mathbf{y}} + z\hat{\mathbf{z}}$. Instead of x, y, z we will often use x^1, x^2, x^3. In this notation, the components F_x, F_y, F_z of an arbitrary vector will be written F^1, F^2, F^3. The reason for the use of numerical *superscripts*, rather than subscripts, will become clear later.

To this force there corresponds a potential energy

$$V(\mathbf{x}) = - Gm \sum_{i=1}^{N} \frac{m_i}{|\mathbf{x} - \mathbf{x}_i|} \qquad [11]$$

with

$$\mathbf{F}(\mathbf{x}) = - \nabla V(\mathbf{x}) \qquad [12]$$

In component notation, we can write Eq. [12] as

$$F^k = - \frac{\partial V}{\partial x^k} \quad k = 1, 2, 3 \qquad [13]$$

The *gravitational field*, which we regard as the carrier of the interaction, is defined as force per unit mass,

$$\mathbf{g}(\mathbf{x}) \equiv \frac{1}{m} \mathbf{F}(\mathbf{x}) \qquad [14]$$

The corresponding *gravitational potential* is defined as

$$\Phi(\mathbf{x}) \equiv \frac{1}{m} V(\mathbf{x}) = - \sum_i \frac{Gm_i}{|\mathbf{x} - \mathbf{x}_i|} \qquad [15]$$

This definition makes the potential negative, as expected for an attractive force. The gravitational potential is sometimes defined with a sign opposite to that in Eq. [15], but we prefer to choose our signs by analogy with electrostatics.

For a continuous mass distribution, the gravitational potential is

$$\Phi(\mathbf{x}) = - \int \frac{G\rho(\mathbf{x}')}{|\mathbf{x} - \mathbf{x}'|} d^3x' \qquad [16]$$

where $\rho(x')$ is the mass density. Eq. [16] implies that $\Phi(\mathbf{x})$ obeys the Poisson equation

$$\nabla^2 \Phi(\mathbf{x}) = + 4\pi G\rho(\mathbf{x}) \qquad [17]$$

■ *Exercise 1*. Derive the Poisson equation from Eq. [16]. ■

Problems in the theory of the Newtonian gravitational potential involve exactly the same mathematics as in the theory of the electrostatic potential. Because of this, we will not spell out all of the details
of the following derivations. One important difference between gravitation and electrostatics is that the mass density, as opposed to the
charge density, can never be negative. This implies that there can be
no shielding of gravitational fields analogous to the shielding of electric fields by conductors.

By ingenious geometrical arguments, Newton proved that a spherically symmetric mass distribution behaves in the same way as a point
particle located at its center, in that (*i*) it produces the same field in
its exterior and (*ii*) it responds in the same way to fields produced by
external sources. This result, known as *Newton's theorem*, can be
obtained very directly by appealing to the uniqueness theorem and the
mean-value theorem of potential theory.

■ *Exercise 2*. Verify that Φ = [constant]$/r$ is a possible solution of the equation
$\nabla^2\Phi = 0$ in the empty space surrounding a mass distribution. The uniqueness theorem for the potential says that if a potential function satisfies the equation [17]
throughout a given volume and has specified values on the boundary of that
volume, then this potential function is the only possible solution of Eq. [17]. Use this
uniqueness theorem to show that in the case of a spherical mass distribution, Φ =
[constant]$/r$ = $- GM/r$ is the only possible solution. ■

■ *Exercise 3*. The potential energy of a body placed in a gravitational potential
$\Phi(\mathbf{x})$ (produced by given external sources) is

$$V = \int \rho(\mathbf{x})\Phi(\mathbf{x}) \, d^3x$$

where $\rho(\mathbf{x})$ is the mass density of the body. The mean-value theorem for the potential says that for any spherical volume placed in a region exterior to the mass distribution that produces the potential, the average value of the potential over the spherical volume coincides with the value of the potential at the center. Use this mean-
value theorem to show that a spherically symmetric body, placed in a potential produced by external sources, has a potential energy $M\Phi(\mathbf{x}_0)$, where $\mathbf{x}_0$ is the coordinate of the center of the sphere, and $M = \int \rho \, d^3x$ is the total mass of the body. ■

■ *Exercise 4*. Use the uniqueness theorem for the potential function to show
that the potential produced by a uniform spherical mass shell is constant in the interior of the shell. ■

The gravitational self-energy of a continuous mass distribution $\rho(\mathbf{x})$ is

$$\frac{1}{2} \int \rho(\mathbf{x}) \Phi(\mathbf{x}) \, d^3x$$

where $\Phi(\mathbf{x})$ is the potential produced by the mass distribution itself (the factor of $\frac{1}{2}$ is needed to eliminate double counting of the potential energies of pairs of mass elements in the distribution). This gravitational self-energy can be expressed in the alternative forms

$$-\frac{1}{2} \int G \, \frac{\rho(\mathbf{x})\rho(\mathbf{x'})}{|\mathbf{x} - \mathbf{x'}|} \, d^3x \, d^3x' \qquad [18]$$

and

$$\int \frac{1}{8\pi G} \, (\nabla \Phi)^2 \, d^3x + \int \rho \Phi \, d^3x \qquad [19]$$

This equation has an interesting interpretation: By analogy with the corresponding formulas for electrostatics, the quantity $(\nabla \Phi)^2/8\pi G$ may be regarded as the energy density of the gravitational field and $\rho \Phi$ may be regarded as an interaction energy density of field and matter. In the case of electrostatics, all of the electric energy may be regarded as field energy; in the case of gravity, this is not possible because field energy is positive, and hence something negative must be added to it in order to obtain a negative total energy as given by Eq. [18].

■ *Exercise 5.* Show that the expressions [18] and [19] are equivalent. ■

1.4 GRAVITATIONAL MULTIPOLES; THE QUADRUPOLE MOMENT OF THE SUN

The gravitational potential has the simple form $-GM/r$ only in the space surrounding a mass distribution with spherical symmetry. An arbitrary mass distribution produces a potential

$$\Phi(\mathbf{x}) = - \int \frac{G\rho(\mathbf{x}')}{|\mathbf{x} - \mathbf{x}'|} \, d^3x' \qquad [20]$$

If the point $\mathbf{x}$ is outside of the region that contains the mass (see Fig. 1.4), then we can construct a multipole expansion for the potential by using the Taylor series expansion of $1/|\mathbf{x} - \mathbf{x}'|$ about $\mathbf{x}' = 0$:

$$\frac{1}{|\mathbf{x} - \mathbf{x}'|} = \frac{1}{[(x - x')^2 + (y - y')^2 + (z - z')^2]^{1/2}}$$

$$= \frac{1}{r} + \sum_k \frac{x^k x'^k}{r^3} + \frac{1}{2} \sum_{k,l} (3x'^k x'^l - r'^2\delta_k^l) \frac{x^k x^l}{r^5} + \dots \quad [21]$$

where $r = \sqrt{x^2 + y^2 + z^2}$.

Fig. 1.4 The shaded region contains some mass distribution. The point $\mathbf{x}$ is outside of this region.

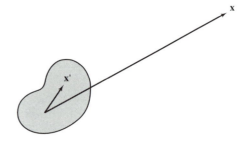

■ *Exercise 6*. Derive Eq. [21]. ■

The integral [20] can therefore be written as

$$\Phi(\mathbf{x}) = - \frac{GM}{r} - \frac{G}{r^3} \sum_k x^k D^k - \frac{G}{2} \sum_{k,l} Q^{kl} \frac{x^k x^l}{r^5} + \dots \qquad [22]$$

where

$$M = \int \rho(\mathbf{x}') \, d^3x' \qquad [23]$$

and

$$D^k = \int x'^k \rho(\mathbf{x}') \, d^3x' \qquad [24]$$

$$Q^{kl} = \int (3x'^k x'^l - r'^2 \delta_l^k) \, \rho(\mathbf{x'}) \, d^3x' \qquad [25]$$

The vector quantity D^k is the mass dipole moment. If the origin of coordinates is chosen to coincide with the center of mass, then $D^k = 0$ (no mass dipole); we will usually assume that this is so. The quantity Q^{kl} is the mass quadrupole tensor. Eq. [22] shows that whenever the quadrupole tensor is nonzero, the potential will contain a term $\propto 1/r^3$, and hence the force will deviate from the inverse-square law by a term $\propto 1/r^4$. Most extended mass distributions have a quadrupole tensor; the obvious exception is a mass distribution with spherical symmetry, for which the quadrupole tensor is always zero.

■ *Exercise 7*. Prove that the quadrupole tensor of a spherical mass distribution is zero. ■

The Earth's polar and equatorial diameters differ by about 3 parts in 10^3. This deviation from spherical shape produces a quadrupole term in the gravitational potential, which causes perturbations in the elliptical Kepler orbits of satellites. The main perturbation is a precession of the Kepler ellipse, that is, a slow rotation of the ellipse around the axis of the Earth (see Fig. 1.5). Such observed perturbations of the orbits of satellites have been used for precise determinations of the multipole moments and the mass distribution in the Earth.

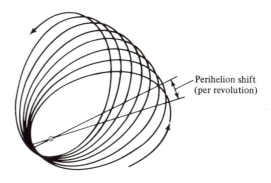

Perihelion shift (per revolution)

Fig. 1.5 Kepler orbit with precession.

The Sun is almost a perfect sphere; its polar and equatorial diameters differ by at most a few parts in 10^5. However, even such a small difference between the polar and equatorial radii, and the consequent small quadrupole term in the potential, could have important consequences when we attempt to compare the observed perihelion precession of the planets with the theoretical predictions. According to Newton's theory, the perihelion precession receives a contribution from the quadrupole term and also a contribution from the gravitational perturbations that the planets exert on each other. This second contribution is actually by far the largest; for instance, for Mercury, the contribution to the perihelion precession from the quadrupole term is at most a few seconds of arc per century, whereas the contribution from interplanetary perturbations is about 500 arcseconds per century. According to Einstein's theory of general relativity, there is an extra contribution of 43 arcseconds to the perihelion precession of Mercury; this extra precession involves relativistic modifications of the gravitational field and of the equation of motion. To test Einstein's theory, we want to isolate this extra contribution, by subtracting the other two contributions from the observed precession. The (large) contribution from interplanetary perturbations poses no difficulties in principle--it can be calculated quite precisely. When this contribution is subtracted from the observed precession, the remainder is 43 arcseconds, in excellent agreement with Einstein's theory. Thus, if there is also a substantial contribution from the quadrupole term, the result of the subtraction would be in disagreement with Einstein's theory.

Since the Sun rotates, we expect it to have a small equatorial bulge, or an oblateness. The observed rotation period at the surface is about 25 days, and under the assumption that the interior rotates at the same rate, we expect that the Sun's equatorial diameter exceeds the polar diameter by about 1 part in 10^5. Such an oblateness would give the Sun a small quadrupole moment. Let us define a dimensionless parameter J_2,

$$J_2 \equiv - \frac{Q^{33}}{2M_\odot R_\odot{}^2} \qquad [26]$$

where $M_\odot$ and $R_\odot$ are the mass and the radius of the Sun, and the z-axis is taken to coincide with the polar axis. This parameter J_2 is a convenient measure of the quadrupole moment. If the Sun were rotating uniformly, the value of J_2 would be about 1×10^{-7}, and the corresponding contribution to the perihelion precession of Mercury would be only a few hundredths of an arcsecond per century; this would be smaller than the uncertainty in the observed value of the perihelion precession, and therefore would not affect the agreement with Einstein's theory.

Unfortunately, direct measurements of the shape of the Sun seem to suggest larger values of J_2. In the scanning method pioneered by Dicke, the Sun's image is thrown on a rotating disk with a slit and a photocell behind it. The photocell measures the intensity of the light as a function of angular position along the rim of the solar image and thereby detects any deviations from roundness. Dicke and Goldenberg (1967) obtained an unexpectedly large value of J_2, which would require that the interior of the Sun rotates much faster than the surface. The large quadrupole moment would produce an extra perihelion precession of 3 arcseconds per century for Mercury. Einstein's theory of gravitation would then disagree with the observed precession by 3 arcseconds per century. Brans and Dicke (1961) proposed a "scalar-tensor" theory of gravitation which avoided any conflict with the observed precession. In this theory, the scalar field replaces the gravitational constant, and this "constant" therefore becomes a function of space and time.

But the results of Dicke and Goldenberg were contradicted by later measurements by Hill et al. (1974) and by Dicke et al. (1985). The inconsistencies among these measurements of the solar shape can possibly be attributed to nonuniformity of the temperature and the brightness of the solar surface. Any excess brightness in some region of the Sun distorts the apparent visual shape; for instance, an excess brightness in the equatorial region simulates an equatorial bulge. Such systematic errors arising from nonuniform brightness made all the visual determinations of the solar oblateness questionable.

A new method for the determination of J_2 became available with the discovery of oscillations of the solar surface. Measurements of the light output, the solar shape, and the Doppler shift of spectral lines all indicate that the body of the Sun vibrates in diverse modes, with periods of several minutes. The analysis of these modes of vibration (solar seismology) can be exploited for an indirect determination of J_2. If the Sun did not rotate, the modes characterized by spherical harmonics Y_l^m with the same given harmonic index l and different azimuthal indices m would be degenerate--they would have the same frequency of vibration. But in a rotating Sun, the frequencies of these modes are split by an amount depending on the rate of rotation. Measurements of these frequencies therefore permit the calculation of the rate of rotation of the solar interior, and consequently the calculation of J_2. Such calculations depend on the details of the model used for the interior of the Sun, and the results for J_2 range from 10^{-7} to 10^{-5} (see Table 1.3).

TABLE 1.3 THE QUADRUPOLE MOMENT OF THE SUN*

Observer	Year	Method	J_2
Dicke and Goldenberg	1967	scanning with photocell	$(2.4 \pm 0.2) \times 10^{-5}$
Hill et al.	1974	scanning with photocell	$(1 \pm 4) \times 10^{-6}$
Hill et al.	1982	solar vibrations	$(5 \pm 1) \times 10^{-6}$
Campbell et al.	1983	solar vibrations	$> 1.6 \times 10^{-6}$
Hill et al.	1984	solar vibrations	4.5×10^{-6}
Duvall et al.	1984	solar vibrations	$(1.7 \pm 0.4) \times 10^{-7}$
Dicke et al.	1985	scanning with photocell	$(7.5 \pm 0.9) \times 10^{-6}$
Hill et al.	1986	solar vibrations	$(8 \pm 2) \times 10^{-6}$
Brown et al.	1989	solar vibrations	1.5×10^{-7}

* For references, see Duvall et al. (1984), Hill et al. (1986), and Brown et al. (1989).

The most recent results favor the low value of J_2, leading to excellent agreement between the observed perihelion precession of Mercury and Einstein's theory.

Another proposed method for the determination of J_2 is based on radar astronomy. The quadrupole moment of the Sun produces not only a secular perturbation of the motion of a planet (a steady, cumulative precession of the orbit), but also periodic perturbations (back-and-forth oscillations of the orbit). Thus, in principle, the quadrupole moment can be determined quite directly by precise observations of the planetary position as a function of time. The high precision required could be achieved by radar tracking of a spacecraft equipped with a transponder parked in an orbit around Mercury. Alternatively, the spacecraft itself could play the role of planet; it would then be advantageous to place the spacecraft in a very tight and very eccentric orbit around the Sun.

The mass distributions we have considered so far are constant in time, and so are their multipole moments. However, if the mass distribution is time dependent, then so are the multipole moments. Such time-dependent quadrupole moments are of much interest in the relativistic theory of gravitation because they act as sources of gravitational radiation. Vibrating and rotating dumbbells (see Fig. 1.6) are simple examples of time-dependent quadrupoles.

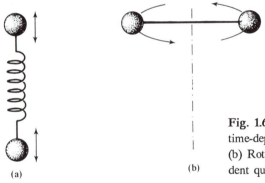

(a)   (b)

Fig. 1.6 (a) Oscillating masses with a time-dependent quadrupole moment. (b) Rotating masses with a time-dependent quadrupole moment.

Note that the values of the elements of the quadrupole tensor depend on the choice of the origin of coordinates. Suppose that the coordinates x^k have their origin at the center of mass. If we shift the origin so we obtain new coordinates

$$x'^k = x^k + b^k \qquad [27]$$

where b^k is a constant, then the quadrupole tensors in the old and new coordinates are related by

$$Q'^{kl} = Q^{kl} + (3b^k b^l - b^2 \delta_k^l)M \qquad [28]$$

■ *Exercise 8.* Derive Eq. [28]. ■

Eq. [28] shows that a change of origin modifies Q^{kl} by only an additive constant. Thus, the time-dependent part of Q^{kl} is independent of the choice of origin. This result will be useful in our study of gravitational radiation (Chapter 5).

1.5 THE EQUIVALENCE OF INERTIAL AND GRAVITATIONAL MASS

It is implicit in Newton's law of universal gravitation that the mass that acts as the source or the receptor of gravitation is the same as the mass that determines the inertia. The equation of motion of a pointlike particle in a given gravitational potential is

$$m \frac{d^2 x^k}{dt^2} = - m \frac{\partial \Phi}{\partial x^k} \qquad [29]$$

Here, the mass on the left side of the equation determines the inertia of the particle and the mass on the right side determines the strength of the gravitational force. If we cancel the masses in this equation, we obtain

$$\frac{d^2 x^k}{dt^2} = - \frac{\partial \Phi}{\partial x^k} \qquad [30]$$

This says that *in a given gravitational field all pointlike particles fall with the same acceleration.* For historical reasons, we will call this statement *Galileo's principle of equivalence.* This statement implies that to some extent gravitational forces behave in the same way as the pseudo-forces that result from the use of a noninertial reference frame. In an accelerated reference frame, free particles appear to accelerate spontaneously and, at a given point, all particles have exactly the same acceleration. To be precise, we should also specify that all the particles are put at the given point with the *same initial velocity.* This is necessary because some pseudo-forces--such as the Coriolis force--are velocity dependent. This restriction on the velocity is also needed for Galileo's principle, since it turns out that the "gravitational force" given by general relativity is velocity dependent.

The equivalence between inertial and gravitational effects can make it rather hard to distinguish between the two. Thus an astronaut in a freely falling spaceship will have a hard time deciding whether she is in free fall in a gravitational field or in unaccelerated motion in a region far away from all fields, if she only performs experiments in the interior of her spaceship.

We must now ask: is it correct to cancel the masses in Eq. [29]? Are the m's on both sides really the same? The mass that appears on the left side of Eq. [29] is the *inertial mass*. For any arbitrary body, the inertial mass is defined by a procedure first proposed by Mach: We take the body and let it interact, somehow, with the standard kilogram. Both the body and the standard will accelerate toward or away from each other. Designating the acceleration of the body and of the standard by a and a_s, respectively, we can then define the inertial mass by

$$m_I \equiv 1 \text{ kg} \times \frac{a_s}{a} \qquad [31]$$

This gives the inertial mass of the body in kilograms.

The *gravitational mass* can be defined in a similar manner. We take a standard body and define its gravitational mass to be one unit; for convenience we will use the standard kilogram as a standard for both inertial and gravitational mass (perhaps an atomic standard would be more relevant). We now place the body at some distance r and let it interact gravitationally with the standard. The body will accelerate toward the standard. We define the gravitational mass of the body in terms of this acceleration and the distance between the bodies:

$$m_G \equiv \lim_{r \to \infty} \left[\frac{a m_I r^2}{1 \text{ kg} \cdot G} \right] \qquad [32]$$

This gives the gravitational mass in kilograms. Note that since $a m_I$ is the gravitational force, or the weight, and $G \cdot 1 \text{ kg}/r^2$ is the gravitational field g generated by the standard mass, Eq. [32] simply says that the gravitational mass is the weight divided by g. Alternatively, we can use Eq. [31] to write Eq. [32] as

$$m_G \equiv \lim_{r \to \infty} \left[\frac{a_s r^2}{G} \right] \qquad [33]$$

The limiting procedure $r \to \infty$ is needed in [32] and [33] in order to eliminate the effects of multipole fields, which depend on the mass distribution of the two bodies. Also, the limiting proceedure $r \to \infty$ eliminates the effect of short-range forces (nuclear forces, van der Waals forces, etc.). At large distances, only the gravitational and electrostatic forces will remain, and the latter can be eliminated by taking the precaution of keeping the standard body neutral.*

---

* A further distinction between *active* and *passive* gravitational mass might be made. The force exerted *by* m' *on* m might be written as $G m p m'_A / r^2$, and the

If we take two identical copies of the standard of mass and let them fall toward each other, the acceleration of each serves to define the constant G:

$$G \equiv \lim_{r \to \infty} \left(\frac{a_{ss} r^2}{1 \text{ kg}} \right) \qquad [34]$$

With these precise definitions of m_I and m_G, the gravitational force between two particles is

$$F = - G \frac{m_G m'_G}{r^2} \qquad [35]$$

and the equation of motion is

$$m_I \frac{d^2 x^k}{dt^2} = - m_G \frac{\partial \Phi}{\partial x^k} \qquad [36]$$

Whether all particles fall with the same acceleration depends on whether all particles have the same value of m_I / m_G. If this ratio is a universal constant, it must have the value $m_I / m_G = 1$ (the standard body has this value by definition). The question is then, is the equation

$$m_I = m_G \qquad [37]$$

satisfied for all bodies? We will call the equality [37] *Newton's principle of equivalence of inertial and gravitational mass*, since Newton first considered the possibility of a difference between mass (inertial) and weight (gravitational).

Before we turn to the experimental evidence, we remark that for a body of appreciable size the acceleration of the center of mass is given by

$$m_I \frac{d^2 x^k}{dt^2} = - \int \rho_G(\mathbf{x}') \frac{\partial \Phi(\mathbf{x}')}{\partial x'^k} d^3 x' \qquad [38]$$

---

force exerted *by m on m'* as $Gm_A m'_P / r^2$; then Eqs. [32] and [33] would define the passive and the active mass, respectively (we might likewise be tempted to introduce active and passive charges in Coulomb's law). But the equality of active and passive mass is required by the equality of action and reaction; an inequality would imply a violation of momentum conservation. We will ignore this possibility.

where the integral is over the volume of the body and ρ_G is the gravitational mass density. Extended bodies fall at the same rate as point particles only if $\partial \Phi / \partial x'^k$ is approximately constant over the volume of the body; in that case we can factor $\partial \Phi / \partial x'^k$ out from the integral, and we then obtain [36].* Hence Newton's principle of equivalence implies equal accelerations only for bodies of sufficiently small size placed in sufficiently homogeneous gravitational fields. Bodies of different size and shape placed in an inhomogeneous field generally fall at different rates.

■ *Exercise 9*. Show that, as a consequence of Newton's principle of equivalence applied to the mass densities (i.e., $\rho_I = \rho_G$), two bodies of the same shape and size and similar (i.e., proportional) mass distribution will fall with the same linear accelerations and angular accelerations when placed in a given gravitational field. ■

According to the derivation that led us to Eq. [30], we see that for point particles the Galileo principle implies the Newton principle and vice versa. This derivation hinges on the validity of Newtonian mechanics; that is, the gravitational fields must be weak ($GM/rc^2 \ll$ 1) and the speeds low ($v \ll c$). If these assumptions are not satisfied, then the Galileo and Newton principles must be regarded as *complementary* rather than equivalent: in general, the Galileo principle is a statement about a special class of particles (point particles) moving with arbitrary velocities in arbitrary fields, whereas the Newton principle is a statement about arbitrary systems moving at low velocity in weak fields. The experiments to be described in the next section constitute evidence for the Newton principle; but since they were performed with particles of low velocities in weak fields, they give only limited support to the Galileo principle.

1.6 THE EMPIRICAL EVIDENCE FOR THE EQUIVALENCE OF m_G AND m_I

The earliest recorded experiments on the equality of rates of fall of different bodies are due to Galileo. He quickly recognized that dropping two bodies from a height was not the most convenient and accurate method for comparing their rates of fall. Instead, he experimented with balls rolling down an inclined plane, and he finally found that comparison of two pendulums gave the best results:

---

* The case of extended bodies with a spherically symmetric mass distribution is exceptional; such bodies behave as point particles regardless of their size (Newton's theorem).

... I took two balls, one of lead and one of cork, the former being more than a hundred times as heavy as the latter, and suspended them from two equal thin strings, each four or five bracchia long. Pulling each ball aside from the vertical, I released them at the same instant, and they, falling along the circumferences of the circles having the strings as radii, passed thru the vertical and returned along the same path. This free oscillation, repeated more than a hundred times, showed clearly that the heavy body kept time with the light body so well that neither in a hundred oscillations, nor in a thousand, will the former anticipate the latter by even an instant, so perfectly do they keep step. (Galileo, 1638)

We can interpret Galileo's observation to mean that the ratio m_I/m_G is the same for lead and for cork, to within a few parts in a thousand.

■ *Exercise 10.* What is the equation for the period of a simple pendulum if the bob has $m_I \neq m_G$? Galileo's claim that the two pendulums showed no deviation after a thousand oscillations is an exaggeration (on pendulums of different weight, the effects of friction will be quite noticeable); his claim for a hundred oscillations is credible. Suppose that after 100 oscillations the pendulums deviate by no more than 1/10 of a cycle. What limit does this set on the difference between the values of m_I/m_G for lead and cork? ■

The first experiments specifically designed to test the equality of inertial and gravitational mass are due to Newton. He also used two pendulums, but he was careful to compensate for the friction:

... I tried the thing in gold, silver, lead, glass, sand, common salt, wood, water, and wheat. I provided two equal wooden boxes. I filled the one with wood, and suspended an equal weight of gold (as exactly as I could) in the centre of oscillation of the other. The boxes, hung by equal threads of 11 feet, made a couple of pendulums perfectly equal in weight and figure, and equally exposed to the resistance of air: and, placing the one by the other, I observed them to play together forwards and backwards for a long while, with equal vibrations. And therefore (by Cor. I and VI, Prop. XXIV, Book II) the quantity of matter in the gold was to the quantity of matter in the wood as the action of the motive force upon all the gold to the action of the same upon all the wood; that is, as the weight of the one to the weight of the other.

And by these experiments, in bodies of the same weight, one could have discovered a difference of matter less than the thousandth part of the whole. (Newton, 1686b)

Table 1.4 summarizes the experimental evidence that has accumulated in favor of the equality of inertial and gravitational mass since the days of Galileo and Newton. These experiments were performed with a large variety of substances. In setting upper limits on $|m_I - m_G|$, one of the substances used by the experimenter is taken as a standard for which $m_I \equiv m_G$.

The most precise results have been obtained with torsion balances, by a method first introduced by Eötvös* around 1890. Eötvös began his experimental investigations on the equality of inertial and gravitational mass in response to a prize offered by the University of Göttingen; the high precision of his results so impressed the jury that they awarded him the prize, even though this prize was originally intended for a theoretical rather than an experimental investigation. The apparatus of Eötvös is shown in Figs. 1.7 and 1.8. Two pieces of matter, labeled "weight," are attached to the arms of a torsion balance. These weights are made of different substances, for instance, copper and platinum (Eötvös used platinum as standard and compared other substances with it). If m_I/m_G has different values for the two substances, then the balance will experience a torque. To understand how this comes about, consider the forces that act on the masses. Seen in the rotating reference frame of the Earth, the forces are of two kinds: there is the gravitational force $m\mathbf{g}$ exerted by the Earth and the centrifugal pseudo-force $m\mathbf{a}$ produced by the rotation of the Earth; the quantity $\mathbf{g}$ is the acceleration of gravity (*without* centrifugal effect**), and $\mathbf{a}$ is the centrifugal acceleration due to the Earth's rotation at the locality of the experiment. Fig. 1.9 shows these forces. In this figure both masses are shown placed directly on the beam of the balance; the asymmetric suspension used in the apparatus of Eötvös is of no relevance to the present experiment (the suspension was originally designed for a different purpose).

In Fig. 1.9, the direction of the z-axis is defined by the direction of $\mathbf{g}$, not the direction of a plumb line (which would include centrifugal effects). The beam of the balance points in the east-west direction. The centrifugal force then has a vertical (opposite to $\mathbf{g}$) component $m_I a_z$ and a horizontal component $m_I a_x$. The torque about the z-axis is obviously

$$\tau = m_I a_x l - m'_I a_x l' \qquad [39]$$

We can eliminate l' by using the equilibrium condition for rotation about the x-axis,

$$(m_G g - m_I a_z)l = (m'_G g - m'_I a_z)l' \qquad [40]$$

---

* Pronounced ötvösh, with the German umlaut ö.

** Careful! Values of "g" quoted in handbooks usually *include* the centrifugal effect.

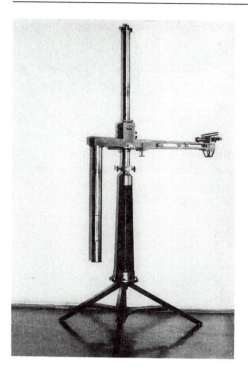

Fig. 1.7 Instrument used by Eötvös and his collaborators for the first gravity survey at Lake Balathon (1902). The instrument used to measure m_I/m_G was of the same type. (Courtesy Hungarian State Geophysical Institute Eötvös Lorand, Budapest.)

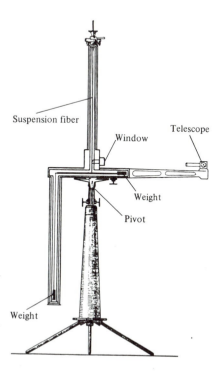

Fig. 1.8 Schematic diagram of the Eötvös apparatus. (After Eötvös, Pekar, and Fekete, 1922.)

TABLE 1.4 EQUALITY OF m_I AND m_G

Experimenter(s)	Year	Method	$\lvert m_I - m_G \rvert / m_I$
Galileo	$\simeq 1610$	pendulum	$< 2 \times 10^{-3}$
Newton	$\simeq 1680$	pendulum	$< 10^{-3}$
Bessel	1832	pendulum	$< 2 \times 10^{-5}$
Eotvos	1890	torsion balance	$< 5 \times 10^{-8}$
Eotvos et al.	1908	torsion balance	$< 3 \times 10^{-9}$
Southerns	1910	pendulum	$< 5 \times 10^{-6}$
Zeeman	1918	torsion balance	$< 3 \times 10^{-8}$
Potter	1923	pendulum	$< 3 \times 10^{-6}$
Renner	1935	torsion balance	$< 2 \times 10^{-10}$
Dicke et al.	1964	torsion balance, Sun	$< 3 \times 10^{-11}$
Braginsky et al.	1971	torsion balance, Sun	$< 9 \times 10^{-13}$
Koester	1976	free fall of neutron	$< 3 \times 10^{-4}$
Keiser et al.	1979	floating mass	$< 4 \times 10^{-11}$
Niebauer et al.	1987	free fall	$< 5 \times 10^{-10}$
Kuroda and Mio	1989	free fall	$< 8 \times 10^{-10}$
Adelberger et al.	1990	torsion balance	$< 1 \times 10^{-11}$
STEP	2000?	free fall in satellite	$< 10^{-17}$?

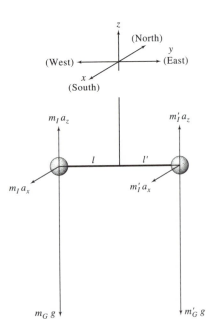

Fig. 1.9 Torsional pendulum suspended near the surface of the Earth at a latitude of about 45^0 N. Note that the centrifugal force vectors $m_I \mathbf{a}$ are shown exaggerated by a factor of about 100.

and we obtain

$$\tau = m_I\, a_x\, l \left(1 - \frac{m'_I}{m_I}\, \frac{m_G g - m_I a_z}{m'_G g - m'_I a_z} \right)$$

$$= m_I\, a_x\, l\, \frac{g(m'_G/m'_I - m_G/m_I)}{g m'_G/m'_I - a_z} \qquad [41]$$

From this it is obvious that a torque exists if and only if $m_G/m_I \neq m'_G/m'_I$.

In equilibrium, the gravitational torque [41] is compensated by a torque produced by the suspension fiber. The presence of the gravitational torque can be detected by rotating the entire apparatus by exactly 180° about the vertical axis (see pivot, Fig. 1.8). If the equilibrium position of the beam was exactly along the east-west direction before this rotation, then it will be slightly off after the rotation; this change in equilibrium position occurs because turning the apparatus around changes the sign of the torque [41].*

The method of Dicke (Dicke, 1964; Roll, Krotkov, and Dicke, 1964) also uses a torsion balance, but detects the torque, if any, produced by the gravitational force of the Sun and the centrifugal force of the Earth's motion around the Sun. If, for the sake of simplicity, we imagine a torsion balance of the type shown in Fig. 1.9 suspended at the north pole of the Earth, then this centrifugal force is entirely horizontal. Assuming equal arms ($l = l'$), the torque about the vertical axis will be

$$\tau = (m_G g - m_I a)l \sin \phi - (m'_G g - m'_I a)l \sin \phi \qquad [42]$$

where g is the acceleration of gravity produced by the *Sun* at the place of the experiment, a is the centrifugal acceleration in the reference frame of the Earth orbiting the Sun,** and ϕ is the angle between the beam of the balance and the Sun.

In Dicke's experiment, the beam is held stationary with respect to the Earth and the torque required to do this is measured. It is obvious from [42] that under these conditions the torque oscillates with a 24-hr period as the Sun angle ϕ increases by 2π. Any other effect ("noise") disturbing the experiment with a period different from 24 hr can be filtered out by Fourier analysis and discarded. Note that Dicke's apparatus actually uses a somewhat more complicated torsion balance with a triangular "beam" holding three masses (see Fig. 1.10). This

---

* Mathematically, turning the apparatus around is equivalent to $l \rightarrow - l$ and $l' \rightarrow - l'$ in Eqs. [39] and [40].

** To a sufficient approximation, $g = a$.

arrangement makes the balance less sensitive to gradients in the gravitational field produced by nearby massive bodies, such as the body of an experimenter. As a further precaution, measurements were carried out by remote control. The experiment of Braginsky and Panov (1971) used the same technique, but the "beam" of his balance was star shaped, with eight arms, and the suspension fiber was much longer.

The recent experiment of Adelberger et al. (1990) combines some of the best features of the Eötvös and the Dicke experiments. Like the Eötvös experiment, it seeks to detect the torque produced by the gravitational force produced by the Earth (which is much stronger than that produced by the Sun), but it also adopts Dicke's method of filtering out unwanted signals by Fourier analysis. For this purpose, the torsion balance is mounted on a turntable which continuously rotates the apparatus about the vertical axis, with a period of about 2 hr. Any noise with a period different from the 2-hr period can then be filtered out and discarded.

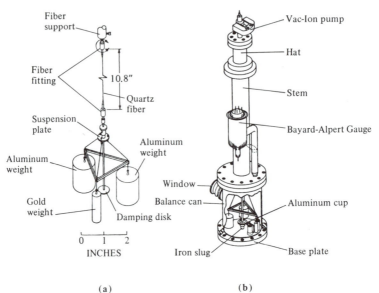

(a)   (b)

Fig. 1.10 The apparatus used by Dicke et al. (a) The torsion balance has a triangular beam with a mass attached at each vertex. (b) The balance is suspended in an evacuated vessel. (From Dicke, *Gravitation and the Universe*. Reprinted with permission of the American Philosophical Society.)

The delicate torsion balances used in these experiments are subject to a wide variety of environmental disturbances. To avoid such disturbances and to attain much higher sensitivity, Barlier et al. (1991) have proposed to perform an experiment with freely falling bodies in a satellite (*S*atellite *T*est of *E*quivalence *P*rinciple, or STEP). Within the satellite several pairs of test masses, in the form of concentric cylindrical shells of different compositions, would float freely. The satellite would be kept effectively drag free by means of thrusters that compensate for atmospheric friction. The masses would therefore remain in free fall for a long time, and even a very small relative acceleration between the two masses in a pair would ultimately yield a detectable displacement. SQUID circuits could detect displacements as small as 10^{-13} cm. It is expected that such an experiment could test the equality of inertial and gravitational mass to 1 part in 10^{17}.

The experiments listed in Table 1.4 have tested the ratio m_I/m_G for a wide variety of materials. Thus, Eötvös et al. compared copper, water, copper sulfate, asbestos, snakewood, etc., with platinum. The results of the experiment may be interpreted to mean that different kinds of energy contribute to the gravitational mass of a system the same amount they contribute to the inertial mass. Forms of energy that appear in the inertial mass (which, as we know from special relativity, is the same thing as total internal energy) of a sample of ordinary matter are listed in the first column of Table 1.5. The second column shows how these different forms of energy gravitate.

TABLE 1.5 DIFFERENT FORMS OF ENERGY AS SOURCES OF GRAVITY*

Form of energy	$\lvert m_I - m_G \rvert / m_I$
Rest mass, protons and neutrons	$< 10^{-11}$
Rest mass, electrons	$< 2 \times 10^{-8}$
Electric fields in nucleus	$< 4 \times 10^{-10}$
Magnetic fields in nucleus	$< 2 \times 10^{-7}$
Strong fields in nucleus	$< 5 \times 10^{-10}$
Weak fields in nucleus	$< 10^{-2}$
Gravitational energy in Earth	$< 2 \times 10^{-3}$

* Based, in part, on Will (1981) and Will (1993).

The nucleus of the atom contains appreciable amounts of energy in its electric, magnetic, and "strong" interaction fields. This permits us to set fairly tight limits on how these kinds of energy contribute to the gravitational mass--they all seem to gravitate in the normal way. The amount of energy in the "weak" interaction fields (which is the kind of interaction responsible for β decay) is much smaller, and the limit we can set on the behavior of this kind of energy is correspondingly worse. Note that the numbers in Table 1.5 hinge on the assumption that there are no fortuitous cancellations among deviations from $m_I = m_G$. Strictly speaking, we cannot make independent statements about, say, electron and proton rest-mass contributions to gravitation, since these particles are present in equal numbers in samples of neutral matter.

Free-fall experiments have been performed with individual neutrons and with individual electrons in the gravitational field of the Earth, but the precision of these experiments is relatively low. For example, measurement of the downward sag of a horizontal beam of low-velocity neutrons emerging from a nuclear reactor has shown that the downward acceleration of these free neutrons agrees with the standard local value of the Earth's gravitational acceleration to within at least 3 parts in 10^4 (Koester, 1976).

The data from the Eötvös experiments do not permit a direct test of the hypothesis that gravitational energy contributes to the gravitational mass. The ostensible macroscopic amounts of gravitational self-energy in masses of laboratory size are much too small to affect the experiments. Theoretical considerations suggest that the rest masses of electrons, protons, and neutrons include large amounts of gravitational self-energy; but we do not know how to calculate these self-energies (for the implications of this, see the next paragraph). If we want to discover whether gravity gravitates, we must examine the behavior of large masses, of planetary size, with a significant and *calculable* amount of gravitational self-energy. Treating the Earth as a continuous, classical mass distribution, we find that its gravitational self-energy is about 4.6×10^{-10} times its rest-mass energy. The gravitational self-energy of the Moon is smaller, only about 0.2×10^{-10} times its rest-mass energy.

If gravitational self-energy does not contribute in the normal way to the gravitational mass, then the Earth and the Moon will fall at different rates in the gravitational field of the Sun. The difference in the rates of fall is effectively equivalent to a uniform extra force field pulling the Moon toward the Sun (if gravitational energy gravitates less than normal) or away from the Sun (if gravitational energy gravitates more than normal). Such an extra force leads to a distortion of the orbit of the Moon relative to the Earth, a distortion called the *Nordvedt effect*. As Fig. 1.11 shows, the orbit is elongated, or polarized, in the direction of the Sun. Although the distortion effect is small, very precise measurements of the Earth-Moon distance have been carried out by the laser-ranging technique already mentioned in Section 1.2. A pulse of light emitted by a laser on the Earth is sent to the Moon, and is reflected back to the Earth by the corner reflectors installed on the Moon during the *Apollo 11* mission. Measurement of the travel time of the pulse determines the distance to within a few centimeters. According to a recent data analyis (Müller et al., 1991), these experiments place a direct limit of 1.5×10^{-3} on the fractional difference between the contributions of gravitational energy to the inertial and to the gravitational mass. Thus, these experiments indicate that gravitational energy gravitates in the normal way.

In setting the above limits on how the strong, electromagnetic, weak, and gravitational energies gravitate, we have ignored the self-energies locked up within the rest masses of electrons, protons, and

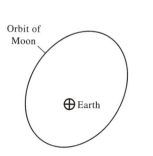

Fig. 1.11 Distortion of the orbit of the Moon produced by a hypothetical uniform extra force field directed toward the Sun. The distortion is shown exaggerated for the sake of clarity (if gravitational energy does not gravitate at all, the lengthening of the orbital diameter would amount to 5 parts in 10^6).

neutrons. This is a rather questionable attitude, since quantum theory suggests that all these particles contain large amounts of strong, electromagnetic, weak, and gravitational self-energies. These particles would therefore not be expected to gravitate in the normal way, unless all these forms of energy gravitate in the normal way. However, from the equal rates of fall of electrons, protons, and neutrons we cannot extract quantitative conclusions for the rates of fall of the self-energies locked up in the rest masses, because we have no way of calculating the magnitudes of these self-energies. A naive calculation of the, say, gravitational self-energy of the electron gives an infinite value; this, of course, proves only that the calculation is wrong, that our understanding of the quantum dynamics is faulty.

In the absence of any precise knowledge of the amount of self-energy contained in particles, we must impose a severe restriction on any putative theory of gravitation: the theory must give equal rates of fall for electrons, protons, and neutrons *regardless* of how much or how little strong, electromagnetic, weak, or gravitational self-energy is contained in the rest masses. Thus, the theory must make all forms of energy gravitate in the same way. If the theory does not satisfy this fundamental requirement of universal free fall, it is incomplete--it cannot make any definite prediction for the rates of fall of electrons, protons, and neutrons. So far only one theory of gravitation has been able to fulfill this requirement: Einstein's theory of gravitation. All the other theories contrived to compete with Einstein fail to meet the fundamental requirement of universal free fall--they all display a Nordvedt effect, with gravitational energy falling at a different rate. Until the inventors of these alternative theories discover some way of calculating the gravitational self-energies of electrons, protons, and neutrons, they cannot predict the rates of fall of these particles. Because of this serious defect of all the alternative theories of gravitation, we will ignore them in this book.*

As concerns antimatter, we do not yet have any direct experimental evidence on the value of m_I/m_G, although an experiment now being developed is soon expected to test the rate of fall of a beam of low-energy antiprotons at CERN (Holzscheiter et al., 1993). Meanwhile we have to rely on some circumstantial evidence in favor of a normal gravitational interaction of antimatter. It has been pointed out by Schiff (1959) that if positrons had a negative gravitational mass ("fell upward"), then one would expect nuclei to have anomalous ratios of m_I/m_G. The reason for this is that a nucleus is surrounded by a fluctuating cloud of virtual electron-positron pairs, a condition known as vacuum polarization. The number of pairs depends on the charge distribution of the nucleus, and if the gravitational mass of positrons were negative, nuclei surrounded by more pairs would appear to have a lower gravitational mass. This effect would lead to a difference in m_I/m_G between platinum and aluminum amounting to a few parts in 10^7, in clear contradiction to Eötvös' experimental results.

A much more precise statement about gravitational interactions of antimatter can be made by examining the famous case of the decay of the neutral K^0 meson. The decay of this meson is very sensitive to quantum-mechanical phase factors, and even a small change in the (small) phase factor $e^{-im_G \Phi t/\hbar}$ arising from the gravitational energy of the meson in the gravitational potential Φ of the Galaxy would presumably lead to drastic changes in the decay (Good, 1961). From the absence of such phase effects in the decay, we can conclude that the gravitational masses of the meson K^0 and the antimeson $\overline{K}^0$ differ at most by a few parts in 10^{10}. (The *inertial* masses of the K^0 and the $\overline{K}^0$ mesons are exactly equal; this is a general theorem for every particle and the corresponding antiparticle.)

The observed equality of gravitation and inertia for all forms of energy means that any system, containing any combination of the different kinds of energy, will have equal gravitational and inertial masses. Thus, we will now postulate Newton's principle of equivalence, $m_I = m_G$, as an exact relation that our gravitational theory must satisfy.

---

* For a catalog of alternative theories of gravitation, see Will (1993). A few of these theories contain several adjustable coupling constants, and by juggling the values of these constants, the Nordvedt effect can be made zero in the lowest order in G; but, barring a miracle, this does not guarantee that the Nordvedt effect will remain zero in higher orders in G. Only Einstein's theory gives us such a guarantee.

1.7 SPECULATIONS ABOUT A FIFTH FORCE

In 1986, Fischbach et al. published a reanalysis of the Eötvös data, which suggested the existence of a new, extra interaction. The data seemed to indicate that the source of this fifth force is baryon number. Fig. 1.12 shows Fischbach's plot of the residual accelerations observed by Eötvös vs. the difference in baryon number per unit mass for pairs of materials used in the Eötvös experiments. The plot reveals a rather striking correlation of relative acceleration and baryon number, and suggests the existence of an extra interaction analogous to the electromagnetic interaction, but with baryon number in the role of "charge." This interaction has been called the "fifth force," or the "fifth interaction" (the gravitational, electromagnetic, strong, and weak interactions are the known four interactions). For two nuclei with baryon numbers B_1 and B_2, respectively, the potential associated with this extra interaction is proportional to

$$\frac{B_1 B_2}{r} e^{-\lambda r} \qquad [43]$$

The sum of the ordinary Newtonian gravitational potential and the extra potential associated with baryon number can then be written as

$$V = -\frac{Gm_1 m_2}{r} + \alpha_5 \frac{B_1}{\mu_1} \frac{B_2}{\mu_2} \frac{Gm_1 m_2}{r} e^{-\lambda r} \qquad [44]$$

where μ_1 and μ_2 are the atomic masses of the materials in the two pieces, and α_5 is a dimensionless constant that characterizes the strength of the fifth force relative to the Newtonian gravitational force. The sign of the potential [44] has been chosen positive, because, by analogy with electromagnetism, we expect that the force between baryon "charges" of equal signs is repulsive (theoretical arguments indicate that if the fifth-force field is a vector field, like the electromagnetic field, then it yields a repulsive force).

Obviously, the potential [44] is similar to the potential [6] that we contemplated in Section 1.2 when examining deviations from the inverse-square law (note that if we write $\alpha = -\alpha_5(B_1/\mu_1)(B_2/\mu_2)$, then [44] becomes formally identical to [6]). However, there is a crucial distinction between the potentials [6] and [44]: the former depends on the masses only, whereas the latter depends on the baryon content, that is, it depends on the chemical composition of the interacting bodies, and it involves a direct violation of the principle of equivalence. This distinction means that we can regard the potential [6] as a modification of the gravitational interaction, but we must regard the fifth interaction as fundamentally different from the gravitational interaction.

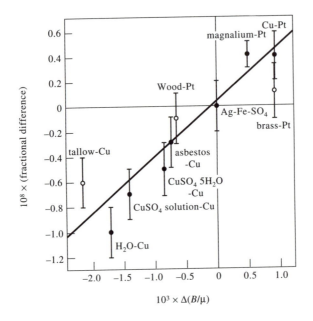

Fig. 1.12 Eötvös' data on the fractional difference between the accelerations of pairs of masses plotted vs. the difference in their baryon numbers.

Although Fig. 1.12 suggests a correlation between relative acceleration and baryon number, it is difficult to draw quantitative conclusions from the data. If there is a fifth force, most of it would arise from topographic irregularities near Eötvös' laboratory--the baryons in a nearby hill would produce more of a deviation in the torsion balance than the baryons in the rest of the Earth. Without detailed topographic information, the Eötvös data are not very helpful for a determination of the coupling strength α_5.

The speculations of Fischbach et al. led to a flurry of experimental activity as several groups of investigators sought to detect the fifth force. Some experimenters used torsion balances to detect the effects of nearby cliffs or hills on masses attached to the arms of the balances (Boynton et al., 1987; Stubbs et al., 1987; Adelberger et al., 1987; Fitch et al., 1988; Adelberger et al., 1990). Others used balls floating in a liquid; such a ball would drift toward the cliff if it were more strongly attracted to the cliff than the liquid (Thieberger, 1987). Others used torsion balances or beam balances with a large mass placed nearby (Speake and Quinn, 1988). And others dropped pairs of masses in a vacuum chamber and monitored the distance between these masses interferometrically during this free-fall motion (Niebauer et al., 1987; Kuroda and Mio, 1989); these experiments have already been mentioned in Table 1.4. The experimental results on the fifth force are summarized in Table 1.6. The limits on the magnitude of the coupling strength set by these experiments depend on the assumed value of the range λ; for purposes of comparison, a nominal value of λ = 100 m has been assumed in Table 1.6 (limits for a wide variety of values of λ are summarized graphically in Adelberger et al., 1990).

TABLE 1.6 SEARCHES FOR THE FIFTH FORCE

Experimenter(s)	Year	Method	Result*
Fischbach et al.	1986	reanalysis of Eotvos	$\|\alpha_5\| \simeq 0.01$
Thieberger	1987	floating ball	$\|\alpha_5\| = 0.01$
Stubbs et al.	1987	torsion balance	negative; $\|\alpha_5\| < 0.0004$
Adelberger et al.	1987	torsion balance	$\|\alpha_5\| \leq 0.001$ if B and L
Niebauer et al.	1987	free fall	negative; $\|\alpha_5\| < 0.07$
Boynton et al.	1987	torsion balance	$\|\alpha_5\| = 0.01$
Fitch et al.	1988	torsion balance	negative; $\|\alpha_5\| < 0.001$
Speake and Quinn	1988	beam balance	negative; $\|\alpha_5\| < 0.03$
Cowsik et al.	1988	torsion balance	negative; $\|\alpha_5\| < 0.003$
Bennett	1989	torsion balance	negative; $\|\alpha_5\| < 0.001$ if B and L
Kuroda and Mio	1989	free fall	negative; $\|\alpha_5\| < 0.01$
Stubbs et al.	1989	torsion balance	negative; $\|\alpha_5\| < 0.001$
Bizetti et al.	1989	floating ball	negative; $\|\alpha_5\| < 0.002$
Adelberger et al.	1990	torsion balance	negative; $\|\alpha_5\| < 0.00001$

* For an assumed range of λ = 100 m and for coupling to baryon number B, unless otherwise noted.

Only two of the experiments showed a positive effect, indicating the presence of a fifth force: the floating-ball experiment of Thieberger (1987) and the torsion-balance experiment of Boynton et al. (1987). But these experiments were contradicted by all the other experiments, which showed no fifth force in excess of the experimental errors. At first it seemed that the contradictions between these results could be resolved by postulating that the source of the fifth force is not baryon number, but a combination of baryon and lepton numbers (for instance, $B - 2L$). Since the experiments were performed with different materials at different locations, with cliffs or hills of different compositions, a suitably contrived combination of baryon and lepton numbers could result in a zero fifth force at some locations and a nonzero force at others. However, later experiments, specifically designed to check the positive results of Thieberger and Boynton, eliminated this loophole (Adelberger et al., 1987; Bizetti et al., 1989; Stubbs et al., 1989).

Thus, in spite of prodigious efforts, the experiments have failed to find any credible evidence for a fifth force. Of course, the experiments cannot *prove* that there is no fifth force; they can only set upper limits on the strength of such a force, consistent with the experimental errors. And a stubborn adherent of the fifth force could always evade such limits by contriving a combination of several Yukawa potentials with different strengths and ranges, maliciously adjusted so as to defeat the available experimental data. The limits given in Table 1.6 assume a single Yukawa potential; if instead we use a combination of two or more Yukawa potentials, the limits become much less stringent. But such artificial attempts at prolonging the life of the fifth force run counter to Newton's First Rule for Reasoning in Philosophy (Newton, 1686a, Book III, Rule I), also known as Occam's razor: "We are to admit no more causes of natural things than such as are both true and sufficient to explain their appearances." Since the experimental results do not require a fifth force, the most reasonable attitude is to disregard this force. The correlation seen in the Eötvös data must then be attributed to some unknown systematic error, or to accidental coincidence.

1.8 TIDAL FORCES

Consider a spacecraft in orbit around the Earth.* The spacecraft is in free fall. The astronauts find themselves in a zero-g environment, they are weightless, they feel no gravitational force. The reference frame attached to the spacecraft simulates an inertial reference frame: a test particle at rest relative to the spacecraft remains at rest, a test particle in motion remains in motion with uniform velocity. However, for an observer at rest relative to the fixed stars, the spacecraft will be in accelerated motion, and such an observer will refute the astronauts' claim that they are in an inertial frame by saying that both the spacecraft and the test particle in it are falling at the same rate, and that the astronauts are being fooled by appearances. Note that the elimination of the Earth's gravitational field by free fall is impossible unless the equivalence of inertial and gravitational mass holds for all bodies; any body with $m_I \neq m_G$ would appear to accelerate spontaneously with respect to the freely falling spacecraft.

---

* We assume that the spacecraft is not spinning about its axis. To be precise: we assume that the rate of spin of the spacecraft has been adjusted so that the axis of a gyroscope carried by the spacecraft remains at rest relative to the spacecraft. In the Newtonian approximation, the axis of spin of a gyroscope will of course maintain its direction relative to the fixed stars. But, as we will see later, relativistic effects produce a precession of the axis of spin; these effects are a generalization of the well-known Thomas precession of special relativity.

We now ask: are the gravitational effects of the Earth *completely* eliminated by free fall? Is there some local experiment that permits the astronauts to find out that they are falling in a gravitational field rather than at rest in some region far away from any attracting mass? The answer is that the astronauts *can detect the gravitational field by the tidal effects* it produces. If the astronauts place a drop of liquid at the center of their spacecraft, they will find that this drop is not exactly spherical, but has two bulges (Fig. 1.13). One bulge points toward the Earth, one away. Since, in the absence of external forces, surface tension would make the drop spherical, the deviation from a sphere indicates the existence of a gravitational field. The bulges result from the inhomogeneity of the gravitational field: the end of the drop nearer the Earth is pulled too much by gravitation, the other end is not pulled enough.

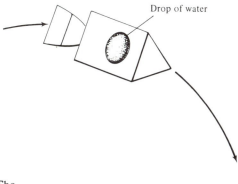

Drop of water

Fig. 1.13 A spacecraft in orbit. The drop of water floating in this spacecraft is distorted by the gravitational pull of the Earth.

Earth

The force that produces the bulges is called the *tidal force*. We can calculate it as follows: Consider a reference point moving in free fall; take this point as the origin of a freely falling coordinate system with the z-axis parallel to the radial line (Fig. 1.14). A particle at position $(0, 0, z)$ in this reference frame experiences a gravitational acceleration

$$- \frac{GM}{(r_0 + z)^2} \qquad [45]$$

where r_0 is the distance from the center of the Earth to the origin of our coordinates, and M is the mass of the Earth. Since the origin has acceleration $-GM/r_0^2$, the acceleration of the particle *relative* to the origin is, in the limit of small z,

$$- \frac{GM}{(r_0 + z)^2} + \frac{GM}{r_0^2} \simeq 2z \frac{GM}{r^3} \qquad [46]$$

Hence, relative to our origin the particle moves as though subjected to a force

$$f_z = 2z \frac{GMm}{r_0^3} \qquad [47]$$

This is a tidal force. Note that it is directly proportional to the distance of the particle from the origin and is repulsive.

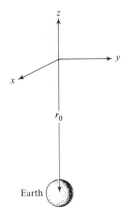

Fig. 1.14 The coordinate system x, y, z belongs to a reference frame in free fall, whose origin is (instantaneously) at a distance r_0 from the Earth.

For a particle at position $(0, y, 0)$, the tidal force points in the $-y$ direction and is given by

$$f_y = -y \frac{GMm}{r_0^3} \qquad [48]$$

This is a tidal restoring force, toward the origin. For a particle at position $(x, 0, 0)$, there is a similar restoring force,

$$f_x = -x \frac{GMm}{r_0^3} \qquad [49]$$

If the particle has simultaneous x, y, and z displacements, then all three forces, [47], [48], and [49], are of course present simultaneously. Note that these expressions for the tidal force only apply if the displacements x, y, z are small compared to r_0.

■ *Exercise 11.* Derive Eqs. [48] and [49]. ■

From these results it is obvious that a drop of liquid, consisting of many particles, will be stretched in the radial direction and compressed in the transverse direction.

In general, given an arbitrary gravitational field, the tidal force in a reference frame whose origin is in free fall can be expressed as follows:

$$f^k = \sum_l x^l \frac{\partial F^k}{\partial x^l} = - \sum_l x^l m \frac{\partial^2 \Phi}{\partial x^l \partial x^k} \qquad [50]$$

where F^k is the gravitational force on the particle and Φ is the potential, and where the derivatives are evaluated at the origin ($x^l = 0$). Eq. [50] is of course valid only near the origin (small x^l).

■ *Exercise 12.* Derive Eq. [50]. ■

We will call the quantity

$$R^k{}_{0l0} \equiv - \frac{1}{mc^2} \frac{\partial F^k}{\partial x^l} = \frac{1}{c^2} \frac{\partial^2 \Phi}{\partial x^k \partial x^l} \qquad [51]$$

the *tidal force tensor.* The extra factor mc^2 and the extra indices 00 have been introduced for later convenience. It turns out that the nine quantities $R^k{}_{0l0}$ are components of the Riemann curvature tensor $R^\mu{}_{\nu\alpha\beta}$; as we will see in Section 7.5, the latter tensor describes the tidal field in the four-dimensional spacetime of general relativity. The tidal force that acts on a particle placed at x^l (with $x^l \ll r_0$) is then

$$f^k = -mc^2 \sum_l R^k{}_{0l0} x^l \qquad [52]$$

which produces an acceleration

$$\frac{d^2 x^k}{dt^2} = - c^2 \sum_l R^k{}_{0l0} x^l \qquad [53]$$

In the special case considered above, the tensor $R^k{}_{0l0}$ has components:

$$R^k{}_{0l0} = \frac{GM}{r_0{}^3 c^2} \begin{pmatrix} 1 & 0 & 0 \\ 0 & 1 & 0 \\ 0 & 0 & -2 \end{pmatrix}$$ [54]

Note that the first index (k) gives the row, and the second (l) the column.

The tidal force given by Eqs. [47], [48], and [49] satisfies the identity

$$\frac{\partial f_x}{\partial x} + \frac{\partial f_y}{\partial y} + \frac{\partial f_z}{\partial z} = - \frac{GMm}{r_0{}^3} - \frac{GMm}{r_0{}^3} + \frac{2GMm}{r_0{}^3} \equiv 0$$

that is, the tidal-force field has zero divergence. This result holds in general for gravitational fields in empty regions of space. We can prove this by using Eq. [50]:

$$\frac{\partial}{\partial x^k} f^k = - \sum_{k,l} \frac{\partial}{\partial x^k} \left[x^l m \frac{\partial^2 \Phi}{\partial x^k x^l} \right]$$

$$= - m \sum_{k,l} \delta^l_k \frac{\partial^2 \Phi}{\partial x^k x^l} = - m \sum_k \frac{\partial^2 \Phi}{\partial x^k \partial x^k} = 0$$ [55]

The last equality holds true because, in empty space, the gravitational potential satisfies the Laplace equation $\nabla^2 \Phi = 0$ (see Eq. [17]). Note that the argument shows that the equations $\nabla \cdot \mathbf{f} = 0$ and $\nabla^2 \Phi = 0$ are equivalent. Hence a gravitational field *in vacuo* may be characterized by the vanishing of the divergence of the tidal force.

■ *Exercise 13.* Show that in the presence of a mass density ρ, the divergence of the tidal force is given by

$$\sum_k \frac{\partial f^k}{\partial x^k} = - 4\pi m G \rho \qquad ■$$ [56]

As we will see later, the tidal forces given by general relativity are somewhat more complicated.* They depend on the velocity of the reference point and on the velocity of the particle with respect to that reference point, and furthermore, the gravitational field is not that

---

* These forces are contained in the equation of geodesic deviation of Section 6.6.

given by Newton. However, the general relativistic tidal force agrees with the above Newtonian value in the limit of weak gravitational fields ($GM/rc^2 \ll 1$) and low speed ($v \ll c$).

The "tidal forces" take their name from the tides they generate on the oceans of the Earth. These tidal forces are exerted by the Moon and, to a lesser extent, by the Sun; we will ignore the Sun for now. For a simple estimate of the height of the tide, consider the Earth as a spacecraft in free fall toward the Moon. According to Eqs. [47], [48], and [49], the tidal force exerted by the Moon on a particle on the surface of the Earth, with coordinates x, y, z (see Fig. 1.15), is

$$f_z = 2z\frac{GMm}{r_0^{\ 3}}, \qquad f_y = -y\frac{GMm}{r_0^{\ 3}}, \qquad f_x = -x\frac{GMm}{r_0^{\ 3}}$$

where M is now the mass of the Moon and r_0 the Earth-Moon distance.

The potential that gives this force is

$$-\frac{GMm}{r_0^{\ 3}}\left[z^2 - \frac{1}{2}x^2 - \frac{1}{2}y^2\right] = -\frac{GMm}{r_0^{\ 3}}\left(-\frac{1}{2}R^2 + \frac{3}{2}z^2\right) \qquad [57]$$

$$= -\frac{GMm}{r_0^{\ 3}}\left(\frac{3\cos^2\theta - 1}{2}\right)R^2 \qquad [58]$$

where $R = (x^2 + y^2 + z^2)^{1/2}$ is the radius of the Earth and θ is the angle shown in Fig. 1.15.

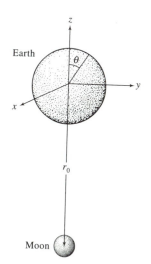

Fig. 1.15 The coordinate system x, y, z is attached to the Earth, which is in free fall toward the Moon.

A particle on the surface of the Earth also feels the gravitational attraction of the Earth. Since we are concerned with particles that remain near the surface, we can write the potential produced by the Earth as mgh, where h is the height above the surface.* The net gravitational potential is then

$$V(h, \theta) = mgh - \frac{GMmR^2}{r_0^3}\left(\frac{3\cos^2\theta - 1}{2}\right) \qquad [59]$$

To find the height of the tide, let us assume that the water is in static equilibrium. Then the surface must be an equipotential

$$gh - \frac{GMR^2}{r_0^3}\left(\frac{3\cos^2\theta - 1}{2}\right) = [\text{constant}] \qquad [60]$$

This gives us the height h of the water as a function of θ. High tides occur at $\theta = 0$ and at $\theta = \pi$; low tides at $\theta = \pi/2$. The difference between high and low tide (the "tidal range") is given by

$$\Delta h = \frac{GMR^2}{gr_0^3}\left[\frac{3\cos^2\theta - 1}{2}\right]_{\theta=\pi/2}^{\theta=0} \qquad [61]$$

The numerical value of the tidal range given by Eq. [61] is 53 cm (the Sun produces an effect about half as large). This is a reasonable order of magnitude estimate for the tide in the open sea. But our results are not accurate because the water does *not* reach static equilibrium with the tidal force. As a consequence of the rotation of the Earth about its axis, the tidal field appears to move once around the Earth every lunar** day, and we must seek the response of water to an oscillating force of period 12 hr 25 min. The water in ocean basins has certain natural periods of oscillation, in much the same way as water in a bathtub has a well-defined period for sloshing back and

---

* To be precise, we should write $(R + h)$ in Eq. [59], rather than R. But for $h \ll R$, this makes little difference.

** The time between successive passages of the Moon through a given meridian is 12 hr 25 min. If the axis of rotation of the Earth were perpendicular to the orbital plane of the Moon, then the two maxima in the tidal potential would simply travel around the equator of the Earth and the period of the tidal driving force would be 12 hr 25 min. Actually, the axis of the Earth makes an angle with the orbit of the Moon and therefore, at any given latitude, the two daily maxima are not exactly equal. The force has a strong Fourier component at 12 hr 25 min ("semidiurnal tide") and a weaker component ("diurnal tide") at 24 hr 25 min. The Sun contributes other components.

forth, and we are therefore dealing with the problem of an oscillator subjected to a periodic driving force. The response of the oscillator depends critically on how the frequency of the driving forces compares with the natural frequency. If the natural frequency is much larger than the driving frequency, the oscillations remain in phase with the driving force and quasistatic conditions prevail (the restoring force is almost in static equilibrium with the driving force). However, for the case of some ocean basins there exist natural modes of oscillation with frequencies *smaller* than 1 cycle per 12 hr 25 min; this leads to an inverted tide, out of phase with the driving force. Basins that have a natural frequency very close to the driving frequency can develop enormous tides by resonance. For example, tides with a range of 15 m are found in the Bay of Fundy.

1.9 THE TIDAL FIELD AS A LOCAL MEASURE OF GRAVITATION

Let us now look somewhat more closely at the detection of gravitational fields by tides. If astronauts in orbit wish to detect the gravitational field of the Earth by measuring the tide produced by the Earth on a drop of water, they will find it desirable to use a very large drop of water. Eq. [61] shows that the height of tide increases with the size of the drop. This suggests that if the astronauts have been ordered to confine their experiments to the interior of a sufficiently small spacecraft, then they will not be able to detect the tide, or the gravitational field. But suppose that the astronauts use as a measure of the tidal field not the height Δh, but rather the ratio $\Delta h/R$, which characterizes the prolateness of the droplets. According to Eq. [61] we have

$$\frac{\Delta h}{R} = \frac{3GMR}{2gr_0{}^3} \qquad [62]$$

where M is now the mass of the Earth, R the radius of the "drop," and g the acceleration of gravity that the "drop" of water produces at its own surface. Strictly speaking, Eq. [62] is applicable only to tides in a thin layer of liquid covering the surface of a dense rigid sphere. For a drop made entirely of liquid, the change in the gravitational field of the drop produced by the change of shape must be taken into account; this modifies [62] by a numerical factor of no great interest to us. Note that we are assuming that the drop is held together by gravitation, rather than by surface tension; this is not realistic for small drops of water, but we will pretend that some very powerful detergent has been added to the water so as to make its surface tension very small.

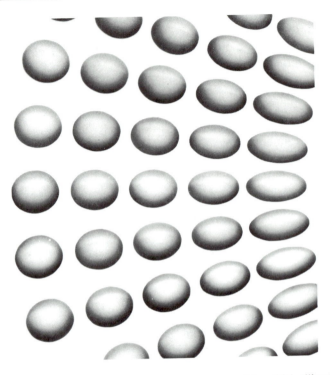

Fig. 1.16 The tidal field of a spherical mass represented by tidal ellipsoids. The mass is beyond the right edge of the drawing.

In terms of the density of the fluid, we can write $g = G(4\pi\rho R^3/3)/R^2$ and

$$\frac{\Delta h}{R} = \frac{9}{8\pi}\frac{M}{\rho r_0^3} \qquad\qquad [63]$$

This shows that the *shape* of the tidal ellipsoid is independent of its size. Even in the limit $R \to 0$, the tidal deformation remains. We can therefore regard the prolateness of the tidal ellipsoid as a *local* measure of the gravitational force.

We can now construct a pretty graphical representation of a gravitational tidal field: at each point of space, draw the tidal ellipsoid that gives the shape of a drop of fluid (with zero surface tension) in free fall at that point. Such a graphical representation of the tidal field of a spherical mass is shown in Fig. 1.16.

■ *Exercise 14.* Suppose that the symmetric tensor $R^k{}_{0l0}$ has been reduced to diagonal form. Show that the components of the tensor cannot all have the same sign. (Hint: Φ cannot have a maximum or a minimum.) ■

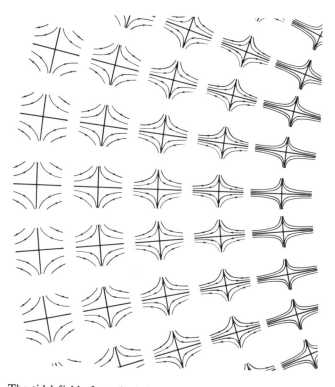

Fig. 1.17 The tidal field of a spherical mass represented by lines of force. Each of the patterns of lines of force is valid only near its center. Note that this two-dimensional drawing does not correctly indicate the density of lines of a three-dimensional field. The density along the radial axis is actually twice as large as the density along the transverse axis.

We can also construct an alternative graphical representation by appealing to Eq. [56]. According to this equation, the tidal force field *in vacuo* has zero divergence; this implies that the tidal force can be represented by lines of force in exactly the same way as an electric field *in vacuo*. Fig. 1.17 shows the pattern of lines of force near points in the space surrounding a spherical mass. Note that we have to draw a separate pattern for the vicinity of each point because our expression for the tidal field is valid only for small displacements.

Although the graphical representations shown in Figs. 1.16 and 1.17 are mathematically equivalent, the picture of tidal ellipsoids has the advantage of emphasizing the local character of the tidal effects. The picture of lines of force suggests that at the center of each pattern (see Fig. 1.17) the tidal effects are absent. We must firmly reject this suggestion. Although some tidal effects do vanish near the origin of a freely falling reference frame, some other effects remain finite, and it is these that ultimately count.

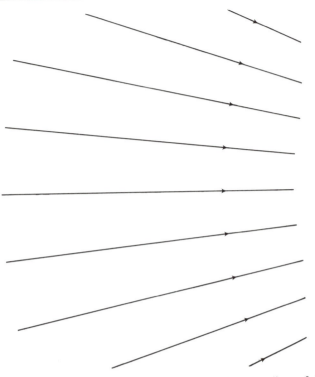

Fig. 1.18 The inverse-square gravitational force represented by lines of force. The diagram is drawn to the same scale as the preceding diagrams.

A more familiar graphical representation of gravitational fields uses the lines of force of the ordinary gravitational force, rather than the tidal force (see Fig. 1.18). This is, of course, very useful to an outside observer who wishes to study the motion of a spacecraft, or whatever, through the field. But this representation fails to describe what happens locally. If we want a description of what the astronauts experience inside their spacecraft, then the picture of tidal forces is much more relevant than that of the ordinary force.

The detection of tidal fields by means of the deformation of drops of water is not a practical method, because for a small drop, the surface tension will prevent the formation of tidal bulges. A somewhat more realistic method is the following: Suppose the astronauts place a freely spinning rigid body in their spacecraft. Fig. 1.19 shows a rigid rod at the origin of the coordinates. The tidal force will exert a torque on the rod and give it an angular acceleration. For the tidal field given by Eqs. [47]-[49], the torque about the x-axis is

$$\tau_x = \int \left[y \left(2z \frac{GM}{r_0^3} \right) - z \left(-y \frac{GM}{r_0^3} \right) \right] dm$$

$$= \frac{3GM}{r_0^3} \int yz \, dm = \frac{3GM}{r_0^3} (-I_{yz}) \tag{64}$$

Here $I_{yz} = I^{23}$ is the y-z component of the moment of inertia tensor,

$$I^{kl} = \int (r^2 \delta_l^k - x^k x^l) \, dm \tag{65}$$

Expressions similar to [64] can be found for the torques about the y- and z-axes.

■ *Exercise 15.* Find τ_y and τ_x. ■

■ *Exercise 16.* Show that the torque exerted by an arbitrary tidal field $R^k{}_{0l0}$ on a body with an inertia tensor I^{ls} is given by

$$\tau^n = c^2 \sum_{k,l} \epsilon^{nkl} R^k{}_{0l0} \left(-I^{ls} + \tfrac{1}{3} \delta_l^s I_{rr} \right) \tag{66}$$

where the quantity ϵ^{nkl} is defined as follows:

$$\epsilon^{123} = \epsilon^{231} = \epsilon^{312} = 1$$
$$\epsilon^{321} = \epsilon^{213} = \epsilon^{132} = -1 \tag{67}$$

with all other components zero. ■

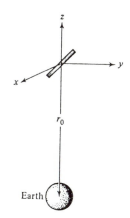

Fig. 1.19 A rigid rod in free fall, instantaneously at rest.

If no other forces act, Eq. [66] implies that the x-component of the spin of the rigid body changes at a rate

$$\frac{dS_x}{dt} = -\frac{3GM}{r_0^3} I_{yz} \qquad [68]$$

Similar formulas can be obtained for the rates of change of the other components of the spin. This rate of change of the spin can serve as a measure of the tidal field.

For a *local* measurement, we must proceed to the limit of small size of the rigid body. The inertia I_{yz} is of the order $I_{yz} \simeq ml^2$ (where l is a typical dimension of the body), and the spin is $S \simeq ml^2\omega$. In the limit $l \to 0$, both torque and spin vanish, but $d\omega/dt$ remains finite. It is easy to see that $d\omega/dt$ depends only on the *shape* of the rigid body, not on the size. We can therefore use the angular acceleration to detect the tidal force. If the body is originally spinning, then the tidal force will cause a precession of the spin. A well-known example is the equinoctial precession of the spin of the Earth, which is caused by the lunar and solar tidal forces acting on the equatorial bulge of the Earth.

■ *Exercise 17.* Show that if a body has the moment of inertia tensor of a sphere, then the torque [66] vanishes. Hence instead of measuring the spin precession [68] relative to the fixed stars, we can measure it locally, relative to a spherical gyroscope. ■

The Stanford gyroscope experiment (Gravity Probe B) provides a good illustration of the importance of tidal torques in a rotating body. This experiment, first proposed by Schiff (1960), has been under development for many years by Fairbanks, Everitt et al., and it is to be sent into orbit with the space shuttle in 1997. The experiment seeks to detect an extremely small relativistic precession predicted by Einstein's theory for a freely spinning gyroscope. Since the experimenters *do not* want their gyroscopes to precess from tidal torques, they had to manufacture spherically symmetric gyroscopes. The shape of these gyroscopes must be spherical to within a few parts in 10^6, otherwise the precession caused by tidal torques would wash out the relativistic effects the experimenters seek to measure (for more on this experiment, see Chapter 7).

There exist other methods for local measurement of the tidal field. The Eötvös balance shown in Fig. 1.7 was originally designed as a gravity gradiometer, to measure gradients in the gravitational field of the Earth by means of the torque they produce on the balance. One of the weights of this balance is placed lower than the other to make the balance sensitive to vertical gradients in the horizontal component of gravitational force. With this arrangement, Eötvös found that tidal fields as small as $R^k{}_{0l0} = (1/c^2)\partial^2\Phi/\partial x^k\,\partial x^l \simeq 10^{-32}/\text{cm}^2$ were detectable.*

The beam of the Eötvös balance was 40 cm long. In order to see how good the balance is at measuring $R^k{}_{0l0}$ *locally*, let us consider the limit $l \to 0$ (where l is a typical dimension of the balance). The tidal torque on the balance will then tend to zero. The equilibrium position of the beam is determined by the condition that the torque exerted by the tidal forces be equal to the restoring torque exerted by the fiber. If we scale down the size of the balance, we must also decrease the torque constant of the fiber, in order to retain the sensitivity of the balance to a given tidal field. This means that the thickness of the fiber must be reduced. (It can be shown that a reduction of the size of the apparatus by a factor of, say, 2 requires a reduction in fiber thickness by a factor of $2^{3/2}$.) There are, of course, practical limitations which prevent us from reducing the size of our apparatus *ad infinitum* and obtaining the value of $R^k{}_{0l0}$ at exactly *one point;* these limitations must, however, not be blamed on the gravitational field, but rather on the properties (ultimately *quantum* properties) of materials.

Compact gravity gradiometers, for use aboard a spacecraft orbiting the Earth, are now being developed by Paik et al. (University of Maryland) and by Fuligni (Frascati). The gradiometer of Paik (Moody, Chan, and Paik, 1986) consists of a pair of accelerometers, each of which contains a superconducting proof mass held by a spring within a housing (see Figs. 1.20 and 1.21). A superconducting coil is placed near the mass. In response to an acceleration of the housing, the proof mass moves toward or away from the coil; by the Meissner effect, this changes the inductance of the coil. Since the magnetic flux linked by any superconducting-coil circuit must remain constant, the current in this circuit changes in proportion to the change in inductance. This change of current is detected by a sensitive SQUID amplifier coupled to the circuit. In the gradiometer, two such superconducting accelerometers are placed end to end, and their circuits are connected in tandem, in such a way that the net output of the SQUID represents the difference between the displacements of the two proof masses, that is, it represents the difference between the gravitational accelerations at the positions of the two masses. The sensitivity attained by this gradiometer depends on the integration time; with a typical integration time of 10 s, the sensitivity demonstrated in a recent test of the gradiometer was about the same as that of the Eötvös balance, that is, about $10^{-32}/\text{cm}^2$, or $10^{-11}/\text{s}^2$ (Paik et al., 1992). Further development of this instrument is expected to improve

---

* In geodesy it is customary to describe tidal fields by $\partial^2\Phi/\partial x^k\partial x^l$, rather than $(1/c^2)\partial^2\Phi/\partial x^k\partial x^l$. A value of $10^{-32}/\text{cm}^2$ for the latter corresponds to $\simeq 10^{-11}/\text{s}^2$, or 10^{-2} E, for the former [the Eötvös unit (E) is defined as $10^{-9}/\text{s}^2$].

the sensitivity by a factor of several hundred.

Such a high-precision orbiting gradiometer could be used to test several aspects of gravitational theory. For instance, measurements of the diagonal components of the tidal-force tensor would serve as a test of the Laplace equation for the gravitational field (see Eq. [55]), that is, a test of the inverse-square law (Paik, 1979). Measurements of the other components of the (relativistic) tidal-force tensor would serve as a test of some of the predictions of Einstein's theory of gravitation.

The center-to-center distance between the two accelerometers making up the gradiometer is about 10 cm; measurements with such an instrument are quite "local." And, in principle, a reduction of the size of the apparatus is possible. As in the case of the Eötvös balance, a reduction of the size requires a reduction of the spring constant.

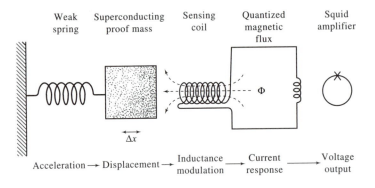

Fig. 1.20 Schematic diagram of the superconducting accelerometer. (After Moody, Chan, and Paik, 1986.)

Fig. 1.21 In this three-axis gradiometer, three pairs of superconducting accelerometers are arranged at right angles, to detect the gradients of the gravitational acceleration in three directions. (Courtesy H. J. Paik, University of Maryland.)

From this survey of actual experiments and *Gedanken*-experiments we see that there exist several methods for measuring the tidal field *locally*, in an arbitrarily small neighborhood of a given point. The limitations on the minimum size of the neighborhood that is needed to perform measurements of a given precision do not arise from any intrinsic properties of the gravitational field; rather, these limitations arise from the quantum nature of matter, which prevents us from constructing an apparatus of arbitrarily small size. The tidal field is no less a local quantity than, say, the electric field. Mathematically, this local nature of the tidal field is perfectly obvious: the tidal field is the derivative of the force field, which means that knowledge of the latter in an arbitrarily small neighborhood of a point is sufficient to determine the former at that point.

Local experiments can distinguish between a reference frame in free fall in a gravitational field and a truly inertial reference frame placed far away from all gravitational fields. *Local* experiments can distinguish between a reference frame at rest in a gravitational field and an accelerated reference frame far away from all gravitational fields. Gravitational effects are *not equivalent* to the effects arising from an observer's acceleration.* It is worth emphasizing this, because in his original paper on the theory of general relativity Einstein wrote:

> [Let K' be a system of reference such that] relative to K' a mass sufficiently distant from other masses has an accelerated motion such that its acceleration and direction of acceleration are independent of its material composition and physical state.
>
> Does this permit an observer at rest relative to K' to draw the conclusion that he is on a "really" accelerated system of reference? The answer is in the negative; for the above mentioned behavior of freely moving masses relative to K' may be interpreted equally well in the following way. The system of reference K' is unaccelerated, but the spacetime region being considered is under the sway of a gravitational field, which generates the accelerated motion of the bodies relative to K'. (Einstein, 1916)

This statement, which is one formulation of the principle of equivalence of gravitation and acceleration, is true only in a limited sense. Gravitation and acceleration are equivalent only as far as the translational motion of point particles is concerned (this amounts to what we called the Galileo principle of equivalence, sometimes also called the "weak" principle of equivalence, or WEP). If the rotational degrees of

---

* It would seem that a perfectly homogeneous gravitational field (zero tidal force) cannot be distinguished from the pseudo-force field of uniform acceleration. This is true, but not very relevant: perfectly homogeneous gravitational fields can exist only under extremely exceptional and unrealistic conditions. It can be shown that uniform fields are possible only in regions (cavities) completely surrounded by a continuous distribution of mass.

freedom of the motion of masses are taken into consideration, then the equivalence fails.

Unfortunately, Einstein's statement has often been generalized to sweeping assertions about all laws of physics being the same in a laboratory freely falling in a gravitational field and in another laboratory far away from any field (this is called the "strong" principle of equivalence, or SEP). Such generalizations are unwarranted since, as we have seen, even quite simple devices will signal the presence of a true gravitational field by their sensitivity to tidal forces and will therefore permit us to discriminate between a gravitational field and the pseudo-force field of acceleration. The confusion surrounding the principle of equivalence led Synge to remark:

> . . . I have never been able to understand this principle. . . . Does it mean that the effects of a gravitational field are indistinguishable from the effects of an observer's acceleration? If so, it is false. In Einstein's theory, either there is a gravitational field or there is none, according as the Riemann tensor* does or does not vanish. This is an absolute property; it has nothing to do with any observer's worldline. . . . The Principle of Equivalence performed the essential office of midwife at the birth of general relativity. . . . I suggest that the midwife be now buried with appropriate honours and the facts of absolute spacetime be faced. (Synge, 1971)

In order to avoid confusion, we will base our further development of gravitational theory on the very precise and unambiguous equality $m_I = m_G$. This equality is necessary and, to a large extent, sufficient for the construction of the relativistic theory.

FURTHER READING

Koestler, A., *The Sleepwalkers* (Grosset and Dunlap, New York, 1963) is a lively history of the beginning of our understanding of planetary motion as well as a study of the psychology of discovery. The book contains fascinating biographies of Copernicus and of Kepler.

North, J. D., *Isaac Newton* (Oxford University Press, Oxford, 1967) is a short but excellent scientific biography. For a wealth of details from Newton's personal life, see More, L. T., *Isaac Newton* (Scribner's, New York, 1934); this is written in the leisurely and elegant manner of the nineteenth century and is full of anecdotes and quotations from Newton's letters. Koyré, A., *Newtonian Studies* (Harvard University Press, Cambridge, 1965) gives a historical analysis of Newton's scientific work. Koyré, A., *The Astronomical Revolution* (Cornell University Press, Ithaca, 1973) covers the contributions of the predecessors of Newton to the theory of planetary motion and gravitation.

---

* Tidal-force tensor.

Mach, E., *Die Mechanik* (Brockhaus, Wiesbaden, 1933), of which a mediocre translation is available under the title *The Science of Mechanics* (Open Court, La Salle, 1960), is a classic historical and critical examination of the foundations of Newtonian physics. It contains profound and surprising insights into many fundamental issues.

The exciting stories of the rediscovery of the minor planet Ceres (through calculations of Gauss) and the discovery of Neptune (through calculations of Adams and LeVerrier) are told by Grosser, M., *The Discovery of Neptune* (Harvard University Press, Cambridge, 1962). Lyttleton, R. A., in an interesting article entitled "The Rediscovery of Neptune," Vistas in Astronomy **3**, 25 (1960), points out that although the calculations of Adams and LeVerrier predicted the position of Neptune to within 1^0 or 2^0, their methods were unnecessarily complicated and gave entirely wrong results for the orbital elements.

A good textbook on Newton's gravitational theory is Danby, J. M., *Fundamentals of Celestial Mechanics* (Macmillan, New York, 1962), which covers both the theory of the potential and the solution of the equations of motion.

MacMillan, W. D., *The Theory of the Potential* (Dover Publications, New York, 1958) is a detailed treatise on the Newtonian potential with many interesting and many irrelevant results; *if* the potentials produced by elliptic cylinders, parallelopipeds, etc., are needed, this is the place to find them.

Mathematically, the theory of the Newtonian potential is equivalent to that of the electrostatic potential; textbooks that treat electrostatics are therefore often helpful, one of the best being Jackson, J. D., *Classical Electrodynamics* (Wiley, New York, 1975).

A broad survey of measurements of the gravitational constant is given by de Boer (1984).

The quantum-mechanical properties of the Yukawa potential are discussed in Wentzel, G., *Quantum Theory of Fields* (Interscience, New York, 1949). (Careful! Wentzel uses the notation h for what is now commonly designated by $\hbar$.)

Some general background for the investigations of deviations from the inverse-square law is provided by Gibbons and Whiting (1981). Reviews of the observational and experimental evidence for such deviations are given by Will (1993, 1992, 1990, 1987), Adelberger et al. (1990, 1991), Talmadge and Fischbach (1987), and Stacey et al. (1987).

A nice description of the measurement of solar oblateness and of the astrophysical background is given in Dicke, R. H., *Gravitation and the Universe* (American Philosophical Society, Philadelphia, 1970). The article Dicke, R. H., "The Oblateness of the Sun and Relativity," Science **184**, 419 (1974) covers the same subjects in a more technical way.

The definitions of inertial mass, gravitational mass, and all manner of other "masses" are examined in their historical context in Jammer, M., *Concepts of Mass* (Harvard University Press, Cambridge, 1961).

The theory of the Eötvös balance and the effect of gravitational gradients on the measurement of m_I/m_G are discussed in "Beiträge zum Gesetze der Proportionalität von Trägheit und Gravität" by Eötvös et al. (1922).

The article "The Equivalence of Inertial and Passive Gravitational Mass" by Roll, Krotkov, and Dicke (1964) is strongly recommended. It is one of the few articles on experimental physics in a research journal that is so well written that it is fun to read. The authors describe the extreme precautions that had to be taken to prevent magnetic contamination, electrostatic effects, temperature gradients, etc., from spoiling their precision experiment. An abbreviated description of the experiment can be found in Dicke, *Gravitation and the Universe*.

A detailed analysis of the implications of the equivalence principle for gravitational theories is given in Will, C. M., *Theory and Experiment in Gravitational Physics* (Cambridge University Press, Cambridge, 1993); some of the same material is included in Will (1992). However, this analysis rests in part on the unrealistic and unjustifiable assumption that the electrons, protons, and neutrons in the ordinary test bodies used in our experiments contain no gravitational self-energies. This assumption leads Will to make a gratuitous distinction between an equivalence principle for bodies with and an equivalence principle for bodies without gravitational self-energy (called the strong equivalence principle and the Einstein equivalence principle, respectively). Furthermore, Will fails to consider the role of tidal forces in the equivalence principle.

Elementary, nonmathematical, but nevertheless interesting descriptions of the tides are given by Clancy, E. P., *The Tides* (Doubleday, New York, 1969) and by Defant, A., *Ebb and Flow* (University of Michigan Press, Ann Arbor, 1958). The static equilibrium theory of Section 1.8 represents the most oversimplified approach conceivable. For a discussion of the dynamics of tidal oscillations and the very important effect produced by Coriolis forces, see Defant, A., *Physical Oceanography*, vol. 2 (Pergamon Press, New York, 1961). The role of tidal forces in the local detection of gravitational fields is discussed in some more detail by Ohanian (1977).

REFERENCES

Adelberger, E. G., Stubbs, C. W., Rogers, W. F., Raab, F. J., Heckel, B. R., Gundlach, J. H., Swanson, H. E., and Watanabe, R., Phys. Rev. Lett. **59**, 849 (1987).

Adelberger, E. G., Stubbs, C. W., Heckel, B. R., Su, Y., Swanson, H. E., Smith, G., Gundlach, J. H., and Rogers, W. F., Phys. Rev. D **42** 3267 (1990).

Adelberger, E. G., Heckel, B. R., Stubbs, C. W., and Rogers, W. F., Ann. Rev. Nucl. Particle Sci. **41**, 269 (1991).

Ander, M. E., et al., Phys. Rev. Lett. (1989).

Barlier, F., Blaser, J. P., Cavallo, G., Damour, T., Decher, R., Everitt, C. W, F., Fuligni, F., Lee, M., Nobili, A., Nordvedt, K., Pace, O., Reinhard, R., and Worden, P., *Satellite Test of the Equivalence Principle*, ESA and NASA Assessment Study Report (1991).

Beams, J. W., Physics Today, May 1971.

Bennett, W. R., Phys. Rev. Lett. **62**, 365 (1989).

Bessel, F. W., Pogg. Ann. **25**, 401 (1832).

Bizetti, P. G., Bizetti-Sona, A. M, Fazzini, T., Perego, A., and Tacetti, N., Phys. Rev. Lett. **62**, 2901 (1989).

Boynton, P. E., Crosby, D., Ekstrom, P., Szumilo, A., Phys. Rev. Lett. **59**, 1385

(1987).

Braginsky, V. B., and Rudenko, V. N., Soviet Physics Uspekhi 13, 165 (1970).

Braginsky, V. B., and Panov, V. I., Zh. Eksp. and Teor. Fiz. 61, 873 (1971); translated in Sov. Phys. JETP 34, 463 (1972).

Brans, C., and Dicke, R. H., Phys. Rev. 124, 925 (1961).

Brown, T. M., Christensen-Dalsgaard, J., Dziembowski, W. A., Goode, P., Gough, D. O., and Morrow, C. A., Astrophys. Journal 343, 526 (1989).

Campbell, L., McDow, J. C., Moffat, J. W., and Vincent, D., Nature 305, 508 (1983).

Cavendish, H., Trans. Roy. Soc. London 18, 383 (1798); ibid., 18, 469 (1798).

Chan, H. A., Moody, M, V., and Paik, H. J., Phys. Rev. Lett. 49, 1745 (1982).

Chen, Y. T., Cook, A. H., and Metherell, A. J. F., Proc. Roy Soc. Lond. A 394, 47 (1984).

Cowsik, R., Krishnan, N., Tandon, S. N., and Unnikrishnan, C. S., Phys. Rev. Lett., 61, 2179 (1988).

Cremieu, M. V., Comptes Rendus 149, 700 (1909).

de Boer, H., "Experiments Relating to the Newtonian Gravitational Constant," in Taylor, B. N. and Phillips, W. D., eds., *Precision Measurement and Fundamental Constants* (Natl. Bur. Stand. U S., Spec. Publ. 617, 1984).

Dicke, R. H., "Experimental Relativity," in DeWitt, C., and DeWitt, B., *Relativity, Groups, and Topology* (Gordon and Breach, New York, 1964).

Dicke, R. H., and Goldenberg, H. M., Phys. Rev. Lett. 18, 313 (1967); see also Dicke, R. H., Science 184, 419 (1974).

Dicke, R. H., et al., Nature 316, 687 (1985).

Duvall, T. L., Dziembowski, W. A., Goode, P. R., Gough, D. O., Harvey, J. W., and Leibacher, J. W., Nature 310, 22 (1984).

Eckhardt, D. H., Jekeli, C., Lazarewicz, A. R., Romaides, A. J., and Sands, R. W., Phys. Rev. Lett. 60, 2567 (1988); for a correction of the data analysis, see Jekeli, C., Eckhardt, D. E., and Romaides, A. J., Phys. Rev. Lett. 64, 1204 (1990).

Einstein, A., Ann. der Phys. 49, 769 (1916); another translation is given in Lorentz, H. A., Einstein, A., Minkowski, H., and Weyl, H., *The Principle of Relativity* (Methuen, London, 1923), p. 114. Quoted with the permission of the Estate of Albert Einstein.

Eötvös, R. V., Math. Nat. Ber. Ungarn 8, 65 (1890).

Eötvös, R. V., Pekar, D., and Fekete, E., Ann. der Phys. 68, 11 (1922).

Facy, L., and Pontikis, C. R., Hebd. Seances Acad. Sci. Ser. B 274, 437 (1972).

Fischbach, E., Sudarsky, D., Szafer, A., Talmadge, C., and Aronson, S. H., Phys. Rev. Lett. 56, 3 (1986).

Fitch, V. L., Isaila, M. V., and Palmer, M. A., Phys. Rev. Lett. 60, 1801 (1988).

Galileo Galilei, *Dialogues Concerning Two New Sciences*. Translated from *Le Opere di Galileo Galilei* (Edizione Nazionale, Florence, 1898), pp. 128, 129.

Gibbons, G. W., and Whiting, B. F., Nature, 291, 636 (1981).

Good, M. L., Phys. Rev. 121, 311 (1961).

Hawking, S. W., and Israel, W. *Three Hundred Years of Gravitation* (Cambridge University Press, Cambridge, 1987).

Heyl, P. R., J. Res. Natl. Bur. Std. (U S.) 5, 1243 (1930).

Heyl, P. R. and Chrzanowski, P., J. Res. Natl. Bur. Std. (U S.) 29, 1 (1942).

Hill, H. A., Clayton, P. T., Patz, D. L., Healy, A. W., Stebbins, R. T., Oleson, J. R., and Zanoni, C. A., Phys. Rev. Lett. **33**, 1497 (1974).

Hill, H. A., Bos, R. J., and Godde, P. R., Phys. Rev. Lett., **49**, 1794 (1982).

Hill, H. A., Int. J. Theor. Phys., **23**, 683 (1984).

Hill, H. A., Rabaey, G. R., and Rosenwald, R. D., "The Sun's Gravitational Multipole Moment Inferred from the Fine Structure of the Acoustic and Gravity Normal Mode Spectra," in Kovalevsky, J., and Brumberg, V. A., eds., *Relativity in Celestial Mechanics and Astrometry* (Reidel, Dordrecht, 1986).

Holzscheiter, M. H., Physica Scripta **46**, 272 (1992); see also Holzscheiter, M. H., et al., Nuclear Physics **A558**, 709c (1993).

Hoskins, J. K., Newman, R. D., Spero, R., and Schultz, J., Phys. Rev. D **32**, 3084 (1985).

Keiser, P. T. and Faller, J. E., "Eötvös Experiment with a Fluid Fiber," in Ruffini, R., ed., *Proceedings of the Second Marcel Grossmann Meeting* (North-Holland, Amsterdam, 1982); see also Keiser, P. T., Faller, J. E., and McLagan, K. H., "New Laboratory Test of the Equivalence Principle," in Taylor, B. N., and Phillips, W. D., eds., Natl. Bur. Stand. (U.S.) Spec. Publ. **617** (1984).

Karagioz, O. V., et al., Phys. Solid Earth **12**, 351 (1976).

Koester, L., Phys. Rev. D **14**, 907 (1976).

Kuroda, K., and Mio, N., Phys. Rev. Lett. **62**, 1941 (1989).

Luther, G. G., and Towler, W. R., Phys. Rev. Lett. **48**, 121 (1982).

Moody, M. V., Chan, H. A., and Paik, H. J., J. Appl. Phys. **60**, 4308 (1986).

Müller, G., Zürn, W., Lindner, K., and Rösch, N., Geophys. J. Int. **101**, 329 (1990).

Müller, J., Schneider, M., Soffel, M., and Ruder, H., Astrophys. J. Lett. **382**, L101 (1991).

Newton, I., *Principia* (1686a), Book III, Proposition VII and Corollary.

Newton, I., *Principia* (1686b), "The System of the World," paragraph 19. Originally published by the University of California Press; reprinted by permission of The Regents of the University of California.

Niebauer, T. M., McHugh, M. P., and Faller, J. E., Phys. Rev. Lett. **59**, 609 (1987).

Ogawa, Y., et al., Phys. Rev. D **26**, 729 (1982).

Ohanian, H. C., Am. J. Phys. **45**, 903 (1977). Reprinted in Worden, P. W., and Everitt, C. W. F., eds., *Gravity and Inertia: Selected Reprints* (American Association of Physics Teachers, Stony Brook, 1983).

Paik, H. J., Phys. Rev. D **19**, 2320 (1979).

Paik, H. J., Canavan, E. R., Kong, Q., and Moody, M. V., preprint, 1992.

Panov, V. I., and Frontov, V. N., Zh. Exsp. Teor. Fiz. **77**, 1701 (1979).

Potter, H. H., Proc. Roy. Soc. (London) **104**, 588 (1923); ibid., **113**, 731 (1927).

Poynting, J. H., article on "Gravitation" in *The Encyclopedia Brittanica*, 11th edition, 1911.

Rapp, R. H., quoted by Stacey et al., 1987.

Reasenberg, R. D., and Shapiro, I. I., "Solar System Tests of General Relativity," in Bertotti, B., ed., *Proceedings of the International Symposium on Experimental Gravitation* (Accad. Naz. Lincei, Pavia, 1976); reprinted in Worden, P. W., and Everitt, C. W. F., *Gravity and Inertia* (American Association of Physics Teachers, Stony Brook, 1983).

Renner, J., Hung. Acad. Sci. **53**, 542 (1935).

Renner, Ya., Determination of the Gravitational Constant in Budapest, in Boulanger, Yu. D., and Sagitov, M. U., eds., *Determination of Gravity Constants and Measurement of Certain Fine Gravity Effects* (Nauka Press, Moscow, 1973).

Roll, P. G., Krotkov, R., and Dicke, R. H., Ann. Phys. (N.Y.) **26**, 442 (1964).

Rose, R. D., Parker, H. M., Lowry, R. A., Kuhltau, A. R., and Beams, J. W., Phys. Rev. Lett. **23**, 655 (1969).

Sagitov, M. U., Dokl. Akad. Nauk. SSSR **245**, 567 (1979); translated in Sov. Phys.-Dokl. (Earth Sci.) **245**, 20 (1981).

Schiff, L. I., Proc. Natl. Acad. Sci. (U.S.) **45**, 69 (1959).

Schiff, L. I., Phys. Rev. Lett. **4**, 215 (1960).

Smith, E., et al., J. Geophys. Res. **90**, 9221 (1985).

Southerns, L., Proc. Roy. Soc. (London) **84**, 325 (1910).

Speake, C. C., and Quinn, T. J., Phys. Rev. Lett. **61**, 1340 (1988).

Spero R., et al., Phys. Rev. Lett. **44**, 1645 (1980).

Stacey, F. D., and Tuck, G. J., Nature **292**, 230 (1981).

Stacey, F. D., Tuck, G. J., Moore, G. I., Holding, S. C., Goodwin, B. D., and Zhou, R., Rev. Mod. Phys. **59**, 157 (1987).

Stubbs, C. W., Adelberger, E., G., Raab, F. J., Gundlach, J., H., Heckel, B., R., McMurry, K. D.., Swanson, H. E., and Watanabe, R., Phys. Rev. Lett. **58**, 1070 (1987).

Stubbs, C. W., Adelberger, E., G., H., Heckel, B., R., Rogers, W. F., Swanson, H. E., Watanabe, R., Gundlach, J. H., and Raab, F. J., Phys. Rev. Lett. **62**, 609 (1989).

Synge, J. L., *Relativity; The General Theory* (North-Holland, Amsterdam, 1971), pp. IX-X. Reprinted with the permission of North-Holland Publishing Company.

Talmadge, C., Berthias, J.-P., Hellings, R. W., and Standish, E. M., Phys. Rev. Lett. **61**, 1159 (1988).

Talmadge, C., and Fishbach, E., "Searching for the Source of the Fifth Force," in De Sabbata, V., and Melnikov, V. N., eds., *Gravitational Measurements, Fundamental Metrology and Constants* (Kluwer, Dordrecht, 1987).

Thieberger, P., Phys. Rev. Lett. **58**, 1066 (1987).

Thomas, J., et al., Phys. Rev. Lett. **63**, 1902 (1989).

Thomas, J., and Vogel, P., Phys. Rev. Lett. **65**, 1173 (1990).

Will, C. M., *Theory and Experiment in Gravitational Physics* (Cambridge University Press, Cambridge, 1981).

Will, C. M., "Experimental Gravitation from Newton's *Principia* to Einstein's general relativity," in Hawking, S. W., and Israel, W., *Three hundred years of gravitation* (Cambridge University Press, Cambridge, 1987).

Will, C. M., Sky and Telescope, November 1990.

Will, C. M., Int. J. Mod. Phys. D **1**, 13 (1992).

Will, C. M., *Theory and Experiment in Gravitational Physics* (Cambridge University Press, Cambridge, 1993).

Yu, H.-T., Ni, W.-T., Hu, C.-C., Liu, F.-H., Yang, C.-H., and Liu, W.-N., Phys. Rev. D **29**, 1813 (1979).

Zahradnicek, J., Phys. Zeits. **34**, 126 (1933).

Zeeman, P., Proc. K. Akad. Amsterdam **20.4**, 542 (1918).

Zumberge, M. A., Hildebrand, J. A., Stevenson, J. M., Parker, R. L., Chave, A. D., Ander, M. E., and Spiess, F. N., Phys. Rev. Lett. **67**, 3051 (1991).

PROBLEMS

1. Consider the gravitational field produced by a sphere with a uniform mass density. Show that if you excavate a concentric spherical cavity in this sphere, the gravitational field will be zero within this cavity (that is, $\mathbf{g} = -\nabla\Phi = 0$). Show that if you excavate a spherical cavity *not* concentric with the original sphere, the gravitational field in this cavity will be constant (that is, $\mathbf{g} = -\nabla\Phi = $ [constant]).

2. Show that the Newtonian gravitational field of an infinitely long cylindrical shell is zero within the shell (that is, $\mathbf{g} = -\nabla\Phi = 0$). Show that the field of a finite cylindrical shell is zero in the equatorial plane within the shell.

3. A pointlike particle generates a Yukawa potential $\Phi = - Gm'e^{-r/\lambda}/r$. In this case, show that the potential outside a spherical distribution of such particles of total mass M and radius R is

$$\Phi(r) = - \frac{GM}{r}e^{-r/\lambda} \times \frac{3[(R/\lambda)\cosh(R/\lambda) - \sinh(R/\lambda)]}{(R/\lambda)^3}$$

Show that for $R \ll \lambda$, this reduces to $\Phi \simeq - GMe^{-r/\lambda}/r$. [Hint: In the interior of the mass distribution, the differential equation for the Yukawa potential is $(\nabla^2 - 1/\lambda^2)\Phi = 4\pi G\rho$.]

4. Under the influence of the gravitational attraction, an electron orbits around a (neutral) meteoroid of mass 1 kg in interstellar space. According to quantum mechanics, what is the Bohr radius for this system? What is the binding energy of the ground state?

5. Suppose that a planet has the shape of an ellipsoid of revolution of uniform density with mass, polar diameter, and equatorial diameter equal to those of the Earth. Evaluate the quadrupole moment tensor of this planet. Compare the magnitudes of the gravitational accelerations produced by the monopole and quadrupole fields of this planet at an altitude of 100 km above the equator.

6. According to the theory proposed by Brans and Dicke, G decreases gradually by perhaps 1 part in 10^{11} or 10^{12} per year. Consider a planet in a circular orbit around the Sun. If G decreases gradually, the orbit will expand and its period will increase. Show that

$$\frac{1}{r}\frac{dr}{dt} = - \frac{1}{G}\frac{dG}{dt}$$

$$\frac{1}{T}\frac{dT}{dt} = - \frac{2}{G}\frac{dG}{dt}$$

What is the rate of increase of the orbital radius and the period for the Earth if

$(1/G)dG/dt = -10^{-11}/\text{yr}$? (Hint: The orbital angular momentum $L = 2\pi mr^2/T$ remains constant as G changes, because the force always remains central.)

7. (a) Express the quadrupole moment tensor Q^{kl} in terms of the moment of inertia tensor I^{kl} and the trace I^{kk}.

(b) The moment of inertia of the Earth about the polar axis is $I^{33} = 0.3306\, MR^2$ and that about an equatorial axis is $I^{11} = I^{22} = 0.3295\, MR^2$, where M is the mass and R the equatorial radius of the Earth. Find all components of Q^{kl}. Evaluate the dimensionless parameter $J_2 = -Q^{33}/2MR^2$.

8. Under the assumption that the Earth is entirely covered with water, estimate the fractional difference between the equatorial and polar radii of the Earth. [Hint: This difference is due to centrifugal effects. Show that in the rotating reference frame of the Earth, the effective potential energy of a particle at a height h above the mean radius R is

$$mgh - \tfrac{1}{2}m\omega^2(R+h)^2 \sin^2\theta$$

where the second term is the centrifugal potential due to the rotation of the Earth; neglect h compared with R in this term. This calculation ignores the change in the gravitational field of the Earth produced by the change of shape and gives only an order of magnitude estimate.]

9. Repeat the preceding problem for the Sun, assuming $\omega = 2\pi$ radians/25 days. If the oblateness of the Sun is as reported by Dicke and Goldenberg, what ought to be the value of ω?

10. Two particles of mass M on the y-axis are connected by a spring (see Fig. 1.6a). Their relative distance undergoes simple harmonic motion: $s = s_0 + b_0 \sin \omega t$. Find the mass quadrupole tensor if the origin of coordinates is at the center of mass. Show that the time-dependent part of this quadrupole tensor is independent of the choice of origin.

11. A symmetric dumbbell rotates about an axis perpendicular to the line joining the masses (see Fig. 1.6b). Treating the masses as particles, find the quadrupole moment tensor (with respect to the center) as a function of time.

12. In spherical coordinates centered on the Earth, what are the radial and tangential components of the centrifugal acceleration due to the Earth's rotation at Budapest? By how much does the plumb line at Budapest deviate from the radial direction? Suppose that one plumb bob has $m_I = m_G$, and another bob has $m_I - m_G = 10^{-9}\, m_I$. What is the angular difference between the corresponding plumb lines?

13. What is the ratio of gravitational self-energy to rest-mass energy for a platinum sphere of radius 10 cm? For the Earth? For the Sun? For a neutron star? Treat the objects as continuous mass distributions of uniform density (ignore the structure of elementary particles). The radius of the neutron star is $\simeq$ 20 km and the mass equal to that of the Sun.

14. An accelerometer is a device that is used to measure gravitational and inertial accelerations. Essentially, it consists of a spring balance with a standard mass attached to it; the balance is calibrated so that when placed at a point it directly reads the acceleration that a free particle would have (in the rest frame of the accelerometer) when released at this point.

In an experimental test, a spacecraft in circular orbit around the Earth carried an accelerometer which was placed at a distance of 2 m from the center of mass (see Fig. 1.22). The acceleration measured by the accelerometer was 10^{-6} m/s² with the spacecraft oriented as shown in the figure.

(a) What is the radius of the orbit? Assume that the spacecraft is not rotating relative to the fixed stars.

(b) Explain how the output from several such accelerometers, placed at different points in the spacecraft, could be used to determine the orientation of the spacecraft relative to the Earth.

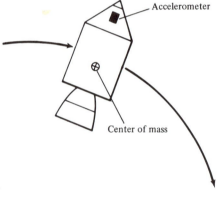

Fig. 1.22 Spacecraft with accelerometer.

Earth

15. Estimate the tidal force that acts on the atoms of a hydrogen molecule placed at the surface of a neutron star of mass $M \simeq 10^{33}$ g, radius $R \simeq 10$ km. (Give a numerical answer, to within a factor of 10 or so.)

16. A meter stick of mass 200 g is in free fall near the surface of the Earth. The meter stick is oriented vertically. Integrate the expression for the tidal force along the meter stick and thereby find the exact value of the tidal "tension" at the midpoint of the meter stick.

17. Calculate the height of the tide produced by the Sun according to our simple equilibrium model. Taking both Sun and Moon into account, what is the height of the tide when the Moon is full or new (spring tide)? In the first quarter or last quarter (neap tide)?

18. Estimate the fractional tidal deformation produced by the Earth on a drop of water of radius $\simeq$ 3 mm orbiting the Earth at an altitude of 100 km. [Hint: Assume that the drop has the shape of a spheroid with a semi-minor axis b, and semi-major axis $a = 3V/4\pi b^2$ (where the constant V is the volume of the drop). Estimate the work dW_t done by tidal forces if the semi-minor axis changes by db. Find the work dW_s done by surface tension. For equilibrium, $dW_t + dW_s = 0$.]

19. The Bay of Fundy is $\simeq$ 250 km long and $\simeq$ 100 m deep. Show that the natural frequency of the lowest mode of oscillation of the water is of the order of 12 hr. The velocity of waves in water of depth h is $v = \sqrt{gh}$ (this formula assumes that the depth is small compared to the wavelength, a condition that is always satisfied for tidal waves).

20. In the estuary of the river La Rance (on the coast of Brittany) the tidal range is as much as 13 m. The tidal bulge is captured behind a dam and as the tide ebbs the water is gradually allowed to run out through turbine generators. The total volume of the flow is 2×10^8 m^3. Estimate the time average electric power (in kW) generated by this tidal power plant. Does the energy come from the Moon or from the Earth? How is angular momentum conserved?

21. The mean angular velocity of the Moon is decreasing at the rate of $\simeq$ (26 ± 3) arcsec/(century)^2 (measured in ephemeris time). Assuming that this deceleration is entirely due to tidal friction on the Earth, find the rate at which the spin angular velocity of the Earth must decrease. (Hint: Treat the Moon's orbit as circular and use conservation of the orbital angular momentum of the Moon plus spin angular momentum of the Earth; neglect the spin of the Moon.) Find the rate at which rotational kinetic energy of the Earth is transformed into orbital energy of the Moon and find the rate at which energy is dissipated by tidal friction. Give numerical answers.

22. Use the result quoted in Table 1.4 to estimate the minimum torque that the Eötvös balance can detect. The balance beam is 40 cm long and the masses are 25 g each. What tidal field would produce such a torque?

23. What is the magnitude of the tidal field of the Earth at a distance equal to the Earth-Moon distance? What is the tidal field of a 1-kg lead sphere at a distance of 100 cm? Compare these numbers with the sensitivity of the Eötvös balance (10^{-32} cm^{-2}).

24. A spherical solid moon of mass m and radius R is orbiting a planet of mass M. Show that if the moon approaches the planet closer than

$$r_{crit} = R \left[\frac{2M}{m} \right]^{1/3}$$

then rocks lying on the surface of the moon will be lifted off by the tidal force. The distance r_{crit} is called *Roche's limit;* a moon approaching a planet closer than this limit will be disrupted by the tidal force.

2. The Formalism of Special Relativity

RAFFINIERT IST DER HERRGOTT, ABER BOSHAFT
IST ER NICHT. [God is cunning, but not malicious.]

Albert Einstein

We will be concerned with fields, that is, functions of space and time. It is therefore quite clear that we must first get to know the structure of spacetime. Unfortunately, because the velocity of light is so large, everyday experience leads us to acquire a certain number of misconceptions about the structure of spacetime. This set of misconceptions serves as the foundation of Newtonian, or Galilean, spacetime. The true structure of spacetime was discovered by Einstein in a study of electrodynamics (1905). As Einstein wrote in his autobiographical notes:

> After ten years of reflection such a principle (the principle of special relativity) resulted from a paradox upon which I had already hit at the age of sixteen: If I pursue a beam of light with the velocity c (velocity of light in a vacuum), I should observe such a beam as an electromagnetic field constant in time, periodic in space. However, there seems to exist no such thing, neither on the basis of experience, nor according to Maxwell's equations .... (Einstein, 1951)

This left Einstein with two alternatives: either Maxwell's equations were wrong in all reference frames except one (the ether frame), or else something was wrong with the structure of spacetime suggested by everyday experience. Einstein had a strong conviction that Maxwell's equations should not permit any instrinsic distinction between two inertial reference frames, and that therefore the equations should have the same form in all such reference frames. Accordingly, he decided for the second alternative: Newton's spacetime is wrong. He was then faced with the task of finding a structure of spacetime compatible with Maxwell's equations. In fact, it is not hard to find what geometry of spacetime will make Maxwell's equations true in all inertial reference frames. The desired geometry follows quite directly from the requirement that the velocity of light be the same in all inertial reference frames. The drastic modification of the geometry of space and time demands a corresponding modification of mechanics. Newton's laws of mechanics are true in all inertial reference frames if, and only if, spacetime is Newtonian. Einstein and his followers had to develop a new set of mechanical equations, designed so as to hold in all inertial reference frames in the relativistic spacetime.

In this chapter we review the special theory of relativity and develop the mathematical formalism of four-vectors and tensors. Detailed discussions of the physical basis of special relativity and of the experiments that compel us to accept the relativistic spacetime may be found in the books listed at the end of the chapter.

2.1 THE STRUCTURE OF SPACETIME

There are three different aspects to the structure of spacetime: the differential structure, the topological structure, and the geometric structure. Crudely speaking, the differential structure tells us how smooth spacetime is, and how many dimensions it has. The topological structure tells us how the different parts of spacetime are connected, that is, which points are in the neighborhood of which. Topology might be called "rubber-sheet geometry," because it is concerned with those properties of a space that are unchanged by arbitrary "smooth" deformations of the space. Finally, the geometric structure tells us how to construct parallel lines, and it tells us the distances between points in spacetime.

Experience teaches us that spacetime is a four-dimensional continuum. All the points in spacetime--such as, say, the point where and when a firecracker exploded, or the points where and when the fragments landed on the ground---can be smoothly parametrized by four real numbers, which we call the coordinates of the points. In Newtonian physics, as well as in special-relativistic physics, this parametrization of all points of spacetime by four coordinates can be carried out globally; if we use rectangular coordinates x, y, z, t, each of which ranges from $-\infty$ to $+\infty$, then these coordinates span the entire spacetime. In general-relativistic physics, where spacetime is curved, such a global parametrization is not possible; attempts at extending the coordinates in all directions usually result in singularities in the coordinates. Such coordinate singularities are similar to what a mapmaker finds when he attempts to use the longitude and latitude angles as coordinates for the curved two-dimensional surface of the Earth; these coordinates develop singularities at the poles of the Earth, where the longitude coordinate fails to be unique. Nevertheless, even for the curved spacetime of general relativity, it is still possible to construct well-behaved four-dimensional coordinate patches locally, for a finite neighborhood of any given point. In the same way, a mapmaker can draw a map with well-behaved coordinates for any part of the Earth; he merely must make sure to avoid latitude and longitude angles when drawing the map of the polar regions, and use instead, say, a rectangular grid. Mathematically, a space that can be covered by coordinate patches with real numbers as coordinates is called a *manifold*. Thus, spacetime is a four-dimensional manifold.

The topology of the spacetime of Newtonian physics, as well as the topology of the spacetime of special-relativistic physics, is that of Euclidean four-dimensional space (R^4). This simple topology is what permits us to parametrize these spacetimes globally, with a single coordinate patch for the entire spacetime. The neighborhood of a given point is the set of all points such that their coordinates differ only little from those of the given point. In view of the direct correspondence between points of spacetime and points of R^4, we can afford to use the careless terminology "the point (t, x, y, z)" rather than the more precise "the point whose coordinates are (t, x, y, z)."

All of this seems trivial, but is not because it is quite possible that the topology of spacetime is not Euclidean. One possible violation could occur at large distances: our universe might be closed and have the topology of a sphere. A violation of Euclidean topology is also possible at very small distances. When dealing with distances of 10^{-13} cm, or 10^{-23} cm, or even less, it is not possible to set up a rigid coordinate system of clocks and meter sticks. We tend to believe that it cannot be done because of the atomic structure of matter, but maybe there is a more fundamental reason why it cannot be done. Perhaps spacetime at a subnuclear level has a very pathological structure, is multiply connected, full of wormholes and bubbles; perhaps it is not even a *continuum*. We can only test for such diseases indirectly: we assume spacetime is a Euclidean continuum and use coordinates (t, x, y, z). If our theories based on the use of such coordinates are successful, then we can feel reasonably confident that we have made the right assumption.

At really small distances, of the order of 10^{-33} cm, a Euclidean topological structure is quite unlikely. At such distances the fluctuations of quantum gravitation will be extremely violent and probably produce an ever-changing, dynamic topology (Wheeler, 1968). The length $\sqrt{\hbar G/c^3}$ = 1.6 × 10^{-33} cm is called the *Planck length*; it gives the characteristic size of the fluctuations in quantum gravitation.

The differential and the topological structures of the spacetimes of Newton and of special relativity are the same. But these spacetimes differ in the definitions of distances, that is, in the metrical structure. Before we write down an expression for the distance between two points, we must make a restriction: choose the t-, x-, y-, z-axes of spacetime in such a way that we obtain an inertial reference frame. The trajectories of free particles in spacetime are supposed to be straight lines. This means that to obtain an inertial reference frame we must choose the axes, and the units along these axes, so that the trajectory of a free particle has the equation of a straight line in four dimensions, that is, $x = a_1 t + b_1$, $y = a_2 t + b_2$, $z = a_3 t + b_3$.

The trajectory of a particle in four-dimensional spacetime is called the *worldline* of the particle. Fig. 2.1 shows some worldlines of free particles in an inertial reference frame.

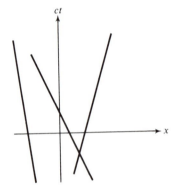

Fig. 2.1 Worldlines of free particles. For such a particle, the worldline, or the plot of position vs. time, is represented by a linear equation in x, y, z, and t.

Consider now two spacetime points separated by a small interval dt, dx, dy, dz (the use of infinitesimal intervals will be helpful for our later work). In order to define the geometry, we write down a quadratic form in dt, dx, dy, dz that specifies the distance. Newtonian spacetime has *two distances*, relativistic spacetime *only one*:

NEWTONIAN SPACETIME:
$$dl^2 = dx^2 + dy^2 + dz^2 \quad \text{(space interval)}$$
$$dt \quad \text{(time interval)} \tag{1}$$

RELATIVISTIC SPACETIME:
$$ds^2 = c^2 dt^2 - dx^2 - dy^2 - dz^2 \quad \text{(spacetime interval)} \tag{2}$$

In Eq. [2], c is the velocity of light in vacuum.

Thus, Newtonian spacetime contains two separate geometries: a three-dimensional Euclidean geometry for space and a one-dimensional geometry for time. Relativistic spacetime contains only one geometry which combines both space and time. As Minkowski expressed it: "Henceforth space by itself, and time by itself, are doomed to fade away into mere shadows, and only a kind of union of the two will preserve an independent reality" (Minkowski, 1908).

Note that the spacetime interval of the relativistic geometry is not always positive. Intervals with a positive, negative, or zero value of ds^2 are called, respectively, timelike, spacelike, and lightlike. Fig. 2.2 shows examples of displacements that have time-, space-, and lightlike intervals; in this figure lightlike displacements make an angle of 45^0 with the spacetime axes. Along the worldline of a massive particle, the interval must be timelike. This follows from Eq. [2] if we take into account that the speed $(dx^2 + dy^2 + dz^2)^{1/2}/dt$ for such a particle is less than c.

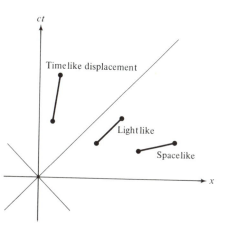

Fig. 2.2 Timelike, lightlike, and spacelike displacements. The lightlike displacements make an angle of 45^0 with the axes since for these displacements $dx = \pm cdt$.

Along the worldline of a light signal, the interval is of course light-like. The set of all worldlines of light signals leaving or arriving at a given point forms a three-dimensional "surface" in the four-dimensional spacetime. If the given point is the origin, this surface has the equation

$$c^2t^2 - x^2 - y^2 - z^2 = 0$$

and it is called the *light cone* corresponding to the given point. Since we cannot make a drawing of a four-dimensional space, let us omit the z coordinate; then the light cone is simply the ordinary cone $c^2t^2 - x^2 - y^2 = 0$ shown in Fig. 2.3. The upper ($t > 0$) and lower ($t < 0$) parts of this cone are called the forward and backward light cones, or the future and past light cones, respectively. Particles that leave the origin must have worldlines inside the forward light cone; particles that arrive at the origin must have worldlines inside the backward light cone.

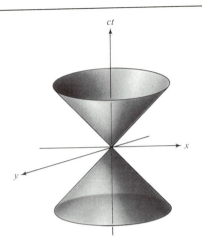

Fig. 2.3 The light cone.

An essential property of distance is that it is independent of the choice of coordinates. This is obvious in Newtonian spacetime, where we routinely measure space intervals with meter sticks and time intervals with clocks, without using coordinates. It is also true in relativistic spacetime, where we can directly measure the spacetime interval s^2 by means of the following general "radar-ranging" procedure: Suppose that P and P' are two arbitrary spacetime points separated by some finite interval. Take a clock with a straight worldline (a freely moving clock) that passes through the point P (see Fig. 2.4), and send a light signal from this clock to the point P', where the light signal is reflected and returns to the clock. If the time registered by the clock is zero when it passes through the point P, t_1 when the light signal is emitted, and t_2 when the reflected signal is received, then the spacetime interval between P and P' is

$$s^2 = c^2 t_1 t_2 \qquad [3]$$

It is instructive to consider two special cases of Eq. [3]. If the worldline of the clock passes through both of the points P and P' (which means that in the reference frame of the clock the spatial distance between P and P' is zero), then Eq. [3] reduces to the self-evident result $s^2 = c^2 t_1^2$. And if the worldline of the clock is at right angles to the displacement from P to P' (which means that in the reference frame of the clock P and P' are simultaneous), then $t_1 = -t_2$, and Eq. [3] reduces to $s^2 = -c^2 t_1^2$. This result implies that the spatial distance is $l = ct_1$, as was to be expected, since in this case our generalized radar-ranging procedure reduces to ordinary radar ranging. Note that, as an alternative, we could calculate the spacetime interval by means of Eq. [2], from separate measurements of the space and time intervals x and t. However, such a calculation is less direct than the radar-ranging procedure, and, furthermore, it requires us to set up a coordinate system, so we can define x and t. The radar-ranging procedure permits us to bypass the coordinate system, and it makes clear that the spacetime interval is independent of the choice of coordinates.

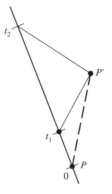

Fig. 2.4 Procedure for the direct measurement of the spacetime interval PP'.

The requirement that coordinate transformations between two inertial reference frames be such as to leave the distances invariant uniquely determines the form that these transformations may take. If the origin of the x', y', z' coordinates moves with velocity V along the x-axis of the x, y, z coordinates, then the Newtonian and relativistic transformations between these coordinates are, respectively:

NEWTONIAN SPACETIME-GALILEAN TRANSFORMATION
  Invariant distances:

$$dl^2 = dl'^2 \text{ (for } dt = 0)$$

$$dt = dt'$$

$$\Rightarrow \quad \begin{cases} t' = t \\ x' = x - Vt \\ y' = y \\ z' = z \end{cases}$$

[4]

RELATIVISTIC SPACETIME-LORENTZ TRANSFORMATION
  Invariant distance:

$$ds^2 = ds'^2 \quad \Rightarrow \quad \begin{cases} t' = \dfrac{t - Vx/c^2}{\sqrt{1 - V^2/c^2}} \\[2mm] x' = \dfrac{x - Vt}{\sqrt{1 - V^2/c^2}} \\[2mm] y' = y \\ z' = z \end{cases}$$

[5]

It can be shown that these are the only *linear* transformations that leave the distances [1] and [2] invariant, except for trivial shifts of origin or rotations of the axes. The transformations must be linear, in order that the worldlines of free particles be straight lines in both reference frames.

■ *Exercise 1.* Derive the transformation equations [4] and [5]. ■

We will now concentrate on relativistic spacetime and the Lorentz transformation. This transformation imposes severe restrictions on acceptable laws of physics. The reason is this: the spacetime structure has been designed so that, as far as the behavior of free particles and light rays is concerned, there is no distinction between different inertial reference frames--a free particle obeys the same laws in all inertial frames. We will now demand that *all* physical systems obey the same laws in all inertial frames, and that therefore no experiment can make an intrinsic distinction between different inertial reference frames. This is a symmetry principle imposed on the laws, called the *principle of relativity.* We can state this symmetry principle as follows:

Principle of relativity: All laws of physics must be invariant under Lorentz transformations.

In this context, "invariant" means that each law retains the same mathematical form, and that all numerical constants retain the same values.

The restrictions placed on the laws of physics by this symmetry principle are so stringent that, for example, in electrodynamics we can derive all the Maxwell equations from Coulomb's law alone (Schwartz, 1972; Ohanian, 1988). The importance of these restrictions is even greater in quantum theory: the principle of relativity and the structure of spacetime tell us that only particles of integer or half-integer spin can exist (for instance, spin $\frac{1}{4}\hbar$ is forbidden), that there must be a connection between spin and statistics, etc. (Streater and Wightman, 1964). What concerns us most is that the principle of relativity also puts very tight restrictions on possible theories of gravitation. We will exploit these restrictions in Section 3.2.

2.2 TENSORS IN SPACETIME

In this section we will develop the definitions of tensors in spacetime. Mathematicians often define tensors as operators that act on vectors. But we prefer to adopt a simpler definition based on the transformation properties of tensors under Lorentz transformations. These transformation properties are essential for the formulation of the principle of relativity.

We begin with some convenient notations. First, we will replace t, x, y, z by

$$x^0 \equiv ct, \quad x^1 \equiv x, \quad x^2 \equiv y, \quad x^3 \equiv z \qquad [6]$$

We can then write Eq. [5] as

$$ds^2 = \sum_{\mu=0}^{3} \sum_{\mu=0}^{3} \eta_{\mu\nu} \, dx^\mu \, dx^\nu \qquad [7]$$

where $\eta_{\mu\nu}$ is a matrix with components

$$\eta_{\mu\nu} = \begin{pmatrix} 1 & 0 & 0 & 0 \\ 0 & -1 & 0 & 0 \\ 0 & 0 & -1 & 0 \\ 0 & 0 & 0 & -1 \end{pmatrix} \qquad [8]$$

This is called the *metric tensor*, or the *Minkowski tensor*, of space-time.* We will use Einstein's summation convention, according to which any repeated Greek super- or subscript appearing in a term of an equation is to be summed from 0 to 3. Then Eq. [7] can also be written as

$$ds^2 = \eta_{\mu\nu} \, dx^\mu \, dx^\nu \qquad [9]$$

It is also convenient to define

$$x_\mu = \eta_{\mu\nu} \, x^\nu \qquad [10]$$

so

$$x_0 = ct, \quad x_1 = -x, \quad x_2 = -y, \quad x_3 = -z \qquad [11]$$

Note that in Eq. [10] the summation convention affects only the repeated index ν, not the index μ; the latter simply tells us that the equation has four components ($\mu = 0, 1, 2, 3$). An index, such as ν, repeated in a given term in an equation is called a *dummy* index because it disappears when the summation is written out in full,

$$\eta_{\mu\nu} x^\nu \equiv \eta_{\mu 0} x^0 + \eta_{\mu 1} x^1 + \eta_{\mu 2} x^2 + \eta_{\mu 3} x^3$$

Hence $\eta_{\mu\nu} x^\nu$, $\eta_{\mu\alpha} x^\alpha$, $\eta_{\mu\beta} x^\beta$, etc., are exactly the same thing. An index such as μ that appears only once in each term of an equation is called a *free* index.

The quantities x^μ and x_μ are called, respectively, the *contravariant* and *covariant* components of the position vector. For the sake of brev-

---

* The metric tensor is sometimes defined with signs opposite to those used here. However, the signs in Eq. [7] are those used by Einstein in his original work (1916), and there is no good reason for going against his convention.

ity, we will often use the phrase "the vector x^μ" or "the vector x_μ" instead of the cumbersome "the vector whose components are x^μ (or x_μ), with $\mu = 0, 1, 2, 3$." The abbreviated phraseology blurs the distinction between a vector and its components; but there is not much harm in that, since a vector can be regarded as a set of components. In terms of x^μ and x_μ, Eq. [9] becomes

$$ds^2 = dx^\mu \, dx_\mu \qquad [12]$$

which resembles the familiar dot product of two vectors.

In general, quantities with Greek superscripts are called contravariant, quantities with subscripts are called covariant. Thus, $\eta_{\mu\nu}$ is the covariant metric tensor. If we invert the relation [10], we obtain

$$x^\nu = \eta^{\nu\mu} x_\mu \qquad [13]$$

where $\eta^{\nu\mu}$ is the matrix inverse to $\eta_{\mu\nu}$. This inverse is defined by

$$\eta^{\nu\mu} \eta_{\mu\alpha} = \eta^\nu_\alpha \qquad [14]$$

where η^ν_α is the Kronecker delta (we also sometimes write δ^ν_α for the Kronecker delta). The matrix $\eta^{\nu\mu}$ is the contravariant metric tensor; the tensor η^ν_μ is *mixed*.

■ *Exercise 2.* Show that $\eta^{\nu\mu}$ has the same numerical value as [7]. ■

In terms of x^μ, the Lorentz transformation [5] can be written

$$x'^0 = \frac{x^0 - Vx^1}{\sqrt{1 - V^2}}$$

$$x'^1 = \frac{x^1 - Vx^0}{\sqrt{1 - V^2}}$$

$$x'^2 = x^2$$

$$x'^3 = x^3 \qquad [15]$$

In these equations we have adopted the convention $c = 1$, which can always be achieved by a clever choice of units. For instance, we can measure time in seconds and distance in light-seconds; then $c = 1$ light-second per second = 1. The convention $c = 1$ amounts to the same thing as omission of all factors of c. The restoration of the missing factors of c is easy: insert as many such factors as is necessary to make the equation dimensionally correct in cgs units. For numerical calculations, we will always use cgs units; that is, the factors of c are to be restored before inserting numerical values into any equations.

A neater way to write the Lorentz transformation is the following:

$$x'^\mu = a^\mu{}_\nu x^\nu \tag{16}$$

where $a^\mu{}_\nu$ is the matrix

$$a^\mu{}_\nu = \begin{pmatrix} 1/\sqrt{1 - V^2} & -V/\sqrt{1 - V^2} & 0 & 0 \\ -V/\sqrt{1 - V^2} & 1/\sqrt{1 - V^2} & 0 & 0 \\ 0 & 0 & 1 & 0 \\ 0 & 0 & 0 & 1 \end{pmatrix} \tag{17}$$

(The first index μ gives the row, the second index ν the column.)

The Lorentz transformation [16] is a particularly simple kind of transformation, in that the velocity is entirely along the x-axis, and therefore has only one component. The matrix $a^\mu{}_\nu$ for a general Lorentz transformation* depends on six parameters: three components of the relative velocity plus three angles which indicate how much the x'-, y'-, z'-axes are rotated with respect to the x-, y-, z-axes. Lorentz transformations with zero rotation angles are called *pure* transformations, or *boosts*. Lorentz transformations with zero velocity are nothing but ordinary rotations of the x-, y-, z-axes.

The matrix $a^\mu{}_\nu$ for a general Lorentz transformation must satisfy the equation

$$a^\alpha{}_\mu \eta_{\alpha\beta} a^\beta{}_\nu = \eta_{\mu\nu} \tag{18}$$

To see this, we substitute $x'^\alpha = a^\alpha{}_\mu x^\mu$ into the equation

$$x'^\alpha \eta_{\alpha\beta} x'^\beta = x^\mu \eta_{\mu\nu} x^\nu \tag{19}$$

which expresses the invariance of the (finite) spacetime interval between the origin and the x^μ. We obtain

$$a^\alpha{}_\mu x^\mu \eta_{\alpha\beta} a^\beta{}_\nu x^\nu = x^\mu \eta_{\mu\nu} x^\nu \tag{20}$$

or

$$x^\mu x^\nu (a^\alpha{}_\mu \eta_{\alpha\beta} a^\beta{}_\nu - \eta_{\mu\nu}) = 0 \tag{21}$$

In Eq. [21], x^μ and x^ν are arbitrary. Hence the coefficient of $x^\mu x^\nu$ must vanish; this implies Eq. [18].

---

* We will deal only with homogeneous transformations; thus, shifts of the origin of coordinates are excluded (no additive constant in Eq. [16]).

We will now give the definitions of tensors. The simplest tensor is a *scalar* (tensor of rank zero); this is a one-component object that remains unchanged under Lorentz transformations. For example, any numerical constant is a scalar. The spacetime interval ds^2 is a scalar since the Lorentz transformation has been designed precisely so as to leave ds^2 invariant. Another closely related scalar is the proper time interval $d\tau$ associated with the worldline of a particle. This is defined as equal to ds along the worldline of the particle, that is,

$$d\tau = \sqrt{dx^\mu dx^\nu} = \sqrt{dt^2 - dx^2 - dy^2 - dz^2} = dt\sqrt{1 - v^2} \qquad [22]$$

where $v = (dx^2 + dy^2 + dz^2)^{1/2}/dt$ is the particle's velocity. In the particle's rest frame, $d\tau = dt$. This gives us the physical interpretation for the proper time: it is the time interval measured by a clock that moves along with the same velocity as the particle.

A *vector* (or tensor of rank one) is a four-component object A^μ that transforms under Lorentz transformations in exactly the same way as x^μ:

$$A'^\mu = a^\mu_{\ \nu} A^\nu \qquad [23]$$

According to this definition, x^μ (and also dx^μ) serves as a prototype for all vectors.

If x^μ is the position vector of a particle, then the quantity

$$u^\mu \equiv \frac{dx^\mu}{d\tau} \qquad [24]$$

is a vector since it consists of the product of a vector dx^μ by a scalar $d\tau$. This vector is called the four-velocity. Written out explicitly, the components of u^μ are

$$u^\mu = \left(\frac{1}{\sqrt{1 - v^2}}, \ \frac{v_x}{\sqrt{1 - v^2}}, \ \frac{v_y}{\sqrt{1 - v^2}}, \ \frac{v_z}{\sqrt{1 - v^2}} \right) \qquad [25]$$

Another important vector is the energy-momentum four-vector, defined as the product of the four-velocity and the particle's rest mass,

$$p^\mu \equiv mu^\mu \qquad [26]$$

This proportionality between p^μ and u^μ looks like the proportionality between Newtonian momentum and velocity, but each side of Eq. [26] is more complicated than in the Newtonian case. Explicitly, the components of the energy-momentum four-vector are

$$p^\mu = \left(\frac{m}{\sqrt{1-v^2}}, \frac{mv_x}{\sqrt{1-v^2}}, \frac{mv_y}{\sqrt{1-v^2}}, \frac{mv_z}{\sqrt{1-v^2}} \right)$$

$$= (E, p_x, p_y, p_z) \tag{27}$$

These components are the energy (E) and the three components of the momentum (p_x, p_y, p_z) of the particle.

The general definition of a tensor* is as follows:

A tensor of rank r is an object $A^{\mu\nu...\kappa}$ with 4^r components that under a Lorentz transformation transforms according to

$$A'^{\alpha\beta...\gamma} = a^\alpha{}_\mu a^\beta{}_\nu ... a^\gamma{}_\kappa A^{\mu\nu...\kappa} \tag{28}$$

Note that the object $A^{\mu\nu...\kappa}$ has r indices, and there is one Lorentz transformation matrix for each index.

As an example of a second-rank tensor, consider the 16-component object

$$x^\mu x^\nu \tag{29}$$

From the transformation law for x^μ, it is easy to see that this object does transform as a tensor.

■ *Exercise 3.* Write down the components of the object [29] as a matrix. Prove that the object transforms as a second-rank tensor. ■

In general, any object constructed by taking the "product" of two vectors is a second-rank tensor. Thus, the objects

$$x^\mu p^\nu - x^\nu p^\mu \tag{30}$$

and

$$mu^\mu u^\nu \tag{31}$$

are sums of such tensors. The first is the angular-momentum tensor of a particle, and the second is related to the energy-momentum tensor (see below).

We saw above that the position vector can be written either in the contravariant or in the covariant form. For an arbitrary contravariant vector A^μ, we define the covariant component A_ν by

---

* Strictly speaking, what follows is the definition of a *Lorentz* tensor. We will deal with more general tensors in Chapter 6.

$$A_\nu = \eta_{\nu\mu} A^\mu \qquad [32]$$

This can be inverted to give

$$A^\mu = \eta^{\mu\nu} A_\nu \qquad [33]$$

■ *Exercise 4.* Show that [32] implies [33]. ■

The operations performed with the metric tensor on the right sides of Eqs. [32] and [33] are called, respectively, *lowering and raising an index*. The same procedure can be applied to one or more indices of a tensor of arbitrary rank. For example,

$$A_\mu{}^\nu = \eta_{\mu\alpha} A^{\alpha\nu}$$
$$A^\mu{}_\nu = \eta_{\nu\alpha} A^{\mu\alpha}$$
$$A_{\mu\nu} = \eta_{\mu\alpha}\eta_{\nu\beta} A^{\alpha\beta}$$
$$A^{\mu\nu} = \eta^{\mu\alpha}\eta^{\nu\beta} A_{\alpha\beta} \qquad [34]$$

■ *Exercise 5.* Write down the components of $u^\mu u^\nu$ as a matrix, expressing everything in terms of the ordinary velocities v_x, v_y, v_z. Repeat for $u_\mu u^\nu$, for $u^\mu u_\nu$, and for $u_\mu u_\nu$. ■

Our rules for raising and lowering indices can also be applied to the Lorentz-transformation matrix $a^\mu{}_\nu$. Thus, Eq. [18] can be written

$$a_{\beta\mu} a^\beta{}_\nu = \eta_{\mu\nu} \qquad [35]$$

If we raise the index μ, we obtain

$$a_\beta{}^\mu a^\beta{}_\nu = \eta^\mu_\nu \qquad [36]$$

This shows that $a_\beta{}^\mu$ is the inverse of $a^\beta{}_\nu$,

$$(a^{-1})^\mu{}_\beta = a_\beta{}^\mu \qquad [37]$$

The matrix $a_\beta{}^\mu$ can be used to write the transformation law for covariant tensors in the form

$$A'_{\alpha\beta\ldots\gamma} = a_\alpha{}^\mu a_\beta{}^\nu \ldots a_\gamma{}^\kappa A_{\mu\nu\ldots\kappa} \qquad [38]$$

This equation is obtained from [28] by lowering all the free indices.

■ *Exercise 6.* Prove this. ■

In particular,

$$x'_\mu = a_\mu{}^\nu x_\nu \qquad [39]$$

Combining this with Eq. [36] yields

$$x_\nu = a^\mu{}_\nu x'_\mu \qquad [40]$$

and this in turn gives

$$x^\nu = a_\mu{}^\nu x'^\mu \qquad [41]$$

■ *Exercise 7.* Derive Eqs. [40] and [41]. ■

Differentiation of Eqs. [16] and [41] gives, respectively,

$$a^\mu{}_\nu = \frac{\partial x'^\mu}{\partial x^\nu}$$

$$a_\mu{}^\nu = \frac{\partial x^\nu}{\partial x'^\mu} \qquad [42]$$

Given a second-rank tensor $A^{\mu\nu}$, the quantity

$$A^\mu{}_\mu \qquad [43]$$

is called the *trace* of the tensor. This is a one-component object which transforms as a scalar. We will sometimes simply use the notation A for the trace of a tensor $A^{\mu\nu}$,

$$A \equiv A^\mu{}_\mu \qquad [44]$$

■ *Exercise 8.* Show that $A^\mu{}_\mu$ remains invariant under Lorentz transformations. Also show that

$$A^\mu{}_\mu = A_\mu{}^\mu \qquad \blacksquare \qquad [45]$$

A tensor of rank higher than two has more than one trace. For example,

$$A^\mu{}_\mu{}^\alpha, \quad A^{\mu\nu}{}_\mu, \quad A^{\mu\nu}{}_\nu \qquad [46]$$

are three different traces of a third-rank tensor $A^{\mu\nu\alpha}$. The operation of taking the trace (for example, [46] or [43]) is often called *contrac-*

tion. Thus, $A^{\mu}{}_{\mu}{}^{\alpha}$ is the contraction of the first two indices of $A^{\mu\nu\alpha}$, and so on. It is obvious that contraction reduces the rank of a tensor by two.

For a tensor of the special type

$$B^{\mu} C^{\nu} \tag{47}$$

(where B^{μ} and C^{ν} are vectors), the contraction gives the scalar

$$B^{\mu} C_{\mu} \tag{48}$$

This is sometimes simply written as $B \cdot C$, and called the *scalar product* of the vectors B^{μ} and C^{ν}.

■ *Exercise 9.* Show that

$$u^{\mu} u_{\mu} = 1 \tag{49}$$

and that

$$p^{\mu} p_{\mu} = m^2 \quad ■ \tag{50}$$

Note that in summations, such as those implied by the repeated dummy indices in Eqs. [46] and [48], one of the repeated dummy indices is always covariant, and one is contravariant. If we were to form the sum $B^{\mu} C^{\mu}$, then the resulting object would not be a scalar--that is, we would get a result that depends on the reference frame used in the evaluation; such an object is uninteresting.

■ *Exercise 10.* Show that $B^{\mu} C^{\mu}$ is *not* a scalar. Show that $A^{\mu\mu\alpha}$ is *not* a vector. ■

A tensor $A^{\mu\nu}$ is said to be *symmetric* in $\mu\nu$ if

$$A^{\mu\nu} = A^{\nu\mu} \tag{51}$$

and it is said to be *antisymmetric* in $\mu\nu$ if

$$A^{\mu\nu} = -A^{\nu\mu} \tag{52}$$

Given an arbitrary tensor $B^{\mu\nu}$, it is always possible to write it as a linear combination of a symmetric and an antisymmetric part as follows:

$$B^{\mu\nu} = \tfrac{1}{2}(B^{\mu\nu} + B^{\nu\mu}) + \tfrac{1}{2}(B^{\mu\nu} - B^{\nu\mu}) \tag{53}$$

Obviously, the first term in parentheses is symmetric, and the second antisymmetric. Eq. [53] can also be compactly written as

$$B^{\mu\nu} = B^{(\mu\nu)} + B^{[\mu\nu]}$$ [54]

with the definitions

$$B^{(\mu\nu)} \equiv \tfrac{1}{2}(B^{\mu\nu} + B^{\nu\mu})$$ [55]

$$B^{[\mu\nu]} \equiv \tfrac{1}{2}(B^{\mu\nu} - B^{\nu\mu})$$ [56]

■ *Exercise 11.* Show that Lorentz transformations transform symmetric tensors into symmetric tensors and antisymmetric tensors into antisymmetric tensors. ■

Finally, given an arbitrary tensor $B^{\mu\nu}$ it is always possible to write it as a combination of a traceless tensor (that is, a tensor of zero trace) and $\eta^{\mu\nu}$ as follows:

$$B^{\mu\nu} = (B^{\mu\nu} - \tfrac{1}{4}\eta^{\mu\nu} B) + \tfrac{1}{4}\eta^{\mu\nu} B$$ [57]

Since $\eta^{\mu}{}_{\mu} = \delta^{\mu}{}_{\mu} = 4$, the term in parentheses is traceless.

2.3 TENSOR FIELDS

The definitions of the preceding section hold for tensors that are constant and also for tensors that vary in space and time; the latter are called *tensor fields*. For example, a *scalar field* is a function $\phi(P)$ that has different values at different points P of spacetime. At each given point of spacetime, the value of the scalar field is left invariant under Lorentz transformations:

$$\phi'(P) = \phi(P)$$ [58]

Instead of describing a given point of spacetime by the symbol P we can, of course, use coordinates. If the point P has coordinates x and x', respectively, in the two coordinate systems, then Eq. [58] becomes

$$\phi'(x') = \phi(x)$$ [59]

Here, we use an abbreviated notation for spacetime variables, without superscripts. In this abbreviated notation, x stands for x^0, x^1, x^2, x^3; and $\phi(x)$ stands for $\phi(x^0, x^1, x^2, x^3)$. If the function $\phi(x)$ is given, then Eq. [59] determines the function $\phi'(x')$, since x can be expressed in terms of x' by Eq. [41]:

$$\phi'(x') = \phi(x(x')) \qquad\qquad [60]$$

Note that $\phi'(x')$ is *not* the same function of x' as $\phi(x)$ is of x. The important thing is that each point of spacetime, ϕ and ϕ' take the same value. This is what is meant by invariance, and the notation [58] makes this clearer than [60] does. We will often simply write $\phi = \phi'$, leaving it understood that the functions are to be evaluated at the same point.

In general, a tensor field is a function of spacetime such that the tensor transformation law [28] holds at each given point of spacetime.

It can be shown that all the fields that describe the state of (continuous) physical systems in classical physics must necessarily be tensor fields. This important theorem rests on a group theoretical analysis of the Lorentz transformations, and therefore we must forgo the proof (Wigner, 1939). Thus, the theorem tells us that fields with more complicated transformation laws are not admissible. It follows that the *only* building blocks that we can use in the construction of a classical field theory are scalars, vectors, tensors, etc.* We will rely heavily on this fact in the next chapter.

The gradient operator in spacetime is

$$\partial_\mu \equiv \left(\frac{\partial}{\partial x^0}, \frac{\partial}{\partial x^1}, \frac{\partial}{\partial x^2}, \frac{\partial}{\partial x^3} \right) \qquad\qquad [61]$$

The chain rule for differentiation and Eq. [42] tell us that

$$\partial'_\mu = \frac{\partial}{\partial x'^\mu} = \frac{\partial x^\nu}{\partial x'^\mu} \frac{\partial}{\partial x^\nu} = a_\mu{}^\nu \partial_\nu \qquad\qquad [62]$$

This shows that ∂_μ transforms as a covariant vector. We will also sometimes use

$$\partial^\mu \equiv \eta^{\mu\nu} \partial_\nu = \left(\frac{\partial}{\partial x^0}, -\frac{\partial}{\partial x^1}, -\frac{\partial}{\partial x^2}, -\frac{\partial}{\partial x^3} \right) \qquad\qquad [63]$$

This operator has the transformation law of a contravariant vector,

$$\partial'^\mu = a^\mu{}_\nu \partial^\nu \qquad\qquad [64]$$

The "Laplacian" operator in spacetime is

---

* In quantum theory, besides tensor fields there exist spinor fields. The tensor fields correspond to quanta of spin 0, 1, 2,...; the spinor fields to quanta of spin 1/2, 3/2,....

$$\partial^\mu \partial_\mu = \left[\frac{\partial}{\partial x^0}\right]^2 - \left[\frac{\partial}{\partial x^1}\right]^2 - \left[\frac{\partial}{\partial x^2}\right]^2 - \left[\frac{\partial}{\partial x^3}\right]^2 \qquad [65]$$

This operator is invariant under Lorentz transformations:

$$\partial'^\mu \partial'_\mu = \partial^\mu \partial_\mu \qquad [66]$$

■ *Exercise 12.* Prove Eqs. [64] and [66]. ■

As a consequence of Eq. [64], $\partial^\mu \phi$ is a vector whenever ϕ is a scalar. It also follows that $\partial^\mu \partial^\nu \phi$ is a tensor and $\partial^\mu \partial_\mu \phi$ is a scalar. In general, when ∂^μ operates on a tensor of rank r, the result is a tensor of rank $r + 1$.

■ *Exercise 13.* Prove that $\partial^\mu A^{\alpha\beta...\kappa}$ is a tensor of rank $r + 1$ whenever $A^{\alpha\beta...\kappa}$ is a tensor of rank r. ■

It is convenient to indicate derivatives by commas, in the style of the following examples:

$$\phi_{,\mu} \equiv \partial_\mu \phi$$
$$\phi^{,\mu} \equiv \partial^\mu \phi$$
$$\phi^{,\mu,\nu} \equiv \partial^\nu \partial^\mu \phi$$
$$\phi^{,\mu}{}_{,\mu} \equiv \partial^\mu \partial_\mu \phi$$
$$A^{\alpha\beta}{}_{,\mu} \equiv \partial_\mu A^{\alpha\beta} \text{ , etc.} \qquad [67]$$

This concludes our discussion of the mathematical aspects of special relativity. More on tensor analysis in four-dimensional spacetime will be found in the books listed at the end of the chapter. We will now turn to some important applications of the formalism.

2.4 THE ENERGY-MOMENTUM TENSOR

In our search for a relativistic theory of gravitation the energy density will play an important role. But it makes little sense to consider the energy density by itself, because what is energy density in one reference frame will be some combination of energy density, energy flux density, and momentum flux density as seen from another reference frame. Hence all these quantities form a single tensor object, and they must be considered together.

It will be best to explain this for a system consisting of a collection of noninteracting particles (a cloud of dust). Suppose that in the neighborhood of a point in this cloud the density of particles is n per unit volume and their velocity is **v**. In this case the *energy density* can be expressed in terms of the density and the velocity of the particles*:

$$T^{00} = \frac{nm}{\sqrt{1 - v^2}}$$ [68]

This is simply the product of the number n of particles per unit volume and the (relativistic) energy per particle. We use the notation T^{00} for energy density for reasons which will soon be obvious.

The *energy flux density* can be defined in the following way: The energy flux in the x direction is the amount of energy transported in unit time across a unit y-z area, that is,

$$\frac{nmv_x}{\sqrt{1 - v^2}}$$ [69]

This is the product of particle current nv_x by the energy per particle. In general, the energy flux density in the k direction is

$$T^{k0} = \frac{nmv^k}{\sqrt{1 - v^2}}$$ [70]

As an alternative notation for [70], we will use T^{0k}, so $T^{k0} \equiv T^{0k}$. Note that T^{k0} can also be regarded as the *density of momentum*.

Finally, let us define the *momentum flux density*. The x-y momentum flux density is defined as the amount of x momentum that flows in the y direction per unit area and unit time. Since the x momentum per particle is

$$\frac{mv_x}{\sqrt{1 - v^2}}$$

the x-y momentum flux density must be

$$\left(\frac{nmv_x}{\sqrt{1 - v^2}} \right) v_y$$ [71]

And the general expression for the k-l momentum flux density is

$$T^{kl} = \frac{nmv^k v^l}{\sqrt{1 - v^2}}$$ [72]

---

* This expression follows the convention $c = 1$. Restoration of factors of c would give $T^{00} = nmc^2/\sqrt{1 - v^2/c^2}$ in cgs units.

It is now easy to show that the 16-component object $T^{\mu\nu}$ given by Eqs. [68], [70], and [72] is a tensor under Lorentz transformations. To show this, we write $T^{\mu\nu}$ as follows:

$$T^{\mu\nu} = n_0 m u^\mu u^\nu \qquad [73]$$

where

$$n_0 \equiv n\sqrt{1 - v^2} \qquad [74]$$

■ *Exercise 14.* Show that Eq. [73] agrees with Eqs. [68], [70], and [72]. ■

The quantity n_0, called the *proper particle density*, is the particle density as measured in a reference frame that moves with the particles. The relation between n_0 and n given by Eq. [74] is a consequence of the well-known volume-contraction effect of special relativity: a volume containing a given number of particles and moving along with them is contracted as measured in the laboratory frame; hence the laboratory density n is increased over the proper density n_0.

Since n_0 is a number measured in the local rest frame, it is a scalar. Hence $n_0 m u^\mu u^\nu$ is the product of a scalar n_0 by the tensor $m u^\mu u^\nu$, and therefore $n_0 m u^\mu u^\nu$ is a tensor. Eq. [73] can also be expressed as

$$T^{\mu\nu} = \rho_0 u^\mu u^\nu \qquad [75]$$

where

$$\rho_0 \equiv n_0 m \qquad [76]$$

is the mass density as measured in the local rest frame of the particles, called the *proper mass density*. Note that the tensor $T^{\mu\nu}$ is symmetric:

$$T^{\mu\nu} = T^{\nu\mu} \qquad [77]$$

The definition of $T^{\mu\nu}$ can be summarized as follows:

$$T^{00} = [\text{energy density}]$$

$$T^{0k} = T^{k0} = [k \text{ momentum density}] = [\text{energy flux density}]$$

$$T^{kl} = T^{lk} = [k \text{ momentum flux density in } l \text{ direction}] \qquad [78]$$

This definition applies in general. For example, if instead of a system of dust particles we have a system consisting of electromagnetic fields, then we can still use Eq. [78] to define the corresponding energy-momentum tensor (see next section).

The energy-momentum tensor will usually be a function of space and time. For example, in Eq. [75] both the particle density and the velocity may vary in space and time. The energy-momentum tensor of a complete system is*

$$\partial_\nu T^{\mu\nu} = 0 \qquad [79]$$

The proof of this is no harder than the well-known proof for the conservation law of charge. Consider the $\mu = 1$ component of this equation:

$$\frac{\partial}{\partial x^0} T^{10} + \frac{\partial}{\partial x^k} T^{1k} = 0 \qquad [80]$$

To derive this from the conservation of momentum, take a small volume in the system. The x momentum in the volume is

$$\int T^{10} \, d^3x \qquad [81]$$

and the rate of decrease of this x momentum is

$$-\int \frac{\partial}{\partial x^0} T^{10} \, d^3x \qquad [82]$$

This momentum decrease must be the result of x momentum flowing out of the volume. The total outflow of x momentum across the boundary of the volume is

$$\int T^{1k} \, dS_k \qquad [83]$$

We can apply Gauss' theorem to the three-component object T^{1k} ($k = 1, 2, 3$) and convert [83] into

$$\int \frac{\partial}{\partial x^k} T^{1k} \, d^3x \qquad [84]$$

---

* By a "complete system" is meant a system not subject to external forces (that is, all forces are already included in the system).

By conservation of momentum, the terms [82] and [84] must be equal. If the volume is small enough, the integrands are nearly constant and the integration can be replaced by a multiplication of integrand by volume:

$$\left[-\frac{\partial}{\partial x^0} T^{10} \right] \Delta x \Delta y \Delta z = \left[\frac{\partial}{\partial x^k} T^{1k} \right] \Delta x \Delta y \Delta z \qquad [85]$$

This reduces to Eq. [80] if we cancel the volumes. The proof for the other components of Eq. [79] is similar.

According to our definition of $T^{\mu\nu}$, the volume integrals of T^{00} and of T^{k0} give, respectively, the total energy and the total momentum of the system:

$$P^0 = \int T^{00} \, d^3x \qquad [86]$$

$$P^k = \int T^{k0} \, d^3x \qquad [87]$$

The four-component object P^μ ought to be a four-vector, but it is not at all obvious that the volume integrals appearing on the right side of Eqs. [86] and [87] do transform correctly under Lorentz transformations. To prove that the four integrals [86] and [87] do transform as a four-vector, we begin with the conservation law

$$\frac{\partial}{\partial x^\mu} T_\nu{}^\mu = 0 \qquad [88]$$

and multiply this by an arbitrary *constant* vector B^ν,

$$\frac{\partial}{\partial x^\mu} (T_\nu{}^\mu B^\nu) = 0 \qquad [89]$$

Integration over a four-dimensional volume gives

$$\int \frac{\partial}{\partial x^\mu} (T_\nu{}^\mu B^\nu) \, dt d^3x = 0 \qquad [90]$$

The left side of this equation has the form

$$\int \frac{\partial}{\partial x^\mu} C^\mu \, dt d^3x$$

where $C^\mu = T_\nu{}^\mu B^\nu$. We can apply the four-dimensional analog of the familiar Gauss theorem to convert the volume integral into a "surface" integral,

$$\int \frac{\partial}{\partial x^\mu} C^\mu \, dt d^3x = \int C^\mu \, dS_\mu \qquad [91]$$

Since the volume is four-dimensional, its "surface" is actually a three-dimensional hypersurface. The vector dS_μ has a magnitude equal to the corresponding three-dimensional volume element and a direction "perpendicular" to the hypersurface; that is, $dS_\mu \delta x^\mu = 0$ whenever δx^μ is any displacement vector within the hypersurface.

■ *Exercise 15.* Prove the four-dimensional Gauss theorem of Eq. [91]. (Hint: A volume of arbitrary shape can be regarded as made up of a large number of rectangular boxes. Hence, it suffices to prove Eq. [91] for the case of a volume in the shape of a rectangular box with sides perpendicular to the coordinate axes.) ■

For simplicity, we will assume that $T_\nu{}^\mu$ differs from zero only in some finite region of space. If we consider this finite region of space at successive instants of time, then in spacetime $T_\nu{}^\mu$ is nonzero in a four-dimensional tube of infinite length in the time direction and of a finite cross section in the space directions. Fig. 2.5 shows this tube in the x^0-x^1 plane: the two vertical lines form the boundaries of the tube, and $T_\nu{}^\mu$ is different from zero only between these boundaries.

Consider now a Lorentz transformation which introduces new coordinates x'^μ according to Eq. [16]. The equations

$$x^0 = [\text{constant}] \qquad [92]$$

and

$$x'^0 = [\text{constant}] \qquad [93]$$

define three-dimensional hypersurfaces in spacetime. These surfaces are distinct (if we express [93] in terms of x coordinates, it becomes $x^0 - vx^1 = [\text{constant}] \times \sqrt{1 - v^2}$, which differs from [92]). In Fig. 2.5, the hypersurfaces [92] and [93] are represented by the heavy lines; in this figure it has been assumed that $[\text{constant}] = 0$. We now concen-

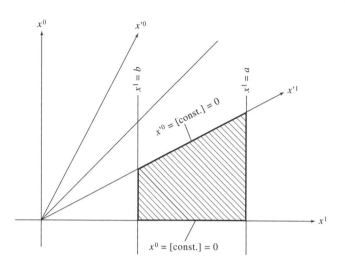

Fig. 2.5 The energy-momentum tensor is different from zero in the region included between the lines $x^1 = a$, $x^1 = b$.

trate on that part of the tube included between the surfaces [92] and [93]; in Fig. 2.5 this four-dimensional region is shown hatched. If we apply Eqs. [90] and [91] to this region, we obtain

$$0 = \int T_\nu{}^\mu B^\nu \, dS_\mu$$

$$= \int_I T_\nu{}^\mu B^\nu \, dS_\mu + \int_{II} T_\nu{}^\mu B^\nu \, dS_\mu + \int_{III} T_\nu{}^\mu B^\nu \, dS_\mu \qquad [94]$$

where the surface I is at the top (Eq. [93]), the surface II is at the bottom (Eq. 92]), and the surface III is the "lateral" surface of the tube (see Fig. 2.5). Since $T_\nu{}^\mu = 0$ on the lateral surface, we obtain

$$\int_I T_\nu{}^\mu B^\nu \, dS_\mu = - \int_{II} T_\nu{}^\mu B^\nu \, dS_\mu \qquad [95]$$

The surface element appearing on the right side is $dS_\mu = (-d^3x, 0, 0, 0)$, and hence the right side reduces to

$$\int_{II} T_\nu{}^0 B^\nu \, d^3x \tag{96}$$

The surface element appearing on the left side of Eq. [95] is more complicated. However, we can take advantage of the fact that the complete integrand is a scalar ($T_\nu{}^\mu$ is a tensor, B^ν and dS_μ are vectors), so

$$\int_I T_\nu{}^\mu B^\nu \, dS_\mu = \int_I T'_\nu{}^\mu B'^\nu \, dS'_\mu \tag{97}$$

Since in the new coordinates $dS'_\mu = (d^3x', 0, 0, 0)$, it follows that the left side of Eq. [95] is simply

$$\int_I T'_\nu{}^0 B'^\nu \, d^3x \tag{98}$$

We have therefore established the equality

$$B'^\nu \int_I T'_\nu{}^0 \, d^3x = B^\nu \int_{II} T_\nu{}^0 \, d^3x \tag{99}$$

which shows that the quantity $B^\nu \int T_\nu{}^0 d^3x$ is a scalar under Lorentz transformations. Since, by hypothesis, B^ν is an arbitrary contravariant vector, we can conclude that $\int T_\nu{}^0 d^3x$ is a covariant vector. This is what we wanted to prove.

2.5 RELATIVISTIC ELECTRODYNAMICS

We will now examine the implications of the principle of relativity for Maxwell's equations. In general, the principle of relativity demands the invariance of the laws of physics under Lorentz transformations. It is obvious that any equation of the form

$$A^{\mu\nu\ldots\kappa} = 0 \tag{100}$$

where $A^{\mu\nu\ldots\kappa} = 0$ is a tensor of rank r, will satisfy the principle of relativity. In fact, multiplication of Eq. [100] by $a^{\alpha}{}_{\mu} a^{\beta}{}_{\nu} ... a^{\gamma}{}_{\kappa}$ gives

$$A'^{\mu\nu...\kappa} = 0 \qquad\qquad [101]$$

which shows that the law [100] is invariant under the Lorentz transformation that is described by the matrix $a^{\alpha}{}_{\mu}$. Our next objective is to show that the equations of electrodynamics have the form of tensor equations, like Eq. [100].

Maxwell's equations of electrodynamics are in cgs (Gaussian) units,*

$$\nabla \cdot \mathbf{E} = 4\pi\rho \qquad\qquad [102]$$

$$\nabla \times \mathbf{B} - \frac{1}{c}\frac{\partial \mathbf{E}}{\partial t} = \frac{4\pi}{c}\mathbf{j} \qquad\qquad [103]$$

$$\nabla \cdot \mathbf{B} = 0 \qquad\qquad [104]$$

$$\nabla \times \mathbf{E} + \frac{1}{c}\frac{\partial \mathbf{B}}{\partial t} = 0 \qquad\qquad [105]$$

These equations do not seem to be of the type [100], and their Lorentz invariance is not at all obvious.

In order to discuss the Lorentz-transformation properties of Eqs. [102]-[105], we begin by rewriting them in four-dimensional component notation as follows:

$$\partial_{\mu} F^{\mu\nu} = \frac{4\pi}{c} j^{\nu} \qquad\qquad [106]$$

$$\partial^{\alpha} F^{\mu\nu} + \partial^{\mu} F^{\nu\alpha} + \partial^{\nu} F^{\alpha\mu} = 0 \qquad\qquad [107]$$

where the four-component object j^{ν} is defined as

$$j^{\nu} \equiv (c\rho, j_x, j_y, j_z) \qquad\qquad [108]$$

and the 16-component object $F^{\mu\nu}$ is defined as

$$F^{\mu\nu} \equiv \begin{pmatrix} 0 & -E_x & -E_y & -E_z \\ E_x & 0 & -B_z & B_y \\ E_y & B_z & 0 & -B_x \\ E_z & -B_y & B_x & 0 \end{pmatrix} \qquad [109]$$

With these definitions, it is a straightforward exercise to check that

---

* For relativity, cgs units have the advantage that the units of $\mathbf{E}$ and $\mathbf{B}$ coincide, which simplifies the Lorentz transformations between these fields.

the components of Eq. [106] are equivalent to Eqs. [102], [103]; and the components of Eq. [107] are equivalent to Eqs. [104], [105]. For instance, if $\nu = 0$, Eq. [106] gives

$$\partial_0 F^{00} + \partial_1 F^{10} + \partial_2 F^{20} + \partial_3 F^{30} = \frac{4\pi}{c} j^0$$

According to Eqs. [108] and [109] this is simply

$$\frac{\partial}{\partial x} E_x + \frac{\partial}{\partial y} E_y + \frac{\partial}{\partial z} E_z = 4\pi\rho$$

which agrees with Eq. [102].

■ *Exercise 16.* Check the other components of Eqs. [106] and [107]. ■

Written in the form [106] and [107], Maxwell's equations look like tensor equations. However, appearances can be deceiving, and unless $F^{\mu\nu}$ is a second-rank tensor and j^ν is a four-vector, these equations will *not* be tensor equations. Before we jump to the conclusion that the equations are tensor equations, we must examine the Lorentz-transformation properties of j^ν and $F^{\mu\nu}$.

To establish that the object j^ν is a four-vector, we recall the definition of charge and current density. Suppose we have a cloud of particles (a fluid), all of the same kind. In the neighborhood of any given point, we describe this distribution of particles by the number per unit volume and by the velocity. Both the number density and the velocity are functions of space and time:

$$n = n(x)$$
$$\mathbf{v} = \mathbf{v}(x) \tag{110}$$

The charge density is then defined by

$$\rho(x) = qn(x) \tag{111}$$

and the current density by

$$\mathbf{j} = qn(x)\mathbf{v}(x) \tag{112}$$

In these equations q is the charge of one particle. For example, $q = -e$ for electrons, $q = +e$ for protons, $q = 0$ for neutrons. If several types of particles are present, then we must combine their charge and current densities additively.

Note that as a consequence of the conservation of the number of particles, the charge is conserved:

$$\nabla \cdot \mathbf{j} + \frac{\partial \rho}{\partial t} = 0 \qquad [113]$$

■ *Exercise 17.* Show this. ■

Conservation of the number of each kind of particle is sufficient, but not necessary, for the validity of Eq. [113]. Obviously all that is necessary is that in any reaction that destroys or creates particles, the net charge should be the same before and after.

In view of Eqs. [111] and [112], we can express the four-component object j^ν as

$$j^\nu = (c\rho, j_x, j_y, j_z)$$

$$= (c\rho n, qnv_x, qnv_y, qnv_z) \qquad [114]$$

We can also write this as

$$j^\nu = n_0 q u^\nu$$

or as

$$j^\nu = \rho_0 u^\nu \qquad [115]$$

where

$$n_0 = n\sqrt{1 - v^2/c^2} \qquad [116]$$

and

$$\rho_0 = qn_0 \qquad [117]$$

Here n_0 is the proper particle density (see Eq. [74]) and ρ_0 the proper charge density*; these proper quantities are scalars. Since j^ν consists of u^ν multiplied by a scalar, it must have the transformation law of a four-vector,

$$j'^\beta = a^\beta{}_\nu j^\nu \qquad [118]$$

The four-vector j^ν is called the electric *current density four-vector.*

In terms of this four-vector, we can write the conservation law [113] as

---

* The symbol ρ_0 has also been used for mass density in Section 2.4. Which meaning is intended will be clear from the context.

$$\partial_\nu j^\nu = 0 \qquad [119]$$

This equation is of the general form [100], and its Lorentz invariance is obvious.

Next, we want to establish the Lorentz-transformation properties of the 16-component object $F^{\mu\nu}$. In contrast to j^ν, for which we could rely on the known Lorentz-transformation properties of the building blocks ρ_0 and u^ν out of which j^ν is constructed, we now have no such building blocks. Instead, we must appeal to the principle of relativity, which tells us that under a Lorentz transformation the law [106] must become

$$\partial'_\alpha F'^{\alpha\beta} = \frac{4\pi}{c} j'^\beta \qquad [120]$$

Since $\partial'_\alpha = a_\alpha{}^\mu \partial_\mu$ and $j'^\beta = a^\beta{}_\sigma j^\sigma$, we can also express Eq. [120] as

$$a_\alpha{}^\mu \partial_\mu F'^{\alpha\beta} = \frac{4\pi}{c} a^\beta{}_\sigma j^\sigma$$

If we multiply this by $a_\beta{}^\nu$ and use Eq. [36], we obtain

$$a_\beta{}^\nu a_\alpha{}^\mu \partial_\mu F'^{\alpha\beta} = \frac{4\pi}{c} j^\nu \qquad [121]$$

We now compare Eqs. [121] and [106]. Agreement between these equations demands

$$a_\beta{}^\nu a_\alpha{}^\mu \partial_\mu F'^{\alpha\beta} = \partial_\mu F^{\mu\nu} \qquad [122]$$

To extract the transformation properties of $F^{\mu\nu}$ from this equation, we note that, on physical grounds, the relationship between $F'^{\mu\nu}$ and $F^{\mu\nu}$ must be linear and homogeneous,* that is, it must be of the form

$$F^{\mu\nu} = C^{\mu\nu}_{\alpha\beta} F'^{\alpha\beta} \qquad [123]$$

where $C^{\mu\nu}_{\alpha\beta}$ is a set of constant numerical coefficients, independent of the values of the electromagnetic fields. If we substitute Eq. [123] into

---

* Linearity is required because an increase of the electric and magnetic fields by, say, a factor of 2 in one reference frame must also result in such an increase in any other reference frame; and homogeneity (no additive constant in Eq. [123]) is required because zero electric and magnetic fields in one reference frame must also result in zero electric and magnetic fields in any other reference frame.

Eq. [122], we obtain

$$a_\beta{}^\nu a_\alpha{}^\mu \partial_\mu F'^{\alpha\beta} = \partial_\mu C^{\mu\nu}_{\alpha\beta} F'^{\alpha\beta}$$

Since $\partial_\mu F'^{\alpha\beta}$ can attain any arbitrary value (at a given point of spacetime), the coefficients multiplying $\partial_\mu F'^{\alpha\beta}$ on both sides of this equation must be equal, which implies

$$a_\beta{}^\nu a_\alpha{}^\mu = C^{\mu\nu}_{\alpha\beta}$$

and consequently

$$F^{\mu\nu} = a_\beta{}^\nu a_\alpha{}^\mu F'^{\alpha\beta} \qquad [124]$$

This says that $F^{\mu\nu}$ is a second-rank tensor. This tensor $F^{\mu\nu}$ is called the *electromagnetic field tensor*.

Now that we have established the tensor character of $F^{\mu\nu}$, we can claim that Eqs. [106] and [107] are tensor equations of the general type of Eq. [100], which are in accord with the principle of relativity.

From the tensor transformation equations for the field tensor $F^{\mu\nu}$ we can readily extract the transformation equations for the electric and magnetic fields, which are components of the tensor $F^{\mu\nu}$ (see Eq. [109]). Since we will have no occasion to use these equations, we will not bother with them.

The equation of motion for a charged particle can be written in terms of the field tensor $F^{\mu\nu}$. The equation of motion is

$$\frac{d\mathbf{p}}{dt} = q\mathbf{E} + \frac{q}{c}\,\mathbf{v} \times \mathbf{B} \qquad [125]$$

where $\mathbf{p} = m\mathbf{v}/\sqrt{1 - v^2/c^2}$ is the relativistic momentum. A simple calculation shows that Eq. [125] is equivalent to

$$\frac{dp^\mu}{d\tau} = \frac{q}{m}\, p_\nu F^{\mu\nu} \qquad [126]$$

The Lorentz invariance of Eq. [126] is obvious.

■ *Exercise 18.* Check that the components of Eq. [126] give [125]. ■

In later chapters, we will need the energy-momentum tensor of the electromagnetic field (and of other systems). The energy-momentum tensor of the electromagnetic field is

$$T^{\mu\nu} = -\frac{1}{4\pi}\left[F^{\mu\alpha}F^{\nu}{}_{\alpha} - \frac{1}{4}\eta^{\mu\nu}F^{\alpha\beta}F_{\alpha\beta}\right] \qquad [127]$$

A complete derivation of this expression is given in Appendix 1. The T^{00} component is the familiar energy density of the electromagnetic field,

$$T^{00} = \frac{1}{8\pi}(\mathbf{E}^2 + \mathbf{B}^2) \qquad [128]$$

and the T^{0k} component is the Poynting vector,

$$T^{0k} = \frac{1}{4\pi}(\mathbf{E} \times \mathbf{B})^k \qquad [129]$$

The trace of $T^{\mu\nu}$ is zero:

$$T^{\mu}{}_{\mu} = 0 \qquad [130]$$

■ *Exercise 19.* Obtain Eqs. [128] and [129] from Eq. [127]. ■

■ *Exercise 20.* Prove Eq. [130]. ■

If we take for granted that the energy density in the electromagnetic field is $(\mathbf{E}^2 + \mathbf{B}^2)/8\pi$, then we can give a very simple derivation of the expression [127]. The argument is this: Suppose Eq. [128] for the energy density is true in every reference frame. Since [127] agrees with [128], it follows that $T^{\mu\nu}$ can differ from the expression [127] only by a (symmetric) tensor $B^{\mu\nu}$ whose B^{00} component vanishes identically in *every reference frame*. It is easy to see that in this case all components of $B^{\mu\nu}$ must vanish.

■ *Exercise 21.* To see that $B^{11} = 0$, take a Lorentz transformation of the type [17]. Then

$$B'^{00} = a^0{}_0 a^0{}_0 B^{00} + a^0{}_1 a^0{}_1 B^{11} + 2a^0{}_1 a^0{}_1 B^{01}$$

and if $B'^{00} = B^{00} = 0$, it follows that

$$0 = a^0{}_1 B^{11} + 2a^0{}_0 B^{01}$$

This cannot be true for all possible choices of $a^0{}_1$ and $a^0{}_0$ unless $B^{11} = B^{01} = 0$. Given that B^{00}, B^{11}, and B^{01} vanish in every reference frame, construct similar arguments to prove that all the remaining components also vanish. ■

Finally, we present the relativistic formulation of the electromagnetic potentials. The usual connection between fields and potentials is

$$\mathbf{E} = -\nabla\phi - \frac{1}{c}\frac{\partial \mathbf{A}}{\partial t} \tag{131}$$

$$\mathbf{B} = \nabla \times \mathbf{A} \tag{132}$$

where ϕ is the "scalar" potential and $\mathbf{A}$ the "vector" potential (these are scalars and vectors only in the three-dimensional sense!). If we define a four-component object

$$A^\mu = (\phi, A_x, A_y, A_z) \tag{133}$$

then we can write Eqs. [131] and [132] simply as

$$F^{\mu\nu} = \partial^\mu A^\nu - \partial^\nu A^\mu \tag{134}$$

■ *Exercise 22.* Check that Eq. [134] is equivalent to [131] and [132]. ■

Since the left side of Eq. [134] is a tensor, it follows that A^μ must be a four-vector, usually called the *four-vector potential*.

If we insert the expression [134] into the Maxwell equations [106] and [107], we find that the second of these is satisfied identically whereas the first gives

$$\partial_\mu \partial^\mu A^\nu - \partial^\nu \partial_\mu A^\mu = \frac{4\pi}{c} j^\nu \tag{135}$$

This is the wave equation for the potential.

We will not attempt the solution of the relativistic Maxwell equations, because the application of these equations to electromagnetic problems is beyond the scope of our discussion. We are interested in electrodynamics only as an example of a relativistic field theory. In the next chapter we will see how the understanding of the electromagnetic field leads to an understanding of the gravitational field.

2.6 DIFFERENTIAL FORMS AND THE EXTERIOR CALCULUS

In the preceding sections we have adopted the point of view that, in four dimensions, a vector or a tensor is an n-tuple of real numbers. In three dimensions an alternative, "geometrical" point of view enjoys much popularity: a vector is thought of as a directed line segment, or an arrow, and it is designated by a vector symbol, such as $\mathbf{A}$. This point of view not only provides us with a nice intuitive mental picture of vectors, but also proves very convenient when we attempt to formulate vector identities, especially identities involving the curl and identities involving vector integrals. For instance, the identities $\nabla \cdot (\nabla \times \mathbf{A}) = 0$, or $\nabla \times \nabla \Phi = 0$, or $\int \mathbf{B} \cdot d\mathbf{l} = \int (\nabla \times \mathbf{B}) \cdot d\mathbf{S}$ (Stokes' theorem) have a very concise form in this geometrical vector language, but they look rather awkward when expressed in component language.

■ *Exercise 23.* Write these vector identities in (three-dimensional) component language. (Hint: In component notation, the cross product of two vectors A^m and B^n can be expressed as $C^k = \epsilon_{kmn} A^m B^n$, where ϵ_{kmn} is the third-rank "alternating" tensor, antisymmetric under exchange of any two indices, whose components are all zero except $\epsilon_{123} = \epsilon_{231} = \epsilon_{312} = 1$, and $\epsilon_{213} = \epsilon_{132} = \epsilon_{321} = -1$.) ■

The connection between the "geometrical" description of a vector and the component description is given by

$$\mathbf{A} = A^1 \mathbf{e}_1 + A^2 \mathbf{e}_2 + A^3 \mathbf{e}_3 \qquad [136]$$

where $\mathbf{e}_1$, $\mathbf{e}_2$, and $\mathbf{e}_3$ are the three unit vectors along the x-, y-, and z-axes, respectively.

Of course, in four dimensions, we could adopt an analogous picture of a vector as a directed line segment in spacetime, and we could express such a "geometrical" vector as a superposition of four unit vectors $\mathbf{e}_0$, $\mathbf{e}_1$, $\mathbf{e}_2$, and $\mathbf{e}_3$. But this point of view is of limited usefulness. First of all, much of the intuitive appeal is lost, because we are unable to visualize line segments in four dimensions. More seriously, the naive geometrical picture of a vector as a directed line segment does not admit of generalization to *curved* spacetime, and is therefore hardly worth bothering with. We can see why curved space leads to trouble for the naive geometrical picture of vectors in the following simple example of a curved two-dimensional space. Pretend that the surface of the Earth is a perfect sphere, and consider a displacement of 4000 km due east along the surface of the sphere and a displacement of 4000 km due north. Can we regard these displacements as vectors, subject to the usual rules for vector addition? We cannot, because the two displacements *do not commute* (see Fig. 2.6a). There are two ways out of this difficulty. We can postulate that the vector addition is to be performed on a flat plane tangent to the sphere at the starting point (see Fig. 2.6b). This is, of course, how we normally deal with the addition of vectors on the surface of the Earth; but it has the serious drawback that such an addition yields a vector that lies outside of the curved two-dimensional surface of the sphere, that is, it yields a vector in (flat) three-dimensional space, rather than a vector in the (curved) two-dimensional space. If we wanted to adopt this procedure for vector addition for a curved four-dimensional spacetime, we would first have to embed the curved spacetime in a flat space, a procedure which is awkward and requires the introduction of unphysical, unobservable extra dimensions of space and time.* Alternatively, we can postulate that vector addition merely means the addition of components. For instance, if we adopt as coordinates on the surface of the sphere the usual polar angles $x^1 = \theta$ and $x^2 = \phi$, then a vector in this curved space has components (A^1, A^2), and the addition of this vector and a second vector (B^1, B^2) simply gives the vector sum $(A^1 + B^1, A^2 + B^2)$, which is obviously commutative. However, such a vector in the curved space is nothing but a pair of numbers, and we cannot associate a mental picture with it.

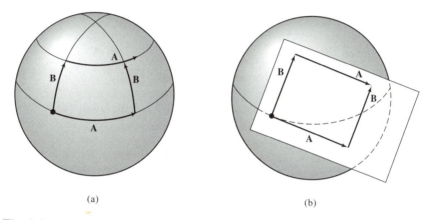

(a) (b)

Fig. 2.6 (a) Eastward and northward displacements on the surface of the Earth do not commute. (b) The displacements commute if the addition is performed on a flat plane tangent to the sphere.

Although it is mathematically consistent to think of a vector as nothing more than a set of components, this view seems to lack something. We expect that if there are components, there should be some mathematical entity such that the components are the "pieces" of this mathematical entity. Of course, we could say that the n-tuple is the entity of which the components are pieces; but this is a rather shallow and unsatisfying point of view, and we expect that it should be possible to construct some deeper, more interesting mathematical entity. Such a construction is provided by Cartan's calculus of differential forms. In this intricate construction, vectors are viewed as linear operators acting on "test" functions, and tensors are viewed as operators acting on vectors. Thus, in Cartan's calculus, everything is operators!

Vectors Cartan's notion of vector hinges on the observation that, in flat space, such as three-dimensional space, displacement vectors are in one-to-one correspondence with directional differential operators. Thus, with the displacement vector **A** we can associate the differential

---

* If we want to regard a curved space of four dimensions as a curved surface in a flat space, we require, in general, a flat space of ten dimensions.

operator $\mathbf{A}\cdot\nabla$, or $A^k \partial/\partial x^k$, which is proportional to the component of
the gradient operator along the direction of $\mathbf{A}$. The algebra of such
differential operators mimics the algebra of vectors--the differential
operators can be added and multiplied by numbers in the same way as
displacement vectors. This means that the differential operators are
"vectors," according to the general mathematical definition of an abstract vector space. But the differential operators have a crucial
advantage over displacement vectors in that they can be generalized to
curved spacetime.

Vectors constructed from differential operators are called *tangent
vectors*. The formal definition of such a vector states that the vector is
a linear operator acting on test functions, that is, infinitely differentiable functions of position. At any given point of spacetime, the vector
produces a real number when it acts on a test function. In mathematical language: the vector is a linear map from test functions to real
numbers. We will designate such vectors by boldface letters, for in-
stance, $\mathbf{u}$. When the vector $\mathbf{u}$ acts on a test function f, it produces the
real number $\mathbf{u}(f)$. When the vector acts on a sum or on a product of
two test functions f and g, the result is given by the following rules:

(*i*) $\mathbf{u}(af + bg) = a\mathbf{u}(f) + b\mathbf{u}(g)$, for any real numbers a and b
(*ii*) $\mathbf{u}(fg) = g\mathbf{u}(f) + f\mathbf{u}(g)$

The first of these rules merely expresses linearity, and the second rule
is the familiar rule for the effect of a first-order differential operator
on a product of functions. Thus, the second rule characterizes the
vector as a differential operator. In component language, these rules
can be shown to imply that $\mathbf{u}$ must be a superposition of the differen-
tial operators $\partial/\partial x^0$, $\partial/\partial x^1$, $\partial/\partial x^2$, and $\partial/\partial x^3$:

$$\mathbf{u} = u^\mu \frac{\partial}{\partial x^\mu} \qquad [137]$$

In this equation, we can regard the operators $\partial/\partial x^\mu$ as basis operators,
or basis vectors. Eq. [137] is then analogous to Eq. [136], with $\partial/\partial x^\mu$
playing the role of the unit vectors $\mathbf{e}_k$. The coefficients u^μ are the
components of the vector $\mathbf{u}$ with respect to the basis vectors $\partial/\partial x^0$,
$\partial/\partial x^1$, $\partial/\partial x^2$, and $\partial/\partial x^3$ (note that in this context, $\partial/\partial x^0$, $\partial/\partial x^1$,
$\partial/\partial x^2$, and $\partial/\partial x^3$ are *not* thought of as components of the gradient
operator $\partial/\partial x^\mu$, but as four separate items).

The vector $\mathbf{u}$ is called a "tangent" vector because the algebra of
such vectors mimics the algebra we would obtain if we were to perform vector addition with ordinary displacement vectors on a flat
plane tangent to the (possibly) curved spacetime at the given point.
The set of all possible vectors at the given point is called the *tangent
space*.

The superscript notation we have adopted in Eq. [137] suggests that the components u^μ should be contravariant components; we can verify that this is indeed the case. If we perform a Lorentz transformation, the new components become u'^μ and the new basis vectors become $\partial/\partial x'^\mu$; however, the vector **u** *remains the same*:

$$\mathbf{u} = u^\mu \frac{\partial}{\partial x^\mu} = u'^\nu \frac{\partial}{\partial x'^\nu} \tag{138}$$

With Eq. [62], this reduces to

$$u^\mu \frac{\partial}{\partial x^\mu} = u'^\nu a_\nu{}^\mu \frac{\partial}{\partial x^\mu}$$

Since the basis vectors are linearly independent, this equation demands that the coefficients of equal basis vectors on both sides be equal, that is,

$$u^\mu = a_\nu{}^\mu u'^\nu \tag{139}$$

This indicates that the components u^μ have the contravariant transformation law, as expected.

As an example of a vector, consider the energy-momentum vector of a particle. In component language, this vector consists of the components

$$(p^0, p^1, p^2, p^3) = (E, p_x, p_y, p_z)$$

In our new language, this vector becomes

$$\mathbf{p} = E \frac{\partial}{\partial x^0} + p_x \frac{\partial}{\partial x^1} + p_y \frac{\partial}{\partial x^2} + p_z \frac{\partial}{\partial x^3} \tag{140}$$

Vectors can be given a geometrical interpretation in terms of directional derivatives along curves in spacetime. Consider a curve specified by a parametric equation $x^\alpha(\lambda)$, where λ is a parameter that varies smoothly along the curve. Then the directional derivative of a test function f along the curve, or the rate of change of f along the curve, is

$$\mathbf{u}(f) = \frac{df}{d\lambda} = \frac{dx^\alpha}{d\lambda} \frac{\partial f}{\partial x^\alpha} \tag{141}$$

Thus, the components of the vector $\mathbf{u} = d/d\lambda$ are $dx^\alpha/d\lambda$; these components agree with what, in component language, we would call the components of a vector tangent to the curve at the given point. By choosing different curves, all passing through the same point but with different directions, we obtain the set of all possible vectors that can be defined at that point. Any vector at this point can then be regarded as a directional derivative, for one of these curves. Note that the basis vectors $\partial/\partial x^0$, $\partial/\partial x^1$, $\partial/\partial x^2$, and $\partial/\partial x^3$ are among the possible vectors we obtain for suitable choices of curves. For instance, $\partial/\partial x^0$ is the directional derivative we obtain for a curve whose direction coincides with the x^0-axis, and whose parameter is simply the time coordinate x^0.

The advantage of describing vectors in terms of directional derivatives along curves is that the curve and its parameter uniquely specify the vector (at a given point), without any need to invoke coordinates explicitly. This is the rationale for the claim, often made by aficionados, that the Cartan definition of a vector is a "geometrical" definition, independent of coordinates. However, this claim is somewhat of an exaggeration. Spacetime is featureless and amorphous; points and curves in spacetime cannot be identified and defined except by their coordinates. We might be tempted to identify a point by describing an event that occurred at this point, for example, "this is the point where the small red Chinese firecracker exploded." But such an anecdotal identification is meaningful only to eyewitnesses of the event; the anecdotal identification lacks the universal communicability that is supposed to be a hallmark of scientific data. If the eyewitnesses try to communicate the position of the event to an investigator, or if they try to make a written record for posterity, they find themselves reduced to describing the position by explicit or implicit coordinates: "the firecracker exploded five meters to the north of the fire hydrant on Mercer Street." In essence, coordinates give us an operational procedure for the identification of points in spacetime, which an anecdotal identification fails to do. Furthermore, the anecdotal identification works only for those points of spacetime where there is an actual event (an actual happening). Most of spacetime is empty, devoid of actual events. The set of spacetime points is a continuous, uncountable set, whereas the set of actual events is a discrete, countable set; it is therefore not possible to regard these sets as identical.

Coordinates lie not only at the root of the definition of a manifold (see Section 2.1) and at the root of any identification of points or curves in spacetime, but they are also essential whenever we want to compare the results of a calculation with experimental measurements. An experimenter cannot measure a Cartan vector; she can only measure components. Thus, the Cartan formalism is no more than a clever mathematical tool for the efficient and elegant derivation of some vector and tensor theorems. It has no advantage in principle over the component formalism.

1-forms In the mathematics of vector spaces it is customary to define a dual vector as a linear map from vectors to real numbers, that is, the dual vector produces a number whenever it acts on a vector. In the Cartan calculus, dual vectors are called *differential forms* or *1-forms*. Since a vector is a differential operator, the 1-form associates numbers with differential operators. We will designate 1-forms by boldface Greek letters, such as $\boldsymbol{\alpha}$. The 1-form $\boldsymbol{\alpha}$ produces a number $\boldsymbol{\alpha}(\mathbf{u})$ when it acts on the vector $\mathbf{u}$. When acting on a superposition of two vectors, the 1-form obeys the rule

$$\boldsymbol{\alpha}(a\mathbf{u} + b\mathbf{w}) = a\boldsymbol{\alpha}(\mathbf{u}) + b\boldsymbol{\alpha}(\mathbf{w}) \qquad [142]$$

As basis for the space of dual vectors, we use the 1-forms ω^0, ω^1, ω^2, and ω^3 defined by their action on the basis vectors $\partial/\partial x^0$, $\partial/\partial x^1$, $\partial/\partial x^2$, and $\partial/\partial x^3$, as follows:

$$\omega^\nu \left[\frac{\partial}{\partial x^\mu} \right] = \delta^\nu_\mu \qquad [143]$$

For instance, ω^0 yields +1 when acting on $\partial/\partial x^0$, and zero when acting on any of the other basis vectors. Any 1-form can then be expressed as a superposition of the basis 1-forms ω^μ:

$$\boldsymbol{\alpha} = \alpha_\mu \, \omega^\mu \qquad [144]$$

Here, the coefficients α_μ are the components of the 1-form with respect to the basis ω^μ.

When a 1-form with components α_μ acts on a vector with components u^μ, the result is

$$\boldsymbol{\alpha}(\mathbf{u}) = \alpha_\mu \, \omega^\mu \left[u^\nu \, \frac{\partial}{\partial x^\nu} \right] = \alpha_\mu u^\nu \, \omega^\mu \left[\frac{\partial}{\partial x^\nu} \right]$$

$$= \alpha_\mu u^\nu \, \delta^\mu_\nu = \alpha_\mu u^\mu$$

In component language, we would call this the scalar product of the contravariant vector u^μ and the covariant vector α_μ (we will see later that the components α_μ are, indeed, covariant components). For example, the energy-momentum can be expressed as a 1-form,

$$\boldsymbol{\pi} = E\omega^0 + p_1\omega^1 + p_2\omega^2 + p_3\omega^3$$

$$= E\omega^0 - p_x\,\omega^1 - p_y\,\omega^2 - p_z\,\omega^3$$

and if we let this 1-form energy-momentum act on the vector energy-momentum (see Eq. [140]), we obtain

$$\pi(p) = (E\omega^0 - p_x\,\omega^1 - p_y\,\omega^2 - p_z\,\omega^3)\left(E\,\frac{\partial}{\partial x^0} + p_x\,\frac{\partial}{\partial x^1} + p_y\,\frac{\partial}{\partial x^2} + p_z\,\frac{\partial}{\partial x^3}\right)$$

$$= (E^2 - p_x{}^2 - p_y{}^2 - p_z{}^2) = m^2$$

Gradient   The *gradient* of a function is a special case of a 1-form. The gradient df of a function f is defined by what it does when it acts on some arbitrary tangent vector $d/d\lambda$:

$$df\left(\frac{d}{d\lambda}\right) = \frac{df}{d\lambda} \qquad [145]$$

Thus, the number that the gradient produces when acting on the tangent vector $d/d\lambda$ associated with a given curve is simply the derivate of the function along the curve. As a simple example of this formula for the gradient, consider the function $f = x^0$, which depends linearly on the coordinate x^0, and suppose that the vector $d/d\lambda$ is one of the basis vectors, $d/d\lambda = \partial/\partial x^\nu$. Then Eq. [145] becomes

$$dx^0\left(\frac{\partial}{\partial x^\nu}\right) = \frac{\partial x^0}{\partial x^\nu} = \delta^0_\nu \qquad [146]$$

More generally, if $f = x^\mu$, so f depends linearly on the coordinate x^μ (and only on this one coordinate), we obtain

$$dx^\mu\left(\frac{\partial}{\partial x^\nu}\right) = \frac{\partial x^\mu}{\partial x^\nu} = \delta^\mu_\nu \qquad [147]$$

Comparing this with Eq. [143], we see that the gradients dx^0, dx^1, dx^2, and dx^3 of the coordinates coincide with the basis 1-forms ω^0, ω^1, ω^2, and ω^3, respectively:

$$dx^\mu = \omega^\mu \qquad [148]$$

Thus, the component-expression [144] for a 1-form can be written as

$$\alpha = \alpha_\mu\,dx^\mu \qquad [149]$$

In particular, the gradient of a function f can be written as

$$df = \frac{\partial f}{\partial x^\mu}\,dx^\mu \qquad [150]$$

The components of this gradient are, of course, what we would normally expect from our experience with component language.

■ *Exercise 24.* Derive Eq. [150]. (Hint: Show that the 1-forms given by Eqs. [150] and [145] yield the same result when acting on each of the basis vectors $\partial/\partial x^0$, $\partial/\partial x^1$, $\partial/\partial x^2$, and $\partial/\partial x^3$.) ■

Keep in mind that although Eq. [150] looks very much like the familiar equation for a small change of a function engendered by a small displacement, it has nothing to do with that; df and dx^μ are *not* small changes, but 1-forms. If we wanted to find the small change in the function f engendered by a small displacement, we would have to write

$$df = dx^\mu \frac{\partial f}{\partial x^\mu} \qquad [151]$$

To interpret this equation in the language of differential forms, we observe that $dx^\mu \, \partial/\partial x^\mu$ is a vector δ which acts on the function f; the small changes dx^μ are the components of this vector δ. Eq. [151] for the small change in the function f then becomes

$$df = \delta f$$

Never confuse this df with $\mathbf{d}f$!

In the above equations we have adopted a subscript notation for the components of a 1-form. This notation suggests that these components should obey the transformation law for covariant components,

$$\alpha_\nu = a^\mu{}_\nu \alpha'_\mu$$

By an argument similar to that given in connection with the components of a vector (see Eqs. [138], [139]), it is easy to show that this is indeed the case.

■ *Exercise 25.* Show that the components α_μ in Eq. [149] are covariant components. (Hint: $dx'^\mu = da^\mu{}_\nu x^\nu = a^\mu{}_\nu dx^\nu$.) ■

Tensors Tensors are defined as linear maps from 1-forms and vectors to real numbers, that is, tensors are operators that yield real numbers when acting on 1-forms and vectors.

For instance, a tensor **T** of third rank, contravariant in the first index and covariant in the next two indices [called a tensor of type $\binom{1}{2}$], produces a real number $\mathbf{T}(\alpha, \mathbf{u}, \mathbf{v})$ when it acts on one 1-form and two vectors. The map is linear in the 1-form and in each of the vectors individually. One way to construct a tensor of this type is by means of the "tensor" product of one vector and two 1-forms, such as

$$s \otimes \zeta \otimes \xi \qquad [152]$$

When this tensor acts on one 1-form α and on two vectors $\mathbf{u}$ and $\mathbf{w}$ it gives, by definition,

$$s \otimes \zeta \otimes \xi \, (\alpha, \mathbf{u}, \mathbf{w}) \equiv \alpha(s) \, \zeta(\mathbf{u}) \, \xi(\mathbf{w}) \qquad [153]$$

Here, the right side is simply a product of three real numbers. Note that the tensor-product symbol $\otimes$ merely indicates that the factors in Eq. [153] are to be treated independently when they act on 1-forms or vectors placed on the right, and only in the last step, after all the 1-forms and vectors have been acted upon and have been converted into numbers, is the product to be taken. The most general tensor of the type $\binom{1}{2}$ is a superposition of tensor products of the kind displayed in Eq. [152], with different choices of s, ζ, and ξ.

If we construct tensor products with the basis vectors and basis 1-forms, we obtain basis tensors, such as

$$\frac{\partial}{\partial x^\sigma} \otimes dx^\mu \otimes dx^\nu \qquad [154]$$

(Note that when such a tensor acts on a vector, we must specify which of the two basis 1-forms acts.) In terms of these basis tensors, an arbitrary third-rank tensor of the type $\binom{1}{2}$ can be expressed as

$$\mathbf{T} = T^\sigma{}_{\mu\nu} \frac{\partial}{\partial x^\sigma} \otimes dx^\mu \otimes dx^\nu \qquad [155]$$

The coefficients $T^\sigma{}_{\mu\nu}$ are the components of this tensor with respect to the basis tensors.

■ ***Exercise 26.*** Show that the transformation law for the components of this tensor is as expected from the super- and subscripts. ■

Note that 1-forms and vectors are themselves tensors of the first rank. Thus, a 1-form is a first-rank tensor of the type $\binom{0}{1}$, which yields a real number when it acts on a vector. And a vector is a first-rank tensor of the type $\binom{1}{0}$, which yields a real number when it acts on a 1-form.

The *contraction* of a tensor is accomplished by allowing it to act on dx^σ and $\partial/\partial x^\sigma$, with summation over σ. For instance, the contraction of the third-rank tensor $\mathbf{T}$ in Eq. [155] results in

$$\overline{\mathbf{T}} = \mathbf{T}\left(dx^\sigma, \frac{\partial}{\partial x^\sigma}, \right) \qquad [156]$$

where the remaining empty slot in the tensor **T** indicates that it has now become a first-rank tensor, of type ($_0^1$), that acts on a tangent vector to produce a number.

In the language of differential forms, the metric tensor is a second-rank tensor of the type ($_2^0$), which can be written as

$$\boldsymbol{\eta} = \eta_{\mu\nu} \, dx^\mu \otimes dx^\nu \qquad [157]$$

The inverse of the metric tensor is a second-rank tensor of the type ($_0^2$):

$$\boldsymbol{\eta}^{-1} = \eta^{\mu\nu} \, \frac{\partial}{\partial x^\mu} \otimes \frac{\partial}{\partial x^\nu} \qquad [158]$$

It is easy to verify that when $\boldsymbol{\eta}$ acts on $\boldsymbol{\eta}^{-1}$, the result is the identity operator,

$$\boldsymbol{\eta}\boldsymbol{\eta}^{-1} = 1 \qquad [159]$$

■ *Exercise 27.* Verify this. ■

When the metric [157] acts on two vectors, it produces a real number:

$$\eta_{\mu\nu} \, dx^\mu \otimes dx^\nu \, (\mathbf{u}, \mathbf{w}) = \eta_{\mu\nu} \, dx^\mu(\mathbf{u}) \, dx^\nu(\mathbf{w})$$

$$= \eta_{\mu\nu} \, dx^\mu \left[u^\sigma \frac{\partial}{\partial x^\sigma} \right] dx^\nu \left[w^\tau \frac{\partial}{\partial x^\tau} \right]$$

$$= \eta_{\mu\nu} u^\mu w^\nu \qquad [160]$$

When the metric tensor acts on a single vector, it produces a 1-form:

$$\boldsymbol{\eta}(\mathbf{u}) = \eta_{\mu\nu} \, dx^\mu \otimes dx^\nu \, (\mathbf{u}) = \eta_{\mu\nu} \, dx^\mu(\mathbf{u}) \, dx^\nu$$

$$= \eta_{\mu\nu} \, dx^\mu \left[u^\sigma \frac{\partial}{\partial x^\sigma} \right] dx^\nu$$

$$= \eta_{\mu\nu} u^\mu \, dx^\nu = u_\nu \, dx^\nu \qquad [161]$$

This action of the metric tensor on a single vector replaces the original vector and its contravariant components u^σ by a 1-form and its covariant components u_ν. Thus, the action of the metric tensor on a vector is effectively equivalent to the operation of lowering an index we

encountered in component language. In fact, all the index manipulations and all the equations in component language can be translated
into the language of differential forms. Sometimes the result of such a
translation is rather cumbersome; for instance, Eq. [161] is rather more
messy than the simple and direct equation $u_\nu = \eta_{\nu\sigma} u^\sigma$ of component
language. But sometimes the translation into the language of differential forms yields elegant and concise results, especially when dealing
with equations containing derivatives, as we will see later.

■ *Exercise 28.* Show that when the inverse metric tensor η^{-1} is acted upon by
a single 1-form with components α_ν, it produces a vector with components $\alpha^\mu =
\eta^{\mu\nu} \alpha_\nu$. ■

p-forms The *exterior product*, or the *wedge product*, of 1-forms is an
antisymmetrized tensor product. The exterior product of two 1-forms
is called a *2-form*. This exterior product is defined by

$$\alpha \wedge \beta = \alpha \otimes \beta - \beta \otimes \alpha \qquad [162]$$

Note that the exterior product is antisymmetric under the exchange of
α and β, that is, $\alpha \wedge \beta = -\beta \wedge \alpha$. As a consequence of this asymmetry, the exterior product of any 1-form with itself vanishes identi-
cally, $\alpha \wedge \alpha \equiv 0$.

In terms of components, the 2-form $\alpha \wedge \beta$ becomes

$$\alpha \wedge \beta = \alpha_\mu dx^\mu \wedge \beta_\nu dx^\nu \qquad [163]$$

We can rewrite the right side of this equation as

$$\alpha \wedge \beta = \tfrac{1}{2}(\alpha_\mu \beta_\nu \, dx^\mu \wedge dx^\nu - \alpha_\mu \beta_\nu \, dx^\nu \wedge dx^\mu)$$

$$= \tfrac{1}{2}(\alpha_\mu \beta_\nu - \alpha_\nu \beta_\mu) \, dx^\mu \wedge dx^\nu \qquad [164]$$

We can regard the coefficients $\tfrac{1}{2}(\alpha_\mu \beta_\nu - \alpha_\nu \beta_\mu)$ as the components of
the 2-form; Eq. [164] shows that the components are antisymmetric in
μ and ν.

More generally, given a set of coefficients $\alpha_{\mu\nu}$, antisymmetric in μ
and ν, we can define the 2-form $\tfrac{1}{2}\alpha_{\mu\nu} dx^\mu \wedge dx^\nu$. An example of
such a 2-form is the electromagnetic field tensor **F** defined by

$$\mathbf{F} = \tfrac{1}{2} F_{\mu\nu} \, dx^\mu \wedge dx^\nu \qquad [165]$$

where the components $F_{\mu\nu}$ have the usual values

$$F_{\mu\nu} = \begin{pmatrix} 0 & E_x & E_y & E_z \\ -E_x & 0 & -B_z & B_y \\ -E_y & B_z & 0 & -B_x \\ -E_z & -B_y & B_x & 0 \end{pmatrix} \qquad [166]$$

If we write Eq. [165] out in full, it becomes

$$\mathbf{F} = E_x\ \mathbf{d}x^0 \wedge \mathbf{d}x^1 + E_y\ \mathbf{d}x^0 \wedge \mathbf{d}x^2 + E_z\ \mathbf{d}x^0 \wedge \mathbf{d}x^3$$

$$- B_z\ \mathbf{d}x^1 \wedge \mathbf{d}x^2 + B_y\ \mathbf{d}x^1 \wedge \mathbf{d}x^3 - B_x\ \mathbf{d}x^2 \wedge \mathbf{d}x^3 \qquad [167]$$

Here, terms with 1-forms in opposite order have been combined.

With this electromagnetic field tensor we can write the equation of motion of an electric charge as follows:

$$\frac{d\mathbf{p}}{d\tau} = \frac{e}{m}\ \mathbf{F}(\mathbf{p}) \qquad [168]$$

where $\mathbf{p}$ is the momentum vector,

$$\mathbf{p} = p^\mu\ \frac{\partial}{\partial x^\mu} \qquad [169]$$

Eq. [168] looks very simple, and this is one of the advantages of the language of differential forms. But, unfortunately, we cannot do anything much with this simple equation of motion. Whenever we want to solve the equation of motion for a particle in a given electromagnetic field with some given initial conditions, we have to disassemble this equation into components, and compute the (numerical) values of these components one by one.

The exterior product can also be defined for more than two 1-forms. The exterior product of p 1-forms is called a p-form. This is an antisymmetrized tensor product, which is antisymmetric in each adjacent pair of factors. For instance, with three 1-forms we can define the 3-form

$$\alpha \wedge \beta \wedge \zeta = \alpha \otimes \beta \otimes \zeta + \beta \otimes \zeta \otimes \alpha + \zeta \otimes \alpha \otimes \beta$$

$$- \beta \otimes \alpha \otimes \zeta - \alpha \otimes \zeta \otimes \beta - \zeta \otimes \beta \otimes \alpha \qquad [170]$$

If we express the 1-forms in components, we can rewrite this as

$$\alpha \wedge \beta \wedge \zeta = \frac{1}{3!}(\alpha_\mu \beta_\nu \zeta_\sigma + \alpha_\nu \beta_\sigma \zeta_\mu + \alpha_\sigma \beta_\mu \zeta_\nu$$

$$- \alpha_\nu \beta_\mu \zeta_\sigma - \alpha_\mu \beta_\sigma \zeta_\nu - \alpha_\sigma \beta_\nu \zeta_\mu)\, dx^\mu \wedge dx^\nu \wedge dx^\sigma \quad [171]$$

With some arbitrary set of coefficients $\alpha_{\mu\nu\sigma}$, antisymmetric in pairs of adjacent indices, we can construct a general 3-form,

$$\frac{1}{3!}\alpha_{\mu\nu\sigma}\, dx^\mu \wedge dx^\nu \wedge dx^\sigma \quad [172]$$

Note that in four-dimensional spacetime, a 5-form, or a 6-form, or any higher p-form, necessarily vanishes,

$$\alpha \wedge \beta \wedge \zeta \wedge \xi \wedge \chi = 0 \quad [173]$$

This is so because, when written with the basis 1-forms dx^μ, the left side of Eq. [173] contains five factors dx^μ, which means that at least one of the basis 1-forms occurs twice--but the exterior product of equal 1-forms is zero. Also, note that, to within an overall numerical factor, there is only one 4-form,

$$\epsilon = dx^0 \wedge dx^1 \wedge dx^2 \wedge dx^3 \quad [174]$$

This 4-form can also be written as

$$\epsilon = \frac{1}{4!}\epsilon_{\mu\nu\sigma\tau}\, dx^\mu \wedge dx^\nu \wedge dx^\sigma \wedge dx^\tau \quad [175]$$

where $\epsilon_{\mu\nu\sigma\tau}$ is the "alternating" symbol, antisymmetric under the exchange of any two adjacent indices, whose components are zero unless all the indices are different, in which case $\epsilon_{0123} = \epsilon_{1230} = \epsilon_{2301} = ... = 1$, $\epsilon_{1023} = \epsilon_{0231} = ... = -1$, etc.

Exterior Derivative   The exterior derivative of a p-form is a $(p + 1)$-form. If the p-form is

$$\alpha = \frac{1}{p!}\alpha_{\mu\nu...}\, dx^\mu \wedge dx^\nu \,... \quad [176]$$

then the exterior derivative is defined by

$$d\alpha = \frac{1}{p!}\, d\alpha_{\mu\nu...} \wedge dx^\mu \wedge dx^\nu \,... \quad [177]$$

Note that the extra 1-form $d\alpha_{\mu\nu...}$ that makes its appearance on the right side of Eq. [177] is simply the gradient of the function $\alpha_{\mu\nu...}$. With the usual formula [150] for the gradient, Eq. [177] becomes

$$d\alpha = \frac{1}{p!}\left(\frac{\partial}{\partial x^\sigma}\alpha_{\mu\nu...}\right)dx^\sigma \wedge dx^\mu \wedge dx^\nu \ ... \qquad [178]$$

The exterior derivative of an exterior derivative is always zero. This is an immediate consequence of Eq. [178]:

$$dd\alpha = \frac{1}{p!}\left(\frac{\partial^2}{\partial x^\tau \partial x^\sigma}\alpha_{\mu\nu...}\right)dx^\tau \wedge dx^\sigma \wedge dx^\mu \wedge dx^\nu \ ... \qquad [179]$$

Here, the second derivative with respect to x^τ and x^σ is symmetric in τ and σ, but the exterior product $dx^\tau \wedge dx^\sigma$ multiplying this is antisymmetric; hence the summation over σ and τ necessarily yields zero. The identity $dd\alpha = 0$ can be regarded as a generalization of the familiar three-dimensional identity $\nabla \cdot (\nabla \times \mathbf{A}) = 0$. Note that the rule that two exterior derivatives in succession give zero also applies to the gradient of a function, for which we readily obtain the identity $ddf = 0$. This identity is a generalization of the familiar three-dimensional identity $\nabla \times (\nabla f) = 0$.

Exterior derivatives permit us to express Maxwell's equations in an elegant way. It is easy to check that the Maxwell equation [107] simply says that the exterior derivative of the electromagnetic field tensor $\mathbf{F}$ is zero:

$$d\mathbf{F} = 0 \qquad [180]$$

■ *Exercise 29.* Check this. ■

The other Maxwell equation, Eq. [106], is somewhat more difficult to translate into the language of differential forms. A clever way to accomplish the translation is to start with the contravariant components of the electromagnetic field tensor and of the current density, and to define a 2-form $^*\mathbf{F}$ and a 3-form $^*\mathbf{j}$ as follows:

$$^*\mathbf{F} \equiv \frac{1}{4}\epsilon_{\mu\nu\sigma\tau}F^{\sigma\tau}dx^\mu \wedge dx^\nu \qquad [181]$$

and

$$^*\mathbf{j} \equiv \frac{1}{3!}\epsilon_{\mu\nu\sigma\tau}j^\tau dx^\mu \wedge dx^\nu \wedge dx^\sigma \qquad [182]$$

These are called the dual electromagnetic field tensor and the dual current density (in this context, the word *dual* has a new, different meaning from that in *dual vector space*). The Maxwell equation [106] then says that the exterior derivative of *F is proportional to *j:

$$d^*F = \frac{4\pi}{c} \,^*j \qquad [183]$$

■ *Exercise 30.* Verify that Eq. [183] implies Eq. [106]. ■

Note that if we take the exterior derivative of Eq. [183], we find that the left side is zero, according to the usual rule for the exterior derivative of an exterior derivative. Hence

$$0 = d^*j \qquad [184]$$

This is the conservation law for electric charge, expressed in the language of differential forms.

■ *Exercise 31.* Verify that Eq. [184] implies $\partial_\mu j^\mu = 0$. ■

Finally, in the language of differential forms, the vector potential is a 1-form,

$$A = A_\mu \, dx^\mu \qquad [185]$$

and the connection between the vector potential and the electromagnetic field tensor is simply

$$F = dA \qquad [186]$$

Equation [180] is then an identity,

$$dF = ddA \equiv 0 \qquad [187]$$

Although all the equations of electrodynamics look deceptively simple when expressed in the language of differential forms, we must keep in mind that the simple notation hides a great many complications. All these complications reemerge when we attempt any numerical calculation, where we must necessarily use components.

These examples of the application of differential forms to some equations of physics give us a glimpse of the elegance of the language of differential forms. But to grasp the full power of this language, we have to look at how it deals with integration of tensor quantities over regions of spacetime. With a suitable definition of the integral of a differential form, it can be shown that Gauss' theorem, and Stokes' theorem, and other integral theorems of this kind are corollaries of a single general theorem that relates the integral of a differential form over a region of spacetime to the integral over the boundary of this region. These results can be found in the references listed at the end of the chapter.

2.7 LOCAL FIELDS VERSUS ACTION-AT-DISTANCE

Although Newton's formulation of universal gravitation suggests action-at-distance, Newton himself had considerable misgivings about this interpretation. In his words:

> It is inconceivable that inanimate brute Matter should without the Mediation of something else which is not material, operate upon, and affect other Matter without mutual Contact. . . . That Gravity should be innate, inherent and essential to Matter so that one Body may act upon another at a Distance *thro'* a *Vacuum* without the Mediation of any thing else, by and through which their Action and Force may be conveyed from one to another, is to me so great an Absurdity that I believe no Man who has in philosophical Matters a competent Faculty of thinking can ever fall into it. . . . (Newton, quoted in Cohen, 1958)

Nevertheless, the followers of Newton, and in particular Laplace and his contemporaries, regarded Newton's gravitational forces as action-at-distance. Because of the tremendous success of the theory, it became fashionable throughout the eighteenth and most of the nineteenth centuries to regard all forces as action-at-distance. This trend was reversed by the work of Faraday and Maxwell in electromagnetism. Faraday's picture of lines of force had irresistible appeal, and with Maxwell's brilliant mathematical formulation of the field equations of electrodynamics, the interpretation of forces by means of fields and action-by-contact gained prevalence. Attempts were then made at describing gravitation as a field with a finite velocity of propagation. Einstein's theory may be regarded as the most successful such attempt.

Nowadays, all the fundamental interactions are regarded as due to local fields. The interaction of two particles that are widely separated is regarded as an action-by-contact of one particle on the field followed by a second action-by-contact of the field on the second particle. Why do we prefer fields to action-at-distance? The answer is simple: we need fields in order to uphold the laws of conservation of energy and momentum.

In nonrelativistic physics, the conservation of momentum is a trivial consequence of Newton's law of equality of action and reaction. In relativistic physics, the conservation of momentum is much harder to achieve. For example, consider two particles that interact but do not come into contact--that is, the interaction takes place while the particles are separated by some distance. The momenta of the particles are $m u_1{}^\mu$ and $m u_2{}^\mu$, and as a result of the interaction these momenta change in some way. The total momentum of the particles is

$$P^\mu = m_1 u_1{}^\mu + m_2 u_2{}^\mu \qquad [188]$$

We will prove that this total momentum is *not* constant during the interaction.

The proof is by contradiction. Suppose that the total momentum *is* conserved. Under a Lorentz transformation, this total momentum transforms as a four-vector, that is,*

$$P'^\mu = a^\mu{}_\nu P^\nu \qquad [189]$$

By conservation, each side of this equation is a constant and it makes no difference at which time t' or t we evaluate P'^μ or P^μ. Let us evaluate P'^μ at the time $t' = 0$ and P^μ at the time $t = 0$. If the worldlines and coordinates are as shown in Fig. 2.7, we obtain

$$m_1 u'_1{}^\mu \Big|_O + m_2 u'_2{}^\mu \Big|_{Q'} = a^\mu{}_\nu m_1 u_1{}^\nu \Big|_O + a^\mu{}_\nu m_2 u_2{}^\nu \Big|_Q \qquad [190]$$

But at the point O, the Lorentz-transformation law for the four-velocity is $u'_1{}^\mu = a^\mu{}_\nu u_1{}^\nu$; and at the point Q', $u'_2{}^\mu = a^\mu{}_\nu u_2{}^\nu$. Hence

$$m_1 a^\mu{}_\nu u_1{}^\nu \Big|_O + m_2 a^\mu{}_\nu u_2{}^\nu \Big|_{Q'} = a^\mu{}_\nu m_1 u_1{}^\nu \Big|_O + a^\mu{}_\nu m_2 u_2{}^\nu \Big|_Q \qquad [191]$$

---

* Since P^μ is the sum of two four-vectors (see Eq. [188]), at first sight it seems obvious that it should be a four-vector. However, there is a catch: in Eq. [188] it is implicitly assumed that the two four-velocities are evaluated at the same time. Since Lorentz transformations do not preserve simultaneity, the meaning of "same time" is different in the two reference frames, and this difference could complicate the transformation law of the sum [188]. However, for the special case of *free* particles, Eq. [189] is obviously correct since in this case the four-velocities are constant, and it is irrelevant whether the velocities are evaluated simultaneously or not. It then follows that Eq. [189] must also be correct for any particles that were free (did not interact) at some time in their past: the momenta P'^μ and P^μ are *constant* by hypothesis, and if they had the transformation law [189] at one time, they must keep it forever.

This reduces to

$$m_2 a^\mu{}_\nu u_2{}^\nu \bigg|_{Q'} = m_2 a^\mu{}_\nu u_2{}^\nu \bigg|_{Q} \qquad [192]$$

Since $a^\mu{}_\nu$ is arbitrary, it follows that

$$m_2 u_2{}^\nu \bigg|_{Q'} = m_2 u_2{}^\nu \bigg|_{Q} \qquad [193]$$

which means that the four-momentum of the particle remains constant as the particle moves from Q to Q'. We have therefore shown that the total momentum is conserved if and only if the *momentum of each particle remains constant*, that is, if and only if the particles *do not interact*. This no-interaction theorem is due to Van Dam and Wigner (1966).

This theorem tells us that in order to preserve the law of conservation of energy-momentum for interacting particles, we must introduce another entity which can serve as a storehouse of energy-momentum. This entity is the field.

To gain some understanding of how the field achieves the balance of momentum, we begin with the observation that an interaction that propagates instantaneously, with infinite speed, is ruled out by the requirements of special relativity. Infinite speed fails to remain invariant under Lorentz transformations (the worldline of a signal of infinite speed coincides with the x-axis; but if we perform a Lorentz transformation, this worldline will fail to coincide with the x'-axis). The only speed that remains invariant under Lorentz transformations is the speed of light (see Problem 6), and hence if gravitational effects propagate with a definite speed, it will have to be the speed of light. Gravitational interactions, like all other interactions, must be retarded; the effects of a change in the source of gravitation propagate outward with the speed of light. Of course, this strongly suggests that what is propagated is a field, but it is also possible to describe the situation by means of "retarded action-at-distance." By the latter is meant that the action of a source of gravitation (or whatever) on some particle is always assumed to be delayed by a time r/c, where r is the distance between source and particle.

As a concrete example, let us take a look at momentum conservation in the Earth-Sun system. In this case Newton's law of action and reaction states that the force exerted by the Sun on the Earth is equal

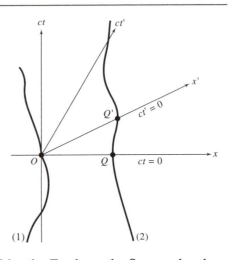

Fig. 2.7 Worldlines of two inter-acting particles. Particle 1 is at the origin O at time $t = t' = 0$. Particle 2 is at the point Q at time $t = 0$, and at the point Q' at time $t' = 0$.

and opposite to the force exerted by the Earth on the Sun; under these conditions the momentum of the Earth-Sun system would be con-served. However, it is easy to see that the equality of action and reac-tion is in contradiction with the requirement that gravitational signals propagate at the speed of light. Consider the following *Gedan-ken*-experiment: at some time the Earth is suddenly given an accelera-tion (for instance, by receiving a series of kicks from the impact of a stream of meteors). This changes the position of the Earth in the field of the Sun and therefore changes the force of the Sun on the Earth. The force of the Earth on the Sun will not change simultaneously. It takes light signals $r/c \simeq 500$ s to travel the Earth-Sun distance r, and hence any signal will be delayed by at least as much. The gravitational signal that "adjusts" the gravitational pull of the Earth moves from Earth to Sun at the speed of light and until it reaches its destination, action and reaction will certainly be out of balance. This conclusion is of course in agreement with the no-interaction theorem.

Our *Gedanken*-experiment is less farfetched than it seems. The Earth is in continuous acceleration as it orbits the Sun, and action and reaction are in fact out of balance all the time. We can give a rough estimate of the difference between action and reaction. The force of the Sun on the Earth is $GM_\odot M/r^2$, where $M_\odot$ and M are the masses of Sun and Earth, respectively. The force of the Earth on the Sun differs from this by a term depending on the acceleration of the Earth. In the time r/c, the radial displacement of the Earth caused by the accelera-tion is

$$\Delta r \simeq \frac{1}{2} \frac{GM_\odot}{r^2} \left(\frac{r}{c}\right)^2 = \frac{1}{2} \frac{GM_\odot}{c^2} \qquad [194]$$

Hence, *without* the acceleration the Earth would have moved off tangentially and increased its distance by this amount. Fig. 2.8 shows the present position of the Earth and the retarded position, that is, the position at a time r/c before the present position. The figure also shows the "extrapolated" position, or the position that the Earth would have reached if it had moved at constant velocity along a straight line beginning at the retarded position. This extrapolated position is nearly on the same radius as the present position, but the radial distance is larger by the amount given by Eq. [194].

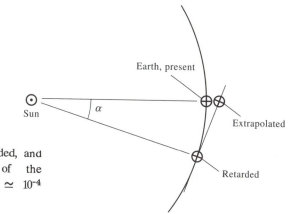

Fig. 2.8 Present, retarded, and extrapolated positions of the Earth. The angle α is $\simeq$ 10^{-4} radian.

As a rough estimate for the force felt at the Sun, we take that force which the Earth would exert if stationary at the extrapolated position. This estimate for the force gives

$$\frac{GM_\odot M}{(r + \Delta r)^2} \simeq \frac{GM_\odot M}{r^2} - \frac{1}{2}\frac{GM_\odot}{c^2}\left(\frac{2GM_\odot M}{r^3}\right) \qquad [195]$$

Of course, Eq. [195] is no more than an educated guess. Essentially, we are assuming that the force depends mainly on the velocity of the Earth at the retarded time, and that to this the acceleration contributes only some insignificant corrections (which have to do with the emission of gravitational radiation). If only the velocity is important, then the force exerted by the Earth is the same as that for a mass that moves with uniform velocity from the retarded position and reaches the extrapolated position at the present time. (For a mass moving with uniform velocity, the center of force will coincide with the position of the mass, as becomes immediately obvious if one asks for the force as seen from the rest frame of that mass. However, the transformation to the rest frame of that mass introduces relativistic corrections which are typically of the same order of magnitude as the extra term in [195].) The difference between action and reaction is then of the order

$$\left(\frac{GM_\odot M}{rc^2}\right)\left(\frac{GM_\odot}{r^2}\right) \qquad\qquad [196]$$

The inequality of action and reaction implies a violation of momentum conservation. The question is now: Where does the momentum go? There is only one place it can go--it must be stored in the gravitational field. What is conserved is not the momentum of the Earth-Sun system, but rather the momentum of the system consisting of Earth, Sun, and gravitational fields. This interpretation is in accord with our estimate [196]. To see this, note that $GM_\odot M/rc^2$ may be interpreted as the "mass" associated with the interaction energy (potential energy) between the Sun and the Earth. This energy is stored in the gravitational fields,* and as the Earth orbits the Sun, so will this extra "mass." To obtain the rate of momentum transfer to the fields, we must take the product of this "mass" by a characteristic acceleration. A reasonable estimate for the acceleration is $GM_\odot/r^2$, and the product then has the form [196].

Note that the momentum transfer to the "convective fields" considered above averages to zero over a period of the motion. However, an orbiting body also loses momentum (and energy) continuously, by radiation of gravitational waves. In this case, the orbiting body continuously transfers momentum and energy to "radiation fields," and these fields propagate outward and carry the momentum and energy away to large distances.

Fields provide us with an elegant way to uphold the law of momentum conservation. In action-at-distance theories, momentum conservation looks quite ugly. At first sight it seems impossible to satisfy momentum conservation at all; when fields are absent there is no place to store the energy and the momentum lost by particles. However, it is possible to get around this by a trick: we can change the definition of momentum of the particles in such a way that it is conserved. This means that we add an extra term (interaction momentum) to the usual expression for momentum. The extra term has a horrible appearance-- it is an integral over the entire history of each particle. In fact this integral is precisely what we would regard as field momentum in a field theory (the integration over history arises from writing the field in terms of the past motion of all the particles that contributed to the field).

The contrast between field theory and action-at-distance theory is brought out sharply by an example that goes somewhat beyond classical physics: the annihilation of particle-antiparticle pairs. If an electron annihilates with a positron, their energy and momentum must continue to exist. According to field theory, the energy-momentum

---

* According to Eq. [1.12], this energy is in part pure field energy and in part interaction energy of field and mass.

exists in the electromagnetic radiation (photons) emitted in the annihilation. According to action-at-distance theory, the energy-momentum exists as an interaction energy-momentum associated with particles; but since the particles have annihilated, we are forced to associate this momentum with the ghosts of particles that once were. This can be done, even though it looks rather artificial; in fact, it can be shown that fields and action-at-distance are two mathematically equivalent formulations of an interaction (Sudarshan, 1972). But fields provide us with the most natural picture of the conservation of energy and momentum, and we therefore prefer to describe electromagnetism, and gravitation, in terms of local fields.

FURTHER READING

As Bertrand Russell said, "Einstein's theory of relativity is probably the greatest synthetic achievement of the human intellect up to the present time." Einstein's recollections of how he arrived at his discoveries are contained in his "Autobiographical Notes," in Schilpp, P. A., ed., *Albert Einstein: Philosopher-Scientist* (Harper Brothers, New York, 1959). A brilliantly written biography of Einstein as a scientist is Hoffmann, B., *Albert Einstein, Creator and Rebel* (Viking Press, New York, 1972).

Einstein's article "Zur Elektrodynamik bewegter Körper," Ann. d. Physik, **17**, 891 (1905), *must* be read. A translation is available in the paperback Lorentz, H., Einstein, A., Minkowski, H., and Weyl, H., *The Principle of Relativity* (Methuen, London, 1923). This is the original article on relativity, and it is amazing for its completeness. It contains just about everything: from the definition of simultaneity and the Lorentz transformations to the transformation of electric and magnetic fields and the relativistic equations for the motion and energy of a particle. It is not unfair to say that the only essential features of special relativity that remained to be added by the followers of Einstein were the geometrical interpretation of spacetime and the theory of the representations of the Lorentz group.

A delightful discussion of spacetime geometry, with a minimum of mathematics, is given in Synge, J. L., *Talking about Relativity* (North-Holland, Amsterdam, 1970).

Qualitative estimates of the effects of quantum fluctuations on the geometry and topology of spacetime will be found in Wheeler, *Einsteins Vision* (Springer-Verlag, Berlin, 1968).

The best elementary introduction to special relativity is Taylor, E. F., and Wheeler, J. A., *Spacetime Physics* (Freeman, New York, 1992). This book stresses that the theory of relativity is a theory of the geometry of spacetime; it contains a wealth of intriguing puzzles and paradoxes and their solutions. Another very good introduction is French, A. P., *Special Relativity* (Norton, New York, 1968). These books lay the physical foundation for the mathematical formalism that we have developed in the present chapter.

A nice discussion of spacetime geometry with emphasis on a graphical view can be found in Ellis, G. F. R., and Williams, R., *Flat and Curved Spacetimes* (Clarendon Press, Oxford, 1988).

More advanced and complete treatments of special relativity may be found in:

Anderson, J. L., *Principles of Relativity Physics* (Academic Press, New York, 1967). (Includes a careful discussion of the spacetime structures underlying Newtonian and relativistic physics.)

Bergmann, P., *Introduction to the Theory of Relativity* (Prentice-Hall, Englewood Cliffs, 1942). (Contains a thorough presentation of general tensor calculus.)

Einstein, A., *The Meaning of Relativity* (Princeton University Press, Princeton, 1955). (A very concise presentation of the mathematical formalism.)

Fock, V., *The Theory of Space, Time, and Gravitation* (Pergamon Press, Oxford, 1964). (Develops special relativity in general coordinates by means of tensor notation, a good preliminary exercise leading into general invariance.)

Lightman, A. P., Press, W. H., Price, R. H., and Teukolsky, S. A., *Problem Book in Relativity and Gravitation* (Princeton University Press, Princeton, 1975). (Nearly 500 interesting problems, and their solutions, covering many puzzling questions in special relativity, general relativity, gravitation, astrophysics, and cosmology.)

Misner, C. W., Thorne, K. S., and Wheeler, J. A., *Gravitation* (Freeman, San Francisco, 1973). (Emphasizes a geometrical view of the spacetime geometry and gives a beautiful introduction to the language of differential forms.)

Møller, C., *The Theory of Relativity* (Clarendon Press, Oxford, 1952). (Very complete, includes a general discussion of relativistic energy, momentum, angular momentum, and center of mass for arbitrary closed systems.)

Naber, G. L., *The Geometry of Minkowski Spacetime* (Springer-Verlag, New York, 1992). (A highly mathematical treatment that includes a large variety of theorems concerning the spacetime geometry.)

Pauli, W., *Theory of Relativity* (Pergamon Press, London, 1958). (A classic written when Pauli was a student of only twenty-one; it was originally written for the *Mathematical Encyclopedia* as a complete review of all the advances in relativity up to 1921, and although some parts are by now outdated, the mathematical material remains valid and informative.)

Rindler, W., *Essential Relativity* (Springer-Verlag, New York, 1977). (Provides excellent explanations of basic concepts.)

Synge, J. L., *Relativity: The Special Theory* (North-Holland, Amsterdam, 1965). (Discusses systems of particles and collisions, and contains a simple definition of the energy-momentum tensor.)

Weinberg, S., *Gravitation and Cosmology* (Wiley, New York, 1972). (A short, neat review of special relativity is given in Chapter 2; it includes some remarks on representations of the Lorentz group.)

The relativistic formulation of Maxwell's equations is discussed in almost all books on relativity and also in many books on electrodynamics such as, for instance, Jackson, J. D., *Classical Electrodynamics* (Wiley, New York, 1975); Schwartz, M., *Principles of Electrodynamics* (McGraw-Hill, New York, 1972); and Ohanian, H. C., *Classical Electrodynamics* (Allyn and Bacon, Boston, 1988). Unfortunately, some books follow the abominable custom of introducing a factor of $\sqrt{-1}$ into the time component of four-vectors. This simplifies the notation somewhat, because it eliminates the metric tensor $\eta_{\mu\nu}$ from all equations; but since the metric is precisely the most important feature of special relativity it makes little sense to pretend that it does not exist.

The historical role of the field concept in gravitation is discussed in the very readable book North, J. D., *The Measure of the Universe* (Oxford University Press, Oxford, 1965). Electrodynamics was the first field theory, and remains the best understood one; the contributions of Maxwell, Faraday, and their predecessors to this theory are presented with a wealth of detail in Whittaker, E., *A History of the Theories of Aether and Electricity* (Harper, New York, 1960). Incidentally: Whittaker expresses the rather curious opinion that the theory of special relativity was invented by Poincaré and Lorentz. The fact is that even the best effort of Lorentz (see his article "Electromagnetic Phenomena in a System Moving with Any Velocity Less than That of Light" in Lorentz et al., op. cit.) amounted to no more than an admittedly very ingenious generalization of the Lorentz-Fitzgerald contraction hypothesis and was in no sense a theory of relativity.

General surveys of classical field theory are given by Soper, D. E., *Classical Field Theory* (Wiley, New York, 1976), and by Davis, W. R., *Classical Fields, Particles, and the Theory of Relativity* (Gordon and Breach, New York, 1970). The first of these is exceptionally clear and concise.

Excellent introductions to differential forms will be found in Schutz, B. F., *Geometrical Methods in Mathematical Physics* (Cambridge University Press, 1980), Schutz, B. F., *A First Course in General Relativity* (Cambridge University Press, Cambridge, 1985), and Misner, Thorne, and Wheeler, op. cit. The latter book develops clever graphical methods for visualizing differential 1-forms and 2-forms and their operation on tangent vectors. The most simple, clear, and elementary introduction to differential forms is Burke, W. L., *Spacetime, Geometry, and Cosmology* (University Science Book, Mill Valley, 1980).

Ingarden, R., and Jamiolkowski, A., *Classical Electrodynamics* (Elsevier, Amsterdam, 1985), show how to apply differential forms to the solution of problems in electrodynamics. Ehlers, J., "General Relativity and Kinetic Theory," in R. K. Sachs, ed., *General Relativity and Cosmology* (Academic Press, New York, 1971), applies differential forms to problems in kinetic theory in curved spacetime.

Synge, J. S., and Schild, A., *Tensor Calculus* (University of Toronto Press, Toronto, 1956).

REFERENCES

Cohen, I. B., ed., *Isaac Newton's Papers and Letters on Natural Philosophy* (Harvard University Press, Cambridge, 1958), Third Letter to Bentley.

Einstein, A., Ann. Physik **49**, 769 (1916); translated in Lorentz, H. A., Einstein, A., Minkowski, H., and Weyl, H., *The Principle of Relativity* (Methuen, London, 1923), p. 119.

Einstein, A., in Schilpp, P. A., ed., *Albert Einstein: Philosopher-Scientist* (Tudor Publishing Company, New York, 1951), p. 53. Reprinted with the permission of the Estate of Albert Einstein and Amiel Book Company.

Marzke, R, F., and Wheeler, J. A., in Chiu, H.-Y., and Hoffmann, W. F., eds., *Gravitation and Relativity* (Benjamin, New York, 1964).

Minkowski, H., in Lorentz, H. A., Einstein, A., Minkowski, H., and Weyl, H., *The*

Principle of Relativity (Methuen, London, 1923) p. 75.

Ohanian, H. C., *Classical Electrodynamics* (Allyn and Bacon, Boston, 1988).

Schwartz, M., *Principles of Electrodynamics* (McGraw-Hill, New York, 1972).

Streater, R. F., and Wightman, A. S., *PCT, Spin, Statistics and All That* (W. A. Benjamin, New York, 1964).

Sudarshan, E. C. G., Fields and Quanta **2**, 175 (1972).

Wheeler, J. A., *Einsteins Vision* (Springer-Verlag, Berlin, Heidelberg, 1968).

Wigner, E. P., Ann. Math., **40**, 39 (1939).

Van Dam, H., and Wigner, E. P., Phys. Rev. **142**, 838 (1966).

PROBLEMS

1. Astronomers on Earth see two novas flare up simultaneously in two galaxies. One of the galaxies is at a distance of 1.0×10^7 light-years in the constellation Draco; the other galaxy is at the same distance in the constellation Tucana, in a direction exactly opposite to that of the first galaxy. According to astronomers in an alien spaceship traveling at $10^{-4}c$ along the line from Draco to Tucana, these novas are not simultaneous. According to these astronomers, which nova happened first? By how many years?

2. The explosion of the supernova 1987A, at a distance of 1.6×10^5 light-years, released a burst of about 10^{58} neutrinos, of which 129 were captured by detectors on the Earth.
(a) All these neutrinos arrived at the Earth within a time interval of 12 s of each other. According to these data, what is the maximum difference between the speeds of these neutrinos?
(b) All these neutrinos arrived within a few hours of the time of arrival of the flash of light from this explosion. According to these data, what is the maximum difference between the speed of the neutrinos and the speed of light?

3. Fig. 2.4 illustrates the direct measurement, by means of a clock, of a timelike interval PP'. Draw similar figures for the direct measurement of a spacelike interval, a lightlike interval, and a timelike interval with P' earlier than P. Verify that Eq. [3] gives the correct values for the spacetime interval in each case.

4. The simplest conceivable clock consists of two facing mirrors between which a light pulse bounces back and forth. Each bounce of the light pulse is one tick of the clock. The two mirrors must be held at a fixed distance, and this would seem to require an extrinsic standard of length (for instance, the mirrors might be attached to the ends of a standard meter bar). However, by a clever argument, Marzke and Wheeler (1964) demonstrated that the clock can be operated without any such extrinsic standard of length. Their clock is called the *geometrodynamic clock*. The essential ingredient in the operation of this clock is the construction of parallel worldlines, since a fixed distance between two bodies, such as mirrors, implies parallel worldlines. Marzke and Wheeler gave the following prescription for constructing parallel worldlines without using any extrinsic standard of length. Fig. 2.9 shows

the worldline PF of a free particle (particle I). To construct a parallel to PF, introduce a second particle with a worldline QF that intersects PF at some point F in
the future. Such a worldline QF can be found by simply shooting the second particle toward the first in such a way that they collide. Arrange for particle I to emit a
light signal at P, and let this light signal be returned to particle I after reflection by
particle II (see dashed line in Fig. 2.9a). Introduce a third particle with worldline
PG that passes through P. Next, arrange for particle I to emit a light signal toward
particle III in such a way that this light signal passes exactly through F after reflection by particle III (for this, it is best to use a trial and error procedure: let particle
I emit many light signals, and discard all of them except for the one signal which
does reach F). Consider now the intersection of PG with the first light signal and
of QF with the second light signal (see Fig. 2.9b). Prove that if a fourth free particle has a worldline AB that passes through both of these points, then AB is parallel
to PF. (Hint: The dashed triangles in Fig. 2.9b are similar.)

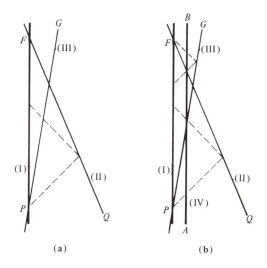

Fig. 2.9 The Marzke-Wheeler construction.

5. The star Capella is at a distance of 46 light-years from the Earth. Suppose that
an astronaut wants to travel from the Earth to Capella in a time of no more than
20 years as reckoned by clocks aboard her spaceship. At what speed would she
have to travel? How long would the trip take as reckoned by clocks on the Earth?

6. The Earth moves around the Sun at a speed of 30 km/s. Assume that in its
own reference frame, the equator of the Earth is a perfect circle of radius 6.4×10^8 cm. In the reference frame of the Sun, the longitudinal diameter of the Earth
suffers length contraction. What is the fractional difference between the longitudinal
and transverse diameters?

7. At 12h 0m 0s Eastern Standard Time a boiler explodes in the basement of the
Museum of Modern Art in New York City. At 12h 0m 0.0003s a similar boiler
explodes in the basement of a soup factory in Camden, N.J., at a distance of 160

km from the first explosion. Find a Lorentz frame in which the first explosion occurs *after* the second.

8. (a) Show that the Lorentz transformation [5] implies the following transformation laws for the (ordinary) velocity:

$$u_x = (u'_x + v)/(1 + vu'_x/c^2)$$

$$u_y = (\sqrt{1 - v^2/c^2}\, u'_y)/(1 + vu'_x/c^2)$$

$$u_z = (\sqrt{1 - v^2/c^2}\, u'_z)/(1 + vu'_x/c^2)$$

where $u'_x = dx/dt'$, etc.

(b) Prove that the only velocity whose magnitude is left invariant by all Lorentz transformations is the velocity of light.

9. Suppose that three sets of coordinates x^μ, x'^μ, and x''^μ are related by pure Lorentz transformations as follows: x'^μ is obtained from x^μ by a boost with velocity v along the x-axis; x''^μ is obtained from x'^μ by a boost with velocity w along the y'-axis.
(a) What is the transformation matrix for the first of these transformations? What is the transformation matrix for the second of these transformations? What is the transformation matrix for the net transformation from x^μ to x''^μ?
(b) Using the transformation matrix for the net transformation from x^μ to x''^μ (or some other method of your choice), show that the x-axis is *not* parallel to the x''-axis and find the angle between these axes.

10. Suppose that there exist particles ("tachyons") that move with a speed *larger* than that of light. Show that you can then send signals into your own past. (Hint: Suppose you are at rest at the origin. Send a tachyon signal, with $v > c$ in your reference frame, towards a spaceship moving away from you. When this spaceship receives the signal, it immediately sends a similar signal, with $v > c$ in the spaceship's frame, back to you. Show that the signal arrives at your position *before* the time at which you sent the original signal.)

11. For an ordinary rotation about the z-axis,

$$a^\mu{}_\nu = \begin{pmatrix} 1 & 0 & 0 & 0 \\ 0 & \cos\theta & \sin\theta & 0 \\ 0 & -\sin\theta & \cos\theta & 0 \\ 0 & 0 & 0 & 1 \end{pmatrix}$$

By explicit matrix multiplication, show that this matrix satisfies Eq. [18].

12. Prove that if a contravariant tensor $A^{\mu\nu}$ is symmetric, then it remains symme-

tric under Lorentz transformations. Is this also true for a mixed tensor $A_\mu{}^\nu$?

13. The *invariant mass* M of a system of particles is defined by

$$M^2 = (p_\mu^{(1)} + p_\mu^{(2)} + \ldots)(p^{(1)\mu} + p^{(2)\mu} + \ldots)$$

where $p^{(1)\mu}$, $p^{(2)\mu}$, etc., are the energy-momentum vectors of particles (1), (2), etc.
    Show that M equals the energy of the system in that reference frame in which the (spatial) momentum is zero (center of momentum, or C.M., frame).

14. Fig. 2.10 shows the worldlines of two clocks that move from A to B. The first clock moves along the straight worldline

$$x = v_0 t$$

where v_0 is a constant. The second clock moves along the curved worldline

$$x = \frac{1}{2} a t^2$$

where a is a constant. For each clock find the elapsed proper time between A and B. Which clock shows the longer proper time?

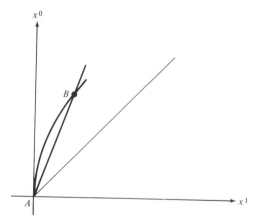

Fig. 2.10 Worldlines of two clocks.

15. Two spacetime points, A and B, have a timelike separation. Consider the straight worldline connecting A and B, and consider curved worldlines. Show that the proper time between A and B,

$$\Delta\tau = \int_A^B d\tau = \int_A^B \sqrt{dx^\mu dx_\mu}$$

is largest when evaluated along the straight worldline. Hence, the worldline of a free particle can be characterized as the spacetime curve of *maximum* proper time.

(Hint: Use a reference frame in which the particle moving along the straight worldline is at rest.)

16. Suppose that $\Phi(x)$ is a scalar field. Show that the integral

$$\int \Phi(x)\, d^3x\, dt$$

evaluated over a given four-dimensional region of spacetime is a scalar. (Hint: Use volume contraction and time dilation to transform $d^3x\,dt$; alternatively, use the Jacobian determinant.)

17. Assume that $A^\mu(x)$ is a vector field with components

$$A^\mu(x) = (0,\ \sin(5x^1 - 3x^0),\ \sin(5x^1 - 3x^0),\ 0)$$

Find the vector field $A'^\mu(x')$ if x and x' are related by the Lorentz transformation [16]. Your answer should be a function of x'.

18. Suppose that the current density $j^\mu(x)$ differs from zero only in a region of finite extent in the x, y, z directions (see Fig. 2.5). Use arguments similar to those given in Eqs. [88] - [99] to show that the *total* charge

$$Q = \int j^0\, d^3x$$

is a Lorentz scalar.

19. Suppose that a vector field $C^\mu(x)$ satisfies the differential equation

$$\partial_\mu C^\mu(x) = 0$$

Assuming that $C^\mu(x)$ is different from zero only in a region of finite extent in the x, y, z directions (see Fig. 2.5), show that

$$\frac{d}{dt}\left[\int C^0(x)\, d^3x\right] = 0$$

20. Express both the energy and momentum of a photon in an electromagnetic wave in terms of the frequency of the wave. Verify that $p_\mu p^\mu = 0$ for the photon.

21. Write down all components of $F_\mu{}^\nu$, $F^\mu{}_\nu$, and $F_{\mu\nu}$ in terms of **E** and **B**. Write down all components of the electromagnetic energy-momentum tensor in terms of **E** and **B**.

22. In the laboratory reference frame, the energy-momentum tensor of a plane electromagnetic wave propagating in the x direction is

$$T^{\mu\nu} = \frac{1}{4\pi} \begin{pmatrix} E^2 & E^2 & 0 & 0 \\ E^2 & E^2 & 0 & 0 \\ 0 & 0 & 0 & 0 \\ 0 & 0 & 0 & 0 \end{pmatrix}$$

where E is the electric field of the wave.

(a) By means of a Lorentz transformation, find the components of the energy-momentum tensor $T'^{\mu\nu}$ in a reference frame moving at speed v in the x direction. Qualitatively, explain why the energy flux T'^{01} is smaller in this new reference frame.

(b) By inspection of the components of the new energy-momentum tensor, what can you say about the magnitude of the electric field of the wave as observed in the new reference frame?

23. Show that $F^{\mu\nu}F_{\mu\nu} = 2(B^2 - E^2)$, where $F^{\mu\nu}$ is the electromagnetic field tensor. Hence, show that $B^2 - E^2$ is a scalar under Lorentz transformations.

24. A very long, charged rod lies along the x-axis. The charge per unit length on the rod is λ, and the electric field at a distance z above the rod is $E_z = 2\lambda/z$ (in Gaussian units). The rod is at rest, and therefore the magnetic field is zero.

(a) What is the electromagnetic field tensor $F^{\mu\nu}$ at the point z?

(b) Consider a new reference frame moving along the x-axis with velocity v. By a Lorentz transformation, find the tensor $F'^{\mu\nu}$ in this new reference frame, at the given distance z above the rod.

(c) From $F'^{\mu\nu}$ find the magnetic field in this new reference frame.

(d) Is your result for the magnetic field in accord with Ampere's law in the new reference frame?

25. Show that the equation of motion [126] has a first integral

$$P_\mu P^\mu = [\text{constant}]$$

26. A plane electromagnetic wave has fields

$$\mathbf{E} = E_0 \, \hat{\mathbf{x}} \, \cos \omega(t - z)$$
$$\mathbf{B} = E_0 \, \hat{\mathbf{y}} \, \cos \omega(t - z)$$

Find all components of the energy-momentum tensor for this wave. Check that the energy flux equals the flux of z momentum. Is this what you expect for a wave made of photons?

27. Show that if the energy of a system of particles in collision is conserved as seen from *all* possible Lorentz frames, then the momentum must necessarily also be conserved. (Hint: Consider the four-vector ΔP^μ that represents the loss of energy-momentum. You are given that $\Delta P^0 = 0$ in *all* reference frames.)

28. Consider a perfect fluid–for instance, a gas–whose state is described by density and pressure (the fluid can support no shear).

(a) Show that in the rest frame of the fluid, in which the macroscopic velocity of the fluid is zero, the energy-momentum tensor is

$$T^{00} = \rho_0, \quad T^{0k} = 0, \quad T^{kl} = p\delta_l^k$$

where ρ_0 is the (macroscopic) mass density measured in this rest frame and p is the pressure.

(b) Show that a Lorentz transformation to a reference frame with respect to which the fluid has a velocity **v** gives

$$T^{\mu\nu} = (\rho_0 + p)u^\mu u^\nu - p\eta^{\mu\nu}$$

29. A slab of solid material has an energy-momentum tensor with $T^{00} = \rho_0$ and all other components equal to zero in its rest frame. What are the components of the energy-momentum tensor in a reference frame with respect to which the slab has a speed v in the x direction?

30. Show that the set of all Lorentz transformations forms a group, and show that scalars, vectors, and tensors form representations of this group.*

31. When a 1-form α acts on the following vectors

$$\mathbf{u} = \frac{\partial}{\partial x^0}, \quad \mathbf{v} = \frac{\partial}{\partial x^1} + \frac{\partial}{\partial x^2}$$

$$\mathbf{w} = \frac{\partial}{\partial x^1} - \frac{\partial}{\partial x^2}, \quad \mathbf{q} = 2\frac{\partial}{\partial x^3}$$

respectively, the results are

$$\alpha(\mathbf{u}) = 1, \quad \alpha(\mathbf{v}) = 0, \quad \alpha(\mathbf{w}) = -2, \quad \alpha(\mathbf{q}) = 1$$

From this, deduce the result that the 1-form produces when it acts on a vector **p** with components (p_0, p_1, p_2, p_3), and deduce the components of the 1-form with respect to the basis dx^μ.

32. Given the 1-forms $\alpha = 2dx^0 + 3dx^1$ and $\beta = -3dx^0 - dx^1 + 4dx^2$, find the 2-form $\alpha \otimes \beta$, and make a list of all its components.

33. For the vector

---

* This problem requires some knowledge of group theory.

$$\mathbf{u} = 3 \; \frac{\partial}{\partial x^1} - 2 \; \frac{\partial}{\partial x^3}$$

evaluate $\boldsymbol{\eta}(\mathbf{u})$, and list the components of the resulting 1-form. Then evaluate $\boldsymbol{\eta}(\mathbf{u}, \mathbf{u})$.

34. The vector potential for a plane electromagnetic wave is

$$\mathbf{A} = \sin \omega (t - z) \; \mathbf{d}x^1$$

Evaluate the electromagnetic field tensor $\mathbf{F} = \mathbf{dA}$ for this plane wave. Evaluate the dual electromagnetic field tensor $^{*}\mathbf{F}$.

3. The Linear Approximation

UN POCO DI VERO FA CREDER TUTTA LA BUGIA.
[A little truth makes the whole lie believable.]
Traditional Italian proverb

We will now seek the relativistic field equations for gravitation. For this purpose, we begin with a linear approximation for the gravitational field, that is, we neglect the effects of the gravitational field on itself. Of course, if Newton's principle of equivalence ($m_I = m_G$) is to hold as an *exact* statement, gravitational energy must gravitate, and the exact field equations must be nonlinear. Although it is true that the most spectacular results of gravitational theory depend in a crucial way on the nonlinearity of the field equations, almost all of the results that have been the subject of experimental investigation can be described by the linear approximation. For example, the deflection of light, the retardation of light, the gravitational time dilation, and gravitational radiation emerge in the linear approximation. Furthermore, this approximation applies to all phenomena that lie in the region of overlap between Newton's and Einstein's theories.

Most discussions of gravitational theory begin by postulating that spacetime is curved and from there proceed to formulate the nonlinear equations that govern the curved spacetime geometry. The linear approximation arises in the end from the full nonlinear equations. The great disadvantage of this approach is that it never makes clear just why anybody would entertain the preposterous notion that our beautiful flat spacetime should be curved, bent, and deformed. The reason why we believe in curved space can best be understood by beginning with flat spacetime and seeking the field equations that describe the gravitational field (in linear approximation) in this flat spacetime. As we will see, the behavior of particles in the gravitational field and the behavior of "clocks" and "meter sticks" are then such that measurements of spacetime indicate that the real geometry is curved. Thus, the flat spacetime with which we begin turns out to be an unobservable, fictitious geometry. But, nevertheless, flat spacetime is a useful starting point in the case of weak fields.

3.1 THE EXAMPLE OF ELECTROMAGNETISM

Since electromagnetism is the linear theory par excellence, it will be instructive to play a game. Let us pretend that we do know that there is a long-range interaction between charges, but we do not know the field equations (Maxwell's equations) that govern this interaction. How could we go about finding the equations?

We will, of course, demand that the equations that describe the interaction between charges obey the principle of relativity. Accordingly, we demand that the equations be tensor equations, and this tells us that an interaction that involves only the charge density is impossible. The charge density ρ and the current density $\mathbf{j}$ taken together form a four-vector (see Eq. [2.114]) and any tensor equation that involves one component of this four-vector must also involve the other components. For example, suppose we have a reference frame in which there are charge densities but no currents. In a new, moving reference frame there will then be both a charge density $j'_0 = a^0{}_0 j^0$ and a current density $j'^k = a^k{}_0 j^0$. Thus, although an observer in the old reference frame attributes the interaction entirely to charges, an observer in the new frame will attribute it to both charges and currents. Any law that pretends to be valid in all reference frames must therefore include both the interaction of charge densities and that of current densities. Hence, if the charge density ρ generates some kind of force field, so will the current density $\mathbf{j}$. To put it briefly: the four-vector j^μ is the source of the electromagnetic field.

The field equations are supposed to establish the link between the field and its source j^μ. As a first step in our search for the field equations, we must decide what kind of field will be used to describe the interaction. According to Section 2.3, our choices are between a scalar, a vector, and a tensor of higher rank. Since the source of the field is the four-vector j^μ, the most natural choice is a vector field, and this is what we will try.

This choice of a vector field can be made plausible by an analogy with Lagrangian mechanics. The equation of motion for a generalized coordinate q^k under the influence of a (generalized) driving force Q^k is typically of the form

$$\frac{d^2 q^k}{dt^2} + ... \propto Q^k$$

where the dots indicate other terms involving the coupling between q^k and other generalized coordinates. In our case, the field plays the role of the generalized coordinate, and the current density plays the role of the driving force. The equation of motion should then have the form

$$\frac{\partial^2 A}{\partial t^2} + ... \propto j^\mu \qquad [1]$$

Here the field variable has simply been written as A, without any index. This quantity A could be a scalar, a vector, or a tensor of higher rank. One obvious defect of our tentative equation of motion is that whereas a Lorentz transformation changes A into A' and j^μ into j'^μ, it does *not* change $\partial^2/\partial t^2$ into $\partial^2/\partial t'^2$; hence the equation is not a tensor equation, and fails to satisfy the principle of special relativity. To achieve invariance, we recall that $\partial^2/\partial t^2$ is only one part of the operator $\partial_\nu \partial^\nu$. If we take the complete operator $\partial_\nu \partial^\nu$, then a Lorentz transformation changes it into $\partial'_\nu \partial'^\nu$, as desired. We therefore arrive at an equation of motion

$$\partial_\nu \partial^\nu A + ... \propto j^\mu \qquad [2]$$

Since the right side of this equation is a four-vector, it follows that A must also be a four-vector. We designate this four-vector by A^μ or by $A^\mu(x)$.*

We now look for the most general second-order linear differential equation for $A^\mu(x)$, with $j^\mu(x)$ acting as source. This means that the equation should have the form of Eq. [2]: on the left side there should appear a linear differential operator acting on the field, and on the right side should appear the current. For example,

$$\partial_\mu \partial^\mu A^\nu + b\partial^\nu \partial_\mu A^\mu + aA^\nu = 4\pi j^\nu \qquad [3]$$

where a and b are constants, is an acceptable equation. In fact, Eq. [3] is the *only* acceptable equation, if we insist that the differential order should be no more than two. To see this, note that Lorentz invariance demands that the left side be a four-vector. But the only four-vectors

---

* Although the analogy with the dynamical equations of mechanics suggests that the field equation should be of second differential order and involve a vector field, it is also possible to write a first-order differential equation involving a second-rank antisymmetric tensor field. It turns out that these alternatives are equivalent (see Eqs. [16] and [17]).

linear in A^μ that can be constructed out of A^μ and no more than two of the differential operators ∂^μ are

$$\partial_\mu \partial^\mu A^\nu \ , \quad \partial^\nu \partial_\mu A^\mu \ , \quad A^\nu \qquad [4]$$

Of course, we would generalize Eq. [3] slightly by inserting an extra arbitrary constant in front of the first term on the left side, but this amounts to no more than a redefinition of A^ν, and is therefore not needed. Likewise, the extra constant that could be inserted in front of j^ν on the right side can always be absorbed in the definition of the current. In fact, the current already contains an adjustable constant q which represents the charge on a particle (see the definition given by Eqs. [2.111] and [2.112]). This is a coupling constant; it characterizes the strength of the coupling between matter and field. The value of this constant q must be determined from experiment. (This can be done only after we discover the field equations. If we do not know Maxwell's equations and the exact form of the law of force between charges, then we cannot give an operational procedure for measuring the constant q that appears in Eqs. [2.111] and [2.112]. But we can write down the expressions [2.111] and [2.112], leaving q as a temporarily unknown parameter.) On the other hand, the values of the constants a and b are completely determined by the following arguments.

As concerns the constant a, we can immediately say that it must be positive or zero. This can best be seen in the special case of vacuum solutions of Eq. [3]; if $j^\nu = 0$, we have

$$\partial_\mu \partial^\mu A^\nu + b\partial^\nu \partial_\mu A^\mu + aA^\nu = 0 \qquad [5]$$

This is a wave equation describing the propagation of the vector field.

■ *Exercise 1.* Show that

$$A^\nu = (0, \cos(kz - \omega t), 0, 0) \qquad [6]$$

with

$$\omega = \sqrt{k^2 + a} \qquad [7]$$

is a possible solution of Eq. [5]. ■

The group velocity of the wave [6] is given by

$$v_g = \frac{d\omega}{dk} = \frac{k}{\sqrt{k^2 + a}} \qquad [8]$$

From this we see that a must be positive, since if a were negative, the group velocity would exceed the speed of light. A positive value of a corresponds to a nonzero mass of the photon; a negative value of a would correspond to an imaginary mass and propagation with speeds larger than the speed of light (tachyons).

To decide what value of a we should adopt, let us look at another special solution of Eq. [3]. Consider a time-independent, spherically symmetric charge distribution. If we consider the $\nu = 0$ component of Eq. [3] and omit all the terms containing time derivatives, we obtain a differential equation for A^0,

$$-\nabla^2 A^0 + aA^0 = 4\pi\rho \qquad [9]$$

In the region exterior to the charge distribution, this has the solution

$$A^0 \propto \frac{e^{\pm\sqrt{a}\,r}}{r} \qquad [10]$$

■ *Exercise 2.* Show this. ■

Only the solution with the negative sign is relevant, and this solution shows that the field decays exponentially with distance outside of the charge distribution. By the rules of our game, we do not know Maxwell's equations, but we do know that the interaction is long range; we therefore will take $a = 0$ (see Section 1.2 for experimental limits on a).

Our field equation then reduces to

$$\partial_\mu \partial^\mu A^\nu + b\partial^\nu \partial_\mu A^\mu = 4\pi j^\nu \qquad [11]$$

The only unknown constant remaining in our field equation is b. To determine the value of this constant, apply ∂_ν to both sides of Eq. [11]:

$$\partial_\nu(\partial_\mu \partial^\mu A^\nu + b\partial^\nu \partial_\mu A^\mu) = 4\pi\partial_\nu j^\nu \qquad [12]$$

If we exchange the (dummy) μ and ν indices in the second term on the left side, we obtain

$$(1 + b)\partial_\mu \partial^\mu(\partial_\nu A^\nu) = 4\pi\partial_\nu j^\nu \qquad [13]$$

The right side of this equation vanishes because the current satisfies the conservation law $\partial_\nu j^\nu = 0$ (Eq. [2.118]). Hence the consistency of our field equation with this conservation law demands that the left side of Eq. [13] vanish. In order to meet this demand, we will take $b =$

-1, so that the left side vanishes *identically*. This is the most elegant way to make the field equations compatible with current conservation. But there are other ways; for instance, we could leave b arbitrary and impose the condition $\partial_\nu A^\nu = 0$; this is a gauge condition (see Eq. [20]).

With $b = -1$, the final form of our field equation becomes

$$\partial_\mu \partial^\mu A^\nu - \partial^\nu \partial_\mu A^\mu = 4\pi j^\nu \qquad [14]$$

This last equation has the same form as the Maxwell equation for the potential that we wrote down in Section 2.5 (see Eq. [2.135]). Obviously, our vector field A^μ is to be identified with the four-vector potential. To obtain the Maxwell equations for the fields, we *define* the tensor $F^{\mu\nu}$ as

$$F^{\mu\nu} \equiv \partial^\mu A^\nu - \partial^\nu A^\mu \qquad [15]$$

and then Eq. [14] becomes

$$\partial_\mu F^{\mu\nu} = 4\pi j^\nu \qquad [16]$$

Furthermore, the tensor [15] identically satisfies the identity

$$\partial^\alpha F^{\mu\nu} + \partial^\mu F^{\nu\alpha} + \partial^\nu F^{\alpha\mu} = 0 \qquad [17]$$

These equations are, of course, exactly the Maxwell equations [2.106] and [2.107] that we formulated in Section 2.5, except that in Section 2.5 we retained the speed of light in these equations, whereas here we assume $c = 1$.

Although in some sense the preceding arguments may be regarded as a derivation of Maxwell's equations, the limitations of this approach should be kept in mind. Clearly, we have had to make quite a few assumptions to reach Eq. [14]. The objective of our game with electrodynamics was to obtain a prescription for finding the field equations in the hope that an analogous prescription will lead us to the field equations for gravitation. The number of ingredients in this prescription need not concern us, as long as each ingredient in the electromagnetic case has an analogous ingredient in the gravitational case. As we will see in the next section, the preceding arguments can indeed be reproduced step-by-step in the gravitational case and lead to a reasonable result.

We will now briefly look at some properties and consequences of the electromagnetic field equation [14]. One of the remarkable properties of this equation is the following: *Whenever A^μ satisfies Eq. [14], then so does $A^\mu + \partial^\mu \Lambda$, where Λ is any arbitrary scalar function.* This is so because in the expression

$$\partial_\mu \partial^\mu (A^\nu + \partial^\nu \Lambda) - \partial^\nu \partial_\mu (A^\mu + \partial^\mu \Lambda) \qquad [18]$$

the terms involving Λ cancel identically, and therefore the differential equation [14] is left unchanged.

The transformation

$$A^\mu \rightarrow A^\mu + \partial^\mu \Lambda \qquad [19]$$

is called a *gauge transformation*. We have shown that the field equation is invariant under this gauge transformation. In general, whenever an object is left unchanged (invariant) by some operation, the operation is said to be a symmetry of the object. Thus, gauge invariance is a symmetry of the field equation.*

The gauge function Λ is completely arbitrary; this means that our equations do not determine the field uniquely. There is no harm in this, because an ambiguity of the form $\partial^\mu \Lambda$ in the field A^μ leads to no observable consequences. The gauge transformations of A^μ leave the field tensor $F^{\mu\nu}$ unchanged; this implies that the observable quantities **E** and **B** are gauge independent.

■ *Exercise 3.* Show that the gauge transformation [19] produces no change in $F^{\mu\nu}$. ■

However, it is helpful to make a particular choice of the gauge function and eliminate some of the ambiguity in A^μ. We can do this by imposing the *Lorentz gauge condition* on the field

$$\partial_\mu A^\mu = 0 \qquad [20]$$

To see how this works, suppose that A^μ does *not* satisfy the condition [20]. Then we can always find a new field $\bar{A}^\mu$ that does. We simply take

$$\bar{A}^\mu = A^\mu + \partial^\mu \Lambda \qquad [21]$$

and choose Λ in such a way that $\partial_\mu \bar{A}^\mu = 0$:

$$0 = \partial_\mu \bar{A}^\mu = \partial_\mu A^\mu + \partial_\mu \partial^\mu \Lambda$$

---

* Our prodedure for the construction of the field equation suggests an intimate connection between current conservation and gauge symmetry, since the gauge invariant form [14] of the field equation emerged from the requirement that the equation be manifestly consistent with current conservation (see Eq. [13]). The study of field theory reveals that, in general, gauge symmetries imply conservation laws. This is Noether's theorem; see, e. g., Soper (1976), Chapter 9.

that is,

$$\partial_\mu \partial^\mu \Lambda = - \partial_\mu A^\mu \qquad [22]$$

This is a differential equation which can be solved for Λ. Eq. [22] always has a solution. In fact, it has *several solutions*, since to any solution Λ, we can always add a solution of the homogeneous equation (accordingly, the Lorentz condition does not remove the ambiguity in the field entirely).

We will from now on assume that the Lorentz condition holds. Note that with this condition the field equation reduces to

$$\partial_\mu \partial^\mu A^\nu = 4\pi j^\nu \qquad [23]$$

This form of the field equation is often the most convenient for finding solutions for a given current.

In vacuo, where $j^\nu = 0$, Eq. [23] is a simple wave equation with a propagation speed equal to the speed of light.

■ *Exercise 4.* Show that for a wave propagating in the z direction, Eq. [23] (with $j^\nu = 0$) and Eq. [20] have the following linearly independent solutions:

$$A^\mu = \epsilon_{(n)}{}^\mu \cos(\omega z - \omega t) \qquad [24]$$

where $\epsilon_{(n)}{}^\mu$ is one of the following four-vectors:

$$\epsilon_{(1)}{}^\mu \equiv (0, 1, 0, 0) \qquad [25]$$
$$\epsilon_{(2)}{}^\mu \equiv (0, 0, 1, 0) \qquad [26]$$
$$\epsilon_{(3)}{}^\mu \equiv (1, 0, 0, 1) \qquad [27]$$

(The subscript n has been enclosed in parentheses to indicate that it is not a tensor index; it is only a label that distinguishes between the four-vectors [25]-[27].)

Show that any other solution of the type $\cos(\omega z - \omega t)$ can be written as a linear combination of the three solutions given above. Find the **E** and **B** fields for each solution, and show that the $\epsilon_{(3)}{}^\mu$ solution gives **E** = **B** = 0; hence only the first two solutions have any physical significance. These two solutions correspond to the familiar two independent directions of polarization of an electromagnetic wave. ■

To complete our electromagnetic theory, we need to find the equation of motion of a charged particle in an electromagnetic field. This equation of motion can be derived from the conservation law of energy-momentum. If we have a charged particle interacting with an electromagnetic field, neither the energy-momentum of the particle nor that of the field is separately conserved; only the *total* energy-momentum is conserved:

$$\partial_\nu (T_{(m)}{}^{\mu\nu} + T_{(em)}{}^{\mu\nu}) = 0 \qquad [28]$$

where $T_{(m)}{}^{\mu\nu}$ and $T_{(em)}{}^{\mu\nu}$ are, respectively, the energy-momentum tensors for particles and for the field.

The expression for $T_{(m)}{}^{\mu\nu}$ is given by Eq. [2.75]. Note that in the present context we want to use this expression to describe a single particle. Hence, we must assume that the mass density has a sharp peak (strictly speaking, a delta function) at the position of the particle and vanishes everywhere else.

The expression for $T_{(em)}{}^{\mu\nu}$ is given by Eq. [2.127]. Unfortunately, according to the rules of our game we must pretend that we do not know Eq. [2.127]. How can we derive this expression for $T_{(em)}{}^{\mu\nu}$?

For a given field equation, there exists a general prescription for constructing the energy-momentum tensor associated with the field ("canonical energy-momentum tensor"). An important feature of this prescription is that it does not make use of the equation of motion of a particle. Since we ultimately want to derive this equation of motion from the energy-momentum tensor, this feature of the canonical prescription is crucial for the consistency of our approach. Other prescriptions for constructing the energy-momentum tensor rely on knowledge of the equation of motion of a particle; such prescriptions would not suit our purposes. In the construction of the canonical energy-momentum tensor we must use the tools of Lagrangian field theory; because of this, the derivation of the expression [2.127] will be left to Appendix 1.

Note that the expression [2.127] has all the properties we expect of an energy-momentum tensor. The first and most obvious property is that for a free field, with $j^\nu = 0$, the differential conservation law $\partial_\nu T_{(em)}{}^{\mu\nu} = 0$ must hold; that this is indeed so follows from Eq. [29] and Exercise 5 (see below). Furthermore, we expect that the tensor is quadratic in the derivatives of the fields. This property is a generalization of what we know to be the case for a system of particles: the (kinetic) energy is quadratic in the time derivatives of the generalized coordinates. Thus we expect time derivatives, and hence (by Lorentz invariance) also space derivatives, of the fields to appear quadratically in the energy-momentum tensor. Finally, the tensor should be symmetric, and its T^{00} component (energy density) should be positive. The expression [2.127] satisfies all these expectations.

Given this expression for $T_{(em)}{}^{\mu\nu}$, a straightforward calculation using Eqs. [16] and [17] shows that

$$\partial_\nu T_{(em)}{}^{\mu\nu} = F^\mu{}_\nu j^\nu \qquad [29]$$

■ *Exercise 5.* Derive Eq. [29]. ■

Hence, Eq. [28] becomes

$$\partial_\nu T_{(m)}{}^{\mu\nu} - F^\mu{}_\nu j^\nu = 0 \qquad [30]$$

This equation tells us how much energy and momentum the electromagnetic field transfers to the matter on which it acts. Thus, the equation determines the rate at which the momentum of a particle changes, and therefore it determines the equation of motion of a particle acted upon by an electromagnetic field.

To derive this equation of motion, we begin by integrating the expression [30] over the volume of the particle (in the present context, "particle" simply means an arbitrary system of small size). The first term gives

$$\int \partial_\nu T_{(m)\mu}{}^\nu \, d^3x = \int \partial_0 T_{(m)\mu}{}^0 \, d^3x + \int \partial_k T_{(m)\mu}{}^k \, d^3x \qquad [31]$$

The second integral on the right side of Eq. [31] can be converted into a surface integral of the form

$$\int T_{(m)\mu}{}^k \, dS_k$$

over a surface surrounding the particle; since $T_{(m)\mu}{}^k$ is zero everywhere outside of the particle, the surface integral is zero. The first integral on the right side of Eq. [31] is simply the rate of change of the momentum of the particle,

$$\frac{d}{dt} \int T_{(m)\mu}{}^0 \, d^3x = \frac{d}{dt} \, p_\mu \qquad [32]$$

To evaluate the volume integral of the second term in [30], we make the assumption that $F^\mu{}_\nu$ is roughly constant over the volume of the particle; this is a good approximation for a sufficiently small particle. We then have

$$- \int F^\mu{}_\nu j^\nu \, d^3x \simeq -F^\mu{}_\nu \int j^\nu \, d^3x \qquad [33]$$

■ *Exercise 6.* Show that integration over the volume of the particle gives

$$\int j^\nu \, d^3x = \int \rho_0 u^\nu \, d^3x = q u^\nu \sqrt{1 - v^2} \qquad [34]$$

(Hint: Assume that u^ν is constant over the volume of the particle. Take the volume contraction into account.) ■

With the results [32] and [34], we obtain the following for the volume integral of [30]:

$$\frac{d}{dt} p_\mu - q\sqrt{1 - v^2}\, F_{\mu\nu} u^\nu = 0 \qquad [35]$$

which is the same as

$$\frac{d}{d\tau} p_\mu = q F_{\mu\nu} u^\nu \qquad [36]$$

This equation of motion agrees with Eq. [2.126], and completes our program of deriving electrodynamics from nothing (or nearly nothing). Note that the equation of motion we have derived is that of a point-particle without electric or magnetic multipole moments. If we drop the assumptions that $F^\mu_{\ \nu}$ and u^ν are constant over the volume of the particle, then we can derive equations for particles with multipole moments.

It may seem strange that the equation of motion of a particle should emerge as a consequence of the electromagnetic field equations. But this becomes much less surprising if one realizes that the equation of motion of a particle is no more than a statement about the exchange of momentum between particle and field. Hence the equation of motion cannot be independent of the conservation law for the energy-momentum tensor of the fields. Note that our equations completely determine the force between charged particles. For example, if we have two positive charges at rest, then we can solve Maxwell's equations (Eqs. [16] and [17]) to obtain the **E** field produced by one of these charges, and then calculate the force (Eq. [36]) that this field exerts on the other charge. The resulting force is of course repulsive. The fact that a vector field, such as A^μ, necessarily leads to a repulsive force between like charges shows that the gravitational field cannot be a vector field. For this and other reasons, we will use a second-rank tensor field as the carrier of the gravitational interaction.

3.2 THE LINEAR FIELD EQUATIONS FOR GRAVITATION

In order to discover the field equations of gravitation, we proceed in a manner entirely analogous to what we did for electromagnetism. This means that we assume that our spacetime is the flat spacetime of special relativity, and we look for a suitable field equation for the gravitational field. The first thing we must do is decide what the source of gravitation is going to be; that is, we must ask what quantity plays a role analogous to that of the electromagnetic source j^μ. The answer is given by Newton's principle of equivalence, according to which the (total) inertial mass of a system gravitates, that is, the energy gravitates. The source of gravitation must therefore be the energy density. However, it is impossible to construct a Lorentz-invariant theory of gravitation in which the energy density is the only source of gravitation. We recall that the energy density T^{00} is only one component of the second-rank tensor $T^{\mu\nu}$. What is pure T^{00} in one reference frame becomes $T'^{\mu\nu} = a^{\mu}{}_0 a^{\nu}{}_0 T^{00}$ in another reference frame and, in general, this is not pure T'^{00}. Thus, what is an energy density in one reference frame will be some combination of energy density, energy flux density, and momentum flux density as seen from another reference frame. If the laws are to have the same form in all Lorentz frames, then *all* these quantities must be sources of gravitation, that is, the tensor $T^{\mu\nu}$ must be the source of gravitation.

Before we proceed, a remark on a possible alternative. We could try to use the invariant trace of the energy-momentum tensor as the source of gravitation. Since

$$T_{\mu}{}^{\mu} = T_0{}^0 + T_1{}^1 + T_2{}^2 + T_3{}^3 \qquad [37]$$

we see that $T_{\mu}{}^{\mu}$ contains T^{00}, and therefore it would seem that the expression [37] satisfies the requirement that the energy density be the main source of gravitation. The trouble with [37] is that for electromagnetic fields, $T_{\mu}{}^{\mu} \equiv 0$ (see Eq. [2.130]). Hence electromagnetic energy would not gravitate, in contradiction with the principle of equivalence.

We must now decide what type of field we will use as carrier of the interaction. Since the source of gravitation is a second-rank, symmetric tensor, arguments analogous to those given in the electromagnetic case suggest that the obvious and natural choice is a second-rank, symmetric tensor field. We will use the notation $h^{\mu\nu}$ for the gravitational field tensor. As an alternative to this second-rank tensor, we might try a third-rank tensor. This gives results equivalent to those obtained by using a second-rank tensor.*

The most general field equation that is linear in $h^{\mu\nu}$, is of second differential order, and contains $T^{\mu\nu}$ as source must have the form

$$\partial_\lambda \partial^\lambda h^{\mu\nu} + a\partial^\mu \partial^\nu h_\sigma{}^\sigma - a'(\partial_\lambda \partial^\nu h^{\mu\lambda} + \partial_\lambda \partial^\mu h^{\nu\lambda})$$
$$+ b\eta^{\mu\nu} \partial_\lambda \partial^\lambda h_\sigma{}^\sigma + b'\eta^{\mu\nu} \partial_\lambda \partial_\sigma h^{\lambda\sigma} = -\kappa T^{\mu\nu} \qquad [38]$$

where a, a', b, b', and κ are constants. We arrive at this form by writing down all possible symmetric tensors that can be constructed out of $h^{\mu\nu}$ and no more than two gradient operators ∂^σ. The only such tensors, linear in $h^{\mu\nu}$, are the following:

---

* It corresponds to the possibility of writing Einstein's equations in terms of $\Gamma^{\mu}{}_{\alpha\beta}$ rather than the usual $g_{\alpha\beta}$; see Chapter 7.

$$\partial_\lambda \partial^\lambda h^{\mu\nu} \,, \quad \partial^\mu \partial^\nu h_\sigma{}^\sigma \,, \quad \partial_\lambda \partial^\nu h^{\mu\lambda} + \partial_\lambda \partial^\mu h^{\nu\lambda} \,,$$

$$\eta^{\mu\nu} \partial_\lambda \partial^\lambda h_\sigma{}^\sigma \,, \quad \eta^{\mu\nu} \partial_\lambda \partial_\sigma h^{\lambda\sigma} \,, \quad h^{\mu\nu} \,, \quad \eta^{\mu\nu} h_\sigma{}^\sigma$$

The left side of Eq. [38] is a general linear combination of these tensors, except that we have omitted the last two, $h^{\mu\nu}$ and $\eta^{\mu\nu} h_\sigma{}^\sigma$. The presence of such terms in Eq. [38] would produce an exponential decrease with distance in the static gravitational field (see Eq. [10]); we will assume that the interaction is long range and that therefore such terms are undesirable. Incidentally: We could attempt to generalize [38] by using $\kappa T^{\mu\nu} + \kappa' \eta^{\mu\nu} T$ on the right side. But this amounts to no more than a redefinition of the constants a, a', b, b'.

Our next task will be to determine the values of the constants a, a', b, b'; we will see that our theory determines all these values unambiguously. The constant κ characterizes the strength of the coupling between matter and gravitation, and its value must be determined by experiment. This constant plays a role analogous to that of the electric charge. The only difference is that the electric charge q is included in the definition of j^μ, whereas the constant κ is written separate from $T^{\mu\nu}$; this difference in notation is sanctified by tradition.

It is convenient to introduce the notation h for the trace of $h^{\mu\nu}$:

$$h \equiv h_\sigma{}^\sigma \tag{39}$$

With this, Eq. [38] becomes

$$\partial_\lambda \partial^\lambda h^{\mu\nu} + a \partial^\mu \partial^\nu h - a'(\partial_\lambda \partial^\nu h^{\mu\lambda} + \partial_\lambda \partial^\mu h^{\nu\lambda})$$
$$+ b \eta^{\mu\nu} \partial_\lambda \partial^\lambda h + b' \eta^{\mu\nu} \partial_\lambda \partial_\sigma h^{\lambda\sigma} = - \kappa T^{\mu\nu} \tag{40}$$

If we operate with ∂_μ on both sides of this equation we obtain

$$\partial_\lambda \partial^\lambda \partial_\mu h^{\mu\nu} + a \partial_\mu \partial^\mu \partial^\nu h - a'(\partial_\lambda \partial_\mu \partial^\nu h^{\mu\lambda} + \partial_\lambda \partial_\mu \partial^\mu h^{\nu\lambda})$$
$$+ b \partial^\nu \partial_\lambda \partial^\lambda h + b' \partial^\nu \partial_\lambda \partial_\sigma h^{\lambda\sigma} = - \kappa \partial_\mu T^{\mu\nu} \tag{41}$$

The right side of this equation vanishes because of the general conservation law for the energy-momentum tensor (Eq. [2.79]). In analogy with the electromagnetic case, we then demand that the left side vanish identically. The constants should therefore satisfy the equations

$$- a' + b' = 0$$
$$1 - a' = 0$$
$$a + b = 0 \tag{42}$$

Hence $a' = 1$, $b' = 1$, and $b = - a$. Our field equation reduces to

$$\partial_\lambda \partial^\lambda h^{\mu\nu} + a\partial^\mu \partial^\nu h - (\partial_\lambda \partial^\nu h^{\mu\lambda} + \partial_\lambda \partial^\mu h^{\nu\lambda})$$
$$- a\eta^{\mu\nu} \partial_\lambda \partial^\lambda h + \eta^{\mu\nu} \partial_\lambda \partial_\sigma h^{\lambda\sigma} = -\kappa T^{\mu\nu} \qquad [43]$$

The value of the constant a is of no physical relevance. Different values of a merely correspond to different ways of describing the same physical gravitational field by means of different, but related, tensors. Thus, if instead of using the tensor $h^{\mu\nu}$ to describe the gravitational field we use a new tensor $\bar{h}^{\mu\nu}$ related to the former by

$$h^{\mu\nu} = \bar{h}^{\mu\nu} - C\eta^{\mu\nu}\bar{h} \qquad [44]$$

(where C is a constant), then the field equation for $\bar{h}^{\mu\nu}$ has exactly the same form as Eq. [43], but a is replaced by $\bar{a}$:

$$\bar{a} = a(1 - 4C) + 2C \qquad [45]$$

■ *Exercise 7.* Substitute Eq. [44] into Eq. [43] and verify Eq. [45]. ■

In particular, if we choose

$$C = \frac{1 - a}{2 - 4a} \qquad [46]$$

then $\bar{a} = 1$, and we obtain the field equation

$$\partial_\lambda \partial^\lambda \bar{h}^{\mu\nu} + \partial^\mu \partial^\nu \bar{h} - (\partial_\lambda \partial^\nu \bar{h}^{\mu\lambda} + \partial_\lambda \partial^\mu \bar{h}^{\nu\lambda})$$
$$- \eta^{\mu\nu} \partial_\lambda \partial^\lambda \bar{h} + \eta^{\mu\nu} \partial_\lambda \partial_\sigma \bar{h}^{\lambda\sigma} = -\kappa T^{\mu\nu} \qquad [47]$$

Note that $a = \frac{1}{2}$ leads to trouble in Eq. [46]. However, this troublesome value of a must in any case be excluded, because for $a = \frac{1}{2}$ our field equation [43] contains insufficient information for determining the field.

■ *Exercise 8.* Show that if $a = \frac{1}{2}$, then

$$h^{\mu\nu} + \eta^{\mu\nu} f$$

is a solution of Eq. [43] whenever $h^{\mu\nu}$ is a solution. The function f is arbitrary. ■

Since the overbars in Eq. [47] serve no further purpose, we omit them and write our final equation as

$$\partial_\lambda \partial^\lambda h^{\mu\nu} + \partial^\mu \partial^\nu h - (\partial_\lambda \partial^\nu h^{\mu\lambda} + \partial_\lambda \partial^\mu h^{\nu\lambda})$$

$$- \eta^{\mu\nu} \partial_\lambda \partial^\lambda h + \eta^{\mu\nu} \partial_\lambda \partial_\sigma h^{\lambda\sigma} = - \kappa T^{\mu\nu} \qquad [48]$$

This is the relativistic field equation for gravitation in the linear approximation. This equation plays a role analogous to that of the Maxwell equations: Eq. [48] determines the gravitational field generated by a given matter distribution. The analogy with electrodynamics is summarized in Table 3.1. Since the gravitational field $h^{\mu\nu}$ is the analog of the four-vector potential A^μ, it would perhaps be more appropriate to call $h^{\mu\nu}$ the gravitational tensor potential; but for convenience we will continue to call it simply the gravitational field since, in the linear approximation, it is the only field quantity we have any need for.

TABLE 3.1 ANALOGY BETWEEN THE ELECTROMAGNETIC AND GRAVITATIONAL FIELD THEORIES

	Electromagnetism	Gravitation (linear approximation)
Source of field	j^ν	$T^{\mu\nu}$
Conservation law	$\partial_\nu j^\nu = 0$	$\partial_\nu T^{\mu\nu} = 0$
Field	A^ν	$h^{\mu\nu}$
Field equation	$\partial_\mu \partial^\mu A^\nu - \partial^\nu \partial_\mu A^\mu$ $= 4\pi j^\nu$	$\partial_\lambda \partial^\lambda h^{\mu\nu} + \partial^\mu \partial^\nu h - \partial_\lambda \partial^\nu h^{\mu\lambda}$ $- \partial_\lambda \partial^\mu h^{\nu\lambda} - \eta^{\mu\nu} \partial_\lambda \partial^\lambda h + \eta^{\mu\nu} \partial_\lambda \partial_\sigma h^{\lambda\sigma}$ $= - \kappa T^{\mu\nu}$
Gauge transformation	$A^\mu \rightarrow A^\mu + \partial^\mu \Lambda$	$h^{\mu\nu} \rightarrow h^{\mu\nu} + \partial^\nu \Lambda^\mu + \partial^\mu \Lambda^\nu$
Preferred gauge condition	$\partial_\mu A^\mu = 0$	$\partial_\mu (h^{\mu\nu} - \frac{1}{2}\eta^{\mu\nu} h) = 0$
Field equation in preferred gauge	$\partial_\mu \partial^\mu A^\nu = 4\pi j^\nu$	$\partial_\lambda \partial^\lambda (h^{\mu\nu} - \frac{1}{2}\eta^{\mu\nu} h) = -\kappa T^{\mu\nu}$
Energy-momentum exchange between field and particle	$\partial_\nu T_{(m)\mu}{}^\nu = F_{\mu\nu} j^\nu$	$\partial_\nu T_{(m)\mu}{}^\nu = \frac{1}{2}\kappa m h_{\alpha\beta,\mu} T_{(m)}{}^{\alpha\beta}$
Equation of motion of particle	$\frac{d}{d\tau} P_\mu = qF_{\mu\nu} u^\nu$	$\frac{d}{d\tau} P_\mu = \frac{1}{2}\kappa m h_{\alpha\beta,\mu} u^\alpha u^\beta$
Energy-momentum of particle	$P_\mu = mu_\mu$	$P_\mu = mu_\mu + m\kappa h_{\mu\alpha} u^\alpha$
Proper time interval	$d\tau^2 = \eta_{\alpha\beta} dx^\alpha dx^\beta$	$d\tau^2 = (\eta_{\alpha\beta} + \kappa h_{\alpha\beta}) dx^\alpha dx^\beta$

The assumption that the gravitational field is entirely described by a single tensor field entered in a crucial way into our construction of the field equation. This is the simplest of all possible choices for the gravitational field and, since all the available experimental data are in agreement with this tensor theory of gravitation, a more complicated choice would be contrary to Newton's First Rule for Reasoning in Philosophy, already mentioned in Chapter 1, "We are to admit no more causes of natural things than such as are both true and sufficient to explain their appearances." To the best of our knowledge, one gravitational field is sufficient to explain all gravitational effects, and more fields would therefore be a wretched excess. If, nevertheless, we were to hypothesize that besides the tensor field there exist one or several *extra* gravitational scalar, vector, or tensor fields, then we could construct no end of endlessly complicated theories. Among these alternatives the least complicated is the scalar-tensor theory of Jordan (1951) and Brans and Dicke (1961) which contains an extra scalar field; qualitatively, this scalar field has the effect of making the gravitational constant dependent on position. The experimental evidence (see Section 4.3) speaks against the scalar-tensor theory; we will therefore not bother with it.

In a region free of matter, the field equation [48] becomes

$$\partial_\lambda \partial^\lambda h^{\mu\nu} + \partial^\mu \partial^\nu h - (\partial_\lambda \partial^\nu h^{\mu\lambda} + \partial_\lambda \partial^\mu h^{\nu\lambda})$$
$$- \eta^{\mu\nu} \partial_\lambda \partial^\lambda h + \eta^{\mu\nu} \partial_\lambda \partial_\sigma h^{\lambda\sigma} = 0$$

This equation describes the propagation of free gravitational waves in empty space; we will investigate such waves in Chapter 5. This equation coincides with the field equation for a massless particle of spin 2, the graviton (Wentzel, 1949). Thus, we could have obtained our field equation for the gravitational tensor field by postulating that the gravitational interaction is carried by massless quanta of spin 2. However, we wanted to formulate the equation in a purely classical context, by relying on an analogy with electromagnetism.

In the electromagnetic case, we were able to obtain a simplified field equation (Eq. [23]) by taking advantage of gauge invariance. We can also do this in the gravitational case. It is easy to show that Eq. [48] is invariant under the gauge transformation

$$h^{\mu\nu} \rightarrow h^{\mu\nu} + \partial^\nu \Lambda^\mu + \partial^\mu \Lambda^\nu \qquad [49]$$

where $\Lambda^\mu(x)$ is an arbitrary vector field.

■ *Exercise 9.* Prove this. ■

Exactly as in the electromagnetic case, gauge invariance implies that the field equations do not determine the field uniquely, but the ambiguity in the field is not serious and leads to no observable consequences.* We find it convenient to eliminate the ambiguity (partially)

---

* Further discussion of the gauge transformation will be given in Section 7.1.

by imposing the following gauge condition on the fields:

$$\partial_\mu (h^{\mu\nu} - \tfrac{1}{2}\eta^{\mu\nu} h) = 0 \qquad\qquad [50]$$

This is known as the *Hilbert condition*. If we are given a field $h^{\mu\nu}$ that does not satisfy Eq. [50], then, by a gauge transformation, we can always obtain a new field $\bar{h}^{\mu\nu}$,

$$\bar{h}^{\mu\nu} = h^{\mu\nu} + \partial^\mu \Lambda^\nu + \partial^\nu \Lambda^\mu \qquad\qquad [51]$$

such that the latter *does* satisfy [50].

■ *Exercise 10.* Show that if Λ^ν is a solution of the differential equation

$$\partial_\mu \partial^\mu \Lambda^\nu = - \partial_\mu h^{\mu\nu} + \tfrac{1}{2}\partial^\nu h \qquad\qquad [52]$$

then $\bar{h}^{\mu\nu}$ will satisfy Eq. [50]. ■

We now assume that the gauge condition holds; then the field equation simplifies to

$$\partial_\lambda \partial^\lambda (h^{\mu\nu} - \tfrac{1}{2}\eta^{\mu\nu} h) = - \kappa T^{\mu\nu} \qquad\qquad [53]$$

■ *Exercise 11.* Derive this from Eqs. [48] and [50]. ■

It is convenient to define a new field variable

$$\phi^{\mu\nu} = h^{\mu\nu} - \tfrac{1}{2}\eta^{\mu\nu} h \qquad\qquad [54]$$

In terms of $\phi^{\mu\nu}$, the field equation becomes

$$\partial_\lambda \partial^\lambda \phi^{\mu\nu} = - \kappa T^{\mu\nu} \qquad\qquad [55]$$

and the gauge condition becomes

$$\partial_\mu \phi^{\mu\nu} = 0 \qquad\qquad [56]$$

■ *Exercise 12.* Show that Eq. [54] implies

$$h^{\mu\nu} = \phi^{\mu\nu} - \tfrac{1}{2}\eta^{\mu\nu} \phi \qquad ■ \qquad [57]$$

Eq. [55] expresses the field equation [48] in a particularly simple and convenient form. The remainder of this chapter and the next chapter are largely concerned with the investigation of different solutions of Eq. [55]. Thus, in Section 4.1 we will find the field surrounding a static, spherically symmetric mass, and in Section 4.7 the field surrounding a rotating mass. In Chapter 4, we will look at the wave solutions of our field equation and investigate the emission of gravitational radiation.

However, it does us little good to calculate the field $h^{\mu\nu}$ (or $\phi^{\mu\nu}$) generated by some matter distribution if we do not know how this field acts on a particle that is placed in it. It is imperative that we try to discover the equation of motion of a particle in a given gravitational field.

3.3 THE INTERACTION OF GRAVITATION AND MATTER

In writing the field equation [48] we assumed that the quantity $T^{\mu\nu}$ is the energy-momentum tensor of matter. In order to obtain a linear field equation we left out the effect of the gravitational field upon itself. Because of this omission, our linear field equation has two (related) defects: (i) According to Eq. [53] matter acts on the gravitational field (changes the fields), but there is no reciprocal action of the gravitational fields on matter; that is, the gravitational field can acquire energy-momentum from matter, but nevertheless the energy-momentum of matter is conserved ($\partial_\nu T^{\mu\nu} = 0$). This is an inconsistency. (ii) Gravitational energy does not act as source of gravitation, in contradiction to the principle of equivalence. Thus, although Eq. [48] may be a fair approximation in the case of weak gravitational fields, it cannot be an exact equation.

The obvious way to correct for our sin of omission is to include the energy-momentum tensor of the gravitational field in $T^{\mu\nu}$. This means that we take for the quantity $T^{\mu\nu}$ in Eq. [53] the total energy-momentum tensor of matter plus gravitation:

$$T^{\mu\nu} = T_{(m)}{}^{\mu\nu} + t^{\mu\nu} \qquad [58]$$

Here $T_{(m)}{}^{\mu\nu}$ and $t^{\mu\nu}$ are, respectively, the energy-momentum tensors of matter and of gravitation. We assume that the interaction energy of matter and gravitation is always included in $T_{(m)}{}^{\mu\nu}$; this is a reasonable convention, since the interaction energy density will differ from zero only at those places where there is matter. For example, consider the energy density $(\nabla\Phi)^2/8\pi G + \rho\Phi$ given by the Newtonian approximation (see Eq. [1.19]). According to the above convention, the first term would be identified with t^{00}, but the second term would be regarded as part of $T_{(m)}{}^{00}$ (together with the density of kinetic and rest-mass energy).

Our field equation now becomes

$$\partial_\lambda \partial^\lambda \phi^{\mu\nu} = -\kappa(T_{(m)}{}^{\mu\nu} + t^{\mu\nu}) \qquad [59]$$

Let us operate with ∂_ν on this equation; the left side is zero because of the gauge condition and therefore

$$0 = \partial_\nu(T_{(m)}{}^{\mu\nu} + t^{\mu\nu}) \qquad [60]$$

This simply expresses the conservation of the total energy-momentum and shows that the modified field equation is free from the inconsistencies mentioned at the beginning of this section. Unfortunately, Eq. [60] does not get us very far because we do not know what $t^{\mu\nu}$ is. We already mentioned that there exists a general procedure for constructing the energy-momentum tensor of a field, if the field equation is known. The trouble with Eq. [59] is that the field equation is not known explicitly, unless we already have some expression for $t^{\mu\nu}$ in terms of $\phi^{\mu\nu}$ and its derivatives. In Chapter 7 we will find ways of solving, or rather bypassing, this difficulty. For now we will adopt a technique of successive approximations.

We begin by writing down the field equation for a *free* gravitational field, in the absence of matter,

$$\partial_\lambda \partial^\lambda \phi^{\mu\nu} = 0 \qquad [61]$$

Given this field equation, we can make use of the canonical prescription to obtain the energy-momentum tensor associated with the field $\phi^{\mu\nu}$. It turns out that the energy-momentum tensor associated with such a free gravitational field is (assuming the gauge condition $\partial_\mu \phi^{\mu\nu} = 0$)

$$t_{(1)}{}^{\mu\nu} = \frac{1}{4}\left[2\phi^{\alpha\beta,\mu}\phi_{\alpha\beta}{}^{,\nu} - \phi^{,\mu}\phi^{,\nu} - \eta^{\mu\nu}\left(\phi^{\alpha\beta,\sigma}\phi_{\alpha\beta,\sigma} - \frac{1}{2}\phi_{,\sigma}\phi^{,\sigma}\right)\right] \qquad [62]$$

Here the subscript $_{(1)}$ serves to indicate that $t_{(1)}{}^{\mu\nu}$ is only a first approximation to the exact energy-momentum tensor $t^{\mu\nu}$. We will leave the derivation of this result to Appendix 1, and only note that [62] has the desirable properties for an energy-momentum tensor: it is quadratic in the field variables, it is symmetric, and, in the static case, the $t_{(1)}{}^{00}$ component agrees with the energy density $(\nabla\Phi)^2/8\pi G$ given by the Newtonian approximation.*

■ *Exercise 13.* Show that, as a consequence of Eq. [61],

$$\partial_\nu t_{(1)}{}^{\mu\nu} = 0 \qquad [63]$$

---

* In general, the energy density $t_{(1)}{}^{00}$ is not positive definite. However, this causes no problems because the integrated energy density $\int t_{(1)}{}^{00} d^3x$ is always positive.

for a *free* gravitational field. ■

If we substitute Eq. [62] into [59] we obtain a new field equation which, although not exact, includes at least some of the nonlinearities of the gravitational field:

$$\partial_\lambda \partial^\lambda \phi^{\mu\nu} = - \kappa(T_{(m)}{}^{\mu\nu} + t_{(1)}{}^{\mu\nu})$$ [64]

The corresponding equation for the conservation of the total energy-momentum is

$$\partial_\nu (T_{(m)}{}^{\mu\nu} + t_{(1)}{}^{\mu\nu}) = 0$$ [65]

We will use this result to derive the equation of motion. The second term on the left side can be calculated explicitly by means of [62] and [64]:

$$\partial_\nu t_{(1)}{}^{\mu\nu} = \frac{1}{4}\left(2\phi^{\alpha\beta,\mu}\phi_{\alpha\beta}{}^{,\nu}{}_{,\nu} - \phi^{,\mu}\phi^{,\nu}{}_{,\nu}\right)$$

$$= -\frac{\kappa}{4}\left(2\phi^{\alpha\beta,\mu}T_{(m)\alpha\beta} - \phi^{,\mu}T_{(m)}\right) + \dots$$ [66]

The dots in Eq. [66] stand for terms cubic in $\phi^{\mu\nu}$ which we omit in this approximation. We then obtain

$$\partial_\nu T_{(m)}{}^{\mu\nu} - \frac{\kappa}{2} T_{(m)\alpha\beta}\left(\phi^{\alpha\beta,\mu} - \frac{1}{2}\eta^{\alpha\beta}\phi^{,\mu}\right) = 0$$ [67]

which we can also write as

$$\partial_\nu T_{(m)\mu}{}^{\nu} - \frac{\kappa}{2} h_{\alpha\beta,\mu} T_{(m)}{}^{\alpha\beta} = 0$$ [68]

This equation is the gravitational analog (see Table 3.1) of the electromagnetic equation [30].

We will next try to derive the equation of motion by the same method as used in the electromagnetic case. One thing that must be kept in mind is that the quantities $T_{(m)\mu}{}^{\nu}$ in Eqs. [30] and [68] are *not* the same; in the gravitational case $T_{(m)\mu}{}^{\nu}$ includes some gravitational interaction energy, and this will make our calculations a bit more difficult. Strictly speaking, we should also have worried about the possibility of interaction energy in the electromagnetic case, but it can be proved that there is no such interaction energy. How we could set up such a proof will become clear from what follows (see also Problem 5).

We begin the derivation by integrating the expression [68] over the volume of the particle. The first term gives

$$\frac{d}{dt} \int T_{(m)\mu}{}^0 \, d^3x \qquad [69]$$

In the electromagnetic case we knew, or we assumed, that $T_{(m)\mu}{}^\nu = \rho_0 u_\mu u^\nu$, and then the volume integral appearing in [69] is simply $p_\mu = mu_\mu$. In the present gravitational case, $T_{(m)\mu}{}^\nu$ contains some interaction part with an as yet unknown dependence on the gravitational field variables; we therefore cannot evaluate the volume integral explicitly. For convenience, let us designate the integral by P_μ and write

$$\frac{d}{dt} \int T_{(m)\mu}{}^0 \, d^3x \ = \ \frac{d}{dt} P_\mu \qquad [70]$$

To evaluate the volume integral of the second term in Eq. [68], we make the assumption that $h_{\alpha\beta,\mu}$ is roughly constant over the volume of the particle. We are then left with

$$- \frac{\kappa}{2} h_{\alpha\beta,\mu} \int T_{(m)}{}^{\alpha\beta} \, d^3x \qquad [71]$$

In order to evaluate this, we note that $T_{(m)}{}^{\alpha\beta}$ has the form

$$T_{(m)}{}^{\alpha\beta} = \rho_0 u^\alpha u^\beta + \dots \qquad [72]$$

where the dots stand for terms that represent the interaction energy between matter and gravitation. These terms are typically of the form $\propto h u^\alpha u^\beta$, and therefore they contribute to [71] only a small correction of order h^2; we will ignore these interaction terms in the evaluation of [71].*

■ *Exercise 14.* Show that integration over the volume of the particle gives

$$\int \rho_0 u^\alpha u^\beta \, d^3x = mu^\alpha u^\beta \sqrt{1 - v^2} \qquad \blacksquare \qquad [73]$$

Therefore [71] becomes

---

* But we cannot ignore these terms in the calculation of [69] because there they contribute corrections of order h, rather than h^2.

$$- \frac{\kappa}{2} \sqrt{1 - v^2} \, mh_{\alpha\beta,\mu} u^\alpha u^\beta \tag{74}$$

With the results [70] and [74] we obtain the following for the volume integral of [68]:

$$\frac{d}{dt} P_\mu - \frac{\kappa}{2} \sqrt{1 - v^2} \, mh_{\alpha\beta,\mu} u^\alpha u^\beta = 0 \tag{75}$$

that is,

$$\frac{d}{d\tau} P_\mu - \frac{\kappa}{2} mh_{\alpha\beta,\mu} u^\alpha u^\beta = 0 \tag{76}$$

This equation has the form we expect for an equation of motion. There is only one remaining difficulty: we do not know how P_μ is related to the particle four-velocity. We expect that P_μ takes the form $P_\mu = mu_\mu + ...$ where the dots stand for an extra term involving both the four-velocity and the gravitational field. Our next concern will be to find a precise expression for this extra term.

3.4 THE VARIATIONAL PRINCIPLE AND THE EQUATION OF MOTION

The next step in our derivation of the equation of motion hinges on the observation that Eq. [76] has the form of an Euler-Lagrange equation. To see what this means, let us recall Hamilton's variational principle, according to which the motion of a particle is such as to make the action between two spacetime points P_1, P_2 stationary. Thus, the action evaluated along the actual worldline of the particle is equal to the action evaluated along any infinitesimally near worldline with the same endpoints P_1, P_2 (see Fig. 3.1).

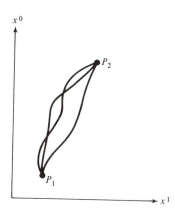

Fig. 3.1 Worldlines between two spacetime points P_1 and P_2. One of these lines, say, the lowest line, represents the actual worldline of the particle; the other lines are obtained by variation. For small variations, the actions along all these worldlines are equal to the action along the actual worldline.

To formulate the condition of stationary action, we need to introduce a parameter that labels the points along the worldlines. In the relativistic case it is convenient to use the proper time of the *actual* worldline as parameter; we will use the usual notation τ for this parameter. It is to be emphasized that although τ is the proper time measured along the actual worldline, it is not the proper time along any of the other nearby worldlines that are considered in the variational principle. Along these other worldlines, τ is to be regarded as a purely mathematical parameter, without any special physical interpretation. The reason why we cannot parametrize the other worldlines by their own proper time is that the parameter used in the formulation of the variational equations must range over exactly the same set of values along all the nearby worldlines between P_1 and P_2, and this cannot be achieved if each worldline is parametrized with its own proper time.

The action can then be written as an integral over this proper time,

$$I = \int_{\tau_1}^{\tau_2} L \ d\tau \qquad [77]$$

where the Lagrangian L is a function of the coordinates x^μ and the corresponding "velocities" $u^\mu = dx^\mu/d\tau$.

If we vary the actual worldline at the point $x_\mu(\tau)$ by an amount $\delta x_\mu(\tau)$, then the velocity will vary by $\delta u^\mu = (d/d\tau)\delta x^\mu$. The change in the action will therefore be

$$\delta I = \int \left(\frac{\partial L}{\partial x^\mu}\delta x^\mu + \frac{\partial L}{\partial u^\mu}\delta u^\mu \right) d\tau$$

$$= \int \left(\frac{\partial L}{\partial x^\mu}\delta x^\mu + \frac{\partial L}{\partial u^\mu}\frac{d}{d\tau}\delta x^\mu \right) d\tau$$

The second term in the integrand can be integrated by parts. Since $\delta x^\mu = 0$ at the endpoints of the integration interval, we obtain

$$\delta I = \int \left(\frac{\partial L}{\partial x^\mu} - \frac{d}{d\tau}\frac{\partial L}{\partial u^\mu} \right) \delta x^\mu \ d\tau$$

If we are to have $\delta I = 0$ for arbitrary variations δx^μ, we need

$$\frac{d}{d\tau}\frac{\partial L}{\partial u^\mu} - \frac{\partial L}{\partial x^\mu} = 0 \qquad [78]$$

This is the *Euler-Lagrange equation*. We can also write this equation as

$$\frac{d}{d\tau}\pi_\mu - \frac{\partial L}{\partial x^\mu} = 0 \qquad [79]$$

where

$$\pi_\mu \equiv \frac{\partial L}{\partial u^\mu} \qquad [80]$$

is the momentum canonically conjugate to x^μ.

By way of illustration, let us consider the case of a free particle. The equation of motion for a free particle is

$$\frac{d}{d\tau}(mu^\mu) = 0 \qquad [81]$$

This equation can be obtained from the action

$$I = \int_{\tau_2}^{\tau_2} \frac{1}{2}m \frac{dx^\mu}{d\tau}\frac{dx_\mu}{d\tau}\, d\tau$$

with the Lagrangian

$$L_0 = \frac{1}{2}m \frac{dx^\mu}{d\tau}\frac{dx_\mu}{d\tau}$$

$$= \frac{1}{2}mu^\mu u_\mu \qquad [82]$$

■ *Exercise 15.* Show that Eq. [82] gives the equation of motion of a free particle and show that the corresponding canonical momentum is mu_μ. ■

Note that the condition $(dx^\mu/d\tau)(dx_\mu/d\tau) = 1$ must not be inserted into Eq. [82]. This condition is valid only along the actual worldline of the particle where the parameter τ coincides with the proper time. The other nearby worldlines that must be considered in the variational principle have their own proper time distinct from τ, and hence along them $(dx^\mu/d\tau)(dx_\mu/d\tau) \neq 1$. It is obvious that if we carelessly insert $(dx^\mu/d\tau)(dx_\mu/d\tau) = 1$ into Eq. [82], we obtain $L_0 = \frac{1}{2}m$, and then no equation of motion results.

The equation of motion for a free particle can also be obtained from a somewhat different variational principle. We can take

$$I = \int_{P_1}^{P_2} \sqrt{dx^\mu \, dx_\mu}$$

as our action. If this integral is evaluated along a given worldline, it gives the proper time that elapses between P_1 and P_2 along that worldline. Again, let us introduce the proper time τ of the actual worldline as a parameter. We can then write the action as

$$I = \int_{\tau_1}^{\tau_2} \sqrt{\frac{dx^\mu}{d\tau} \frac{dx_\mu}{d\tau}} \, d\tau \qquad [83]$$

and the Lagrangian as

$$L_0 = \sqrt{u^\mu u_\mu} \qquad [84]$$

This yelds an equation of motion

$$\frac{d}{d\tau} \frac{u_\mu}{\sqrt{u^\nu u_\nu}} = 0$$

Since this equation refers to the actual worldline, it is valid to insert the condition $u^\nu u_\nu = 1$ into it. We then recover the equation of motion [81] of a free particle. Thus, the motion of a free particle is such as to make [83] stationary; this means that along the worldline of a free particle the proper time is stationary. It is easy to show that the proper time is actually at a maximum.

■ *Exercise 16.* Show this. (Hint: See Problem 2.13.) ■

This concludes our review of the variational principle; we now return to our discussion of the equation of motion of a particle in a gravitational field. A glance at Eq. [76] shows that it has the form of an Euler-Lagrange equation. Comparison shows that Eqs. [76] and [78] will be identical if

$$\frac{\partial L}{\partial x^\mu} = \frac{\kappa}{2} m h_{\alpha\beta,\mu} u^\alpha u^\beta \qquad [85]$$

and

$$\frac{\partial L}{\partial u^\mu} = P_\mu \qquad [86]$$

We can regard Eq. [85] as a differential equation for L. If we integrate this equation with respect to x_μ, we obtain

$$L(x, u) = \frac{\kappa}{2} m h_{\alpha\beta} u^\alpha u^\beta + f(u) \qquad [87]$$

where $f(u)$ is an arbitrary function of the velocity u^μ, but not a function of the coordinate x^μ. The function f is determined by the condition that in the limit $h_{\alpha\beta} = 0$, the Lagrangian must reduce to that of a free particle. The latter is given by Eq. [82],

$$L_0 = \tfrac{1}{2} m \eta_{\alpha\beta} u^\alpha u^\beta \qquad [88]$$

and therefore

$$L = \tfrac{1}{2} m (\eta_{\alpha\beta} + \kappa h_{\alpha\beta}) u^\alpha u^\beta \qquad [89]$$

According to Eq. [86], the quantity P_μ is then

$$P_\mu = \frac{\partial L}{\partial u^\mu} = m u_\mu + m \kappa h_{\mu\alpha} u^\alpha \qquad [90]$$

With this expression for P_μ, we can now return to Eq. [76] and obtain a differential equation for the four-velocity (the masses cancel in this equation):

$$\frac{d}{d\tau} (u_\mu + \kappa h_{\mu\alpha} u^\alpha) - \frac{\kappa}{2} h_{\alpha\beta,\mu} u^\alpha u^\beta = 0 \qquad [91]$$

This is the equation of motion for a particle in a (weak) gravitational field. Note that according to Eq. [91] the motion of a particle in a gravitational field is independent of the mass of the particle, in agreement with the Galileo principle.

We can write Eq. [91] in several ways. For example, we can use

$$\frac{d}{d\tau} h_{\mu\alpha} u^\alpha = h_{\mu\alpha} \frac{d}{d\tau} u^\alpha + h_{\mu\alpha,\beta} u^\beta \qquad [92]$$

Since the acceleration of the particle is of order h, the first term on the right side of Eq. [92] is of order h^2, and can be neglected. Therefore Eq. [91] is approximately equivalent to

$$\frac{du_\mu}{d\tau} + (\kappa h_{\mu\alpha,\beta} u^\alpha u^\beta - \frac{\kappa}{2} h_{\alpha\beta,\mu} u^\alpha u^\beta) = 0 \qquad [93]$$

The quantity P_0 was originally defined as the volume integral of the energy density associated with the particle (see Eq. [70]). Hence P_0 is the kinetic plus potential energy of the particle in the gravitational field:

$$P_0 = mu_0 + m\kappa h_{0\alpha} u^\alpha \qquad [94]$$

According to Eq. [91], this energy is a constant of the motion provided the gravitational field is time independent ($h_{\alpha\beta,0} = 0$). If the gravitational field is time dependent, the particle energy is not constant.

■ *Exercise 17.* Show that if

$$T_{(m)\mu}{}^\nu = \rho_0 u_\mu u^\nu + \kappa\rho_0 h_{\mu\alpha} u^\alpha u^\nu \qquad [95]$$

then the integral $\int T_{(m)\mu}{}^0 d^3x$ agrees with the expression given by Eq. [90]. The second term on the right side of Eq. [95] corresponds to the term indicated by dots in Eq. [72]. ■

It is a general theorem of Lagrangian mechanics that

$$u^\mu \frac{\partial L}{\partial u^\mu} - L \qquad [96]$$

is a constant of the motion regardless of any time dependence of the potential. In our case, this integral of the motion takes the form

$$\frac{1}{2} m(u^\mu u_\mu + \kappa h_{\mu\nu} u^\mu u^\nu) = [\text{constant}] \qquad [97]$$

■ *Exercise 18.* Show that the left side of Eq. [97] agrees with [96] and use the Lagrangian equation of motion to show that the derivative $d/d\tau$ applied to [96] or [97] gives zero. ■

The value of the constant in Eq. [97] is $\frac{1}{2} m$. This becomes obvious if we suppose that the particle was initially outside of the gravitational field (where $h_{\mu\nu} = 0$ and $u^\mu u_\mu = 1$) or, equivalently, that the gravitational field was initially zero and reached its present value by gradually changing over time. Hence Eq. [97] becomes

$$(\eta_{\mu\nu} + \kappa h_{\mu\nu})u^\mu u^\nu = 1 \qquad [98]$$

Alternatively, we can express this integral of the motion in terms of the (ordinary) momentum $p^\mu = mu^\mu$:

$$(\eta_{\mu\nu} + \kappa h_{\mu\nu})p^\mu p^\nu = m^2 \qquad [99]$$

Note that with $u^\mu = dx^\mu/d\tau$, Eq. [98] gives us a new relation between the parameter $d\tau$ and dx^μ:

$$d\tau^2 = (\eta_{\mu\nu} + \kappa h_{\mu\nu})dx^\mu dx^\nu \qquad [100]$$

This equation differs from the relation $d\tau^2 = \eta_{\mu\nu}dx^\mu dx^\nu$ valid in the absence of gravitation. We will examine the implications of Eq. [100] in Section 3.6. For now, we remark only that in the evaluation of Eq. [94], we must use the expression [100] for $d\tau$, rather than the familiar $d\tau = \eta_{\mu\nu}dx^\mu dx^\nu$.*

From Eq. [100] it follows that

$$u_0 = \frac{dt}{d\tau} = \cfrac{dt}{\sqrt{\left[1 - v^2 + \kappa h_{\mu\nu}\dfrac{dx^\mu}{dt}\dfrac{dx^\nu}{dt}\right]dt^2}}$$

$$\simeq \frac{1}{\sqrt{1 - v^2}} - \frac{1}{2}\frac{1}{(1 - v^2)^{3/2}}\kappa h_{\mu\nu}\frac{dx^\mu}{dt}\frac{dx^\nu}{dt}$$

and

$$P_0 \simeq \frac{m}{\sqrt{1 - v^2}} - \frac{1}{2}\frac{m}{(1 - v^2)^{3/2}}\kappa h_{\mu\nu}\frac{dx^\mu}{dt}\frac{dx^\nu}{dt} + m\kappa h_{0\alpha}u^\alpha \qquad [101]$$

The first term on the right side has the form of the usual rest-mass and kinetic energy; the other terms represent a gravitational interaction energy (potential energy).

---

* Note that in order to derive Eq. [76] from Eq. [75] we used $d\tau^2 = dx_\mu dx^\mu = (1 - v^2)dt^2$. We now recognize that this is only an approximation; it was a satisfactory approximation in that context.

3.5 THE NONRELATIVISTIC LIMIT AND NEWTON'S THEORY

Our linear gravitational theory will be a good approximation if the gravitational field is weak. We can express this condition more precisely if we compare the linear and nonlinear terms in Eq. [64]. The nonlinear terms are contained in $\kappa t_{(1)}{}^{\mu\nu}$ on the right side; apart from derivatives and tensor indices, these terms are typically of the form $\kappa\phi^{\mu\nu}\phi_{\mu\nu}$ (see Eq. [62]). The linear terms are on the left side of Eq. [64]; apart from derivatives they are of the form $\phi^{\mu\nu}$. These two kinds of terms, therefore, typically differ by a factor $\kappa\phi^{\mu\nu}$. The nonlinearities are therefore insignificant if

$$\kappa\phi^{\mu\nu} << 1$$

As we will see in Section 4.1, the field surrounding a static mass M is of the order of $\phi^{\mu\nu} \simeq GM/\kappa rc^2$ (see Eq. [4.8]), and hence the above condition becomes

$$\frac{GM}{rc^2} << 1$$

This is the same as the condition for the validity of the Newtonian theory (see Section 1.1), that is, both the Newtonian and the linear relativistic theories hinge on the assumption that the field is weak. However, Newtonian theory also requires the assumption that the speeds are low ($v << c$, or $v << 1$); this assumption is not required in the relativistic theory. Thus the relativistic theory is an improvement because we can treat time-dependent fields, and we can treat the motion of relativistic particles in (weak) gravitational fields. For instance, we can calculate the motion of light (photons) in the field of the Sun; we cannot do this in Newtonian theory.

We will now prove that in the special case of static fields and low speeds, Newton's gravitational theory emerges, as it should, from the relativistic theory. In the nonrelativistic limit, $v^k \equiv dx^k/dt << 1$, and we can therefore neglect terms of order v^2. We must also decide what to do with terms of order $\kappa h \times v$. For a particle orbiting a static mass M, these terms are of order $\simeq (GM/r) \times \sqrt{GM/r}$, which is somewhat larger than the terms of order $v^2 \simeq GM/r$. In the lowest approximation, we neglect both of these kinds of terms. In the next approximation, we neglect the terms of order v^2, but we retain the terms of order $\kappa h \times v$. We will see that the lowest approximation corresponds to Newton's gravitational theory.

In the context of this lowest approximation, we can substitute $du^k/d\tau \simeq dv^k/dt$ in the first term in Eq. [93], and we can substitute $u^0 \simeq 1$, $u^k \simeq 0$ in the term in parentheses. This leads to a simplified equation of motion

$$\frac{d}{dt}v^k = -\kappa\left(h^k_{0,0} - \frac{1}{2}h_{00},^k\right)$$

or, if we assume that the field is time independent,

$$\frac{d}{dt}v^k = \frac{1}{2}\kappa h_{00},^k$$

We can compare this with the equation of motion given by Newton's theory of gravitation (see Eq. [1.30]),

$$\frac{d}{dt}v^k = \Phi,^k$$

This establishes that Newton's equation of motion is a consequence of our theory in the nonrelativistic limit. The Newtonian potential is proportional to h_{00}:

$$\Phi = \frac{1}{2}\kappa h_{00} \tag{102}$$

To complete the proof, we must show that the field equation of Newton's theory,

$$\nabla^2\Phi = 4\pi G\rho \tag{103}$$

is a consequence of our relativistic field equations. In the nonrelativistic limit, the dominant component of the energy-momentum tensor is T^{00}, since this is the component that contains the rest-mass density. We therefore have

$$T^{00} = \rho$$
$$T^{k\mu} \simeq 0 \tag{104}$$

The ^{00} component of the field equation [48] is

$$\partial_\lambda\partial^\lambda h^{00} + \partial^0\partial^0 h - 2\partial_\lambda\partial^0 h^{0\lambda} - \partial_\lambda\partial^\lambda h + \partial_\lambda\partial_\sigma h^{\lambda\sigma} = -\kappa T^{00} \tag{105}$$

To eliminate the last two terms on the left side, we take the trace of the field equation [48], which gives

$$-2\partial_\lambda\partial^\lambda h + 2\partial_\lambda\partial_\sigma h^{\sigma\lambda} = -\kappa T \tag{106}$$

We then subtract one-half times Eq. [106] from Eq. [105], with the result

$$\partial_\lambda \partial^\lambda h^{00} + \partial^0 \partial^0 h - 2\partial_\lambda \partial^0 h^{0\lambda} = -\kappa(T^{00} - \tfrac{1}{2}T)$$

For a time-independent field, this reduces to

$$-\nabla^2 h^{00} = -\kappa(T^{00} - \tfrac{1}{2}T) \qquad [107]$$

With the energy-momentum tensor given by Eq. [104], we obtain $T^{00} - \tfrac{1}{2}T \simeq \tfrac{1}{2}\rho$, and therefore

$$\nabla^2 h_{00} = \tfrac{1}{2}\kappa\rho \qquad [108]$$

Eq. [108] has exactly the same form as Eq. [103] and shows that the Newtonian field equation is a consequence of the relativistic field equation.

When seeking out the values of a, a', b, b' in Section 3.2 we could have saved ourselves some labor by demanding that our equations approach those of Newtonian theory in the nonrelativistic limit. However, it is very satisfying that this correspondence should have emerged spontaneously.

Since, in the nonrelativistic limit, the equations of our theory are the same as those of Newtonian theory, the signs of the gravitational force acting between two particles are the same in both theories. Interaction by means of a tensor field automatically gives an attractive force, just as interaction by a vector field automatically gives a repulsive force (between like charges).

To obtain the numerical value of the constant κ, substitute Eq. [102] into [108]:

$$\frac{2}{\kappa} \nabla^2 \Phi = \frac{1}{2}\kappa\rho \qquad [109]$$

Comparison of Eqs. [109] and [103] then gives*

$$\kappa = \sqrt{16\pi G} \qquad [110]$$

In cgs units this becomes

$$\kappa = \sqrt{16\pi G/c^4} = 2.04 \times 10^{-24} \text{ s/(g·cm)}^{1/2} \qquad [111]$$

To examine the equation of motion and the field equation in the next approximation, we must retain the terms of order $\kappa h \times v$. This

---

* We could equally well take $\kappa = -\sqrt{16\pi G}$. The force between two particles is attractive, no matter what the sign of κ.

means that in such terms $u^0 \simeq 1$, but $u^k \simeq v^k \neq 0$, except in quadratic terms, where $u^k u^l \simeq 0$. In a static field, with $h_{\alpha\beta,0} = 0$, the equation of motion [93] then becomes

$$\frac{du_\mu}{dt} + \kappa(h_{\mu 0,l} - h_{0l,\mu})v^l - \frac{\kappa}{2} h_{00,\mu} = 0 \qquad [112]$$

Here, it is convenient to define a new field quantity

$$f_{\alpha\beta} = \frac{\kappa}{2}(h_{0\beta,\mu} - h_{0\mu,\beta}) \qquad [113]$$

This quantity is a 4×4 antisymmetric matrix (note that $f_{\alpha\beta}$ is not a genuine tensor quantity; its Lorentz transformation properties are not those of a second-rank tensor, but those of particular 0 components of a third-rank tensor, as indicated by Eq. [113]). We can write the matrix $f_{\alpha\beta}$ as

$$f_{\alpha\beta} = \begin{pmatrix} 0 & g_x & g_y & g_z \\ -g_x & 0 & -b_z/2 & b_y/2 \\ -g_y & b_z/2 & 0 & -b_x/2 \\ -g_z & -b_y/2 & b_x/2 & 0 \end{pmatrix} \qquad [114]$$

where $g_x = f_{01} = -(\kappa/2)h_{00,1}$, $b_x = 2f_{32} = \kappa(h_{02,3} - h_{03,2})$, etc. The three-component quantities $\mathbf{g}$ and $\mathbf{b}$ appearing here are three-dimensional vectors (but they are not vectors with respect to Lorentz transformations). In terms of $f_{\alpha\beta}$, the equation of motion [112] becomes

$$\frac{dv_k}{dt} = 2f_{kl}v^l - f_{0k}$$

or, in vector notation,

$$\frac{d\mathbf{v}}{dt} = \mathbf{g} + \mathbf{v} \times \mathbf{b} \qquad [115]$$

This equation is analogous to the usual equation for the Lorentz force in electrodynamics. The gravitational field $\mathbf{g}$, which is simply the Newtonian gravitational field (see Eq. [1.14]), is analogous to the electric field $\mathbf{E}$; and the field $\mathbf{b}$ is analogous to the magnetic field $\mathbf{B}$. Because of this analogy, the field $\mathbf{b}$ is called the *gravimagnetic field*. Of course, the notation for $\mathbf{g}$ and $\mathbf{b}$ used in Eq. [114] was chosen in anticipation of this analogy.

The analogy between gravitational and electromagnetic quantities also applies to the field equations for **g** and **b** (Braginsky et al., 1977). As a consequence of Eq. [113], the quantity $f_{\alpha\beta}$ identically satisfies the differential equation

$$\partial_\alpha f_{\mu\beta} + \partial_\mu f_{\beta\alpha} + \partial_\beta f_{\alpha\mu} = 0 \qquad [116]$$

Note that this equation is analogous to Eq. [2.107] for the electromagnetic field. Since Eq. [114] is analogous to Eq. [2.109] (except for factors of $\frac{1}{2}$), we immediately recognize that the vectors **g** and **b** must satisfy equations analogous to the last two Maxwell equations, Eqs. [104] and [105]:

$$\nabla \cdot \mathbf{b} = 0 \qquad [117]$$

$$2\nabla \times \mathbf{g} + \frac{\partial \mathbf{b}}{\partial t} = 0 \qquad [118]$$

It is also easy to verify these equations by direct substitution of Eq. [114] into [116].

The other two "Maxwell equations" emerge from the gravitational field equation, Eq. [48]. As we saw above, for a time-independent field, the ^{00} component of the field equation reduces to

$$-\nabla^2 h^{00} = -\kappa(T^{00} - \tfrac{1}{2}T) \qquad [119]$$

In terms of $\mathbf{g} = -(\kappa/2)\nabla h^{00}$, this becomes

$$\nabla \cdot \mathbf{g} = -4\pi G(2T^{00} - T) \qquad [120]$$

Likewise, for a time-independent field, the ^{0k} component of the field equation [48] is

$$-\nabla^2 h^{0k} - \partial_l \partial^k h^{0l} = -\kappa T^{0k} \qquad [121]$$

It is easy to check that this equation is equivalent to the k component of the following equation for the curl of **b**:

$$\nabla \times \mathbf{b} = -16\pi G\mathbf{S} \qquad [122]$$

where **S** is the momentum density, or the energy flux density (the Poynting vector), with components $S^k = T^{0k}$. Eqs. [120] and [122] are analogous to the Maxwell equations [2.102] and [2.103]. According to this analogy, $2T^{00} - T$ plays the role of a gravitational charge density (source of the gravitational field **g**), and the momentum density T^{0k} plays the role of a gravitational current density (source of the gravimagnetic field **b**). Note that in Eq. [122], the gravitational current density appears with a factor of 16π, in contrast to the factor of 4π that appears in the analogous Maxwell equation. Thus, although the analogy between static gravitational fields and static electrodynamic fields is surprisingly good, it is not perfect.

The analogy breaks down if the fields are not static. The gravitational equations then contain several terms involving time derivatives not present in the electromagnetic equations (see Harris, 1991).

We could further improve on the approximation by retaining further terms in our equations. The systematic procedure for developing the equation of motion (and also the field equations, including their nonlinearities) in a series of terms involving powers of the speed and powers of $h_{\mu\nu}$ is called the post-Newtonian formalism. The gravitational field is of the order GM/r, and the speed of an orbiting particle is of the order $\sqrt{GM/r}$; hence the terms of the series all are powers of $\sqrt{GM/r}$, or, in cgs units, $\sqrt{GM/rc^2}$, which is a dimensionless quantity. The post-Newtonian formalism is used for accurate predictions of the motion of the planets [currently, in the most accurate numerical calculations, terms up to order $(\sqrt{GM/r})^2$ are retained]. The post-Newtonian formalism is also often used to compare different gravitational theories with each other and with experiment. For this purpose, adjustable parameters are included with the coefficients of the powers of $\sqrt{GM/r}$, so that different theories can be characterized by different values of these parameters. This *parametrized post-Newtonian formalism* (PPN) leads to very general, but very messy, equations for the motion and for the fields (Will, 1993). Since we will deal exclusively with Einstein's gravitational theory, we will have no occasion to use the PPN formalism.

3.6 THE GEOMETRIC INTERPRETATION; CURVED SPACE-TIME

According to Eq. [100], the proper time interval for a particle moving in a gravitational field is given by

$$d\tau^2 = (\eta_{\mu\nu} + \kappa h_{\mu\nu})dx^\mu dx^\nu \qquad [123]$$

This differs from the expression

$$d\tau^2 = \eta_{\mu\nu}dx^\mu dx^\nu \qquad [124]$$

that is valid in the (flat) relativistic spacetime. In fact, [123] is the spacetime interval for a *curved spacetime*.

In Chapter 6 we will engage in a detailed discussion of curved spacetime. For now we only state briefly that in a curved spacetime, the constant Minkowski tensor $\eta_{\mu\nu}$ is replaced by a function $g_{\mu\nu}(x)$ of space and time, and the spacetime interval becomes

$$ds^2 = g_{\mu\nu} dx^\mu dx^\nu \qquad [125]$$

This expression determines the spacetime distances. For example, a coordinate displacement dx^1 along the x-axis has a length $\sqrt{-g_{11}} dx^1$, that is, the measured distance differs from dx^1 by a factor $\sqrt{-g_{11}}$. Likewise, a coordinate time (t time) displacement dx^0 has a duration $\sqrt{g_{00}} dx^0$ when measured by the proper time of a clock at rest. Because $g_{\mu\nu}$ determines the spacetime distances, it is called the *metric tensor* of the curved spacetime. The exact connection between $g_{\mu\nu}$ and the curvature of the spacetime will be established in Chapter 6. Under exceptional circumstances, the spacetime interval [125] may merely represent flat spacetime in curvilinear coordinates. However, there is a simple test for discriminating between flat and curved spacetime: If the spacetime interval [125] represents flat spacetime in curvilinear coordinates, then a coordinate transformation to a suitable set of rectangular coordinates will transform $g_{\mu\nu}$ into $\eta_{\mu\nu}$; if the spacetime interval represents a genuine curved spacetime, then there exists no coordinate transformation that accomplishes this.

Comparison with Eq. [125] shows that the metric tensor corresponding to the spacetime interval [123] is

$$g_{\mu\nu} = \eta_{\mu\nu} + \kappa h_{\mu\nu} \qquad [126]$$

This suggests a geometric interpretation of gravitation: a gravitating mass distorts spacetime, giving it a curvature. Furthermore, the motion of a particle in the field of the gravitating mass can be interpreted as a free motion along the geodesics, or the straightest possible worldlines, in the curved geometry. Such straightest possible worldlines are defined by the condition that the invariant length between two points remains stationary with respect to arbitrary (but small) variations of the worldline, a condition which can be expressed by the variational principle

$$\delta \int_{P_1}^{P_2} d\tau = 0 \qquad [127]$$

This definition is motivated by what we know about a straight worldline in flat spacetime: it is the curve of maximum length connecting two points (see Exercise 16). We can verify that the differential equation for a geodesic implied by the variational principle agrees with our equation of motion for a particle, that is, Eq. [91].

To obtain an explicit differential equation for the geodesic, we note that the proper time measured along a worldline connecting points P_1 and P_2 is

$$\int_{P_1}^{P_2} d\tau = \int_{P_1}^{P_2} \sqrt{g_{\mu\nu}\, dx^\mu\, dx^\nu} = \int_{\tau_1}^{\tau_2} \sqrt{g_{\mu\nu} \frac{dx^\mu}{d\tau} \frac{dx^\nu}{d\tau}}\; d\tau$$

This means that the Lagrangian function is

$$L = \sqrt{g_{\mu\nu} \frac{dx^\nu}{d\tau} \frac{dx^\nu}{d\tau}} \qquad\qquad [128]$$

and the Euler-Lagrange equation is

$$0 = \frac{d}{d\tau} \frac{\partial L}{\partial(dx^\mu/d\tau)} - \frac{\partial L}{\partial x^\mu}$$

$$= \frac{d}{d\tau}\left[\frac{1}{L} g_{\mu\nu} \frac{dx^\mu}{d\tau}\right] - \frac{1}{2L} g_{\alpha\beta,\mu} \frac{dx^\alpha}{d\tau} \frac{dx^\beta}{dt} \qquad [129]$$

Since this equation refers to the geodesic, we can insert the numerical value $g_{\mu\nu}(dx^\mu/d\tau)(dx^\nu/d\tau) = 1$, and we obtain*:

$$\frac{d}{d\tau}\left[g_{\mu\nu} \frac{dx^\nu}{d\tau}\right] - \frac{1}{2} g_{\alpha\beta,\mu} \frac{dx^\alpha}{d\tau} \frac{dx^\beta}{d\tau} = 0 \qquad\qquad [130]$$

This is called the *geodesic equation*.

With $g_{\mu\nu} = \eta_{\mu\nu} + \kappa h_{\mu\nu}$, we see that the geodesic equation takes the form

$$\frac{d}{d\tau}\left[\eta_{\mu\nu} \frac{dx^\nu}{d\tau} + \kappa h_{\mu\nu} \frac{dx^\nu}{d\tau}\right] - \frac{\kappa}{2} h_{\alpha\beta,\mu} \frac{dx^\alpha}{d\tau} \frac{dx^\beta}{d\tau} = 0 \qquad [131]$$

which is the same as Eq. [91]. Our equation of motion for a particle in a gravitational field therefore says that the particle moves along the geodesics in a curved spacetime with a metric $g_{\mu\nu} = \eta_{\mu\nu} + \kappa h_{\mu\nu}$. This means we can replace the effects of the gravitational field and the gravitational force by purely geometrical effects associated with motion in a curved geometry.

However, this geometric interpretation of the gravitational field hinges on the physical interpretation of the parameter τ defined by

---

* Because the numerical value of $g_{\mu\nu}(dx^\mu/d\tau)(dx^\nu/d\tau)$ is 1, we can construct many different Lagrangians that give the same differential equation. For example, $L = (1/2)g_{\mu\nu}(dx^\mu/d\tau)(dx^\nu/d\tau)$ will do just as well (this should be compared with Eq. [89]).

Eq. [123]. In the equation of motion, this parameter certainly seems to play the role of proper time. We therefore expect that it actually is the proper time measured by a clock carried along by the particle. Our expectation is based on what we know to be the case in Newtonian physics: the parameter t that appears in the Newtonian equation of motion does indeed agree with the time measured by clocks. However, relativistic physics is more complicated. We know that in special relativity, the parameter τ in the equation of motion does not agree with the time measured by clocks, but only with the proper time, that is, the time measured by a preferred clock that accompanies the particle during its motion. And it is conceivable that in the presence of gravitation, the situation is even more complicated, and that the parameter τ has some new meaning involving some recalibration of clocks by the gravitational field. In a purely mathematical sense, the variable τ enters the equation of motion [131] merely as a parameter, and Eq. [123] relates this parameter to spacetime displacements. Once the equation of motion has been solved, the parameter τ can be eliminated from the solution, leaving **x** as a function of t. Thus, the value of the parameter does not affect the spacetime trajectory of the particle. It might, therefore, be questioned whether we necessarily have to regard τ as the proper time of the particle. Could it be that the variable τ given by Eq. [123] is a purely mathematical parameter having nothing to do with proper time? If so, then there would be no need to regard spacetime as curved.

In order to confirm the physical interpretation of Eq. [123], we must check that a clock placed in a gravitational field shows a proper time $\sqrt{g_{\mu\nu}dx^\mu dx^\nu}$ rather than $\sqrt{\eta_{\mu\nu}dx^\mu dx^\nu}$. For this, we must analyze the behavior of clocks in a gravitational field. The essential ingredient in such an analysis is the principle of equivalence, which tells us that a clock in free fall in a gravitational field will behave in exactly the same way as though the gravitational field were absent. Whether the clock is a mechanical clock (regulated by a balance wheel), an electronic clock (regulated by a vibrating quartz crystal), or an atomic clock (regulated by the vibrations of atoms) makes no difference--in any clock in free fall in a gravitational field, all the parts of the clock fall with the same acceleration, and hence the gravitational field cannot exert any direct effect on the rate of the clock. The only direct effect of the gravitational field on a clock arises from tidal forces; but these are insignificant if the clock is of sufficiently small size.

Although the gravitational field cannot exert any direct effect on the rate of a clock, it exerts an indirect effect on the relative rates of clocks placed at different positions in this gravitational field. To discover this indirect effect, we must compare the rates of clocks at different positions. Since the behavior of freely falling clocks is completely predictable from the principle of equivalence, we will use freely falling clocks for all our measurements in the gravitational field, even measurements at a fixed position, for instance, a measurement at a fixed position on the surface of the Earth. For this purpose, we use a freely falling clock, instantaneously at rest at the fixed position. As soon as this clock has fallen too far from our fixed position and acquired too much speed, we must replace it by a new clock, instantaneously at rest. Whenever we speak of the time as measured by "a clock located at a fixed position" in a gravitational field, this phrase must be understood as shorthand for a complicated measurement procedure, involving many freely falling, disposable clocks, used in succession.

We are now ready to compare the rates of two clocks at different fixed positions. Suppose that clock 1 is located at position P_1 and clock 2 at position P_2, and suppose that the gravitational field is time independent (if the gravitational field is time dependent, then the comparison of clock rates is ambiguous--the result depends on how and when the comparison is made). We can use either of two methods for the comparison of the rates of the clocks at these different fixed positions: we can send signals from P_1 to P_2, or we can transport a third, auxiliary clock from clock 1 to clock 2, and check each of the two clocks against this auxiliary clock. Since we understand the behavior of freely falling clocks, it will be best to use the method of transport of an auxiliary clock, and to perform this transport by free fall. Suppose that the auxiliary clock has a speed v_1 when it passes clock 1 and a speed v_2 when it passes clock 2. The rest frame of the auxiliary clock is a (local) inertial frame, and relative to this reference frame the first clock has a speed v_1 and the second clock has a speed v_2. Thus, relative to the rest frame of the auxiliary clock, both of the other clocks will suffer a time dilation. Accordingly, when the first clock passes by the auxiliary clock, we find that rate of the first clock differs from that of the auxiliary clock by a factor

$$\frac{d\tau_1}{d\tau_{aux}} = \sqrt{1 - v_1^2} \qquad [132]$$

Likewise, when the second clock passes by the auxiliary clock, we find that

$$\frac{d\tau_2}{d\tau_{aux}} = \sqrt{1 - v_2^2} \qquad [133]$$

The ratio of these two rates gives us the rate of the first clock relative to the rate of the second:

$$\frac{d\tau_1}{d\tau_2} = \sqrt{\frac{1 - v_1{}^2}{1 - v_2{}^2}} \qquad [134]$$

Although this result is exact, it becomes more useful if we make some approximations. The speeds v_1 and v_2 are reckoned in a local inertial reference frame, and in a curved spacetime such a local speed is not simply equal to the rate of change calculated from the coordinate and time displacements dx^k and dt. However, to lowest order in the gravitational constant, we can ignore this fine distinction, and we can approximate $v \simeq \sqrt{dx^2 + dy^2 + dz^2}/dt$. Furthermore, for any freely falling body, we can approximate the difference between $v_1{}^2$ and $v_2{}^2$ by its Newtonian value

$$v_1{}^2 - v_2{}^2 \simeq 2\Phi_2 - 2\Phi_1 \qquad [135]$$

With these approximations we obtain

$$\frac{d\tau_1}{d\tau_2} \simeq 1 - \frac{1}{2}(v_1{}^2 - v_2{}^2) \simeq 1 - (\Phi_2 - \Phi_1)$$

$$\simeq \frac{1 + \Phi_1}{1 + \Phi_2} \qquad [136]$$

This shows that if clock 2 is at a lower potential than clock 1 ($\Phi_2 < \Phi_1$), then $d\tau_2$ is smaller than $d\tau_1$. Thus, the clock that is deeper in the gravitational potential runs slower.

Eq. [136] is the gravitational time-dilation formula, or redshift formula; we will further discuss it in Section 4.2. The equation is valid to lowest order in the gravitational field strength.

If clock 2 is outside of the gravitational potential, at infinity, where $\Phi_1 = 0$, then its proper time coincides with the conventional time coordinate t, and $d\tau_1 = dt$. Then Eq. [136] becomes

$$d\tau_2 = (1 + \Phi_2)dt \qquad [137]$$

If we express this in terms of h_{00} and drop the subscript, it reduces to

$$d\tau = (1 + \tfrac{1}{2}\kappa h_{00})dt \qquad [138]$$

This result shows that the proper time measured by a clock at rest in a gravitational field differs from dt by a factor $(1 + \tfrac{1}{2}\kappa h_{00})$. Since $(1 + \tfrac{1}{2}\kappa h_{00})^2 \simeq (1 + \kappa h_{00})$, this time-dilation factor agrees with g_{00}, as given by Eq. [126]. The agreement between the time-dilation factor obtained from our analysis of measurements by clocks and the factor appearing in Eq. [126] verifies the physical interpretation of $(\eta_{\mu\nu} + \kappa h_{\mu\nu})$ as metric of a curved spacetime, at least as concerns the $_{00}$ component of the metric.

To verify the physical interpretation of the other components of the metric, we must investigate the measurement of intervals of space. Such measurements can be conveniently performed by the "radar-ranging" procedure described in Section 2.1. For a spatial displacement, the general procedure described in Section 2.1 reduces to ordinary radar ranging. Given a spatial displacement (dx, dy, dz), we place a clock at one end of this displacement. We then send a light pulse from this end to the other end, and we reflect the light pulse back to the starting point (see Fig. 3.2). The clock tells us the travel time of the light pulse, that is, it tells us the proper time elapsed at the clock. If half the travel time is $d\tau_{clock}$, then the distance from one end of the displacement to the other must be $dl = d\tau_{clock}$. Note that this radar-ranging procedure is in accord with the modern definition of the unit of length, which states that the meter is the distance traveled by light in 1/299,792,458 s. And note that the radar-ranging procedure implicitly relies on the principle of equivalence: in the local inertial reference frame of the freely falling clock, light propagates as in the absence of the gravitational field, at its standard speed.

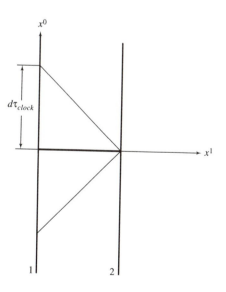

Fig. 3.2 Measurement of the length of a spatial displacement by radar ranging. The light signal starts at one end of the spatial displacement, reflects off the other end, and returns to the first end.

Since we understand the behavior of clocks in the gravitational field, we should be able to relate the result of such a distance measurement to the field tensor $h_{\mu\nu}$. First we need to examine what effect the field tensor $h_{\mu\nu}$ has on the propagation of light. Since we already know the equation of motion of particles in the gravitational field, it is best to think of light as consisting of particles, or photons, of zero mass. For the case of zero mass, Eq. [99] becomes

$$(\eta_{\mu\nu} + \kappa h_{\mu\nu})p^\mu p^\nu = 0 \qquad [139]$$

The direction of propagation of a photon coincides with the direction of its momentum. Hence, along the worldline of the photon, $dx^\mu \propto p^\mu$, and therefore

$$(\eta_{\mu\nu} + \kappa h_{\mu\nu})dx^\mu dx^\nu = 0 \qquad [140]$$

This establishes that along the worldline of a light signal, $d\tau^2$ is zero,

$$d\tau^2 = (\eta_{\mu\nu} + \kappa h_{\mu\nu})dx^\mu dx^\nu = 0 \qquad [141]$$

Of course, this result is the obvious generalization of what we know to be true in the absence of gravitation, where $d\tau^2 = \eta_{\mu\nu} dx^\mu dx^\nu = 0$ for a light signal.

Eq. [141] tells us how the increments dx^μ of the conventional space and time coordinates are related along the worldline of a light signal. To exploit this equation in our radar-ranging procedure, suppose that $h_{0k} = 0$, for the sake of simplicity. (If $h_{0k} \neq 0$, then it is awkward to express the radar-ranging procedure in terms of dx^k, because the light signal takes different amounts of t time going out and coming back. The condition $h_{0k} = 0$ can always be achieved by a gauge transformation; it therefore does not really restrict the gravitational field in any serious way.) Eq. [141] then tells us

$$\sqrt{1 + \kappa h_{00}}\, dt = \sqrt{-(\eta_{kl} + \kappa h_{kl})dx^k dx^l} \qquad [142]$$

From our analysis of the behavior of clocks, we know that $\sqrt{1 + \kappa h_{00}}\, dt$ is the proper time registered by a clock placed at a fixed position in the gravitational field. If this clock is used in the radar-ranging procedure, the measured length of the spatial displacement (dx, dy, dz) is equal to half the elapsed proper time registered by the clock; thus, $dl = d\tau_{clock} = \sqrt{1 + \kappa h_{00}}\, dt$, and according to Eq. [142] we can then conclude that

$$dl = \sqrt{-(\eta_{kl} + \kappa h_{kl})dx^k dx^l} \qquad [143]$$

or

$$dl^2 = -(\eta_{kl} + \kappa h_{kl})dx^k dx^l \qquad [144]$$

This result expresses the measured length in terms of the coordinate increments, and verifies the physical interpretation of the spatial components of the tensor $(\eta_{\mu\nu} + \kappa h_{\mu\nu})$ as metric of a curved space.

If we introduce the usual notation $g_{kl} = \eta_{kl} + \kappa h_{kl}$, we obtain

$$dl^2 = -g_{kl} dx^k dx^l \qquad [145]$$

We can then combine our results for the measurement of space and time intervals to find a general expression for an *arbitrary* spacetime interval measured by means of clocks placed in a gravitational field:

$$ds^2 = d\tau_{clock}^2 - dl^2$$

$$= g_{00} dt^2 + g_{kl} dx^k dx^l$$

$$= g_{\mu\nu} dx^\mu dx^\nu \qquad [146]$$

This result is in complete agreement with the expression for ds^2 that we found from our examination of the equation of motion. Thus, all measurements based on clocks indicate that the metric of spacetime is $\eta_{\mu\nu} + \kappa h_{\mu\nu}$, rather than $\eta_{\mu\nu}$, and we therefore conclude that the real observable geometry is curved. The flat spacetime geometry which we took as the starting point of our calculations in this chapter is unobservable, except, of course, if gravitational fields are absent.

It is to be emphasized that the geometric interpretation of gravity is not optional. The above arguments concerning measurements with clocks in the gravitational field *compel* us to accept this interpretation. From these arguments we see that the formalism of the tensor theory of gravity implicitly carries within it its own physical interpretation.

For our further discussion of the linear approximation, we will take for granted the geometric interpretation, with the metric tensor $\eta_{\mu\nu} + \kappa h_{\mu\nu}$. For some parts of our discussion this is not really essential, since some experiments involve only measurements with clocks placed outside of the gravitational field, where we do not directly detect the curved geometry. The deflection of light by the gravitational field of the Sun (see Section 4.3) is a typical experiment of this kind; the deflection angle is measured in a region that is far away from the deflecting body.

Note the crucial role played by the equivalence principle in several stages of the above arguments. Both the comparison of clocks at different positions and the radar-ranging procedure hinge on the principle of equivalence. This means that any deviation from the principle of equivalence--such as the deviations produced by tidal forces-- would have serious consequences. For instance, we might ask: Can the tidal forces disturb an atomic clock? A simple estimate (see Exercise 19) shows that not even the strong tidal forces found on the surface of a neutron star can change the frequency of atomic vibrations by very much; the internal forces of the atom dominate and resist the tidal distortions.

■ *Exercise 19.* Estimate the change in the frequency of an atomic transition produced by tidal forces. Show that for an atom in free fall at the surface of a neutron star (of mass $M \simeq M_\odot$ and radius $R \simeq 10$ km), the change in proper frequency amounts to no more than 1 part in $\simeq 10^{23}$. {Hint: For a rough estimate, you may regard the electron as bound in the atom by a harmonic force with spring constant k_0. Show that the effect of the tidal force (Eqs. [147]-[149]) can be described as a change in spring constant to $\simeq (k_0 \pm GMm_e/r^3)$. The frequency of oscillation is then changed from $\omega_0 = \sqrt{k_0/m_e}$ to

$$\sqrt{\frac{k_0 \pm GMm_e/r^3}{m_e}} \simeq \sqrt{\frac{k_0}{m_e}}\left(1 \pm \frac{GMm_e}{2k_0 r^3}\right) \simeq \omega_0\left(1 \pm \frac{GM}{2\omega_0^2 r^3}\right)$$

For optical transitions, the value of ω_0 is typically $\omega_0 \simeq 4 \times 10^{15}$ radians/s.} ■

Atomic clocks are usually not found in free fall. They are usually at rest in the gravitational field (for instance, at rest on the surface of the Earth or a star), and this means that they are accelerated relative to freely falling clocks. We might worry about the disturbances caused by this acceleration. These disturbances can be estimated (see Exercise 9), and they are negligible. Incidentally: There is direct experimental evidence that confirms that even very large accelerations have an insignificant effect on the frequency of a well-built clock. In the most severe experimental test of the effects of acceleration, nuclei were used as clocks. The nuclear vibration frequency, as revealed by the emission of gamma rays via the Mössbauer effect, played the role of clock frequency. When a sample of nuclei was placed in an ultracentrifuge and whirled around with an acceleration of 10^8 cm/s², the nuclear frequency was unaffected, except for the ordinary time dilation predicted by special relativity (Hay et al., 1960; Kündig, 1963; Champeney et al., 1963).

■ *Exercise 20.* An atom, or nucleus, at rest in a gravitational field is held in this position by an electric force generated by the atoms that provide the support. Compare the magnitude of the electric force needed to support an electron against the Earth's gravity with the magnitude of the typical internal electric forces in an atom. Compare the magnitude of the electric force needed to support a typical nucleus against the Earth's gravity with the magnitude of the electric fields that the nucleus generates at its own surface. Repeat the calculation for the case of an atom or nucleus placed on the surface of a neutron star. ■

FURTHER READING

Classical field theory is discussed in Soper, D. E., *Classical Field Theory* (Wiley, New York, 1976), and Davis, W. R., *Classical Fields, Particles, and the Theory of Relativity* (Gordon and Breach, New York, 1970), already mentioned in the preceding chapter. Quantum field theory is discussed in Wentzel, G., *Quantum Theory of Fields* (Interscience, New York, 1949).

Derivations of the equations of electrodynamics in the context of special relativity have been given by Schwartz, M., *Principles of Electrodynamics* (McGraw-Hill, New York, 1972); Ohanian, H. C., *Classical Electrodynamics* (Allyn and Bacon, Boston, 1988); Kobe (1986); and Neuenschwander, D. E., and Turner, B. N., Am. J. Phys. **60**, 35 (1992).

An approach to the linear field equations for gravitation quite similar to that of Section 3.2, but different in some details, is offered by Bowler, M. G., *Gravitation and Relativity* (Pergamon, Oxford, 1976).

Lagrangian field theory provides an alternative, and very elegant, derivation of the linear field equations for gravitation. This approach is presented in detail in Thirring, W. E., "An Alternative Approach to the Theory of Gravitation," Ann. Phys. **16**, 96 (1961), and more briefly in Misner, C. W., Thorne, K. S., and Wheeler, J. A., *Gravitation* (Freeman, San Francisco, 1973), Chapter 7.

The most fundamental justification of the linear equations emerges from the analysis of the representations of the Lorentz group. This approach is discussed by Weinberg, S., in the incisive article "The Quantum Theory of Massless Particles," in Deser, S., and Ford, K. W., eds., *Lectures on Particles and Fields* (Prentice-Hall, Englewood Cliffs, 1965); see also Weinberg, S., Phys. Rev. B **138**, 988 (1965).

Nonlinear corrections to the field equations, and the method of successive approximations that can be used to find them, are discussed in the article by Thirring, op. cit., and in the article by Gupta, S. N., "Einstein's and Other Theories of Gravitation," Rev. Mod. Phys. **29**, 334 (1957). Both of these articles also explain what is wrong with scalar and vector theories of gravitation.

The traditional approach is to derive the linear approximation from the exact Einstein equations, rather than vice versa (in Chapter 7 we will use a very simple method to derive the exact nonlinear Einstein equations from the linear equations). This traditional approach may be found in Landau, L. D., and Lifshitz, E. M., *The Classical Theory of Fields* (Addison-Wesley, Reading, 1962); Misner, Thorne, and Wheeler, op. cit.; and Weinberg, S., *Gravitation and Cosmology* (Wiley, New York, 1972). The article by L. Halpern, "On Alternative Approaches to Gravitation," Ann. Phys. **25**, 387 (1963), compares the merits of these alternatives.

REFERENCES

Brans, C., and Dicke, R. H., Phys. Rev. **124**, 925 (1961).
Braginsky, V. B., Caves, C. M., and Thorne, K. S., Phys. Rev. D **15**, 2047 (1977).
Champeney, D. C., Isaak, G. R., and Khan, A. M., Phys. Lett. **7**, 241 (1963).
Harris, E. G., Am. J. Phys., **59**, 421 (1991).
Hay, H. J., Schiffer, J. P., Cranshaw, T., and Egelstaff, P. A., Phys. Rev. Lett. **4**,

165 (1960).

Jordan, P., *Schwerkraft und Weltall* (Vieweg, Braunschweig, 1951).

Kobe, D. H., Am. J. Phys. **54**, 631 (1986).

Kündig, W., Phys. Rev. **129**, 2371 (1963).

Soper, D. E., *Classical Field Theory* (Wiley, New York, 1976).

Wentzel, G., *Quantum Theory of Fields* (Interscience Publishers, New York, 1949).

Will, C. M., *Theory and Experiment in Gravitational Physics* (Cambridge University Press, Cambridge, 1993).

PROBLEMS

1. Show that if the gravitational field $h_{\mu\nu}$ is constant in some region, then this gravitational field can be reduced to zero by a gauge transformation in that region.

2. Find a gauge transformation that leaves invariant the equation [43]. Show that in the limiting case $a \to 1$, your gauge transformation reduces to that given by Eq. [49].

3. Show that for an arbitrary gravitational field the energy density $t_{(1)}{}^{00}$ given by Eq. [62] is not always positive.

4. In Section 4.1 we will see that for a static gravitational field, with $\phi_{\mu\nu,0} = 0$, the only nonzero component of the tensor $\phi_{\mu\nu}$ is ϕ_{00}. Show that in this case, the energy density $t_{(1)}{}^{00}$ is positive.

5. In Eq. [36] it has been assumed that $p_\mu = mu_\mu$. To construct a proof of this, first show that Eq. [36] can be written as

$$\frac{d}{d\tau} P_\mu = qA_{\nu,\mu} u^\nu$$

where $P_\mu = p_\mu + qA_\mu$. Then check that this equation of motion can be obtained from the Lagrangian

$$L = m\sqrt{u_\mu u^\mu} + qA_\mu u^\mu$$

if and only if $p_\mu = mu_\mu$. This proves that the momentum p_μ that appears in Eq. [36] is in fact the ordinary momentum mu_μ.

6. Show that in the nonrelativistic limit ($v \ll 1$), the particle energy given by Eq. [101] agrees with the Newtonian energy $\frac{1}{2}mv^2 + m\Phi$.

7. Show that the familiar Coriolis pseudoforce associated with a rotating reference frame mimics a gravimagnetic force. What is the corresponding gravimagnetic field **b**?

4. Applications of the Linear Approximation

TWINKLE, TWINKLE LITTLE STAR
HOW I WONDER WHERE YOU ARE.
"1.75 SECONDS OF ARC FROM WHERE I SEEM TO BE
FOR $ds^2 \simeq (1 - 2GM/r^2)dt^2 - (1 + 2GM/r)dr^2 - r^2 d\theta^2 - r^2 sin^2 \theta \, d\phi^2$."

Source unknown

In this chapter we will obtain some simple time-independent solutions of the linear field equations, such as the solution for the field produced by a static spherical mass and by a steadily rotating mass. For a given solution of the field equations, the equation of motion then permits us to predict the trajectories of particles or of light signals, which we can compare with observational or experimental data.

Most of the experimental tests that have so far been performed on the relativistic theory of gravitation involve only the linear approximation. For instance, the gravitational time dilation predicted by the linear approximation has been tested and confirmed by experiments with clocks in the gravitational field of the Earth. Some other experimental tests exploit the motion of light signals in the gravitational field of the Sun. Since light signals are necessarily relativistic, their motion provides a severe test of the relativistic features of our equations.

The gravitational field equations and the equation of motion predict a deflection of light rays and also a retardation of light signals passing through the gravitational field of a mass. The observational data on the deflection and on the retardation of light by the gravita-

tional field of the Sun confirm the predictions of the linear aproxima-
tion. The deflection of light was one of the early predictions of Einstein, and the observational confirmation of this effect in 1919, just
after the Great War, was viewed as a spectacular success of his theory
of general relativity. By now, the deflection of light has become a tool
for astronomers, who use it in the study of the multiple images produced by "gravitational lenses." Such multiple images of distant quasars
or distant galaxies are produced when the rays of light are deflected
by the gravitational field of a foreground galaxy, which acts as lens.

4.1 THE FIELD OF A SPHERICAL MASS

The gravitational field surrounding a spherically symmetric, static
mass distribution is as important in the relativistic theory as it is in
Newton's theory. In the exterior of the mass, $T^{\mu\nu} = 0$, and we must
therefore construct the spherically symmetric, time-independent solution of the field equations

$$\partial_\lambda \partial^\lambda \phi^{\mu\nu} = 0 \qquad [1]$$

$$\partial_\mu \phi^{\mu\nu} = 0 \qquad [2]$$

This special solution describes the gravitational field produced by the
Sun or by the Earth, provided we ignore the rotational motion and any
oblateness of these bodies.

With the assumption that the field is time independent, Eq. [1]
reduces to

$$-\nabla^2 \phi^{\mu\nu} = 0 \qquad [3]$$

The spherically symmetric solution of this equation is

$$\phi^{\mu\nu} = \frac{C^{\mu\nu}}{r} \qquad [4]$$

where $C^{\mu\nu}$ is a constant matrix. We could also include an additive
constant in the solution [4], but we will assume that $\phi^{\mu\nu} \to 0$ as $r \to$
∞, and therefore there is no such additive constant. Substitution of
Eq. [4] into Eq. [2] yields

$$C^{k\nu} \partial_k \left(\frac{1}{r} \right) = 0$$

that is,

$$C^{k\nu}\left[\frac{-x^k}{r^3}\right] = 0 \tag{5}$$

This can hold for all values of x^k only if $C^{k\nu} = 0$. Hence the only nonzero component of $C^{\mu\nu}$ is C^{00}. It then follows that

$$h_{00} = \phi_{00} - \frac{1}{2}\eta_{00}\phi = \frac{C_{00}}{r} - \frac{1}{2}\frac{C_{00}}{r} = \frac{C_{00}}{2r} \tag{6}$$

This determines the value of C_{00}, because we already know that h_{00} is proportional to the Newtonian potential (see Eq. [3.102]), that is,

$$h_{00} = \frac{2}{\kappa}\left[-\frac{GM}{r}\right] \tag{7}$$

Hence $C_{00} = -4GM/\kappa$ and

$$\phi_{\mu\nu} = \begin{pmatrix} -4GM/\kappa r & 0 & 0 & 0 \\ 0 & 0 & 0 & 0 \\ 0 & 0 & 0 & 0 \\ 0 & 0 & 0 & 0 \end{pmatrix} \tag{8}$$

The corresponding expression for $h_{\mu\nu}$ is

$$h_{\mu\nu} = \begin{pmatrix} -2GM/\kappa r & 0 & 0 & 0 \\ 0 & -2GM/\kappa r & 0 & 0 \\ 0 & 0 & -2GM/\kappa r & 0 \\ 0 & 0 & 0 & -2GM/\kappa r \end{pmatrix} \tag{9}$$

Note that since the Newtonian potential of the mass M is $\Phi = -GM/r$, Eq. [9] can also be written as

$$h_{\mu\nu} = \begin{pmatrix} 2\Phi/\kappa & 0 & 0 & 0 \\ 0 & 2\Phi/\kappa & 0 & 0 \\ 0 & 0 & 2\Phi/\kappa & 0 \\ 0 & 0 & 0 & 2\Phi/\kappa \end{pmatrix} \tag{10}$$

In this form, the equation is also valid for mass distributions other than spherical. A general mass distribution can be regarded as a collection of point masses, and the linearity of our equations therefore tells us that the net field $h_{\mu\nu}$ of the mass distribution is the sum of the fields of the point masses. Thus, the net field $h_{\mu\nu}$ of the mass distribution is related to the Newtonian potential of this mass distribution by Eq. [10]. For instance, if we want to find the field $h_{\mu\nu}$ in the interior of the Sun, we can simply substitute the Sun's Newtonian potential into Eq. [10].

We could use Eq. [9] and our relativistic equation of motion to study the motion of planets in the gravitational field of the Sun. However, we will not bother with this because the relativistic deviations from the Newtonian orbits are extremely small. The most noticeable deviation between the Newtonian and the relativistic orbital motions is the perihelion precession. The only other deviation that we might hope to detect is a difference in the predicted instantaneous velocities along the orbit. Ordinary telescopic observation cannot detect these differences. With the development of radar astronomy, it has become possible to measure the instantaneous velocity of the (inner) planets to 1 part in $\simeq 10^6$ by means of the Doppler shift suffered by radar signals sent to these planets and reflected by them back to Earth (Shapiro et al., 1966). But not even these radar-echo experiments are quite precise enough to detect the velocity deviations between Newtonian and relativistic gravitation. At present, the only available test of the relativistic equation of motion of massive particles comes from the perihelion precession. Although we could calculate the perihelion precession from our linear theory, it turns out that this precession receives comparable contributions from the linear approximation for $h^{\mu\nu}$ and from nonlinear corrections to $h^{\mu\nu}$ [note that the right side of Eq. [9] is of order GM; the nonlinear corrections are of order $(GM)^2$]. Such corrections come from the nonlinearities of the exact field equations. Let us try to understand how these nonlinearities affect the motion of planets. Consider the equation of motion [3.76],

$$\frac{d}{d\tau} P_\mu = \frac{\kappa}{2} mh_{\alpha\beta,\mu} u^\alpha u^\beta$$

The right side of this equation may be regarded as a force; since $u^0 \simeq 1$ and $u^k \simeq v^k$, it contains terms of the types $\kappa mh_{00,\mu}$ and $\kappa mh_{kl,\mu} v^k v^l$. If we use the linear approximation [9], then the term $\kappa mh_{00,\mu}$ is simply the Newtonian force; this yields no perihelion precession. The term $\kappa mh_{kl,\mu} v^k v^l$ represents a deviation from the Newtonian force and produces a perihelion precession. Since the speed of a planet in orbit is roughly $v \simeq \sqrt{GM/r}$, this term is of order $(GM)^2$. We now notice that nonlinear corrections to h_{00} will change $\kappa mh_{00,\mu}$ by a term of order $(\kappa h_{00})^2 \propto (GM)^2$. Thus, this nonlinear correction will be roughly of the same importance as $\kappa mh_{kl,\mu} v^k v^l$. Hence, for the (approximate) evaluation of the perihelion precession, we need the terms of order $(GM)^2$ in κh_{00}, although we need only the terms of order GM in κh_{kl}. We actually could calculate the required corrections by solving the nonlinear field equation [3.64], and we could then find the correct perihelion precession (see Problems 7 and 7.14). But we prefer to set the whole question aside until after we have written down and solved the exact Einstein equations.

However, there are three interesting observable effects that we can calculate from the linear approximation: the gravitational time dilation, the deflection of light by the gravitational field of the Sun, and the retardation of light signals.

4.2 THE GRAVITATIONAL TIME DILATION

According to Section 3.6, the tensor $\eta_{\mu\nu} + \kappa h_{\mu\nu}$ we obtained in the study of the equation of motion is the metric tensor $g_{\mu\nu}$ that is measured by means of clocks. In the particular case of a spherically symmetric mass distribution, the metric tensor is therefore approximately given by

$$g_{\mu\nu} = \eta_{\mu\nu} + \kappa h_{\mu\nu} \qquad [11]$$

$$= \begin{pmatrix} 1-2GM/r & 0 & 0 & 0 \\ 0 & -(1+2GM/r) & 0 & 0 \\ 0 & 0 & -(1+2GM/r) & 0 \\ 0 & 0 & 0 & -(1+2GM/r) \end{pmatrix} \qquad [12]$$

and the spacetime interval is given by

$$ds^2 = g_{\mu\nu} dx^\mu dx^\nu$$

$$= \left(1 - \frac{2GM}{r}\right) dt^2 - \left(1 + \frac{2GM}{r}\right)(dx^2 + dy^2 + dz^2) \qquad [13]$$

This interval corresponds to a curved spacetime; we will calculate the curvature in Chapter 6.

Although the arguments of Section 3.6 establish that Eq. [13] describes the geometry surrounding, say, the Earth and that therefore we are living in a curved spacetime, it would be good to have some direct experimental evidence for this curvature. In 1821 Gauss attempted to find a curvature of space by laying out a triangle between mountain peaks, about 100 km on one side. Gauss laid out this triangle with light rays, and he used surveying instruments to measure the sum of interior angles. He found no deviation from 180°, to within his experimental error ($\simeq 0.7$ arcsec). But even if he had found a deviation, could this not have been interpreted as a simple bending of the path of light in the Earth's gravitational field with no geometrical implications? In view of our definition of c as the local standard of velocity, the answer must be in the negative. A deflection of light implies that the effective wave velocity defined by reference to standards of length and time kept outside of the gravitational field is a function of position. Hence this effective velocity disagrees with the locally measured velocity, a paradox that can be resolved only if spacetime is curved (this will be discussed in more detail in Section 4.4).

The gravitational time dilation, or slowing down of clocks in a gravitational field, serves as an even more direct test of the curvature of spacetime. We already obtained the time-dilation formula in our general discussion of measurement procedures in Section 3.6. There, we found a time dilation when we compared two clocks at different positions by means of an auxiliary clock allowed to fall freely from one position to the other. In practice, most time-dilation experiments have been performed with light or radio signals sent from one clock to the other. If a regular succession of identical signals (messenger rockets, cannonballs, periodic light pulses, or whatever) is sent from one fixed position to the other, then each signal takes exactly the same amount of t time to complete this trip. This is an immediate consequence of the time independence of the gravitational field and the repetitive character of the signals--the worldlines of successive signals are copies of each other, merely shifted forward in t time (see Fig. 4.1). But if each signal takes the same amount of time to complete the trip, and if the t-time difference between one signal and the next is dt at the point of departure, then the t-time difference must also be dt at the point of arrival.

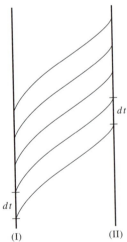

Fig. 4.1 Worldlines of successive, identical signals plotted in r, t coordinates. The t-time difference between departure of one signal and the next is dt, and the t-time difference between the arrival of one signal and the next is also dt.

(I)   (II)

However, t time is not the time that clocks placed in the gravitational field measure for us. Rather, t time is the time indicated by a clock placed at large distance from the gravitating body. For such a distant clock, $r \to \infty$ and $g_{00} \to 1$; hence, the the proper time indicated by such a clock is $d\tau = \sqrt{g_{00}}\,dt = dt$. In contrast, for clocks 1 and 2 placed at finite distances r_1 and r_2, the proper times are, respectively,

$$d\tau_1 = \sqrt{g_{00}(1)}\,dt$$

$$d\tau_2 = \sqrt{g_{00}(2)}\,dt \qquad [14]$$

Given that the delay dt between successive signals traveling from r_1 to r_2 is the same at both positions, the ratio for the proper times that the two clocks indicate between such signals is simply

$$\frac{d\tau_2}{d\tau_1} = \frac{\sqrt{g_{00}(2)}}{\sqrt{g_{00}(1)}} \qquad [15]$$

$$= \frac{\sqrt{1 - 2GM/r_2}}{\sqrt{1 - 2GM/r_1}} \simeq 1 - \frac{GM}{r_2} + \frac{GM}{r_1} \qquad [16]$$

Since $GM/r_1 - GM/r_2 = -\Phi_1 + \Phi_2$, the ratio $d\tau_2/d\tau_1$ we obtain by sending signals from one clock to the other agrees with the ratio we obtained in Section 3.6 by transport of an auxiliary clock from one to the other--both calculations give us the same value for the gravitational time dilation.

The experiments on gravitational time dilation may be regarded as direct measurements of the g_{00} component of the metric tensor. The experiments show that in the vicinity of a massive body the metric does not have the flat spacetime value $g_{00} = 1$, but rather the value $g_{00} = 1 - 2GM/r$. We are constrained to accept this purely geometric interpretation of the time-dilation effect because, according to the principle of equivalence, it is impossible for the gravitational field to influence the rate of a (freely falling) clock in any other way. A modification of the clock rate by a change in the spacetime geometry is the only possible alternative.

For comparison with the experiments, it is convenient to express Eq. [16] as a fractional deviation between the times registered by the two clocks,

$$\frac{\Delta \tau}{\tau} \simeq \frac{d\tau_2 - d\tau_1}{dt} \simeq GM\left(\frac{1}{r_1} - \frac{1}{r_2}\right) \qquad [17]$$

In terms of the clock rate, or frequency, this deviation amounts to a shift of frequency. If a clock registers a large elapsed time, it must be ticking fast, that is, it must have a high frequency. Thus, $\Delta \nu$ must be proportional to $\Delta \tau$:

$$\frac{\Delta \nu}{\nu} = \frac{\nu_2 - \nu_1}{\nu} = \frac{\Delta \tau}{\tau} \simeq GM\left(\frac{1}{r_1} - \frac{1}{r_2}\right) \qquad [18]$$

If $r_2 > r_1$, then $\Delta \nu > 0$, and hence a periodic signal emitted in step with the ticking of the clock at the smaller radius r_1 will display a low frequency when it arrives at, and is compared with, the ticking of the clock at the larger radius r_2. In particular, the spectral lines in the light emitted by an atom placed deep in a gravitational potential will display a *redshift* when compared with the spectral lines emitted by a similar atom placed outside of the potential.

   If the clocks are near the surface of the Earth, we can make the approximation

$$GM\left(\frac{1}{r_1} - \frac{1}{r_2}\right) \simeq GM\,\frac{\Delta r}{R^2} = g\Delta r$$

where Δr is the difference in height between the clocks and R is the radius of the Earth. With this approximation Eq. [18] becomes, in cgs units,

$$\frac{\Delta \tau}{\tau} \simeq g\,\frac{\Delta r}{c^2} \qquad [19]$$

For example, if $\Delta r = 10$ km, which is the typical height that can be attained by an aircraft, the time dilation is

$$\frac{\Delta \tau}{\tau} \simeq 10^{-12} \qquad [20]$$

   Although the time-dilation effect predicted by Eq. [20] is small, it is not beyond the accuracy of atomic clocks. For a direct test of the time dilation, Häfele and Keating (1972) carried cesium-beam clocks on flights around the world in commercial aircraft. On each flight, four cesium clocks were carried, and they were continuously inter-compared so that small spontaneous jumps in rate, to which the individual clocks are prone, could be identified and corrected for. The clocks carried to high altitude (about 10 km) by the aircraft were found to have gained of the order of 100 nanoseconds when brought back to a laboratory clock that stayed on the ground. To be precise, the clocks were found to have gained time after a kinematic correction was subtracted to take into account the special-relativistic time dilation. The kinematic correction and the predicted redshift were calculated from the flight data of the aircraft. The experiment verified the gravitational time dilation to within about 10%.

A similar, but more precise test was performed by Alley (1979), who kept an aircraft equipped with atomic clocks flying in a holding pattern above a ground station, also equipped with atomic clocks. The clocks in the aircraft and on the ground were compared during flight by laser signals, and the aircraft was tracked continuously by radar, to determine height and speed.

The development of highly stable hydrogen-maser clocks has permitted the most precise test to date. Vessot et al. (1980) used a Scout rocket to launch a hydrogen-maser clock to an altitude of about 10,000 km. Radar signals were used to compare the rate of the clock on the rocket with a similar clock on the ground. Furthermore, separate radar signals sent to the rocket and returned by a transponder were used to monitor the height and instantaneous speed of the rocket, by the Doppler shift (if a radar signal travels up and down in the gravitational field, it would return to the ground with unchanged frequency if the rocket were instantaneously stationary; any frequency shift of such radar signal is a Doppler shift, which reveals the speed of the rocket). The data from this experiment verified the prediction for the gravitational time dilation to within 2 parts in 10^4.

The accuracy of atomic clocks is not yet sufficient for a laboratory test of the time dilation, over a height of 10 m or so. The first (and only) laboratory test of the gravitational time dilation was performed with a "nuclear" clock. Pound and Rebka (1960) placed an emitter of γ rays (^{57}Fe) at ground level and detected the γ rays with an absorber placed on top of a tower, 22.6 m above the emitter (see Fig. 4.2). For $\Delta r = 22.6$ m, Eq. [19] predicts a fractional shift in frequency of

$$\frac{\Delta \nu}{\nu} \simeq \frac{\Delta \tau}{\tau} = g \frac{\Delta r}{c^2} = 2.46 \times 10^{-15} \qquad [21]$$

The measurement of such a small frequency shift is possible because resonant absorption of γ rays in an absorber of ^{57}Fe is very sensitive to the frequency of the incident γ rays. The experiment relied on the Mössbauer effect to prevent loss of any of the γ-ray energy by recoil of the nucleus during emission or absorption. The most precise version of this experiment (by Pound and Snider, 1964) agreed with the predicted value to within the experimental error, about 1%.

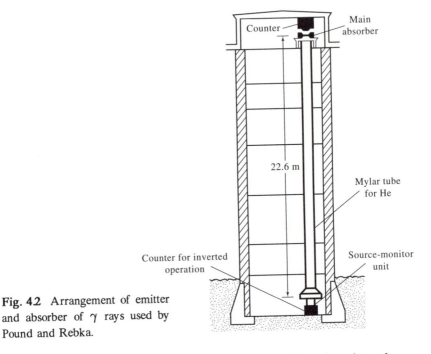

Fig. 4.2 Arrangement of emitter and absorber of γ rays used by Pound and Rebka.

The experiments with flying clocks, in aircraft and rockets, have a crucial conceptual advantage over the γ-ray experiment, in that the former experiments show in the most direct way that clocks in a gravitational potential run slower. The latter experiment does not give quite as direct an indication of whether the frequency shift between the absorbed γ rays and the natural nuclear oscillations is due to a slower oscillation rate of the emitter or a loss of frequency suffered by the γ ray as it climbs upward in the Earth's field. However, we can rule out the possibility of a simple frequency loss during propagation of the light wave by noting that if pulses of light, or whatever, are injected into a static medium (for instance, a gravitational field) at one point and received at another point, then, under steady-state conditions, the rate of injection reckoned in t time must be identical with the rate of reception, also reckoned in t time. This is obvious because under steady-state conditions, pulses cannot accumulate in the space between the two points. (In optics, this corresponds to the well-known result that if a medium has an index of refraction that varies with position, then light propagates with a changing wavelength but a constant frequency.) The fact that the frequency of a stationary nuclear clock disagrees with the frequency of a γ ray sent to it by another stationary nuclear clock shows that these clocks do not have the same frequency, as reckoned in t time. Thus the rates of clocks depend on the gravitational field, that is, the spacetime geometry depends on the gravitational field.

Gravitational redshifts have also been measured for light emitted by atoms on the surface of stars. For example, with respect to oscillations of atoms at a large distance, oscillations of atoms on the surface of the Sun have a frequency shift of

$$\frac{\Delta \nu}{\nu} = - \frac{GM_\odot}{R_\odot c^2} = -2.12 \times 10^{-6} \qquad [22]$$

relative to the oscillations of identical atoms far from the Sun. It has been difficult to obtain clear-cut experimental results for this redshift because there are strong convection currents in the solar atmosphere, and the spectral lines are subjected to Doppler shifts arising from the motion of the gas (Bertotti et al., 1962). However, measurements by Brault (1962) and Snider (1971) found redshifts in good agreement with the value given by Eq. [22]. These measurements were performed on spectral lines of sodium and potassium which are believed to originate above the layers of the photosphere where convection currents produce significant Doppler shifts.

For light emitted by white dwarfs, the redshift is larger than that given by Eq. [22] by a factor of up to $\simeq 100$, because the radius is smaller than $R_\odot$ by approximately this factor, whereas the mass is about equal to $M_\odot$. The spectral lines of white dwarfs tend to be very broad and diffuse, and meaningful measurements have been possible only on the two brightest white dwarfs, Sirius B (the companion of Sirius A) and 40 Eridani B. Both of these white dwarfs are in binary systems, and this makes possible the determination of their masses from the observational data on their orbital motion. The radius is determined from the absolute luminosity and surface temperature.

Between 1930 and 1960 the angular separation between Sirius A and Sirius B was too small for any measurements (Bertotti et al., 1962). The observations made before this period suffered seriously from scattering of the light of Sirius A by the atmosphere of Sirius B. Observations made after this period (Greenstein et al., 1971) took advantage of a clever diffraction technique to eliminate the undesirable light from Sirius A and gave a redshift $\Delta \nu / \nu = -(30 \pm 5) \times 10^{-5}$, in excellent agreement with the value $-(28 \pm 1) \times 10^{-5}$ predicted from the measured radius. In the case of 40 Eridani B, the separation from its companion is large and the measured redshift, $\Delta \nu / \nu = -(5.7 \pm 1) \times 10^{-5}$, is in good agreement with the expected value, $-(7 \pm 1) \times 10^{-5}$.

Table 4.1 lists time-dilation experiments; the last column of this table gives the ratio of experimental result to theoretical prediction.

Finally, we present two simple derivations of the redshift. The first argument is based on energy conservation. Suppose an atom at rest in a gravitational field emits a photon, and this photon is absorbed by a second atom at rest at a different position. Then the mass lost by the

TABLE 4.1 TIME-DILATION EXPERIMENTS

Experimenter(s)	Method	$\Delta\nu_{exp}/\Delta\nu_{theor}$
Adams, Moore (1925, 1928)	redshift of H lines on Sirius B	0.2 to 0.5
Popper (1954)	redshift of H lines on 40 Eridani B	1.2 ± 0.3
Pound and Rebka (1960)	redshift of γ rays	1.05 ± 0.10
Brault (1962)	redshift of Na lines on Sun	1.0 ± 0.05
Pound and Snider (1964)	redshift of γ rays	1.00 ± 0.01
Greenstein et al. (1971)	redshift of H lines on Sirius B	1.07 ± 0.2
Snider (1971)	redshift of K lines on Sun	1.10 ± 0.06
Hafele and Keating (1972)	cesium clocks on aircraft	0.9 ± 0.1
Alley (1979)	cesium and rubidium clocks on aircraft	1 ± 0.02
Vessot and Levine (1979)	hydrogen-maser clock on rocket	1 ± 0.0002

first atom by emission of the photon is*

$$\Delta m_1 = h\nu_1 \qquad\qquad [23]$$

and the mass gained by the second atom by absorption of the photon is

$$\Delta m_2 = h\nu_2 \qquad\qquad [24]$$

where ν_2 is the frequency of the photon as received at the position of the second atom. The *total* energy cannot change in the emission-absorption process:

$$-\Delta m_1 + \Delta m_2 + (\Phi_2 \Delta m_2 - \Phi_1 \Delta m_1) = 0 \qquad\qquad [25]$$

The term in parentheses represents the change in gravitational potential energy. Using Eqs. [24] and [25], we obtain

$$-\nu_1 + \nu_2 + \Phi_2\nu_2 - \Phi_1\nu_1 = 0 \qquad\qquad [26]$$

---

*   Here, h is Planck's constant.

or

$$\frac{\Delta \nu}{\nu} = \frac{\nu_2 - \nu_1}{\nu_1} \simeq - (\Phi_2 - \Phi_1) \qquad [27]$$

which agrees with our old result [18], except for sign (the sign change arises because in Eq. [27] ν_2 is not the frequency of the second atom, but the frequency of a photon that is received at the position of the second atom).

What is involved in this argument is not just energy conservation, but also the principle of equivalence, since in writing the potential-energy term in Eq. [25] we assumed that the energy change Δm produces a corresponding change in the *gravitational* mass.* It would therefore be wrong to say that energy conservation by itself implies the redshift. Also, the argument should not be construed as a denial of curved spacetime. Rather, the argument shows that energy conservation and the principle of equivalence imply the redshift and hence, as we saw in Section 3.6, imply that spacetime is curved.

Another simple derivation of the redshift makes use of the principle of equivalence to replace the gravitational field by the pseudo-force field generated in an accelerated reference frame. A uniform gravitational field g is simulated by placing the two clocks in a reference frame ("elevator") accelerating upward at the rate g (in the *absence* of any gravitational field; see Fig. 4.3). An inertial observer then describes the situation as follows: Clocks 1 and 2, separated by a

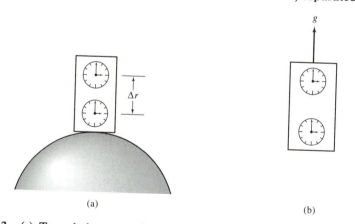

(a)

(b)

Fig. 43 (a) Two clocks at rest in the gravitational field of the Earth. (b) Two clocks in an accelerated reference frame.

---

* Also, we assumed that the usual connection $E = h\nu$ holds in the freely falling reference frame.

vertical distance Δr,* are both accelerating upward with acceleration g. Clock 1 is emitting light pulses at the rate ν_1 signals per second. In the time $\simeq \Delta r/c$ that these light pulses take to reach clock 2, the velocity of clock 2 increases by $\Delta v = g\Delta r/c$. This means that the rate ν_2 at which pulses are received by clock 2 is decreased by the Doppler shift:

$$\nu_2 \simeq \nu_1 \left(1 - \frac{\Delta v}{c}\right) = \nu_1 \left(1 - \frac{g\Delta r}{c^2}\right) \tag{28}$$

that is,

$$\frac{\Delta \nu}{\nu} \simeq -\frac{g\Delta r}{c^2} \tag{29}$$

which agrees with Eq. [27].

4.3 THE DEFLECTION OF LIGHT

To find out how light propagates in a gravitational field, we adopt the view that light consists of particles (photons), since this permits us to use our relativistic equation of motion for a particle in the gravitational field. However, since the mass of the photon is zero, we must first rewrite the equation of motion somewhat. For this purpose, we multiply the equation of motion [3.93] by $md\tau$:

$$d(mu_\mu) + \left[\kappa h_{\mu\alpha,\beta}\, mu^\alpha u^\beta\, d\tau - \frac{\kappa}{2}\, h_{\alpha\beta,\mu}\, mu^\alpha u^\beta\, d\tau\right] = 0 \tag{30}$$

If we substitute $p_\mu = mu_\mu$ and $dx^\beta = u^\beta\, d\tau$, we obtain

$$dp_\mu + \left[\kappa h_{\mu\alpha,\beta}\, p^\alpha - \frac{\kappa}{2}\, h_{\alpha\beta,\mu}\, p^\alpha\right] dx^\beta = 0 \tag{31}$$

In this equation neither m nor $d\tau$ appears explicitly, and hence the equation can be applied to a photon. It tells us how the momentum of the photon is changed by the gravitational field, and permits us to calculate the deflection of a ray of light passing through the gravitational field. For such a calculation we can treat the factor p^α in the second term in Eq. [31] as constant, since any nonconstant portion in p^α is of

---

* We neglect the length contraction; hence, Δr could be measured either with respect to the accelerated reference frame or with respect to the Earth.

order κ, and would make a negligible contribution of order κ^2 in the second term in Eq. [31]. The net change in p_μ produced while the photon passes through the gravitational field is therefore given a simple integral:

$$\Delta p_\mu = -\kappa p^\alpha \int_{-\infty}^{\infty} \left(h_{\mu\alpha,\beta} - \frac{1}{2} h_{\alpha\beta,\mu} \right) dx^\beta \qquad [32]$$

where it is understood that the integration is performed along the (approximately) straight trajectory of the photon. The first term on the right side of Eq. [32] can be omitted, since $\int h_{\mu\alpha,\beta} dx^\beta = h_{\mu\alpha}(\infty) - h_{\mu\alpha}(-\infty)$, which is zero because the field $h_{\mu\alpha}$ is zero at infinite distance from the gravitating mass. Thus,

$$\Delta p_\mu = \frac{\kappa}{2} p^\alpha \int_{-\infty}^{\infty} h_{\alpha\beta,\mu} dx^\beta \qquad [33]$$

Consider a ray of light passing by the Sun with impact parameter b (see Fig. 4.4). Taking the z-axis along the direction of incidence, we see that the displacement along the (approximately) straight ray and the momentum vector are, respectively,

$$dx^\beta \simeq (dt, 0, 0, dz) = (dz, 0, 0, dz) \qquad [34]$$

$$p^\alpha \simeq (p^3, 0, 0, p^3) \qquad [35]$$

Inserting Eqs. [34] and [35] into Eq. [33], we obtain

$$\Delta p_1 = \frac{\kappa}{2} p^3 \int_{-\infty}^{\infty} (h_{00,1} + h_{03,1} + h_{30,1} + h_{33,1}) dz \qquad [36]$$

For a spherically symmetric gravitational field, the terms h_{03} and h_{30} vanish and $h_{00} = h_{33} = -2GM_\odot/\kappa r$. Hence

$$\Delta p_1 = \frac{\kappa}{2} p^3 \int_{-\infty}^{\infty} \frac{\partial}{\partial x} (h_{00} + h_{33}) dz$$

$$= -\frac{\kappa}{2}\, p^3 \int_{-\infty}^{\infty} \frac{\partial}{\partial x}\, \frac{4GM_\odot}{\kappa r}\bigg|_{x=b} dz \qquad [37]$$

$$= 2GM_\odot p^3 \int_{-\infty}^{\infty} \frac{b}{(z^2 + b^2)^{3/2}}\, dz$$

$$= \frac{4GM_\odot}{b}\, p^3$$

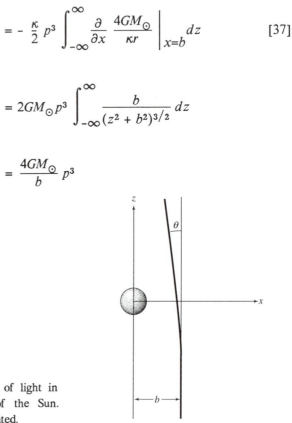

Fig. 4.4 Path of a ray of light in the gravitational field of the Sun. The deflection is exaggerated.

The deflection angle for the light ray is therefore

$$\theta \simeq \frac{\Delta p_x}{p_z} = -\frac{\Delta p_1}{p^3} = -\frac{4GM_\odot}{b} \qquad [38]$$

or, expressed in cgs units,

$$\theta = -\frac{4GM_\odot}{bc^2} \qquad [39]$$

■ **Exercise 1.** Show that, within the same approximations,

$$\Delta p_0 = 0$$
$$\Delta p_3 = 0 \qquad ■$$

Note that Eq. [37] can be expressed in terms of the Newtonian potential $\Phi = - GM_\odot/r$ of the gravitating mass,

$$\Delta p_x = - \Delta p_1 = - 2p^3 \int_{-\infty}^{\infty} \left. \frac{\partial \Phi}{\partial x} \right|_{x=b} dz$$

Written in this form, the equation is valid for the deflection produced by some general mass distribution with some general potential Φ, since any such general mass distribution can be regarded as a collection of point masses, which produce additive contributions to the potential Φ and to the deflection angle. Accordingly, the deflection angle for a ray passing by (or through) a general mass distribution is

$$\theta = -2 \int_{-\infty}^{\infty} \frac{\partial \Phi}{\partial x} \, dz \qquad [40]$$

where it is understood that the derivative is to be evaluated at the impact parameter of the ray. This general formula will be useful later in this chapter.

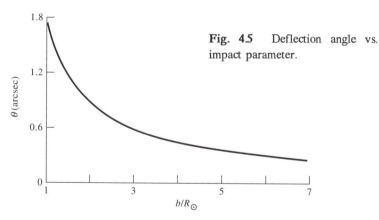

Fig. 4.5 Deflection angle vs. impact parameter.

Fig. 4.5 shows a plot of deflection versus impact parameter for rays passing by the Sun. For the case of a light ray grazing the Sun, we must take $b = R_\odot = 6.96 \times 10^{10}$ cm and $M_\odot = 1.99 \times 10^{33}$ g, which gives

$$\theta = - \frac{4GM_\odot}{R_\odot c^2} = - 1.75 \text{ arcsec} \qquad [41]$$

Such a deflection of the light reaching the Earth from a star results in an (apparent) shift in the position of the star as seen from the Earth (see Fig. 4.6). Observations of this effect are difficult to perform. The stars near the Sun are visible only during a total eclipse of the Sun, and even then the brightness of the solar corona restricts observations to impact parameters $b > 2R_\odot$. The experimental procedure involves taking a photograph of the star field surrounding the eclipsed Sun and comparing this photograph with another one taken at night several months before (or after) when the Sun is not in this star field. The

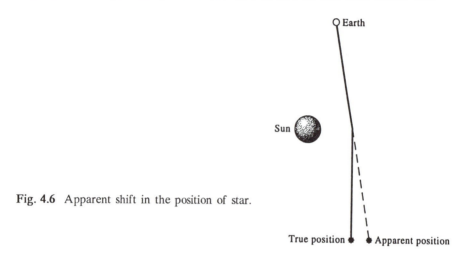

Fig. 4.6 Apparent shift in the position of star.

Fig. 4.7 Eclipse instruments at Sobral. The two separate telescopes are mounted horizontally and mirrors (center left) are used to throw the Sun's image into them. (From Eddington, *Space, Time, and Gravitation*. Reprinted with permission of Cambridge University Press.)

technical difficulties are that the eclipse photograph must be taken in the very short time interval of totality and, furthermore, disturbances to the instrument must be prevented (or corrected for) during the waiting period between the eclipse photograph and the night photograph. In a temporary field station set up in the path of the eclipse in some remote part of the world all of this is hard to achieve.

In response to Einstein's prediction of light deflection, Eddington and Dyson organized two expeditions to observe the eclipse of May 29, 1919, on the islands of Sobral (off Brazil) and Principe (in the Gulf of Guinea). Fig. 4.7 shows the instruments installed at Sobral. The experimental results obtained by these expeditions were in reasonable agreement with the prediction.

Since then, observations have been carried out during many solar eclipses. Table 4.2 (based on the review by von Klüber, 1960) is a summary of these observations. The numbers given in this table are the deflections extrapolated to an impact parameter $b = R_{\odot}$; they should be compared with Eq. [41]. Evidently, the observations of the light deflection are afflicted with large experimental errors, and in spite of fifty years of efforts, the errors in the most recent observations are about as large as those in the first observations. On the basis of these results, all we can say is that a deflection of approximately 2 arcsec is present, but no precise test of the predicted deflection is possible.

TABLE 4.2 EXPERIMENTAL RESULTS ON THE DEFLECTION OF LIGHT*

Observatory	Eclipse	Site	θ (arcsec)
Greenwich	May 29, 1919	Sobral	1.98 ± 0.16
Greenwich	" "	Principe	1.16 ± 0.40
Adelaide-Greenwich	Sept. 21, 1922	Australia	1.77 ± 0.40
Victoria	" "	Australia	1.42 to 2.16
Lick	" "	Australia	1.72 ± 0.15
Lick	" "	Australia	1.82 ± 0.20
Potsdam	May 9, 1929	Sumatra	2.24 ± 0.10
Sternberg	June 19, 1936	U.S.S.R	2.73 ± 0.31
Sendai	"	Japan	1.28 to 2.13
Yerkes	May 20, 1947	Brazil	2.01 ± 0.27
Yerkes	Feb. 25, 1952	Sudan	1.70 ± 0.10
U. of Texas	June 30, 1973	Mauritania	1.66 ± 0.19

* For the first eleven entries, see von Klüber (1960); for the last entry, see Jones (1973).

More precise results have been obtained by the use of radio waves rather than optical light. In this case, it is not necessary to wait for an eclipse; rather one must wait for the Sun's limb to approach some radio source in the sky. The quasistellar source 3C279 is occulted by the Sun on the eighth of October of each year, and radio interferometers, consisting of two or more antennas, have been used to measure the apparent shift of the position of this source. The shift is measured relative to the quasistellar sources 3C273 and 3C48, which are about 10° farther away and therefore suffer a much smaller shift. The deflection of the radio waves must be corrected for refraction by the plasma in the solar corona. This is accomplished by measuring the deflections at two or more different radio frequencies; the slight differences between the deflections at different frequencies permit a calculation of the plasma density. The experimental results listed in Table 4.3 have been expressed in terms of a ratio of the observed deflection and the predicted deflection.

TABLE 4.3 EXPERIMENTAL RESULTS ON THE DEFLECTION
OF RADIO WAVES

Radio telescope	Observer(s)	Baseline (km)	$\theta_{exp}/\theta_{theor}$
Owens Valley	Seielstadt et al. (1970)	1.07	1.01 ± 0.11
Goldstone	Muhleman et al. (1970)	21.56	$1.04 \left\{ \begin{array}{l} +0.15 \\ -0.10 \end{array} \right.$
National RAO	Sramek (1971)	2.7	0.9 ± 0.05
Mullard RAO	Hill (1971)	$\simeq 1$	1.07 ± 0.17
Cambridge	Riley (1973)	4.57	1.04 ± 0.08
Westerbork	Weiler et al. (1974)	1.44	0.96 ± 0.05
Haystack and National RAO	Counselman et al. (1974)	845	0.99 ± 0.03
National RAO	Fomalont and Sramek (1975)	35	1.015 ± 0.011
National RAO	Sramek (1974)	35	0.97 ± 0.08
Westerbork	Weiler et al. (1975)	$\simeq 1$	1.04 ± 0.03
National RAO	Fomalont and Sramek (1976)	35	1.007 ± 0.009
VLBI	Roberston and Carter (1984)	$\simeq 10,000$	1.004 ± 0.002
VLBI	Robertson et al. (1991)	$\simeq 10,000$	1.0001 ± 0.0001

The most precise result has been obtained by Very Long Baseline Interferometry (VLBI), which combines measurements taken simultaneously with several radio telescopes separated by intercontinental distances. The signals are recorded on tapes, tagged with precise time markers provided by synchronized atomic clocks at each radio telescope. The tapes can later be played back in precise synchronization, and the superposition of their signals gives us the interference pattern, as though the separate radio telescopes had operated as a single radio interferometer. According to Rayleigh's criterion, the resolving power of a radio interferometer is limited by the width of the diffraction peak

$$\Delta\theta \simeq [\text{wavelength}]/[\text{baseline}] = \lambda/d \qquad [42]$$

With an intercontinental baseline of $\simeq 10,000$ km and a wavelength of 3 cm, this theoretical diffraction limit is $\Delta\theta \simeq 0.0006$ arcsec. The uncertainty of ± 0.0001 arcsec in the best available result for the deflection angle is actually below this diffraction limit because radio astronomers can plot the shape of the diffraction peak, and determine the location of its center without much regard to its width.

   Incidentally: The results of the radio-wave deflection experiments give evidence against the scalar-tensor theory of gravitation (see Section 3.2). In this alternative theory of gravitation, the predicted deflection is smaller than about 1.66 arcsec, which disagrees with the observational data.

4.4 THE RETARDATION OF LIGHT

Another observable effect that we can obtain from the linear approximation is the time delay suffered by a radar signal sent from the Earth to a planet and reflected back to Earth. As first recognized by Shapiro, the gravitational field of the Sun contributes a measurable increment to this time delay, because it reduces the speed of propagation of light signals.

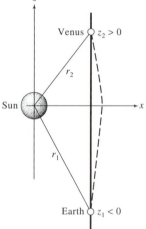

Fig. 4.8 Path of a light ray between the Earth (at z_1) and a planet (at z_2). The true path is indicated by the dashed line. The straight line is an approximation.

The speed of propagation can be deduced from Eq. [31]. Consider a light signal that moves along a straight line connecting the Earth and another planet (see Fig. 4.8; we take the z-axis parallel to this line). Of course, because of the deflection of the path of light, the true path will not be exactly straight. But this has next to no effect on the travel time because the difference in length between the straight and curved paths shown in Fig. 4.8 is only of order θ^2 (where θ is the deflection angle); this difference is therefore a second-order correction which we can ignore. Along the straight line, the change in p_3 given by Eq. [31] and Eqs. [34] and [35] is

$$dp_3 = -\kappa h_{33,3} p^3 dz + \frac{\kappa}{2} h_{33,3} p^3 dz + \frac{\kappa}{2} h_{00,3} p^3 dz \qquad [43]$$

With $h_{00} = h_{33} = -2GM_\odot/\kappa r$, the terms on the right side of Eq. [43] cancel; thus $dp_3 = 0$, and p_3 is constant. The change in p_0 is

$$dp_0 = -\kappa h_{00,3} p^0 dz \simeq -\kappa h_{00,3} p^3 dz \qquad [44]$$

which leads to

$$p_0 = [\text{constant}] - \kappa h_{00} p^3$$

$$= (1 - \kappa h_{00}) p^3 \qquad [45]$$

Since $p^\mu = m dx^\mu / d\tau$, the velocity of the photon can be expressed as

$$\frac{dz}{dt} = \frac{dx^3}{dx^0} = \frac{p^3}{p^0}$$

$$= \frac{p^3}{(1 - \kappa h_{00}) p^3} \simeq (1 + \kappa h_{00}) \qquad [46]$$

or as

$$\frac{dz}{dt} = \left[1 - \frac{2GM_\odot}{r} \right] \qquad [47]$$

Note that this velocity is independent of the frequency; hence there is no dispersion, and the group velocity (signal velocity) coincides with the phase velocity. The result [47] shows that the gravitational field of the Sun decreases the speed of propagation of light. In fact, the deflection of light, which we calculated earlier, can be regarded as a refraction of the wavefront: the part of the wavefront nearer the Sun is slowed down more than the part farther away, and hence the direction of advance of the wave is deflected toward the Sun.

Fig. 4.8 shows the Earth (at $z_1 < 0$) and the target planet (at $z_2 > 0$) in the gravitational field of the Sun. The travel time for a light or radio signal from z_1 and z_2 is

$$\Delta t = \int_{z_1}^{z_2} \frac{dz}{dz/dt}$$

$$\simeq \int \left[1 + \frac{2GM_\odot}{r} \right] dz = \int \left[1 + \frac{2GM_\odot}{\sqrt{z^2 + b^2}} \right] dz$$

$$= z_2 + |z_1| + 2GM_\odot \ln \frac{\sqrt{z_2^2 + b^2} + z_2}{\sqrt{z_1^2 + b^2} - |z_1|} \qquad [48]$$

The second term on the left side of Eq. [48] represents the extra time delay produced by the gravitational field. The extra delay is large when the Earth and planet are on nearly opposite sides of the Sun ("superior conjunction"). Then b/z is small and the extra time delay becomes approximately

$$2GM_\odot \ln \left(\frac{4|z_1||z_2|}{b^2} \right) \qquad [49]$$

■ *Exercise 2.* Derive Eq. [49] from [48]. ■

Clearly, b cannot be smaller than the radius of the Sun (grazing light ray). If we set $b = R_\odot = 6.96 \times 10^{10}$ cm, $|z_1| =$ [radius of Earth orbit] $\simeq 14.9 \times 10^{12}$ cm, and $z_2 =$ [radius of Mercury orbit] $\simeq 5.8 \times 10^{12}$ cm, we find from Eq. [49] that the maximum possible extra time delay is $\simeq 1.1 \times 10^{-4}$ s; the extra delay for the signal to go to Mercury and return is twice as large.

Before we can compare this with experiment, we must make a correction because the clocks on the Earth used to measure the time delay give the proper time rather than the t time which appears in Eq. [48]. According to Section 4.2, the connection between proper time and t time for a clock at rest in the gravitational field of the Sun is

$$\Delta\tau = \sqrt{g_{00}}\Delta t = \sqrt{1 + \kappa h_{00}}\Delta t \simeq \left(1 - \frac{GM_\odot}{r}\right)\Delta t \qquad [50]$$

To convert the expression [48] to proper time at the Earth, we must therefore multiply it by $1 - GM_\odot/r$, where r is the distance between Earth and Sun. Since the Earth is moving, the time-dilation effect of special relativity must also be taken into account. If the orbit of the Earth were circular, these corrections would merely give a constant, uninteresting factor; but because the orbit of the Earth is elliptical, with a variable distance and speed, these corrections add a variable term to the time delay, which alters the measured time delay by as much as 10% for the case of Mercury. Further (nongravitational) corrections must be made for the change in the velocity of propagation of the radio signal caused by the solar corona and interplanetary plasma, and for distortion of the return signal caused by topographic features on the reflecting surface of the target planet.

The delays of radar echoes from Mercury and Venus were measured by Shapiro et al. (1968, 1971), using the Haystack and Arecibo radio telescopes. In order to separate the *extra* time delay due to gravitation from the total measured delay, it is necessary to know the position of the reflecting surface very precisely. This means that both the position and the size of the reflecting planet must be known with an accuracy that is beyond what can be achieved by ordinary observa-

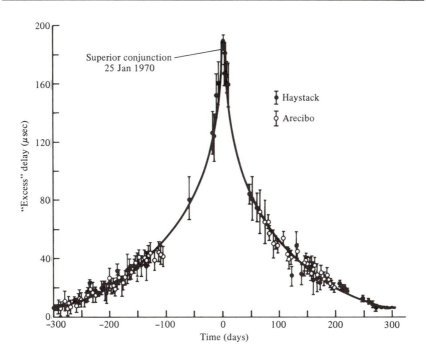

Fig. 4.9 Results of Earth-Venus time-delay measurements. The solid curve gives the theoretical prediction. (From Shapiro et al., 1971.)

tional astronomy. This difficulty was resolved by performing a large number of time-delay measurements over a period of several years (beginning in 1966). The orbital elements of the planets, their sizes, topographical features, masses, and finally the gravitational delay were then obtained from these data. The curve in Fig. 4.9 (from Shapiro et al., 1971) is a plot of the predicted extra time delay for Venus produced by gravitational effects as a function of time. The data points obviously fit this curve very well--the theoretical and experimental values agree to within 2%.

Time delays were also measured by Anderson et al. (1975) for signals transmitted to the *Mariner* 6 and 7 spacecraft in orbit around the Sun; signals received by the spacecraft were retransmitted to the Earth by a transponder on the spacecraft. The experimental delays were found to be in excellent agreement with the theoretical values. Unfortunately, these spacecraft are subject to unknown random forces produced by outgassing of the attitude-control systems and by fluctuations in the solar radiation pressure. These unknown perturbations introduce uncertainties in the positions and consequent uncertainties in the delays; a realistic estimate of the total uncertainty is 3%.

More precise results were obtained for signals transmitted to the *Viking* orbiters and landers, in orbit around Mars and on the ground on Mars. Under these conditions, the spacecraft are "anchored" to Mars, and the uncertainties in their positions are minimized. Analysis of the *Viking* data has confirmed the theoretical value of the time delay to within 0.1%. Table 4.4 summarizes the results of time-delay measurements.

TABLE 4.4 EXPERIMENTAL RESULTS ON THE RETARDA-TION OF LIGHT

Target	Observer(s)	$\Delta\tau_{exp}/\Delta\tau_{theor}$
Mercury and Venus	Shapiro et al. (1968, 1971)	1.02 ± 0.05
Mariner 6	Anderson et al. (1975)	1.00 ± 0.03
Mariner 7	Anderson et al. (1975)	1.00 ± 0.03
Mariner 9 (orbiting Mars)	Reasenberg and Shapiro (1977)	1.00 ± 0.02
Mariner 9 (orbiting Mars)	Anderson et al. (1978)	1.00 ± 0.02
Viking (on Mars)	Shapiro and Reasenberg (1977)	1.000 ± 0.005
Viking (on Mars)	Reasenberg and Shapiro (1979)	1.000 ± 0.001
Voyager 2	Krisher et al. (1991)	1.00 ± 0.02

A comment on just what is meant by the "slowing down" of the speed of propagation of light (Eq. [47]) may be helpful. Although a light signal that passes close to the Sun will be delayed, this does not mean that an observer who measures the speed of the signal at a point near the Sun will obtain a result different from the usual speed of light. If this observer uses "good" clocks and "good" meter sticks, he will always find a speed $c = 3 \times 10^{10}$ cm/s. The speed given by Eq. [47] is not a locally measured speed but an effective speed measured with standards of time and length that are far away, outside of the gravitational field.

An idealized situation is shown in Fig. 4.10a. We have two points P_1, P_2 connected by a light signal. The points P_1, P_2, are outside of the gravitational field (alternatively, the field is so weak at these points that it can be neglected for our purposes), but the light ray connecting them passes through a region where the field is strong. In order to measure the speed of the light signal along P_1P_2 we could try to lay out our meter sticks along P_1P_2; but since, according to Eq. [13], the gravitational field affects the measurement of space and time, let us not do this. An alternative method is to use surveying instruments to lay out the lines P_1A, AB, BP_2, where all the angles are right angles (the figure is drawn under the simplifying assumption that there is no appreciable bending of the light rays). All these lines are outside of the gravitational field and their lengths can be measured without trouble. According to Euclidean geometry the distance P_1P_2 equals AB. Furthermore, clocks at P_1, A, B, P_2 can be synchronized by the procedure familiar from special relativity. We may then say that

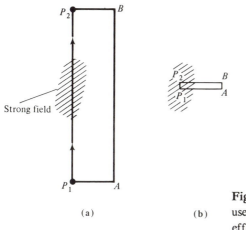

(a)   (b)

Fig. 4.10 Light path and survey used for the measurement of the effective speed.

the effective speed along P_1P_2 is the distance measured along AB divided by the time elapsed between departure from P_1 and arrival at P_2. It is this effective speed that is smaller than c.

Of course this procedure gives us only the average effective speed. If we want to obtain the instantaneous effective speed, then we must make the segments P_1P_2 and AB infinitesimal, and then both P_1 and P_2 will be in the region where the gravitational field is appreciable (see Fig. 4.10b). In this case surveying instruments cannot be used directly to lay out the rectangle of Fig. 4.10b, because the light rays on which these instruments rely will be bent in the gravitational field. But if we assume that the formula for the amount of bending is known, we can correct for it and lay out the necessary straight lines. Furthermore, when measuring the time elapsed between departure of a light signal from P_1 and arrival at P_2 we will not want to use the local clocks of Fig. 4.10b, but rather a clock outside of the gravitational field. This can be done by signaling the departure and arrival to a clock at a large distance and using that clock to measure the time interval. Note that essentially this is an operational procedure for setting up the coordinates x, y, z, t by reference to the coordinates at large distance, the speed dz/dt (Eq. [47]) being defined in terms of the changes in these coordinates.

The disagreement between local speed c measured along P_1P_2 and the effective speed measured indirectly by the construction of P_1ABP_2 shows quite clearly that something is foul with flat spacetime, or our measurement instruments, or both. In a curved spacetime, the disagreement is not at all surprising because then the distance P_1P_2 need not equal AB. In fact, the effective speed of propagation can be obtained immediately from the curved spacetime interval [13]. The propagation of light is characterized by a lightlike interval, that is,

$$0 = ds^2 = g_{\mu\nu}dx^\mu dx^\nu \qquad [51]$$

From Eq. [13], we obtain

$$0 = \left[1 - \frac{2GM_\odot}{r}\right]dt^2 - \left[1 + \frac{2GM_\odot}{r}\right](dx^2 + dy^2 + dz^2)$$

Hence the speed of propagation of the light signal is

$$v = \frac{\sqrt{dx^2 + dy^2 + dz^2}}{dt} = \sqrt{\frac{1 - 2GM_\odot/r}{1 + 2GM_\odot/r}}$$

$$\simeq \left[1 - \frac{2GM_\odot}{r}\right] \qquad [52]$$

This agrees with Eq. [47].

Since both the light deflection and the time delay can be obtained from the speed formula [47] or [52], we could have derived these effects without using the equation of motion [31] explicitly. Note, however, that this alternative derivation is not entirely independent of the equation of motion because the interpretation of $g_{\mu\nu}$ as metric hinges on that equation (see Section 3.6).

In the above calculation we adopted the view that light is made of particles. Of course, we can also adopt the view that light is a wave, and we expect that the wave speed matches the particle speed. In this context, Eq. [45] poses somewhat of a puzzle, since, with the familiar proportionality $\hbar\omega = p_0$, it suggests that the frequency of the wave varies during propagation. But such a variation of the frequency of the wave is in conflict with a general property of wave propagation in a static medium--in such a medium, the wavelength of the wave varies, but the frequency always remains constant. The answer to this puzzle is that the familiar proportionality of ω and p_0 fails. The frequency ω of the wave is not proportional to the "kinetic" energy p_0, but rather to the "canonical" energy P_0,

$$\hbar\omega = P_0 = p_0 + \kappa h_{0\alpha}p^\alpha = (1 + \kappa h_{00})p_0 \qquad [53]$$

(More generally, the wave vector k_μ is proportional to the canonical energy-momentum vector P_μ.) Eq. [53] simply means that the frequency is proportional to the *total* energy of the photon in the gravitational field. Since, according to Eq. [45], $p_0 = (1 - \kappa h_{00})p^3 \simeq p^3/(1 + \kappa h_{00})$, we see that ω is indeed constant as the wave propagates through the gravitational field, that is, the light wave propagates without change in frequency. As we learned in Section 4.2, the redshift of light reaching us from an emitter in a gravitational field is due to a slowing down of clocks when placed in a gravitational field. The redshift is not due to a gradual change of frequency as the light propagates through the gravitation field; rather, it is due to a reduction of the frequency of oscillation of the emitter when placed deep in a gravitational potential (in this context, "frequency" means frequency as reckoned by a clock placed at large distance from the gravitational field, and therefore not affected by the field).

TABLE 4.5 EVIDENCE FOR CURVED SPACETIME

Experimental observation	Implication
Equality of inertial and gravitational mass	Leads to the construction of the tensor field equation for gravitation and the equation of motion, which compels us to accept the geometric interpretation of the tensor field.
Deflection of light and retardation of light	Shows that speed of light c measured with instruments placed in the gravitational field disagrees with speed measured by instruments at infinity. This contradicts flat spacetime.
Gravitational time dilation	Shows that clocks placed in the gravitational field run slow as compared with clocks at infinity. This contradicts flat spacetime.

Thus, the experimental observations of the deflection of light and of the retardation of light signals constitute direct evidence for the curvature of spacetime. From a purely phenomenological point of view, these experiments may be regarded as measurements of the ratio g_{00}/g_{kk} since this is what enters into the speed formula, Eq. [52]. The experiments, of course, lend support to the values of g_{00} and g_{kk} given by the linear approximation.

Table 4.5 summarizes the several lines of evidence that converge to the conclusion that spacetime is curved.

4.5 GRAVITATIONAL LENSES

When rays of light from a distant source are deflected by the gravitational field of a mass, the rays that pass on opposite sides of the mass intersect at some large distance beyond (see Fig. 4.11). An observer placed at such a large distance will simultaneously see the source at two locations in the sky, that is, she will see two images of the source. For rays deflected by the Sun, the nearest point at which these rays intersect, and two images become visible, is about 50 light-years; hence, on the Earth, we cannot see any of the multiple images produced by light deflection in the gravitational field of the Sun. However, we can expect to see the multiple images produced by light deflection in the gravitational field of distant stars or distant galaxies.

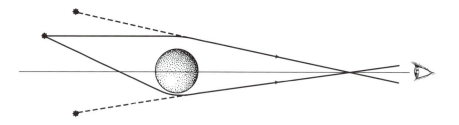

Fig. 4.11 The observer sees the source at two distinct locations.

Stars or galaxies that produce multiple images are called *gravitational lenses*. However, such "lenses" do not have the optical properties of genuine lenses. The magnitude of the deflection angle of a ray passing through a genuine lens increases in direct proportion to the impact parameter (with the sign convention used in Eq. [38], the deflection angle is negative for a convex lens, and positive for a concave lens). When a bundle of (almost) parallel rays from a very distant source is incident on a genuine lens, the rays either converge to a point (convex lens, real image) or they diverge so their backward extrapolations appear to come from a point (concave lens, virtual image). But the magnitude of the deflection angle of a ray passing through the gravitational field of a star *decreases* with impact parameter (see Eq. [38]). Hence such a gravitational lens has no well-defined focal length, and it cannot produce genuine images, real or virtual. When a bundle of parallel rays passes through this gravitational lens, the rays diverge, but their backward extrapolation has no unique point of intersection (see Fig. 4.12), and therefore rays do not form a well-defined image point in the sky. Thus, the multiple "images" generated by gravitational lenses are nothing but multiple directions of incidence of rays on the eye or on the telescope of the observer. The "images" are like the rainbow produced by raindrops, which also corresponds merely to a direction of incidence of light on the observer, not a genuine image in the sky. If the observer shifts position, the apparent position of the rainbow also shifts. For the raindrops, as for gravitational lenses, the only genuine images are those produced by the lens of the observer's eye or a telescope on her retina or on a photographic film, where each of the multiple directions of incidence of rays registers as an image.

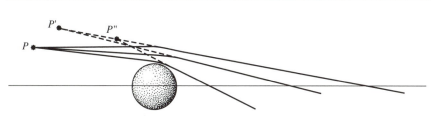

Fig. 4.12 Backward extrapolation of the rays fails to lead to a unique point of intersection.

If the source of light, the deflecting star, and the observer are aligned exactly, all the rays that pass at the appropriate impact parameter around the deflecting star, at any azimuth, reach the position of the observer. Under these exceptional circumstances, the observer sees an infinite number of images, which form a ring around the deflecting star, called the *Einstein ring* (such rings were first discussed by Chwolson, 1924). Let us assume that the source of light is much farther from the deflecting star than the observer. The rays incident on the deflecting star are then nearly parallel to the line of alignment, and the deflection angle required for the ray to reach the observer is b/D, where D is the distance from the deflecting star to the observer. Thus,

$$\frac{b}{D} = \frac{4GM}{b}$$

which can also be written as

$$\frac{b}{D} = \sqrt{\frac{4GM}{D}} \qquad [54]$$

This tells us the angular radius of the Einstein ring. For a typical star in our galaxy, with $M \simeq M_\odot$ and $D \simeq 10^4$ light-years, the angular radius of the Einstein ring is

$$\frac{b}{D} \simeq 8 \times 10^{-9} \text{ radian} \simeq 2 \times 10^{-3} \text{ arcsec} \qquad [55]$$

This is the characteristic angle for the gravitational-lens phenomenon produced by stars. This angle gives us not only the size of the Einstein ring, but also the typical angular separation between the two images, and the maximum permissible deviation from alignment if two images are to be seen (Refsdal, 1964; Liebes, 1964).

An angle of 10^{-3} arcsec is beyond the limit of resolution of the existing optical telescopes. Thus, the only observable effect of the alignment of a deflecting star and a pointlike source--such as another, more distant star--and the Earth would be the increase in the intensity that we would perceive when the light from all the azimuthal directions is deflected toward us. But the chances for a sufficiently close alignment of two stars are not good. Furthermore, for the observational demonstration of the lens effect, it is not enough to measure the intensity at one time--we need to measure the intensity as a function of time, during the transit of one star in front of another. Such a transit takes years, and we cannot afford to monitor the intensity of a distant star over many years in the hope that a foreground star will align with it and display the lens effect.

A much better opportunity for observation of the gravitational-lens effect is provided by galaxies. Light or radio waves passing outside of a galaxy will be deflected in much the same way as light passing by an individual star. In fact, the deflection produced by a galaxy can be calculated by performing a (vector) sum of the deflections produced by all the stars or mass elements in the galaxy (see the next section). Since the mass distribution of the galaxy is not spherical, the details of the deflection of light are somewhat more complicated than for the spherically symmetric gravitational field of a single star. However, the simple formula [54] gives us an order-of-magnitude estimate of the deflection. For a typical galaxy at a typical cosmological distance, with $M \simeq 10^{11} M_\odot$ and $D \simeq 10^{10}$ light-years, we obtain

$$\frac{b}{D} \simeq 1 \text{ arcsec} \qquad [56]$$

Such an angle is at the limit of resolution of optical telescopes, but it is well within the resolution attainable by radio telescopes. Thus, the Einstein ring as well as multiple images should be easily observable with radio telescopes. Furthermore, since an alignment to within a few arcseconds suffices for the production of multiple images, the probabilities for alignments of a galaxy lens are much better than for a star lens. With the current estimates for the overall density of galaxies in the universe, it is expected that at least 1 galaxy in 10 is sufficiently well aligned with another background galaxy to produce multiple images of the latter (Press and Gunn, 1973).

When a galaxy acts as a gravitational lens, it not only produces deflections of the rays passing outside it, but also of rays passing *through* the mass distribution of the galaxy. Since the distances between the stars in a galaxy are much larger than the diameters of the stars, the chance of interception of a ray of light by the body of a star is negligible, and the galaxy is effectively transparent to light rays. It is therefore possible for rays to reach the observer by several paths,

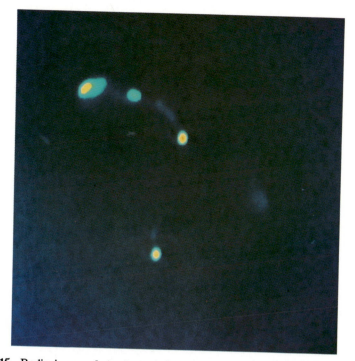

Fig. 4.15 Radio image of the "twins" Q0957+561 A and B discovered by Walsh et al. (1979). The foreground galaxy G1 that acts as gravitational lens can be seen faintly just above the lower image B. The radio lobe that extends outward from the upper image A fails to produce a twin radio lobe because it is too far from the lens galaxy. (J. N. Hewitt, Massachusetts Institute of Technology.)

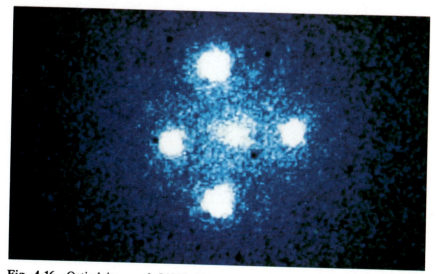

Fig. 4.16 Optical image of Q2237+031, called the "Einstein Cross," obtained with the Hubble Space Telescope. Four images of the quasar surround the central, fainter image of the galaxy that acts as gravitational lens. (F. D. Macchetto, NASA/ESA.)

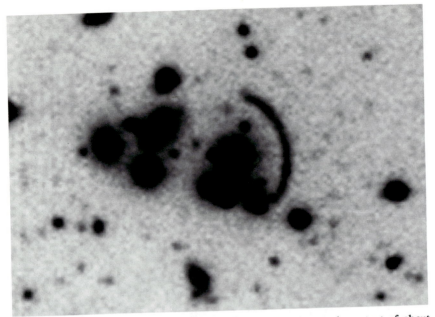

Fig. 4.17 A large blue arc in Cl 2244-02. The arc has an angular extent of about 110°. It is the image of a distant galaxy distorted by a foreground cluster of galaxies acting as gravitional lens. (V. Petrosian, Stanford University, and R. Lynds, NOAO.)

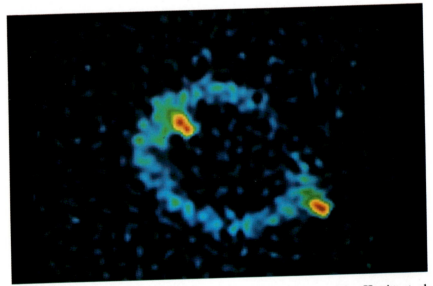

Fig. 4.18 The spectacular radio ring MG1131+0456 discovered by Hewitt et al. (1988). This is the image of a compact radio source distorted by a foreground galaxy (not seen on this radio map) acting as gravitational lens. The good alignment of the radio source and the lensing galaxy gives rise to a nearly complete Einstein ring. The picture on the front jacket shows the same radio ring, imaged at a different wavelength. (J. N. Hewitt, Massachusetts Institute of Technology.)

Fig. 4.19 Numerous small blue arcs encircle the core of the rich cluster Abell 2218. The yellowish galaxieş are members of the cluster. The blue arcs (note especially the two in the upper right quadrant) are the distorted images of much more distant blue galaxies located behind the cluster. (V. Petrosian, Stanford University, and R. Lynds, NOAO.)

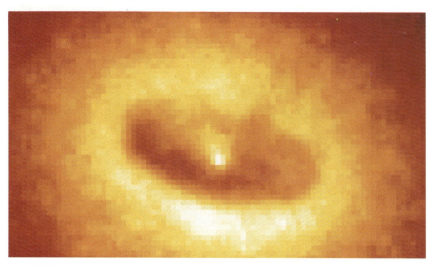

Fig. 8.30 Large disk of cool dust and gas orbiting around the nucleus of the galaxy NGC 4261. The bright spot at the center is believed to arise from thermal emission by a much smaller hot accretion disk belonging to a black hole in the nucleus of the galaxy. (W. Jaffe, Leiden Observatory, H. Ford, JHU/STScI, and NASA.)

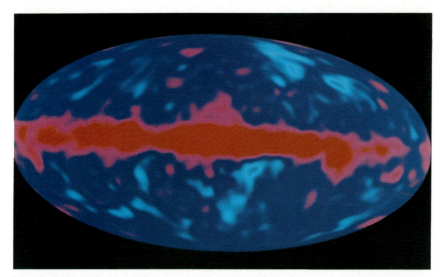

Fig. 9.10 (a) This is a microwave map of the whole sky, prepared from one year of data taken by the COBE differential radiometer. The $\cos \theta$ contribution due to the motion of the Earth has been subtracted. The intense red color indicates a temperature excess of 0.01% and the dark blue color a temperature deficit of 0.01%. The warm (red) band along the equator of this map is due to emission by our Galaxy.

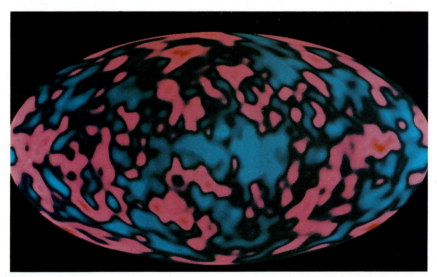

(b) In this map, both the $\cos \theta$ contribution and the galactic contribution have been subtracted. The residual variations revealed by the red and blue patches include both instrumental noise and also the actual temperature variations in the sky. Most of the patches in this map are the result of instrumental noise, rather than actual hot and cold spots. However, statistical analysis shows that the variations in this map are somewhat larger than the expected instrumental noise; thus, a faint cosmic signal is hidden within the patches. (G. F. Smoot, Lawrence Berkeley Laboratory.)

passing outside or inside the galaxy. The observer then sees multiple images of the source.

The number of images depends on the details of the mass distribution, and we will investigate the optical properties of the gravitational lenses with arbitrary, complicated mass distributions in the next section. But in all cases, the number of images produced by a transparent mass distribution is an odd number. This general theorem is most easily proved by appealing to wave optics. When the wavefronts pass through the gravitational field, they are deformed, because of the slowing down of the speed of light in the field. The deformed wavefront subsequently folds on itself (see Fig. 4.13). When the folded wavefront reaches the observer, it gives rise to the multiple images. Each layer in the folded wavefront has a different direction of propagation (different wave vector), and when the folded, layered wavefront passes over the observer, each layer is perceived as arriving from a different direction--hence the observer sees images in different directions. Since the portions of the wavefront far from the gravitational field are more or less undisturbed, the number of layers in each folded wavefront is necessarily odd*--and the observer sees an odd number of images. Note that this theorem does not apply to the

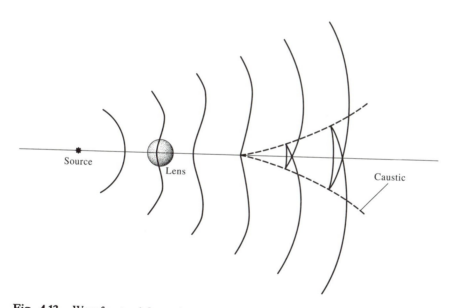

Fig. 4.13 Wavefronts deformed by the gravitational field fold upon themselves. The number of layers of any of the folded wavefronts passing over the observer is odd. The edge of the fold of the wavefront traces out a surface in space (dashed lines).

---

* You can easily convince yourself of this by crumpling and folding the central part of a (stretchy) bedsheet while keeping the edges of the sheet more or less fixed.

images produced by an individual star, since the star is not transparent and blocks out some rays (that is, the star cuts holes in the wavefront, and we then cannot make any self-evident assertion about the number of layers in the wavefront). However, the theorem appplies to the case of a galaxy, except perhaps a galaxy with heavy dust clouds or a dense galactic nucleus, which might cut holes into the wavefront.

The edge of the fold of the traveling wavefront traces out a singular surface in space, called the *caustic surface* (see Fig. 4.13). The layers of the wavefront are tangent to each other at the fold, that is, their rays are parallel, and they are tangent to the caustic surface. Hence, the caustic surface is a surface of tangency of rays, that is, it is an envelope of rays.

On the caustic surface, the rays are concentrated and the intensity of light is exceptionally high. Fig. 4.14 illustrates this concentration of rays. Consider a narrow conical bundle of rays emitted by the source on the left. Initially, the cross section of this bundle is circular, but it stretches into into an ellipse as the bundle propagates through the lens. The ellipse continues to stretch until it reaches a caustic surface, where the ellipse collapses into a line segment lying flat on the surface. Thus, at each caustic surface, the bundle of rays becomes a ribbon, tangent to the surface. Note that at each caustic surface, the gravitational lens achieves this concentration of rays for only one of the transverse dimensions of the bundle. In contrast, a genuine lens, at its real image point, achieves simultaneous concentration for both of the transverse dimensions of such a bundle. The real image point formed by a genuine lens is a caustic point, which can be regarded as an extreme, degenerate form of a caustic surface, a caustic surface that has contracted to a point.

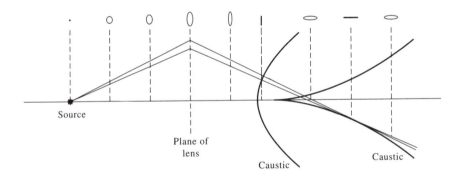

Fig. 4.14 A narrow conical bundle of rays is emitted by the source on the left. In the case illustrated here, the gravitational lens has two distinct caustic surfaces (for a more complete three-dimensional view of the shape of the caustic surfaces of a typical gravitational lens, see Fig. 4.22). The bundle of rays first converges tangentially, forming a radial (vertical) ribbon at the first caustic surface; it then converges radially, forming a tangential (horizontal) ribbon at the second caustic surface. The diagrams at the top indicate the cross sections of the bundle at different locations.

The caustic surfaces of a gravitational lens separate the regions of space where the observer sees different numbers of images; each time the observer crosses a caustic surface, the number of images she sees increases or decreases by two. For instance, if the observer approaches the caustic surface illustrated in Fig. 4.13 from inside, she will see that two of the three images merge, become exceptionally bright on the caustic, and then disappear.

The most favorable configuration for the observation of multiple images is the alignment of a quasar with a galaxy. Quasars are intense, almost pointlike sources, and they are therefore likely to yield sharp, nonoverlapping multiple images. In fact, all of the instances of multiple images discovered so far are images of quasars. The first such multiple image was discovered in 1979 by Walsh, Carswell, and Weyman (Walsh et al., 1979), who noticed that the two quasars Q0957+561 A and B had identical optical and radio spectra, with the same redshifts. (The redshift is a Doppler shift of the light, resulting from the motion of expansion of the universe. A large redshift means a large distance. For more about this, see Chapter 9.) Walsh et al. proposed that these "twins" were two images of a single quasar formed by some invisible foreground galaxy acting as gravitational lens (Fig. 4.15; see color plate). Subsequently, a faint galaxy was discovered in roughly the right position for gravitational lensing. This galaxy is part of a cluster of more than 100 galaxies, all of which also contribute to the gravitational-lens effect of the "twins." The positions and the intensities of the images can be accounted for with a plausible theoretical model for the mass distribution of the galaxies (for example, Young et al., 1981). The theoretical model predicts that a third image should also be present. But this image is faint and overlaps the lensing galaxy; it is therefore not readily detectable.

Several more instances of gravitational lensing of quasar images have been discovered, most as a result of a systematic search program with the VLA radio telescope. We now know of about a dozen confirmed cases of double and multiple images and another dozen probable and tentative cases. Good optical images have been obtained for a few of these (Fig. 4.16; see color plate). Table 4.6 lists some of the confirmed cases. Note that in almost all of these cases of gravitational lensing, the number of observed images is even, in contradiction to the general theorem that the number should be odd. There are several possible explanations for this discrepancy: the missing image or images might be too faint to be seen, or an image might merge with another image (when near a caustic), or an image might be blocked out by a cloud of dust in the lensing galaxy.

If the source is not a pointlike quasar, but a galaxy or a radio source with a disk of noticeable extent, the gravitational lens distorts the shape and the size of the source. The distortion becomes severe near the caustic. For instance, a disklike source will be distorted into one or several arcs; and if the alignment is exceptional, the arcs spread and merge to form an annular Einstein ring.

Several cases of such arcs and annular Einstein rings have been discovered. Fig. 4.17 shows a large arc and Fig. 4.18 (see color plate) shows an almost complete Einstein ring. These represent cases of alignment of an extended source with a foreground galaxy or cluster of galaxies that acts as lens. Several cases of swarms of small arcs, or arclets, have also been discovered. Fig. 4.19 (see color plate) gives one example of such a swarm. The small blue arcs are the images of a multitude of distant blue galaxies more or less aligned with a rich, massive foreground cluster of galaxies that acts as lens.

Long before the discovery of any gravitational lenses, Zwicky (1937) speculated on their possible applications to investigations of distant galaxies. Today, gravitational lenses are quickly becoming an important tool for the investigation of galaxies, quasars, and cosmology. Among the interesting results that have already been obtained is a determination of the Hubble constant by means of time delays measured in light signals in the "twins," an estimate of quasar size by means of fluctuations observed in the "twins," and a determination of the distribution of dark mass in clusters of galaxies.

The light from the "twin" quasar travels to the Earth along two different paths, of different lengths for each image. Furthermore, where the paths penetrate through the gravitational field of the lens, the effective speeds of light are different. Hence, the travel times for the light arriving from the two images are different. To measure the difference in travel time, astronomers exploit the spontaneous intensity changes that sometimes occur in quasars. As observed on the Earth, the intensity changes seen in the two images will be separated by a time delay equal to the difference in the travel times. The observed time delay between the images A and B in the "twins" is about 540 days (Hewitt, 1993; Schild, 1990; Vanderriest et al., 1989; for objections, see Falco et al., 1990), with image A leading image B. If the overall geometry of the light rays is assumed known from a theoretical model of the mass distribution of the lensing galaxies, then the difference in travel times fixes the overall scale of the distances to the lensing galaxies and to the source quasar. The scale of distances is the essential ingredient for the calculation of the Hubble constant (the Hubble constant equals the speed of recession of a galaxy or a quasar divided by its distance; the speed of recession is easily determined from the redshift). The value of the Hubble constant implied by the data from the "twins" is in the range from 35 to 90 km/s·Mpc* (Hewitt, 1992). The large range of values arises from the uncertainties in the mass distribution of the lensing galaxies. Unfortunately, the gravitational lens that produces the "twins" is rather complicated, and its mass distribution is not determined unambiguously by the configuration of the images. It is hoped that if and when time delays are observed in other multiple quasar images, with simpler gravitational

---

* The Mpc, or megaparsec, is a unit of distance, 1 Mpc $= 3.26 \times 10^6$ light-years $= 3.08 \times 10^{24}$ cm.

lenses, we will be able to extract a more precise value for the Hubble constant.

An image formed by rays passing through a galaxy is expected to exhibit intensity fluctuations when stars moving within this galaxy pass close to the rays and produce extra, time-dependent deflections in excess of the deflection contributed by the average, smoothed mass distribution of the galaxy (Chang and Refsdal, 1979; Young, 1981). Such fluctuations caused by individual stars are called *microlensing*, in contrast to deflection caused by the average mass distribution, called *macrolensing*. An event of microlensing has been observed in the "twin" quasar, whose image B suffered a fluctuation of intensity while image A did not (Irwin et al., 1989; the time delay of 540 days must of course be taken into account when comparing the intensities of the images). Microlensing is strongly suppressed if the size of the disk of the source exceeds the size of the Einstein ring of the microlensing star, since in that case the light merely gets rearranged within the disk, with no net loss of light from the disk. Hence, from the observation of a fluctuation of intensity, astronomers have obtained the first direct estimate for the size of the disk of a quasar--the upper limit for the size is about 10^{15} cm.

In general, the intensity fluctuations caused by microlensing depend on the distribution of stars in the lensing galaxy, their masses, their speeds, and on the size of the luminous disk of the source. By monitoring the intensity fluctuations, we can extract information about all the parameters that affect the intensity. The hypothetical dark mass in intergalactic space (see Chapter 9) also contributes to the gravitational-lens effect, and, if the dark mass has lumps, it will cause intensity fluctuations. Observation of intensity fluctuations originating from such microlensing in intergalactic space would give us direct observational evidence for dark matter.

We already have direct observational evidence for dark matter from the study of the swarms of "arclets" seen in some gravitational lenses. As mentioned above, these arclets are distorted images of background galaxies formed by a cluster that acts as lens. From the analysis of the amount of distortion as a function of radial distance from the center of the cluster, it is possible to calculate the cluster mass (Tyson et al., 1990). The mass deduced in this way is larger than the sum of the masses of the luminous galaxies in the cluster; thus, there must be dark matter in the cluster. The amount of dark matter deduced from this analysis of gravitational lensing is in good agreement with other, dynamical methods for the determination of the masses of clusters.

TABLE 4.6 GRAVITATIONAL LENSES*

Name	Source redshift**	Deflector redshift	Number of images	Max. image separation
Double Images				
Q0957+561	$z = 1.41$	$z = 0.36$	2	6.1 arcsec
Q0142-100	$z = 2.72$	$z = 0.49$	2	2.2
Multiple Images				
Q1115+080	$z = 1.72$	?	4	2.3 arcsec
Q2016+112	$z = 3.27$	$z = 1.01$	3	3.8
Q2237+031	$z = 1.69$	$z = 0.039$	4	1.8
H1413+117	$z = 2.55$	?	4	1.1
Q0414+053	$z = 2.63$	?	4	3.0
Arcs				
Abell 370	$z = 0.72$	$z = 0.37$	large arc and many arclets	
Cl 2244-02	$z = 2.23$	$z = 0.33$	large arc	
Abell 963	$z = 0.77$	$z = 0.21$	two large arcs	
Abell 2218	$z = 0.70$	$z = 0.17$	many arclets and a ring	
Cl 0024+17	?	$z = 0.39$	two large arcs and many arclets	
Abell 2390	$z = 0.92$	$z = 0.23$	"straight" arc and several arclets	
Radio Rings				
MG1131+0456	?	?		2.2 arcsec
MG1643+1346	$z = 1.75$	$z = 0.25$		2.1
PKS1830-211	?	?		1.0
MG1549+3047	?	$z = 0.11$		1.8
0218+357	?	?		0.3

* Based on Fort (1990), Blandford and Narayan (1992), and Hewitt (1993).
** The redshift factor z listed in this column is the fractional change of the wavelength of light, $z = \Delta\lambda/\lambda$.

4.6 OPTICS OF GRAVITATIONAL LENSES

To determine the optical properties of gravitational lenses we must examine the deflection of a ray that passes by or through an arbitrary mass distribution. Suppose that the source is located at a distance $z = D_1$, the deflecting mass at $z = 0$, and the observer at $z = -D_2$, on the z-axis (see Fig. 4.20).* The transverse coordinates of the ray are x, y in the plane of the source and x', y' in the plane of the deflecting mass. The ray suffers simultaneous deflections α_x and α_y in the x and the y directions. These deflections can be expressed in terms of the gradients of the Newtonian potential, as in Eq. [40]:

$$\alpha_x = -2 \int_{-\infty}^{\infty} \frac{\partial \Phi}{\partial x'}\, dz' \qquad \alpha_y = -2 \int_{-\infty}^{\infty} \frac{\partial \Phi}{\partial y'}\, dz' \qquad [57]$$

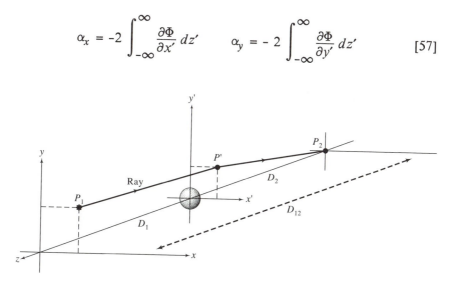

Fig. 4.20 Source, deflector, and observer with z coordinates $z = D_1$, $z = 0$, and $z = -D_2$, respectively. The observer is at P_2 on the z-axis; the source is at P_1 off the z-axis, with coordinates x, y.

For convenience, we combine these two components into a complex angle (Bourassa and Kantowski, 1975):

---

* For large, cosmological distances, it is necessary to take into account the overall deviations of the background geometry from flat Euclidean geometry and also the expansion of the universe. It can be shown that the formulas of this section remain valid even in a curved, expanding universe, provided the distances D_1 and D_2 are interpreted as "angular-diameter" distances. For instance, the distance D_2 is defined as the actual diameter of the deflecting mass divided by its apparent angular size as we see it in the sky.

$$\alpha = \alpha_x + i\alpha_y \qquad [58]$$

If we then define a complex scattering function $I(x', y')$ as

$$I(x', y') \equiv \mathrm{Re}(I) + i\mathrm{Im}(I) \equiv \int_{-\infty}^{\infty} \left(\frac{\partial \Phi}{\partial x'} - i\, \frac{\partial \Phi}{\partial y'} \right) dz' \qquad [59]$$

we obtain a simple formula for the complex angle α:

$$\alpha = - 2[\mathrm{Re}(I) - i\mathrm{Im}(I)]$$

$$= - 2I^*$$

The scattering function $I(x', y')$ can be expressed as an integral over the density of the deflecting mass. From the familiar formula for the Newtonian potential,

$$\Phi(\mathbf{x}') = - G \int \frac{\rho(\xi, \varsigma, \chi)}{[(x' - \xi)^2 + (y' - \varsigma)^2 + (z' - \chi)^2]^{1/2}}\, d\xi d\varsigma d\chi \qquad [60]$$

we obtain

$$\frac{\partial \Phi}{\partial x'} = G \int \frac{(\xi - x')\rho(\xi, \varsigma, \chi)}{[(x' - \xi)^2 + (y' - \varsigma)^2 + (z' - \chi)^2]^{3/2}}\, d\xi d\varsigma d\chi \qquad [61]$$

and when we integrate this over z', we find

$$\int_{-\infty}^{\infty} \frac{\partial \Phi}{\partial x'}\, dz' = G \int \frac{(\xi - x')\rho(\xi, \varsigma, \chi)}{(x' - \xi)^2 + (y' - \varsigma)^2}\, d\xi d\varsigma d\chi \qquad [62]$$

Furthermore, when we integrate the mass density $\rho(\xi, \varsigma, \chi)$ over χ, we obtain the projected surface mass density in the plane of the deflecting mass:

$$\int \rho(\xi, \varsigma, \chi)\, d\chi = \sigma(\xi, \varsigma)$$

Therefore

$$\int_{-\infty}^{\infty} \frac{\partial \Phi}{\partial x'}\, dz' = 2G \int \frac{(x' - \xi)\sigma(\xi, \varsigma)}{(x' - \xi)^2 + (y' - \varsigma)^2}\, d\xi d\varsigma \qquad [63]$$

With a similar expression for $\int (\partial \Phi / \partial y')dz'$, we arrive at the following result for the scattering function I:

$$I(x', y') = 2G \int \frac{(x' - \xi) - i(y' - \varsigma)}{(x' - \xi)^2 + (y' - \varsigma)^2}\, \sigma(\xi, \varsigma)\, d\xi d\varsigma$$

$$= 2G \int \frac{\sigma(\xi, \varsigma)}{(x' - \xi) + i(y' - \varsigma)}\, d\xi d\varsigma \qquad [64]$$

For a given projected surface mass density, the evaluation of this integral is straightforward in principle, but tedious in practice. Explicit formulas for I can be obtained for some fairly simple mass distributions with spheroidal symmetry (Bourassa and Kantowski, 1975).

From the deflection angle α, we can calculate the angular displacement of the image relative to the source. According to Fig. 4.20, the deflection angle is the difference between the angles of the lines $P_1 P'$ and $P_2 P'$:

$$\alpha_x = \frac{x - x'}{D_1} - \frac{x'}{D_2} = \frac{x}{D_1} - \frac{x' D_{12}}{D_1 D_2}$$

where $D_{12} = D_1 + D_2$. This gives

$$\frac{x'}{D_2} = \frac{x}{D_{12}} - \alpha_x \frac{D_1}{D_{12}} \qquad [65]$$

and, likewise,

$$\frac{y'}{D_2} = \frac{y}{D_{12}} - \alpha_y \frac{D_1}{D_{12}} \qquad [66]$$

Here x/D_{12}, y/D_{12} and x'/D_2, y'/D_2 are the angular positions at which the observer sees the (undeflected) source and the image, respectively. Thus, as seen in the sky, the complex angular displacement of the image relative to the source is $- \alpha D_1/D_{12}$. In terms of the scattering function $I(x', y')$, the angular displacement of the image relative to the source is $+ 2I^* D_1/D_{12}$. Fig. 4.21 illustrates this angular displacement.

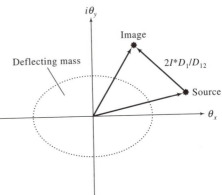

Fig. 4.21 In this diagram, the angular positions of source and the image in the sky are represented by complex angles of the form $\theta_x + i\theta_y$. The angular displacement of the image relative to the source is $2I^*D_1/D_{12}$.

The scattering function tells us not only the position of the image (or images) relative to the source, but also the brightness of the image relative to the brightness of the source. Suppose that when the gravitational lens is absent, the source is a small (infinitesimal) disk whose surface brightness (energy per unit solid angle and unit time as seen from the position of the observer) is uniform. Suppose that the displacement from the center of this disk to a point at its edge is δx, δy. For small displacements, Eqs. [65] and [66] give linear relations among δx, δy and $\delta x'$, $\delta y'$.

$$\frac{\delta x'}{D_2} = \frac{\delta x}{D_{12}} - \delta x' \frac{\partial \alpha_x}{\partial x'} \frac{D_1}{D_{12}} = \frac{\delta x}{D_{12}} + 2\delta x' \frac{\partial}{\partial x'} \mathrm{Re}(I) \frac{D_1}{D_{12}} \qquad [67]$$

$$\frac{\delta y'}{D_2} = \frac{\delta y}{D_{12}} - \delta y' \frac{\partial \alpha_y}{\partial y'} \frac{D_1}{D_{12}} = \frac{\delta x}{D_{12}} - 2\delta y' \frac{\partial}{\partial y'} \mathrm{Im}(I) \frac{D_1}{D_{12}} \qquad [68]$$

We can solve these equations for the angular displacements $\delta x'/D_2$, $\delta y'/D_2$ of the edge of the image relative to its center. In matrix form, the result is

$$\begin{pmatrix} \delta x'/D_2 \\ \delta y'/D_2 \end{pmatrix} = \frac{1}{F_1^2 - |F_2|^2} \begin{pmatrix} F_1 + \mathrm{Re}(F_2) & -\mathrm{Im}(F_2) \\ -\mathrm{Im}(F_2) & F_1 - \mathrm{Re}(F_2) \end{pmatrix} \begin{pmatrix} \delta x/D_{12} \\ \delta y/D_{12} \end{pmatrix} \qquad [69]$$

where

$$F_1 = 1 - \left[\frac{\partial I}{\partial x'} + i \frac{\partial I}{\partial y'} \right] \frac{D_1}{D_{12}}$$

$$F_2 = \left[\frac{\partial I}{\partial x'} - i \frac{\partial I}{\partial y'} \right] \frac{D_1}{D_{12}} \qquad [70]$$

■ *Exercise 3.* Obtain Eq. [69] from Eqs. [67] and [68]. ■

The eigenvalues of the matrix appearing on the right side of Eq. [69] tell us by what factor the semiaxes of the image ellipse are enlarged relative to the radius of the source disk. These eigenvalues are

$$\frac{1}{F_1 - |F_2|} \quad \text{and} \quad \frac{1}{F_1 + |F_2|} \tag{71}$$

■ *Exercise 4.* Verify that the eigenvalues of the matrix are as stated in Eq. [71]. ■

The product of these eigenvalues is proportional to the solid angle subtended by the image. Thus, the solid angle subtended by the image is enlarged relative to the solid angle subtended by the source by a factor of

$$\frac{1}{F_1^2 - |F_2|^2} \tag{72}$$

The surface brightness (energy per unit solid angle and unit time) of the image is the same as the surface brightness of the source.* Hence the enhancement [72] of solid angle leads to a corresponding enhancement of the intensity of the image relative to the intensity of the source. When we observe quasars, the sources are almost pointlike, and so are the images produced by the gravitational lens. The distortions of the shapes of the images are therefore not directly observable; however, the sizes of the images affect their relative intensities, which are observable.

If the denominator in Eq. [72] vanishes, the intensity predicted by this equation becomes infinite. The equation $F_1^2 - |F_2|^2 = 0$ determines when the observer is on a caustic surface. Although our simple calculation, based on geometrical optics, predicts an infinite intensity on the caustic surface, a more careful calculation, based on wave optics, establishes that the intensity on the caustic surface is limited by diffraction effects, and it remains finite (Ohanian, 1983; Chang, 1984; Schneider, Ehlers, and Falco, 1992a, Chapter 7).

In the calculations of this section we have assumed that the observation point is fixed on the z-axis, whereas the source point has variable coordinates x, y, and $z = D_1$. Hence the equation $F_1^2 - |F_2|^2 = 0$ actually gives the locus of those source points that place the observation point on a caustic. We will call the locus of such source points the *conjugate caustic surface*. The conjugate caustic is actually the caustic

---

* This can be shown by a simple application of the Liouville theorem.

that the lens would form if we were to place a light source at the position of the observer. This follows from optical reversibility--if the gravitational field is time independent, we can retrace the rays in Fig. 4.20 from left to right, and the rays that diverge from the position of the observer on the old caustic will then reconverge at the position of the source. Thus, we have a reciprocal relationship: if the observer is on the caustic generated by the source, then the source is on the caustic generated by a source placed at the position of the observer (this is analogous to the reciprocal relationship between object and image points for lenses in optics: if we place a source at the image point, the lens forms an image at the object point). If the source and the observer are symmetrically located in relation to the lens, then the conjugate caustic surface has the same shape and size as the caustic surface--the equations of the two surfaces are related by the simple exchange $z \leftrightarrow - z$; but if the source and observer are not symmetrically located, then the shapes and sizes of the caustic and the conjugate caustic surfaces are different. In discussions of gravitational lenses, it has become customary to examine the conjugate caustic, rather than the caustic. Unfortunately, it has also become customary to call this conjugate caustic simply the "caustic," which can be somewhat confusing.

The shape of the conjugate caustic surface or surfaces can be calculated from the mass distribution in the lens and the equation $F_1^2 - |F_2|^2 = 0$. In general, these surfaces are complicated, with several intersecting smooth curved pieces. Fig. 4.22 illustrates the conjugate caustic surfaces associated with an ellipsoidal mass distribution of uniform density (such a mass distribution is a simple, but crude, model for a galaxy). For this mass distribution there are two caustic surfaces. At the far left, the larger of these surfaces is roughly an ellipse in cross section; the smaller surface is an astroid with four spikes. These two caustic surfaces correspond to the two ways that a bundle of rays can be focused: tangentially or radially (compare Fig. 4.14).

Fig. 4.22 Conjugate caustic surfaces for a lens consisting of an ellipsoidal mass. (After Blandford and Narayan, 1986.)

Although the shape of the conjugate caustic surface in Fig. 4.22 is rather complicated, it consists of several basic building blocks, which can be identified and classified by the mathematical theory of catastrophes. The basic building blocks are called the *elementary catastrophes*. Fig. 4.22 includes three of the five possible elementary catastrophes that can occur in caustics. The simplest and most pervasive elementary catastrophe is the *fold*, which makes up all the smooth, curved lateral surfaces in Fig. 4.22, with the exception of the points or lines of intersection of one smooth surface with another. The terminology for the fold is in accord with the discussion in the preceding section, where we saw that the smooth surfaces correspond to those points where the wavefront has a fold, that is, the wavefront is folded back on itself (a double layer of wavefronts). Another elementary catastrophe discernible in Fig. 4.22 is the *cusp*, where two fold catastrophes merge into each other. At the cusp, two folded wavefronts merge together, which amounts to a more complicated folding of the wavefront upon itself (a triple layer of wavefronts). The third elementary catastrophe in Fig. 4.22 is the *hyperbolic umbilic*, where a cusp of the astroid merges with the fold of the ellipsoid (a quadruple layer of wavefronts). The two other possible elementary catastrophes (not shown in Fig. 4.22) are the *swallow tail*, which involves the merging of two cusps at a point; and the *elliptic umbilic*, which involves the merging of three cusps at a point.

As remarked in the preceding section, the caustic surfaces divide space into regions in which the observer sees different numbers of images. The conjugate caustics divide space into regions in which the source must be located to produce different numbers of images. Whenever the source crosses one of the conjugate caustic surfaces, the number of images the observer sees increases or decreases by two. For the case illustrated in Fig. 4.22, the observer sees one image if the source is outside of the large caustic; she sees three images if the source is between the large and the small caustic; and she sees five images if the source is inside the small caustic. More complicated mass distributions give rise to more caustic surfaces and more multiple images. But, if the mass distribution is transparent, then the number of images must always be odd, in agreement with the general qualitative argument given in the preceding section.

The result [72] for the enhancement of the intensity is valid only for a small (infinitesimal) source. For an extended source, the shape of the image must be calculated by mapping the boundary of the source point by point into the image plane, by means of Eq. [66]. If the source is at or near a conjugate caustic, the image suffers a severe distortion, and its solid angle becomes much larger than that subtended by the source. For example, Fig. 4.23a shows the image seen when the

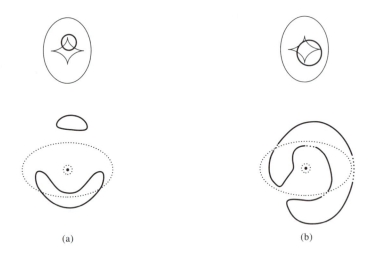

(a)   (b)

Fig. 4.23 Diagrams at the top show the position of the source disk in relation to the conjugate caustic. Diagrams at the bottom show the corresponding images. (a) If a disklike source is near the cusp of the conjugate caustic, the image is an elongated arc. (b) If the source is almost centered on the conjugate caustic, the image is an almost complete annular ring. (From Blandford and Narayan, 1992.)

source, in the shape of a disk, overlaps the cusp of the conjugate caustic; the image displays one small and one large arc, similar to the arc seen in the radio source Cl 2244-02. Fig. 4.23b shows the image seen when the source is almost centered on the conjugate caustic; the image then displays an almost complete annular Einstein ring, similar to the ring seen in the radio source MG1131+0456.

4.7 THE FIELD OF A ROTATING MASS; LENSE-THIRRING EFFECT

The gravitational field surrounding a rotating mass differs from that surrounding a nonrotating mass. We can understand this by analogy with the case of a rotating, uniformly charged sphere; such a sphere produces both electric and magnetic fields whereas a nonrotating sphere produces only an electric field.

Since the rotations of the Sun and of the Earth are quite slow (nonrelativistic), the effects of the rotation on the gravitational fields and on the motion of particles in these fields are very small. However, a high-precision experiment now under development will attempt to detect the effect of the rotation of the Earth on a gyroscope placed near the Earth. To study this effect, we need the solution for the gravitational field surrounding a nonrelativistic rotating mass. The general solution with relativistic rotation will be given in Section 8.4.

We assume that the rotation is steady, so the field is time independent. It is convenient to begin with the general solution of Eq. [3.55]. In the time-independent case, this equation reads

$$- \nabla^2 \phi^{\mu\nu}(\mathbf{x}) = - \kappa T^{\mu\nu}(\mathbf{x}) \tag{73}$$

and has the solution

$$\phi^{\mu\nu}(\mathbf{x}) = - \frac{\kappa}{4\pi} \int \frac{T^{\mu\nu}(\mathbf{x})}{|\mathbf{x} - \mathbf{x}'|} d^3x' \tag{74}$$

■ *Exercise 5.* Show that [74] solves Eq. [73]. ■

In order to evaluate the integral appearing in Eq. [74], we use the Taylor series expansion for $1/|\mathbf{x} - \mathbf{x}'|$ (see Eq. [1.21]) and keep only the first two terms:

$$\frac{1}{|\mathbf{x} - \mathbf{x}'|} \simeq \frac{1}{r} + \frac{x^k x'^k}{r^3} \tag{75}$$

This gives

$$\phi^{\mu\nu}(\mathbf{x}) \simeq - \frac{\kappa}{4\pi r} \int T^{\mu\nu}(\mathbf{x}') d^3x' - \frac{\kappa x^k}{4\pi r^3} \int x'^k T^{\mu\nu}(\mathbf{x}') d^3x' \tag{76}$$

The first integral in Eq. [76] has the value

$$\int T^{00}(\mathbf{x}') d^3x' = M \tag{77}$$

$$\int T^{k\mu}(\mathbf{x}') d^3x' = 0 \tag{78}$$

where M is the mass of the system. Eq. [77] is obvious and requires no comment. To derive Eq. [78], we begin with the conservation law

$$\frac{\partial}{\partial x'^0} T^{0\mu}(\mathbf{x}') + \frac{\partial}{\partial x'^l} T^{l\mu}(\mathbf{x}') = 0 \tag{79}$$

Since the matter distribution is assumed to be time independent, this reduces to

$$\frac{\partial}{\partial x'^l} T^{l\mu}(\mathbf{x}') = 0 \tag{80}$$

Next, we multiply Eq. [80] by x'^k and integrate over the volume of the system:

$$\int x'^k \frac{\partial}{\partial x'^l} T^{l\mu}(\mathbf{x}) \, d^3x' = 0 \qquad [81]$$

An integration by parts then gives Eq. [78].

■ *Exercise 6.* Check this. ■

The second integral appearing in Eq. [76] has the value

$$\int x'^k T^{00}(\mathbf{x}') \, d^3x' = 0 \qquad [82]$$

$$\int x'^k T^{l0}(\mathbf{x}') \, d^3x' = \tfrac{1}{2} \epsilon^{kln} S^n \qquad [83]$$

$$\int x'^k T^{ln}(\mathbf{x}') \, d^3x' = 0 \qquad [84]$$

where ϵ^{kln} is the object defined in Eq. [1.67], and S^n is the spin angular momentum of the system.

Eq. [82] simply states that the origin of coordinates is at the center of mass; this is an *assumption* which we make for the sake of simplicity.

The derivation of Eq. [83] proceeds as follows: We multiply Eq. [80] by $x'^k x'^n$ and we integrate over the volume of the system:

$$\int x'^k x'^n \frac{\partial}{\partial x'^l} T^{l\mu} \, d^3x' = 0 \qquad [85]$$

With an integration by parts, we obtain the identity

$$- \int x'^n T^{k\mu} \, d^3x' - \int x'^k T^{n\mu} \, d^3x' = 0 \qquad [86]$$

The spin angular momentum is the integral of the cross product of the position vector and the density of momentum. Thus, the x component of the spin angular momentum is defined as

$$S^1 = \int (x'^2 T^{30} - x'^3 T^{20}) \, d^3x' \qquad [87]$$

The other components of the spin angular momentum are defined similarly.

■ *Exercise 7.* Show that in general

$$S^n = \int \epsilon^{nkl} x^k T^{l0} \, d^3x'. \ ■ \qquad [88]$$

In view of the identity [86], each of the two terms in the integrand of Eq. [87] contributes the same amount, and therefore

$$S^1 = 2 \int x'^2 T^{30} \, d^3 x' \qquad [89]$$

This equation is exactly the same as the $k = 2$, $l = 3$ component of Eq. [83].

■ **Exercise 8.** Supply the proof for the other components of Eq. [83]. ■

Finally, to derive Eq. [84] we begin with the identity

$$T^{ln}(\mathbf{x}') = \frac{1}{2} \frac{\partial}{\partial x'^m} [x'^n T^{lm}(\mathbf{x}') + x'^l T^{nm}(\mathbf{x}')] \qquad [90]$$

■ **Exercise 9.** Derive this identity. (Hint: Use the conservation law and the symmetry of the energy-momentum tensor.) ■

Hence, with an integration by parts,

$$\int x'^k T^{ln}(\mathbf{x}') \, d^3 x' = \frac{1}{2} \int x'^k \frac{\partial}{\partial x'^m} (x'^n T^{lm} + x'^l T^{nm}) \, d^3 x'$$

$$= -\frac{1}{2} \int \delta^k_m (x'^n T^{lm} + x'^l T^{nm}) \, d^3 x'$$

$$= -\frac{1}{2} \int (x'^n T^{lk} + x'^l T^{nk}) \, d^3 x' \qquad [91]$$

The last expression vanishes as a consequence of the identity [86].

If we insert the results [77], [78], and [82]-[84] into Eq. [76], we obtain

$$\phi^{00}(\mathbf{x}) = -\frac{\kappa M}{4\pi r}$$

$$\phi^{l0}(\mathbf{x}) = \phi^{0l}(\mathbf{x}) = -\frac{\kappa}{8\pi r^3} \epsilon^{kln} x^k S^n$$

$$\phi^{ln}(\mathbf{x}) = 0 \qquad [92]$$

The corresponding expressions for $h_{\mu\nu}$ are

$$h_{00}(\mathbf{x}) = -\frac{2GM}{\kappa r}$$

$$h_{l0}(\mathbf{x}) = h_{0l}(\mathbf{x}) = \frac{2G}{\kappa r^3} \epsilon^{kln} x^k S^n$$

$$h_{ln}(\mathbf{x}) = -\frac{2GM}{\kappa r} \delta_l^n \qquad\qquad [93]$$

If the direction of the spin angular momentum coincides with the z-axis of our coordinate system, we obtain

$$h_{\mu\nu} = \begin{pmatrix} -2GM/\kappa r & -2GS_z\,y/\kappa r^3 & 2GS_z\,x/\kappa r^3 & 0 \\ -2GS_z\,y/\kappa r^3 & -2GM/\kappa r & 0 & 0 \\ 2GS_z\,x/\kappa r & 0 & -2GM/\kappa r & 0 \\ 0 & 0 & 0 & -2GM/\kappa r \end{pmatrix} \quad [94]$$

Comparing this with Eq. [9], we see that the field of a rotating mass differs from that of a nonrotating mass by the presence of the off-diagonal components h_{0l}.*

The spins of the Sun and Earth are fairly small, and the effects of the off-diagonal components of the gravitational field on the orbits of planets and ordinary satellites are not directly detectable. However, in the case of the Earth, artificial satellites of a special design may make it possible to detect the effects of the off-diagonal components. The main effect is a precession of the orbital angular momentum, that is, a precession plane of the orbit, around the axis of spin of the Earth. This is called the Lense-Thirring precession (Lense and Thirring, 1918).

The Lense-Thirring precession can be calculated by examining the equation of motion of a particle in the gravitational field [94]. However, the calculation is rather complicated, and it is much simpler to obtain the answer by exploiting the analogy between the gravitational and the electrodynamic equations discussed in Section 3.5. This analogy cannot give us the full expression for the field tensor, since the fields $\mathbf{g}$ and $\mathbf{b}$ involve only the components h^{00} and h^{0k} of the field tensor, and they do not involve any of the h^{kl} components. But, for a nonrelativistic (slow) particle, the equation of motion [3.115] depends only on $\mathbf{g}$ and $\mathbf{b}$, and we do not need to know h^{kl}. The Lense-Thirring precession is analogous to the precession of the orbital angular momentum of a charged particle orbiting around another charged particle endowed with a magnetic dipole moment. The time-average magnetic moment $\mathbf{m}'$ associated with the angular momentum of the

---

* When the rotation is very fast, there is a further difference: the total mass M of Eq. [94] includes the kinetic energy of rotation.

orbiting particle then couples to the magnetic dipole moment **m** of the central particle, via the magnetic field, and this results in an effective torque on the orbital angular momentum. The magnetic field of the dipole **m** is given by the familiar formula

$$\mathbf{B} = \frac{3\mathbf{n} \; \mathbf{n} \cdot \mathbf{m} - \mathbf{m}}{r^3} \qquad [95]$$

where **n** is the unit vector in the radial direction, $\mathbf{n} = \mathbf{r}/r$. The torque exerted by this magnetic field on the dipole **m′** is

$$\tau = \mathbf{m}' \times \mathbf{B} = \mathbf{m}' \times \frac{3\mathbf{n} \; \mathbf{n} \cdot \mathbf{m} - \mathbf{m}}{r^3} \qquad [96]$$

By analogy, the gravimagnetic field of the Earth is

$$\mathbf{b} = -4G \frac{3\mathbf{n} \; \mathbf{n} \cdot \mathbf{S} - \mathbf{S}}{2r^3} \qquad [97]$$

Here, we have included an extra factor of 2 in the denominator, because the analog of the magnetic moment **m** is **S**/2, rather than **S** (the definition $\mathbf{m} = \frac{1}{2} \int \mathbf{x}' \times \mathbf{j} \; d^3x'$ includes a factor of $\frac{1}{2}$, whereas the definition of **S**, Eq. [88], includes no such factor). The torque exerted by the gravimagnetic field on the orbital angular momentum is

$$\tau = \frac{\mathbf{L}}{2} \times \mathbf{b} = -G\mathbf{L} \times \frac{3\mathbf{n} \; \mathbf{n} \cdot \mathbf{S} - \mathbf{S}}{r^3} \qquad [98]$$

In this expression, the term **n** $\mathbf{n} \cdot \mathbf{S}$ must be replaced by its average value over the orbit, since it is only in an average sense that we can regard the orbiting particle as a spinning body for which the dipole-dipole interaction is valid. For a circular or nearly circular polar orbit, the time-average value $\langle \mathbf{n} \; \mathbf{n} \cdot \mathbf{S} \rangle$ is $\frac{1}{2}\mathbf{S}$, and for an equatorial orbit it is zero.

■ *Exercise 10.* Obtain these results for the time-average values of **n** $\mathbf{n} \cdot \mathbf{S}$. ■

The rate of change of the orbital angular momentum is then

$$\frac{d\mathbf{L}}{dt} = -G \frac{\mathbf{L} \times (3\langle \mathbf{n} \; \mathbf{n} \cdot \mathbf{S} \rangle - \mathbf{S})}{r^3} \qquad [99]$$

From this, we can identify the precession angular velocity as

$$\Omega = G \, \frac{3\langle \mathbf{n} \; \mathbf{n \cdot S}\rangle - \mathbf{S}}{r^3} \qquad\qquad [100]$$

For a polar orbit, with $\langle \mathbf{n} \; \mathbf{n \cdot S}\rangle = \tfrac{1}{2}\mathbf{S}$, the direction of the precession is in the same direction as that of the rotation of the Earth; it is east-ward. For an equatorial orbit, with $\langle \mathbf{n} \; \mathbf{n \cdot S}\rangle = 0$, it is westward.

   If the polar orbit is of low altitude ($r \simeq R_E$), the magnitude of the Lense-Thirring precession amounts to about 0.05 arcsec per year. Unfortunately, satellite orbits are subject to much larger precessions due to the quadrupole moment of the Earth. The quadrupole preces-sion vanishes if the orbit is *exactly* polar, but this does not help-- there are practical limitations on our ability to measure the angle of the plane of the orbit with ground-based instrumentation, and we cannot be sure that the orbit is *exactly* polar. As a clever way out of this difficulty, it has been proposed that two satellites be placed in similar nearly polar orbits, so the plane of the orbit of one satellite is slightly on one side of the polar axis, and the plane of the orbit of the other satellite is slightly on the other side (see Fig. 4.24). The quadru-pole moment of the Earth then produces precessions of opposite direc-tions for the two satellites, whereas the Lense-Thirring effect pro-duces a westward precession for both. To separate these two effects, we must measure not only the precession rates of the two satellites, but also the small angle α between their orbits. The latter measurement could be done with the required precision by means of Doppler radars on the satellites. The sum of the rates of precession caused by the action of the quadrupole moment on the two orbits can be shown to be proportional to the small angle α and to the magnitude of the qua-drupole moment. Thus, from the measurement of α and from the known magnitude of this quadrupole moment, we can predict the sum of the precession rates caused by the quadrupole moment. Comparison of this predicted value with the actual value found by observation of the orbits determines the magnitude of the Lense-Thirring precession.

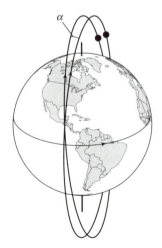

Fig. 4.24 Two satellites in similar nearly polar orbits. The angle between these orbits is α.

An alternative method to measure the Lense-Thirring precession proposed by Ciufolini (1989) would make use of a satellite already in orbit (LAGEOS) and require only one new satellite (LAGEOS X). Since the orbit of LAGEOS is far from polar, the precession contributed by the quadrupole moment of the Earth is large, and the uncertainty in our knowledge of the quadrupole moment precludes a sufficiently accurate prediction of the precession rate. However, if the orbit of the new LAGEOS X is inclined at an exactly opposite angle with respect to the axis of the Earth (see Fig. 4.25), then in the sum of the precession rates of LAGEOS and LAGEOS X, the contributions from the quadrupole moment cancel exactly, and only the Lense-Thirring effect remains. This proposal has the advantage of simplicity and relatively low cost, since only one new satellite, without complicated Doppler-radar equipment, is required. But the experiment is subject to an uncertainty of at least 10% arising from errors in the initial inclination of the second satellite, measurement errors in the orbital parameters, etc.

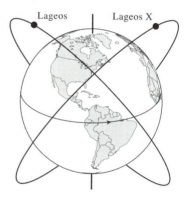

Fig. 4.25 Orbits of the LAGEOS satellite and of the proposed LAGEOS X satellite of opposite inclination.

Although the Lense-Thirring effect was originally conceived as an orbital precession, it applies equally well to the precession of a gyroscope placed near the Earth. The spin of such a gyroscope is, again, coupled to the spin of the Earth via the gravimagnetic field. The gyroscope experiences a torque, and, if it is suspended in frictionless mountings, it will precess at the rate given by Eq. [100]. For a gyroscope placed on the surface of the Earth, the friction in the mountings is always much larger than the gravimagnetic torque; but the Stanford gyroscope experiment (Gravity Probe B), already mentioned in Section 1.9 and further discussed in Section 7.8, seeks to detect this torque with a gyroscope placed in a satellite in orbit, where the gyroscope is weightless, and requires no suspension. However, such an orbiting gyroscope is subject to an additional precession effect, arising from its motion through the gravitational field of the Earth. This additional precession, called the de Sitter precession, is about 100 times larger than the Lense-Thirring precession, but the two kinds of precession can be separated experimentally because they differ in their dependences on the direction of the axis of the gyroscope. The Stanford gyroscope experiment is designed to measure both of these precession effects.

This concludes our discussion of the time-independent solutions of the linear field equations. In the next chapter we will investigate the time-dependent fields in the radiation zone, that is, the fields at large distance from some time-dependent mass distribution. We could also investigate the fields in the convection zone, that is, the fields in the immediate vicinity of a time-dependent mass distribution. In fact, we could work out a set of "Liénard-Wiechert" potentials for gravitation entirely in analogy with electromagnetism. But although the close similarity between linear gravitation and electromagnetism may tempt us to solve the same mathematical problems, we must keep in mind that there are few gravitational problems that are accessible to experiment. We cannot manufacture noticeable gravitational fields as easily as we manufacture electric and magnetic fields. With the exception of the Cavendish experiment, the gravitational fields that we use in our experiments are generated by processes that are beyond our control.

FURTHER READING

A detailed survey of the experimental and observational tests of the relativistic theory of gravitation is presented in Will, C. M., *Theory and Experiment in Gravitational Physics* (Cambridge University Press, Cambridge, 1993) and more concisely in Will, C. M., "Experimental Gravitation from Newton's Principia to Einstein's General Relativity," in Hawking, S. W., and Israel, W., eds., *Three Hundred Years of Gravitation* (Cambridge University Press, Cambridge, 1987).

The simple derivation of the redshift given at the end of Section 4.2 was found by Einstein long before he developed his theory of general relativity. His article "Über den Einfluss der Schwerkraft auf die Ausbreitung des Lichtes," Ann. der Physik [translated in Lorentz, H. A., Einstein, A., Minkowski, H., and Weyl, H., *The Principle of Relativity* (Methuen, London, 1923)] is therefore of great historical interest. Note that the light-deflection formula derived in this article is off by a factor of 2 because Einstein did not (yet) take into account the curved spacetime geometry.

von Klüber, H., "The Determination of Einstein's Light Deflection in the Gravitational Field of the Sun," Vistas in Astronomy **3**, 47 (1960) gives a thorough and interesting review of the light-deflection observations (up to 1952), including a discussion of the experimental difficulties in obtaining reliable results, by an astronomer who participated in several expeditions. Eddington, A. S., *Space, Time, and Gravitation* (Cambridge University Press, Cambridge, 1920) tells the story of the first expedition that set out to measure the light deflection.

In the last few years, gravitational lensing has become a topic of high interest for astronomers and astrophysicists, with an ever-increasing number of papers published each year. The most comprehensive treatise on both the theoretical and observational aspects of gravitational lenses is Schneider, P., Ehlers, J., and Falco, E. E., *Gravitational Lenses* (Springer-Verlag, Berlin, 1992). It includes a historical introduction, case-by-case discussions of the observed lenses, a general treatment of caustics, intensities of images, microlensing, and the use of lenses as astrophysical tools. Some other reviews and collections of papers dealing with gravitational lenses are the following:

Blandford, R., and Narayan, R., Astrophys. J. **310**, 568 (1986). (Exploits Fermat's principle of extremum time for calculations with gravitational lenses.)

Blandford, R. D., and Kochanek, C. S., "Gravitational Lenses," in Bahcall, J., Piran, T., and Weinberg, S., eds., *Dark Matter in the Universe* (World Scientific, Singapore, 1987). [Offers a concise introduction to the theory of gravitational lenses and discusses alternative mathematical formalisms: vector equations for ray deflections (as in Section 4.6), Fermat's principle, and "optical scalar equations."]

Blandford, R. D., Kochanek, C. S., Kovner, I., and Narayan, R., Science **245**, 824 (1989).

Blandford, R. D., and Narayan, R., Ann. Rev. Astronomy and Astrophysics **30**, 311 (1992). (Deals with cosmological applications of gravitational lenses.)

Kayser, R., and Schramm, T., eds., *Gravitational Lenses* (Springer-Verlag, Berlin, 1992).

Mellier, Y., Fort, B., and Soucail, G., eds., *Gravitational Lensing* (Springer-Verlag, Berlin, 1990).

Moran, J. M., Hewitt, J. N., and Lo, K. Y., eds., *Gravitational Lenses* (Springer-Verlag, Berlin, 1989).

Schneider, P., Ehlers, J., and Falco, E. E., eds., *Gravitational Lensing* (Springer-Verlag, Berlin, 1992).

Turner, E. L., "Gravitational Lenses and Dark Matter: Observations," in Kormandy, J., and Knapp, G. R., *Dark Matter in the Universe* (Reidel, Dordrecht, 1987).

A gravitational lens can be simulated optically by a disk of glass with its surface shaped in such a way that light rays are deflected by the same angle as in a gravitational field. The construction of such a lens is described in Dyer, C. C., and Roeder, R. C., J. Roy. Astron. Soc. Can. **75**, 227 (1981).

A general introduction to catastrophe theory and caustics is provided by Berry, M. V., "Waves and Thom's Theorem," Advances in Physics **25**, 1 (1976).

REFERENCES

Alley, C. O., "Relativity and Clocks," in Proceedings of the 33rd Annual Symposium on Frequency Control (Electronic Industries Association, Washington, 1979).

Anderson, J. D., Esposito, P. B. Martin, W., Thornton, C. L. and Muhleman, D. O., Astrophys. J. **200**, 221 (1975).

Anderson, J. D., Keesey, M. S. W., Lau, E. L., Standish, E. M., and Newhall, XX, Acta Astronaut. **5**, 43 (1978).

Bertotti, B., et al., in Witten, L., ed., *Gravitation: An Introduction to Current Research* (Wiley, New York, 1962).

Blandford, R. D., and Narayan, R., Ann. Rev. Astronomy and Astrophysics **30**, 311 (1992).

Bourassa, R. R., and Kantowski, R., Astrophys. J. **195**, 13 (1975).

Bourassa, R. R., Kantowski, R., and Norton, T. D., Astrophys. J. **185**, 747 (1973).

Brault, J. W., Princeton University Thesis, 1962. A brief description of this measurement is given in Dicke (1964).

Chang, K., and Refsdal, S., Nature **279**, 561 (1979); see also Astron. Astrophys. **132**, 168 (1984).

Chang, K., Astron. Astrophys. **130**, 157 (1984).

Ciufolini, I., Int. J. Mod. Phys. A **4**, 3083 (1989).

Counselman, C. C., et al., Phys. Rev. Lett. **33**, 1617 (1974).

Chwolson, O., Astr. Nachrichten, **221**, 329 (1924).

Dicke, R. H., in de Witt, C., and de Witt, B., eds., *Relativity, Groups and Topology* (Gordon and Breach, New York, 1964), p. 170.

Dirac, P. A. M., Nature **139**, 323 (1937).

Einstein, A., Ann. d. Physik **35**, 898 (1911).

Falco, E. E., et al., in Mellier, Y., et al., eds., *Gravitational Lensing* (Springer-Verlag, Berlin, 1990).

Fomalont, E. B., and Sramek, R. A., Astrophys. J. **199**, 749 (1975).

Fomalont, E. B., and Sramek, R. A., Phys. Rev. Lett. **36**, 1475 (1976).

Fort, B., "Clusters of Galaxies: A New Observable Class of Gravitational Lenses," in Mellier, Y., Fort, B., and Soucail, G., eds., *Gravitational Lensing* (Springer-Verlag, Berlin, 1990).

Greenstein, J. L., et al., Astrophys. J. **169**, 563 (1971).

Häfele, J. C., and Keating, R. E., Science **177**, 168 (1972).

Hewitt, J. N., in Akerlof, C. W., Srednicki, M. A., eds, *Texas PASCOS 92: Relativistic Astrophysics & Particle Cosmology* (New York Academy of Sciences, New York, 1993).

Hill, J. M., Mon. Not. Roy. Astron. Soc. **153**, 7P (1971).

Irwin, M. J., Webster, R. L., Hewett, P. C., Corrigan, R. T., and Jedrzejewski, R. I., Astron. J. **98** (1989).

Jones, B. F., Astron. J. **81**, 455 (1976).

Krisher, T. P., et al., Astrophys. J., **373**, 665 (1991).

Lense, J., and Thirring, H., Phys. Zeitsch. **19**, 156 (1918).

Liebes, S., Phys. Rev. B **133**, 835 (1964).

Muhleman, D. O., et al., Phys. Rev. Lett. **24**, 1377 (1970).

Ohanian, H. C., Astrophys. J., **271**, 551 (1983).

Pound, R. V., and Rebka, G. A., Phys. Rev. Lett. **4**, 337 (1960).

Pound, R. V., and Snider, J. L., Phys. Rev. Lett. **13**, 539 (1964); Phys. Rev. B **140**, 788 (1965).

Press, W. H., and Gunn, J. E., Astrophys. J., **185**, 397 (1973).

Reasenberg, R. S., and Shapiro, I. I., in Bertotti, ed., *Proceedings of the International Symposium on Experimental Gravitation* (Pavia, 1976); reprinted in Worden, P. W., and Everitt, C. W. F., eds., *Gravity and Inertia: Selected Reprints* (American Assoc. of Physics Teachers, Stony Brook, 1983).

Shapiro, I. I., Reasenberg, R. D., et al., J. Geophys. Res. **82**, 4329 (1977).

Reasenberg, R. D., Shapiro, I. I., et al., Astrophys. J. **234**, L219 (1979).

Refsdal, S., Mon. Not. R. Astron. Soc. **128** 295 (1964).

Riley, J. M., Mon. Not. Roy. Astron. Soc. **161**, 11P (1973).

Robertson, D.S., and Carter, W.E., Nature **310**, 572 (1984).

Robertson, D.S., et al., Nature **349**, 768 (1991).

Rudenko, V. N., Soviet Physics Uspekhi **21**, 893 (1978); reprinted in Worden, P. W., and Everitt, C. W. F., eds., *Gravity and Inertia: Selected Reprints* (American Assoc. of Physics Teachers, Stony Brook, 1983).

Schneider, P., Ehlers, J., and Falco, E. E., *Gravitational Lenses* (Springer-Verlag, Berlin, 1992).

Seielstadt, G. A., et al., Phys. Rev. Lett. **24**, 1373 (1970).

Shapiro, I. I., et al., Phys. Rev. Lett. **17**, 933 (1966).

Shapiro, I. I., et al., Phys. Rev. Lett. **20**, 1265 (1968).

Shapiro, I. I., et al., Phys. Rev. Lett. **26**, 1132 (1971).

Shapiro, I. I., Reasenberg, R. D., et. al., J. Geophys. Res. **82**, 4329 (1971).

Shapiro, I. I., Reasenberg, R. D., Chandler, J. F., and Babcock, R. W., Phys. Rev. Lett. **61**, 2643 (1988).

Schild, R., in Mellier, Y., Fort, B., and Soucail, G., eds., *Gravitational Lensing* (Springer-Verlag, Berlin, 1990).

Snider, J. L., Phys. Rev. Lett. **28**, 853 (1972).

Sramek, R.A., Astrophys. J. **167**, L55 (1971).

Sramek, R. A., "The Gravitational Deflection of Radio Waves," in Bertotti, B., ed., *Experimental Gravitation* (Academic Press, New York, 1974).

Tyson, A., "Lensing the Background Population of Galaxies," in Mellier, Y., Fort, B., and Soucail, G., eds., *Gravitational Lensing* (Springer-Verlag, Berlin, 1990).

Tyson, J. A., Valdes, F., and Wenk, R. A., Astrophys J. Lett., **349**, 1 (1990).

Vanderriest, C., Schneider, J., Herpe, G., Chevreton, M., Wlérik, G., and Moles, M., in Moran, J. M., Hewitt, J. N., and Lo, K. Y., eds., *Gravitational Lenses* (Springer-Verlag, Berlin, 1989).

Vanderriest, C., et al., Astron. and Astrophys. **215**, 1 (1989).

Vessot, R. F. C., and Levine, M. W., J. Gen. Rel. and Grav. **10**, 181 (1979); see also Vessot, R. F. C., et al., Phys. Rev. Lett., **45**, 2081 (1980).

von Klüber, H., "The Determination of Einstein's Light Deflection in the Gravitational Field of the Sun," Vistas in Astronomy **3**, 47 (1960).

Walsh, D., Carswell, R. F., and Weymann, R. J., Nature **279**, 381 (1979).

Weiler, K. W., et al., Astron. and Astrophys. **30**, 241 (1974).

Weiler, K. W., et al., Phys. Rev. Lett. **35** 134 (1975).

Young, P., Deverill, R. S., Gunn, J. E., Westphal, J. A., and Kristian, J., Astrophys. J. **244**, 723 (1981).

Young, P., Astrophys. J. **244**, 756 (1981).

Zwicky, F., Phys. Rev. Lett. **51**, 290, 679 (1937).

PROBLEMS

1. According to Eqs. [8] and [10], the gravitational field of any static mass distribution is

$$\phi^{\mu\nu} = \begin{pmatrix} 4\Phi/\kappa & 0 & 0 & 0 \\ 0 & 0 & 0 & 0 \\ 0 & 0 & 0 & 0 \\ 0 & 0 & 0 & 0 \end{pmatrix}$$

where Φ is the Newtonian potential. Show that for such a gravitational field, the energy density $t^{00}{}_{(1)}$ is positive and agrees with the energy density of Newtonian theory (see the first term in Eq. [1.19]).

2. A sphere of radius R and mass M has a uniform density.
(a) From the linear gravitational field equations, find the metric tensor $g_{\mu\nu} = \eta_{\mu\nu} + \kappa h_{\mu\nu}$ produced by this sphere in the region $r > R$ and in the region $r < R$.
(b) Pretend that the Earth is a sphere of uniform density. According to the geometric interpretation of the gravitational field discussed in Section 3.6, calculate the difference (in cm) between the measured circumference and 2π times the measured radius of the Earth.

3. In the linear approximation, the spatial part of the geometry surrounding a spherical mass has a length interval

$$dl^2 = \left(1 + \frac{2GM}{r}\right)(dx^2 + dy^2 + dz^2)$$

(a) If we stretch a very long, thin, massless string so it passes by this mass with an "impact parameter" b (see Fig. 4.26), what will be the angle α between the extremes of this string?
(b) Does the angle α agree with the angle of deflection of a light ray of impact parameter b? Explain.

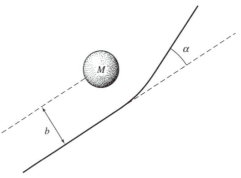

Fig. 4.26 Stretched string "bent" by a mass.

4. Some stars are surrounded by thin spherical shells of mass (planetary nebulas). Suppose that the mass of a star is M_1 and the mass of the shell is M_2. The radius of the star is R_1 and the radius of shell is R_2; the region between the star and the shell is empty. Find the solution of the linear gravitational field equations in the region $r > R_2$ and in the region $R_1 < r < R_2$.

5. A small spherical mass (spherical in its own rest frame) moves with uniform velocity v along the z-axis, passing the origin at time $t = 0$. Find the gravitational field $h_{\mu\nu}$ at the point $x = b$ on the x-axis as a function of time. Plot h_{00}, h_{11}, and h_{22} as a function of laboratory time for $v = 0.5$ and for $v = 0.95$. (Hint: Perform a Lorentz transformation on Eq. [9].)

6. An infinitely long, thin rod has a mass of λ grams per centimeter in its rest frame. The rod lies along the x-axis and is moving in the x direction with a uniform relativistic velocity w.
(a) Find all the components of the gravitational field tensor $h_{\mu\nu}$ as a function of the radial distance r from the rod. Ignore additive constants in the field tensor.
(b) Suppose that a particle is moving radially away from the rod, along the y-axis. The particle has an instantaneous four-velocity u_y. What is the instantaneous acceleration $du_x/d\tau$ of this particle?

7. *In vacuo,* the nonlinear field equation [3.64] is

$$\partial_\lambda \partial^\lambda \phi^{\mu\nu} = - \kappa t_{(1)}{}^{\mu\nu}$$

where $t_{(1)}{}^{\mu\nu}$ is given by Eq. [3.62]. Check that this nonlinear equation has the approximate solution

$$\phi^{\mu\nu} = \begin{pmatrix} -\dfrac{4GM}{\kappa r} + \dfrac{G^2 M^2}{\kappa r^2} & 0 & 0 & 0 \\[2mm] 0 & -\dfrac{G^2 M^2 x^2}{\kappa r^4} & -\dfrac{G^2 M^2 xy}{\kappa r^4} & -\dfrac{G^2 M^2 xz}{\kappa r^4} \\[2mm] 0 & -\dfrac{G^2 M^2 xy}{\kappa r^4} & -\dfrac{G^2 M^2 y^2}{\kappa r^4} & -\dfrac{G^2 M^2 yz}{\kappa r^4} \\[2mm] 0 & -\dfrac{G^2 M^2 xz}{\kappa r^4} & -\dfrac{G^2 M^2 yz}{\kappa r^4} & -\dfrac{G^2 M^2 z^2}{\kappa r^4} \end{pmatrix}$$

Take into account terms of order M and M^2, but neglect terms of order M^3. This solution gives the second-order corrections for the field surrounding a spherically symmetric mass; the perihelion precession can be calculated from this (see Problem 7.14).

8. Two very large, thin, parallel sheets of mass have a uniform mass density σ (g/cm^2) in their own rest frames. One of the sheets moves toward the right at speed w, the other moves toward the left at the same speed w (see Fig. 4.27).
(a) Show that, with the x and y coordinates as in the figure, the components of the gravitational field tensor are

$$h^{\mu\nu} = \begin{pmatrix} 0 & -16\pi G\sigma\gamma^2 wy/\kappa & 0 & 0 \\ -16\pi G\sigma\gamma^2 wy/\kappa & 0 & 0 & 0 \\ 0 & 0 & 0 & 0 \\ 0 & 0 & 0 & 0 \end{pmatrix} + \text{[constant]}$$

where $\gamma = 1/\sqrt{(1 - w^2)}$.

(b) A particle moving in this gravitational field has an instantaneous four-velocity u^μ. Find the instantaneous accelerations $du_x/d\tau$, $du_y/d\tau$, and $du_z/d\tau$ of this particle.

(c) Qualitatively, what is the shape of the orbit of this particle?

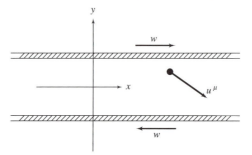

Fig. 4.27 Two large, thin, parallel sheets in motion.

9. When scientists want to make extremely precise comparisons between the atomic clocks at the Institute for Standards and Technology (Boulder, Colorado) and those at the Bureau International de l'Heure (Paris), they must take into account that Boulder and Paris are at different altitudes and that the clocks suffer different amounts of gravitational time dilation. Boulder is at 1600 m above sea level, and Paris is at (approximately) sea level. What time difference (in seconds) will accumulate between identical clocks at Boulder and Paris in one year as a consequence of gravitational time dilation?

10. According to the new definition of the meter adopted in 1983, the meter is the distance traveled by light in a time interval of 1/299,792,458 second. The second is the time required for 9,162,631,770 vibrations of a cesium atom. Suppose that two laboratories located at different altitudes on the equator use these definitions to manufacture meter bars. Taking into account the gravitational time dilation and the gravitational slowing down of the speed of light, calculate by what factor the coordinate lengths dr of the two meter bars will differ. Express your answer in terms of the radial positions r_1 and r_2 of the two laboratories. Evaluate the fractional difference between the coordinate lengths of the meter bars if one of the laboratories is at sea level and the other is at an altitude of 1000 m. Ignore the rotation of the Earth.

11. In the test of the gravitational time dilation by Vessot et al. (see Section 4.2), a hydrogen-maser clock was launched in a rocket to an altitude of 10,000 km above the surface of the Earth. Assume that the trajectory was exactly radial, with a launch point at the equator. The frequency of the maser clock was 1420 MHz. If at apogee this clock sends radio signals toward a similar clock on the ground, what is the frequency shift (in Hz) observed on the ground? In your calculation, ignore the rotational velocity of the Earth (the corresponding eastward velocity is about the same for the clock on the Earth and in the rocket).

12. Consider a clock sitting at sea level at the north pole of the Earth and another identical clock sitting at the equator. Relative to an inertial reference frame, the clock at the north pole is at rest, whereas the clock at the equator circles the axis of the Earth at 460 m/s; hence the equatorial clock suffers time dilation relative to the polar clock. On the other hand, since the Earth has an equatorial bulge, the equatorial clock is at a higher gravitational potential than the polar clock; hence the polar clock suffers gravitational time dilation relative to the equatorial clock. Assuming that the surface of the sea is in hydrostatic equilibrium, show that these two time-dilation effects are (approximately) of the same magnitude, so there is no net relative time dilation between these clocks.

13. Approximately by what factor does the rate of a clock on the Earth differ from that of a clock in intergalactic space (assumed to be flat)? Take into account the potential of the Earth, Sun, and Galaxy.

14. A satellite in a circular orbit of radius r around the Earth carries a radio transmitter which emits a signal of constant frequency ω_0 (measured in the rest frame of the satellite). Find the frequency ω measured by a receiver on the surface of the Earth when the satellite passes overhead and its radio waves come to the Earth along the vertical (no longitudinal Doppler shift!). Take the gravitational time dilation into account and also the special-relativistic time dilation resulting from the velocity of the satellite and the velocity of rotation of the surface of the Earth. Express your answer in terms of r (radius of satellite orbit), ω_0 (transmitter frequency), M (mass of Earth), R (radius of Earth), and Ω (angular velocity of rotation of Earth). Assume that both transmitter and receiver are in the Earth's equatorial plane.

15. Consider a particle of mass m that moves toward the Sun along a straight line of impact parameter b. Assume that the velocity of the particle is nonrelativistic, yet large enough so the orbit is nearly a straight line. Show that the increment in the proper time of the particle can be expressed (approximately) as

$$d\tau \simeq dt + \left(-\frac{E}{m} - \frac{2GM_\odot}{r} \right) dt$$

where E is the Newtonian energy. Since E is a constant of the motion, the effect of the gravitational field of the Sun is entirely contained in the term $-(2GM_\odot/r)dt$. Integrate this term from $r = r_1$ to $r = b$ to $r = r_2$, and find the total time dilation that may be attributed to the Sun. Evaluate numerically for a particle that moves from $r_1 = 1$ light-year to $r_2 = 1$ light-year with an initial velocity $v = 10^9$ cm/s and an impact parameter $b = r_\odot$.

16. The Earth is in an elliptical orbit around the Sun and therefore has a (small) radial velocity. Use the orbital parameters of the Earth to calculate the maximum Doppler shift of the light emitted by the Sun. Compare this Doppler shift with the gravitational redshift. Is it necessary to take the Doppler shift into account when making a redshift measurement?

17. Suppose the Earth were transparent, so we could see the light emitted by the hot, glowing center of the Earth. What would be the redshift of this light when received at the surface? For this calculation, pretend that the Earth is a sphere of uniform density, and ignore the rotation of the Earth.

18. We send a radar signal from the Earth to the surface of the Sun, where it is reflected and returns to the Earth. Calculate the retardation that the gravitational field of the Sun produces in this signal.

19. The famous twin "paradox" of special relativity can be stated as follows: Terra and Stella are initially on the Earth (regarded as an inertial frame). Terra stays on the Earth. Stella boards a spaceship and flies away at speed u. After a while, she decelerates the spaceship, reverses the velocity, and returns to Earth. If we assume that the accelerated portions of the trip last a very short time compared with the unaccelerated portions, then in the inertial reference frame of Terra, the usual time-dilation formula tells us that the elapsed proper times of Terra and Stella are related by

$$\Delta t_{Terra} = \frac{1}{\sqrt{1 - u^2}} \Delta t_{Stella}$$

Thus, Terra has aged more than Stella. If we naively apply the time-dilation formula in the reference frame of Stella, we would reach the opposite conclusion. However, it is not legitimate to apply the time-dilation formula in a noninertial reference frame and hence there actually is no paradox.

If we want to derive the relationship between Δt_{Terra} and Δt_{Stella} in the reference frame of Stella, then we must take into account the gravitational time-dilation effect. When Stella decelerates, there is an apparent gravitational field in her reference frame, such that Terra is at the higher gravitational potential and Stella at the lower. Show that the consequent advance of the clocks of Terra more than compensates for the time dilation that these clocks suffer during the unaccelerated portion of the trip and show that the net result is again given by the above formula. Assume that the accelerated portion of Stella's trip lasts a very short time compared with the unaccelerated portion, so the acceleration essentially occurs at a fixed distance from the Earth. (Ignore the accelerations that occur near the Earth at the beginning and at the end of the trip.)

20. A mass M is distributed uniformly over a sphere of radius R. Suppose that a (massless) neutrino is released at the center of this mass. The neutrino travels outward along a radius, from $r = 0$ to $r = R$. Calculate the time delay (in t time) that the neutrino suffers during this trip because of the gravitational field. Evaluate the delay numerically if the mass is 10^{34} g and the radius is 10^{11} cm. (Hint: The motion of neutrinos is the same as that of photons.)

21. On the Earth, rays of light are used for surveying. Consider a horizontal ray of light propagating from one hilltop to another, 10 km away. What is the gravitational deflection of this ray?

22. Fig. 4.28 shows a triangle laid out in the space within and near the Sun. The angle at the center is a right angle, the horizontal and vertical sides are straight radial lines, and the diagonal side is a light ray. By how much does the sum of interior angles of this triangle differ from 180^0?

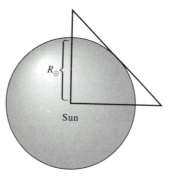

Fig. 4.28 Triangle of light rays laid out in and near the Sun.

23. Repeat the preceding problem if, instead of light rays, the diagonal side of the triangle is a tightly stretched massless string.

24. A right triangle is inscribed in the Earth, as shown in Fig. 4.29. The angle at the center is a right angle, and all the sides are light rays. Assume that the Earth is a transparent sphere of uniform density. Calculate by how much the sum of the interior angles differs from 180^0.

Fig. 4.29 Triangle of light rays within the Earth.

25. Suppose that three artificial satellites are placed in circular orbits of radius $r = 2R_{\oplus} = 2 \times 6.38 \times 10^8$ cm around the Earth. The satellites aim laser beams at each other, and these beams inscribe the Earth in an equilateral triangle (see Fig. 4.30).

Calculate by how much the sum of interior angles of this triangle differs from 180^0. Ignore the refraction of the laser beams by the atmosphere.

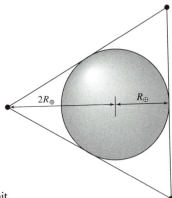

Fig. 4.30 Three satellites in orbit around the Earth.

26. Consider two initially parallel rays of light that pass by the Moon grazing opposite ends of a diameter. How far from the Moon will these rays intersect?

27. Suppose that a galaxy acting as a "gravitational lens" produces two "images" of a distant quasar. Suppose that the galaxy is at a distance r, midway between the quasar and the observer, as illustrated in Fig. 4.31. Suppose that the impact parameters of the two light rays passing by the galaxy are b_1 and b_2. Treating the galaxy as a pointlike mass M and using the linear approximation for the relativistic gravitational field, find an expression for the difference in the travel times along the two light rays. Evaluate the time delay numerically for $b_1 = 5 \times 10^4$ light-years, $b_2 = 1 \times 10^5$ light-years, $r = 1 \times 10^{10}$ light-years, and $M = 10^{12} M_\odot$. (Hint: Take into account the difference in path length between the rays and the difference in the effective speed of light in the gravitational field; in the lowest order of approximation, these two contributions can be combined additively. Ignore cosmological effects, that is, pretend that the background spacetime is flat and static.)

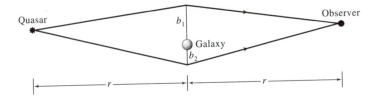

Fig. 4.31 A "gravitational lens" located midway between a quasar and the observer.

28. Rays of light from a distant quasar pass by a spherical galaxy and are deflected. The mass of the galaxy is 10^{45} g, and its diameter is 10^{23} cm.
(a) Consider two parallel rays of light that pass by the galaxy on opposite ends of a

diameter. How far from the galaxy will the rays intersect? An observer at the point of intersection sees the light of the quasar as forming an Einstein ring around the galaxy. What is the angular diameter of the ring he sees?

(b) If another observer is twice as far from the galaxy as the first, what is the angular diameter of the ring she sees?

29. A spherical mass distribution of uniform density has a total mass M and a radius R. Assume the mass is transparent.

(a) Show that the deflection angle of a light ray with impact parameter $b < R$ is

$$\theta = \frac{4GM}{b} \left[1 - \frac{(R^2 - b^2)^{3/2}}{R^3} \right]$$

(b) Show that the deflection can also be expressed as $\theta = 4GM(b)/b$, where $M(b)$ is that part of the mass that is contained in a cylinder of radius b centered on the mass distribution. In this form the result is quite general and applies to any spherically symmetric mass distribution.

(c) Plot θ as a function of b, for $0 < b < 2R$.

30. Show that the function F_1 defined in Eq. [70] reduces to

$$F_1 = 1 - 4\pi G\sigma(x', y') \frac{D_1}{D_{12}}$$

[Hint: Begin by establishing that

$$\left(\frac{\partial}{\partial \xi} + i \frac{\partial}{\partial \zeta} \right) \frac{1}{x' - \xi + i(y' - \zeta)} = 2\pi\delta(x' - \xi)\delta(y' - \zeta) \ ]$$

31. A *relativistic* particle of mass m and velocity v passes by the Sun with an impact parameter b. Find the angular deflection suffered by the particle. Make the same approximations as were used in the calculation of the deflection of light (weak gravitational field, small deflection angle). Show that for a nonrelativistic particle, your result approaches $2GM/b$, and that for an extremely relativistic particle your result approaches $4GM/b$.

32. According to Eq. [47] (or [52]), the gravitational field of the Sun may be regarded as a medium with an effective index of refraction

$$n = \left(1 + \frac{2GM_\odot}{r} \right)$$

Use this expression for n and Huygens' principle for the propagation of wavefronts to derive the deflection formula [38].

33.   Find an expression for the deflection angle of a light ray in the field of a rotating mass. Assume that the light ray moves in the equatorial plane of the rotating mass. Consider both the case of a light ray whose orbital angular momentum is parallel to the spin of the central mass and the case of angular momentum antiparallel to the spin. Numerically estimate the magnitude of the effect produced by the spin of the Sun for an impact parameter $b = R_\odot$.

5. Gravitational Waves

THE EARTH IS JUST A SILLY BALL
TO THEM, THROUGH WHICH THEY SIMPLY PASS,
LIKE DUSTMAIDS DOWN A DRAFTY HALL
OR PHOTONS THROUGH A SHEET OF GLASS.

John Updike

Gravitational effects cannot propagate with infinite speed. This is obvious both from the lack of Lorentz invariance of infinite speed and from the causality violations that are associated with signal speeds in excess of the speed of light. Since the speed of light is the only Lorentz-invariant speed, we expect that gravitational effects propagate in the form of waves at the speed of light.

As a concrete example, consider an apple that hangs on a tree. At some time, the stem of the apple breaks and the apple falls to the ground, which means there is a sudden change in the terrestrial mass distribution. The gravitational field surrounding the Earth must then adapt itself to this new mass distribution. The change in the field will not occur simultaneously throughout the universe--at any given point of space the change will be delayed by a time equal to the time needed for a light signal to travel from the Earth to that point. Hence the disturbance in the gravitational field propagates outward at the speed of light. Such a propagating disturbance is a gravitational wave.

The existence of gravitational waves is an immediate consequence of special relativity; and, to some extent, the experimental discovery of gravitational waves would merely confirm the obvious. However, although the existence of waves is ensured by general arguments, the strength and type of wave depend on the details of the gravitational theory, and hence the experimental investigation of the properties of the waves would serve as a test of the theory. Even more important: gravitational-wave astronomy would be a useful adjunct to optical, radio, and X-ray astronomy. Gravitational waves would permit us to "look" into the very cores of quasars and other regions of strong gravitational fields. The energy, pulse shape, and polarization of bursts of gravitational radiation could reveal to us the astrophysical processes in which these bursts are generated. Table 5.1 [adapted from Press and Thorne (1972)] lists gravitational-wave frequency bands and their likely origins.

TABLE 5.1 FREQUENCY BANDS FOR GRAVITATIONAL
WAVES

Designation	Frequency	Typical sources
Extremely low frequency	10^{-7} to 10^{-4} Hz	Slow binaries, black holes ($>10^8 \, M_\odot$)
Very low frequency	10^{-4} to 10^{-1} Hz	Fast binaries, black holes ($<10^8 \, M_\odot$), white-dwarf vibrations
Low frequency	10^{-1} to 10^2 Hz	Binary pulsars, black holes ($<10^5 \, M_\odot$)
Medium frequency	10^2 to 10^5 Hz	Supernovas, pulsar vibrations
High frequency	10^5 to 10^8 Hz	Man-made?
Very high frequency	10^8 to 10^{11} Hz	Blackbody, cosmological?

The most promising frequency band is that of medium frequency, from 10^2 to 10^5 Hz. There are several probable sources of gravitational waves in this band and, fortunately, detectors that respond to waves in this band can be built. There is little doubt that gravitational waves are incident on the Earth; the question is, can we build a detector sufficiently sensitive to feel them?

5.1 PLANE WAVES

According to Eqs. [3.55] and [3.56], the linear field equation *in vacuo* is

$$\partial_\lambda \partial^\lambda \phi^{\mu\nu} = 0 \qquad [1]$$

with the gauge condition

$$\partial_\mu \phi^{\mu\nu} = 0 \qquad [2]$$

Let us look for plane-wave solutions of the form

$$\phi^{\mu\nu} = \epsilon^{\mu\nu} \cos k_\alpha x^\alpha \qquad [3]$$

where $\epsilon^{\mu\nu}$ is a constant tensor, and k_α is a constant vector; these are called, respectively, the *polarization tensor* and the *wave vector*. This plane wave will satisfy Eqs. [1] and [2] provided that

$$k^\alpha k_\alpha = 0 \qquad [4]$$

$$\epsilon^{\mu\nu} k_\mu = 0 \qquad [5]$$

■ *Exercise 1.* Show this. ■

The frequency of the wave is $\omega = k^0$. According to Eq. [4],

$$k^0 = (k_x{}^2 + k_y{}^2 + k_z{}^2)^{1/2} = |\mathbf{k}| \qquad [6]$$

and hence $\omega/|\mathbf{k}| = 1$, so the speed of the wave is that of light.

The tensor $\epsilon^{\mu\nu}$ must be symmetric, in accord with the general requirement that $h^{\mu\nu}$ and $\phi^{\mu\nu}$ be symmetric. A symmetric second-rank tensor has ten independent components, but Eq. [5] imposes an extra four conditions on these ten components. Hence $\epsilon^{\mu\nu}$ has only six independent components, that is, there are six linearly independent tensors that will solve Eq. [5]. Let us suppose that the wave propagates in the z direction, with

$$k^\alpha = (\omega, 0, 0, \omega) \qquad [7]$$

and

$$\phi = \epsilon^{\mu\nu} \cos (\omega t - \omega z) \qquad [8]$$

For this special case, the linearly independent solutions of Eq. [5] may be taken to be the following:

$$\epsilon_{(1)}{}^\mu \epsilon_{(1)}{}^\nu - \epsilon_{(2)}{}^\mu \epsilon_{(2)}{}^\nu \qquad [9]$$

$$\epsilon_{(1)}{}^\mu \epsilon_{(2)}{}^\nu + \epsilon_{(1)}{}^\nu \epsilon_{(2)}{}^\mu \qquad [10]$$

$$\epsilon_{(1)}{}^\mu k^\nu + \epsilon_{(1)}{}^\nu k^\mu \qquad [11]$$

$$\epsilon_{(2)}{}^\mu k^\nu + \epsilon_{(2)}{}^\nu k^\mu \qquad [12]$$

$$k^\mu k^\nu \qquad [13]$$

$$\epsilon_{(1)}{}^{\mu}\epsilon_{(1)}{}^{\nu} + \epsilon_{(2)}{}^{\mu}\epsilon_{(2)}{}^{\nu} \qquad [14]$$

where the vectors $\epsilon_{(1)}{}^{\mu}$ and $\epsilon_{(2)}{}^{\mu}$ are defined as follows:

$$\epsilon_{(1)}{}^{\mu} = (0, 1, 0, 0) \qquad [15]$$

$$\epsilon_{(2)}{}^{\mu} = (0, 0, 1, 0) \qquad [16]$$

■ *Exercise 2.* Show that [9]-[14] satisfy Eq. [5]. Write each of the tensors [9]-[14] as a matrix, and show that they are linearly independent. ■

■ *Exercise 3.* Show that the tensors [9]-[13] are traceless,

$$\epsilon_{\mu}{}^{\mu} = 0 \qquad [17]$$

and show that [14] is *not* traceless. ■

The polarization tensors [9] and [10] are called *transverse*. They have the form

$$\epsilon_{\oplus}{}^{\mu\nu} = \epsilon_{(1)}{}^{\mu}\epsilon_{(1)}{}^{\nu} - \epsilon_{(2)}{}^{\mu}\epsilon_{(2)}{}^{\nu} = \begin{pmatrix} 0 & 0 & 0 & 0 \\ 0 & 1 & 0 & 0 \\ 0 & 0 & -1 & 0 \\ 0 & 0 & 0 & 0 \end{pmatrix} \qquad [18]$$

and

$$\epsilon_{\otimes}{}^{\mu\nu} = \epsilon_{(1)}{}^{\mu}\epsilon_{(2)}{}^{\nu} + \epsilon_{(1)}{}^{\nu}\epsilon_{(2)}{}^{\mu} = \begin{pmatrix} 0 & 0 & 0 & 0 \\ 0 & 0 & 1 & 0 \\ 0 & 1 & 0 & 0 \\ 0 & 0 & 0 & 0 \end{pmatrix} \qquad [19]$$

These are the only polarizations that correspond to physical gravitational waves. The other four polarizations, [11]-[14], carry no energy and no momentum. In fact, waves of the other four types can always be completely canceled by a gauge transformation. For instance, consider a wave of the type [11],

$$\phi^{\mu\nu} = (\epsilon_{(1)}{}^{\mu}k^{\nu} + \epsilon_{(1)}{}^{\nu}k^{\mu}) \cos k_{\alpha}x^{\alpha} \qquad [20]$$

The gauge transformation [3.49] for $h^{\mu\nu}$ implies the gauge transformation

$$\phi^{\mu\nu} \rightarrow \phi'^{\mu\nu} = \phi^{\mu\nu} + \partial^{\nu}\Lambda^{\mu} + \partial^{\mu}\Lambda^{\nu} - \eta^{\mu\nu}\partial_{\alpha}\Lambda^{\alpha} \qquad [21]$$

for $\phi^{\mu\nu}$. It is easy to see that this gives $\phi'^{\mu\nu} = 0$ if we take

$$\Lambda^\mu = - \epsilon_{(1)}{}^\mu \sin k_\alpha x^\alpha \qquad [22]$$

Thus, the waves of the types [11]-[14] merely represent different choices of the gauge function; they are "gauge waves," not physical waves.

If we substitute the wave solution [3] into the energy-momentum tensor $t_{(1)}{}^{\mu\nu}$ of Eq. [3.62], we obtain

$$t_{(1)}{}^{\mu\nu} = \frac{1}{2}\left(\epsilon^{\alpha\beta}\epsilon_{\alpha\beta} - \frac{1}{2}\epsilon^\alpha{}_\alpha \epsilon^\beta{}_\beta \right)k^\mu k^\nu \sin^2 k_\sigma x^\sigma \qquad [23]$$

The transverse waves, with amplitudes $A_\oplus$ and $A_\otimes$,

$$\phi_\oplus{}^{\mu\nu} = A_\oplus \epsilon_\oplus{}^{\mu\nu} \cos k_\sigma x^\sigma \qquad [24]$$

$$\phi_\otimes{}^{\mu\nu} = A_\otimes \epsilon_\otimes{}^{\mu\nu} \cos k_\sigma x^\sigma \qquad [25]$$

give, respectively, energy-momentum tensors

$$t_{(1)}{}^{\mu\nu} = (A_\oplus)^2 k^\mu k^\nu \sin^2 k_\sigma x^\sigma \qquad [26]$$

$$t_{(1)}{}^{\mu\nu} = (A_\otimes)^2 k^\mu k^\nu \sin^2 k_\sigma x^\sigma \qquad [27]$$

The $t_{(1)}{}^{03}$ component of these tensors gives the energy flux in the z direction. For the average energy flux, we must replace $\sin^2 k_\sigma x^\sigma$ by $\frac{1}{2}$, and we obtain, respectively,

$$t_{(1)}{}^{03} = \frac{1}{2c}(A_\oplus)^2\omega^2 \qquad [28]$$

$$t_{(1)}{}^{03} = \frac{1}{2c}(A_\otimes)^2\omega^2 \qquad [29]$$

where we have inserted a factor of $1/c$, needed to express the energy flux in cgs units.

Each of the "gauge waves" of types [11]-[14] has $t_{(1)}{}^{\mu\nu} = 0$. Furthermore, if $\epsilon^{\mu\nu}$ in Eq. [23] is an arbitrary superposition of the polarization tensors [9]-[14], then the value of $t_{(1)}{}^{\mu\nu}$ is exactly what it is if the terms involving [11]-[14] are omitted from the superposition; that is, the value of $t_{(1)}{}^{\mu\nu}$ depends on the amplitudes $A_\oplus$ and $A_\otimes$ only (see Problem 3).

■ *Exercise 4.* Show that gravitational waves, just as electromagnetic waves, carry not only energy but also momentum. Show that the flux of momentum is equal to the energy flux (in our units). ■

It is often convenient to make use of the complex waves

$$\epsilon_{\oplus}{}^{\mu\nu}\, e^{ik_{\alpha}x^{\alpha}} \qquad\qquad [30]$$

$$\epsilon_{\otimes}{}^{\mu\nu}\, e^{ik_{\alpha}x^{\alpha}} \qquad\qquad [31]$$

in which case it is understood that the real part is to be regarded as the physical wave.

We can also define circularly polarized waves as

$$(\epsilon_{\oplus}{}^{\mu\nu} - i\epsilon_{\otimes}{}^{\mu\nu})\, e^{ik_{\alpha}x^{\alpha}} \qquad\qquad [32]$$

$$(\epsilon_{\oplus}{}^{\mu\nu} + i\epsilon_{\otimes}{}^{\mu\nu})\, e^{ik_{\alpha}x^{\alpha}} \qquad\qquad [33]$$

It can be shown that such circularly polarized waves carry angular momentum. The amount of angular momentum is proportional to the amount of energy carried by the wave:

$$[\text{angular momentum in wave}] = \frac{2}{\omega}\,[\text{energy in wave}] \qquad [34]$$

Waves with the polarization [32] have angular momentum parallel to the energy flux and are said to have *positive helicity*; those with the polarization [33] have angular momentum antiparallel to the energy flux and are said to have *negative helicity*. The result [34] cannot be obtained directly from our solution [3]; this solution ignores the boundaries of the wave in the transverse direction and it is precisely the boundary region which is crucial for the transport of angular momentum.

The quantum-mechanical interpretation of Eq. [34] is that the quanta of the gravitational field, or gravitons, have spin $2\hbar$. The ratio of angular momentum to energy per quantum is

$$\frac{2\hbar}{\hbar\omega} = \frac{2}{\omega}$$

in agreement with Eq. [34]. Incidentally: Our result that there exist only two physical states of polarization for a gravitational wave of given momentum corresponds to the well-known result of relativistic quantum theory that a particle of mass zero must have its spin along the direction of motion (positive helicity) or opposite to the direction of motion (negative helicity); any other direction for the spin is forbidden.

What is the effect of a gravitational wave on a particle? The equation of motion of a particle is (see Eq. [3.93])

$$\frac{du_\mu}{d\tau} = - \kappa \left[h_{\mu\alpha,\beta} - \tfrac{1}{2} h_{\alpha\beta,\mu} \right] u^\alpha u^\beta \qquad [35]$$

For a particle initially at rest, $u_0 = 1$ and $u_k = 0$, and hence

$$\frac{du_\mu}{dt} = - \kappa \left(h_{\mu 0,0} - \tfrac{1}{2} h_{00,\mu} \right) \qquad [36]$$

But for each of the physical gravitational waves [24] and [25] we have $h_{\mu 0} = 0$; hence the right side of Eq. [36] vanishes. Thus, a particle initially at rest remains at rest. Although this means that the position of the particle, as measured by our coordinates, does not change, we must be careful not to jump to the conclusion that there is no physical effect at all. Consider two particles placed on the x-axis, one at $x = x_0$ and the other at $x = - x_0$, and suppose they are subjected to a gravitational wave of the type [24]. According to Eq. [36] the coordinate interval between these particles remains fixed at the value $2x_0$. But the physical distance, as measured by a meter stick or by radar ranging, depends not only on Δx, but also on the metric of spacetime. We already saw in Section 3.6 that the metric is not $\eta_{\mu\nu}$, but rather

$$g_{\mu\nu} = \eta_{\mu\nu} + \kappa h_{\mu\nu} \qquad [37]$$

If so, then the measured distance between the two particles is

$$\Delta l^2 = - g_{11} \Delta x^2 = (1 - \kappa h_{11})(2x_0)^2$$

$$= (1 - \kappa A_\oplus \cos \omega t)(2x_0)^2$$

In the context of the linear approximation, the amplitude $A_\oplus$ is small, and hence

$$\Delta l \simeq \left(1 - \frac{\kappa}{2} A_\oplus \cos \omega t \right) (2x_0) \qquad [38]$$

Thus, the distance between the particles oscillates. Expressed in another way, the particles remain at rest relative to the coordinates, but the coordinates oscillate relative to our meter sticks, and hence we perceive that the particles move.

■ *Exercise 5.* Show that the distance between two particles on the y-axis, at $y = \pm y_0$, exposed to the gravitational wave [24], varies as follows:

$$\Delta l \simeq \left[1 + \frac{\kappa}{2} A_{\oplus} \cos \omega t\right](2y_0) \quad \blacksquare \qquad\qquad [39]$$

Comparison of Eqs. [38] and [39] shows that the distance between the first pair of particles is at minimum when the distance between the second pair is at maximum, and vice versa.

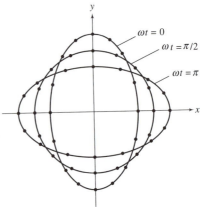

Fig. 5.1 Deformation of a circular "necklace" of particles by a gravitational wave with polarization of type ⊕.

Fig. 5.1 displays the deformation produced by the gravitational wave [24] on a "necklace" of *free* particles originally placed in a circle in the x-y plane. Fig. 5.2 shows the corresponding result for the gravitational wave [25]. It is obvious that the only difference between these figures is a rotation by 45°. In fact, the polarization tensor [18] can be obtained from [19] by a 45° rotation.

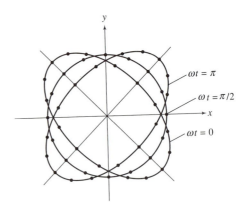

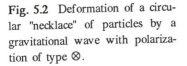

Fig. 5.2 Deformation of a circular "necklace" of particles by a gravitational wave with polarization of type ⊗.

■ *Exercise 6.* Prove this. ■

Finally, Figs. 5.3 and 5.4 show the deformations produced by the circularly polarized waves [32] and [33]. In these two cases, only the deformation, or bulge, rotates in the direction shown; the particles on the necklace do not rotate; they only oscillate in and out around their initial positions.

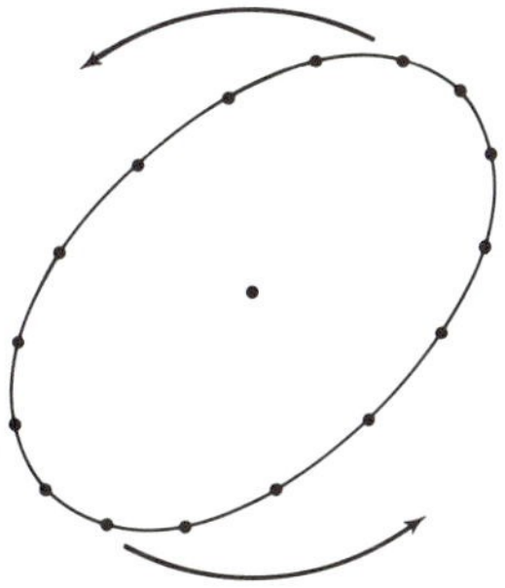

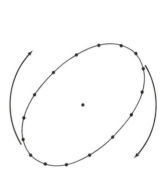

Fig. 5.3 Deformation produced by a wave of positive helicity. The deformation rotates in the counterclockwise direction if the wave is propagating out of the plane of the page, toward you.

Fig. 5.4 Deformation produced by a wave of negative helicity. The deformation rotates in the clockwise direction.

The deformations shown in all these figures are essentially tidal effects, similar to those of Section 1.8, but time dependent rather than static.

Let us calculate the tidal force needed to produce the motion given in Eq. [38]. The acceleration of each particle toward the origin is

$$\frac{d^2}{dt^2}\frac{\Delta l}{2} = -\frac{\kappa}{2}A_\oplus x_0 \frac{d^2}{dt^2}\cos \omega t$$

$$= \frac{\kappa}{2}A_\oplus \omega^2 x_0 \cos \omega t \qquad [40]$$

Hence the tidal force on a particle of mass m on the x-axis is

$$f_x = \left[m \frac{\kappa}{2} A_\oplus \omega^2 \cos \omega t \right] x_0 \tag{41}$$

Likewise, the tidal force on a particle on the y-axis, at $y = y_0$, is

$$f_y = - \left[m \frac{\kappa}{2} A_\oplus \omega^2 \cos \omega t \right] y_0 \tag{42}$$

These expressions should be compared with our earlier results (see Eqs. [1.48], [1.49]) for the tidal force in a static gravitational field. In both cases the strength of the tidal force increases directly with the distance. Note that in Eqs. [41] and [42], x_0 and y_0 may be regarded, to a sufficient approximation, as the distances of the particles from the origin. The difference between the true distance Δl and the coordinate interval Δx is crucial in Eq. [38], but it is not important in Eqs. [41] and [42], since there it leads only to higher-order corrections.

Incidentally: The forces f_x and f_y associated with a gravitational wave can be written in the form

$$f_x = g(t)x_0 \qquad f_y = - g(t)y_0$$

where we are now assuming that both x_0 and y_0 are nonzero, so both the forces [41] and [42] are present simultaneously. From this it is easy to see that

$$\frac{\partial f_x}{\partial x_0} + \frac{\partial f_y}{\partial y_0} = 0 \tag{43}$$

Hence the divergence of the force $\mathbf{f}$ is zero, which tells us that the tidal force field can be represented graphically by field lines. The field lines for a gravitational wave of type $\oplus$ are drawn in Fig. 5.5, for $t = 0$. Obviously, such a force field will tend to produce the deformations shown in Fig. 5.1. The field lines for a gravitational wave of type $\otimes$ are drawn in Fig. 5.6.

■ *Exercise 7.* Show that the radial component of the tidal force is

$$f_r = \left[m \frac{\kappa}{2} A_\oplus \omega^2 \cos \omega t \right] r_0 \cos 2\phi \tag{44}$$

where $r_0 = (x_0^2 + y_0^2)^{1/2}$ is the radial coordinate of the particle, and ϕ is the azimuthal coordinate. ■

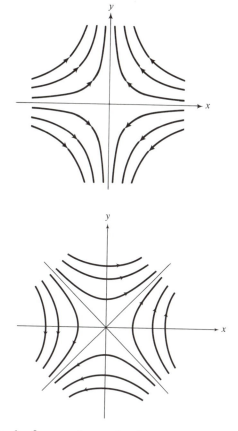

Fig. 5.5 The tidal field lines for a gravitational wave of type $\oplus$. The field lines reverse direction every half period.

Fig. 5.6 The tidal field lines for a gravitational wave of type $\otimes$.

Note that our results are valid only for weak gravitational waves. To be precise, the wave must have a small amplitude, so $\kappa A \ll 1$.

As an illustration of these results, consider the effect of a gravitational wave on the Earth-Moon system. If the frequency of the wave is much larger than the frequency of the orbital motion of the Moon, then we can regard the Earth and the Moon as two free particles. In response to the gravitational wave, incident transversely, the Earth-Moon distance will oscillate as indicated by Eq. [38]. The quantity $2x_0$ may be interpreted as the Earth-Moon distance measured when the wave is absent ($2x_0 \simeq 3.8 \times 10^{10}$ cm).

For a strong astrophysical source of gravitational radiation--say, a supernova at a distance of 10 kpc*--the gravitational wave might have an amplitude of about $\kappa h_{11} = \kappa A_\oplus \simeq 10^{-18}$ (see Table 5.4). This means that the Earth-Moon distance would vary by about 1 part in 10^{18}, or about 10^{-8} cm. By means of a laser pulse sent to the Moon and reflected back to the Earth, the distance can nowadays be measured to within $\simeq 10$ cm. Clearly, this laser-ranging technique is not suffi-

---

* The parsec (pc) equals 3.26 light-years.

ciently precise for detection of the gravitational-wave effect.

Much more precise measurement of distance can be carried out with interferometers of laboratory size. Several interferometers with arm lengths of tens of meters have been built to detect the changes of distance produced by an incident gravitational wave. These interferometers can detect distance changes of about 1 part in 10^{-17}, and larger interferometers now under construction are expected to detect changes of about 1 part in 10^{-22} (see Section 5.7).

The expressions for the change in distance (Eqs. [38] and [39]) and for the tidal force (Eqs. [41] and [42]) hinge, of course, on the interpretation of [37] as the metric of spacetime. But some of the important results of the present chapter can be derived without relying on this geometric interpretation. In particular, the expressions for the cross section for absorption of gravitational waves (see Eqs. [127], [130]) can be obtained by arguments that rely only on general theorems of scattering theory (Weinberg, 1972) or on the principle of detailed balance (Rees, Ruffini, and Wheeler, 1974). However, for the sake of simplicity we will base our derivation on the above expressions for the tidal force.

5.2 THE EMISSION OF GRAVITATIONAL RADIATION

If we have some system with an energy-momentum tensor distribution that changes in time, then gravitational waves will be emitted by the system. The calculation of the radiation fields in the linear approximation proceeds in much the same way as the calculation of the electromagnetic radiation fields associated with a time-dependent charge distribution. We begin with the linear field equation [3.55],

$$\partial_\lambda \partial^\lambda \phi^{\mu\nu}(t, \mathbf{x}) = - \kappa T^{\mu\nu}(t, \mathbf{x}) \tag{45}$$

where it is understood that $T^{\mu\nu}$ is the energy-momentum tensor of matter. The retarded solution of Eq. [45] is

$$\phi^{\mu\nu}(t, \mathbf{x}) = - \frac{\kappa}{4\pi} \int \frac{T^{\mu\nu}(t - |\mathbf{x} - \mathbf{x}'|, \mathbf{x}')}{|\mathbf{x} - \mathbf{x}'|} d^3x' \tag{46}$$

■ *Exercise 8.* Verify that this is a solution of Eq. [45]. [Hint: Use $\nabla^2(1/|\mathbf{x} - \mathbf{x}'|)$ $= - 4\pi\delta(\mathbf{x} - \mathbf{x}')$ and proceed by analogy with the calculation given in textbooks of electromagnetism.] ■

We will assume that the field point $\mathbf{x}$ is in the radiation zone, far away from the matter system (see Fig. 5.7); this means that we can replace $|\mathbf{x} - \mathbf{x}'|$ by $|\mathbf{x}|$ in the denominator of the integrand in Eq. [46].

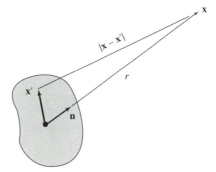

Fig. 5.7 The field point **x** is at a large distance from the region (shaded) containing the distribution of energy momentum.

We will further assume that $T^{\mu\nu}$ does not change very quickly in time; if $T^{\mu\nu}$ does not have an excessively strong time dependence, then it will be a good approximation to replace $t - |\mathbf{x} - \mathbf{x}'|$ by $t - |\mathbf{x}|$ in the numerator of the integrand in Eq. [46].* With the notation $r \equiv |\mathbf{x}|$ we then obtain

$$\phi^{\mu\nu}(t, \mathbf{x}) = - \frac{\kappa}{4\pi r} \int T^{\mu\nu}(t - r, \mathbf{x}') \, d^3x' \qquad [47]$$

In the linear approximation, $T^{\mu\nu}$ satisfies the condition

$$\partial_\nu T^{\mu\nu} = 0 \qquad [48]$$

Let us separate Eq. [48] into space and time components:

$$\frac{\partial}{\partial t} T^{k0} = - \partial_l T^{kl} \qquad [49]$$

$$\frac{\partial}{\partial t} T^{00} = - \partial_l T^{0l} \qquad [50]$$

In consequence of Eq. [49], we have the following identity:

---

* The condition for **x** in the radiation zone can be expressed as $r \gg \lambda$. The condition for "slow" change in $T^{\mu\nu}$ can be expressed in the alternative forms $\omega b/c \ll 1$, $b/\lambda \ll 1$, or $v/c \ll 1$, where ω is the frequency of the emitted radiation, λ the wavelength, b the typical dimension of the system, and v the typical speed of the particles in the system. Corrections to our approximation give multipole radiation of higher order.

$$\int T^{kl} \, d^3x = \frac{1}{2} \frac{\partial}{\partial t} \int (T^{k0} \, x^l + T^{l0} x^k) \, d^3x \qquad [51]$$

where the integration volume includes all of the matter system.

■ *Exercise 9.* Verify this identity. (Hint: Use Eq. [49] on the right side of Eq. [51], and integrate by parts.) ■

Similarly, as a consequence of Eq. [50], we have the identity

$$\int (T^{k0} \, x^l + T^{l0} x^k) \, d^3x = \frac{\partial}{\partial t} \int T^{00} x^k \, x^l \, d^3x \qquad [52]$$

■ *Exercise 10.* Verify this identity. (Hint: Use Eq. [50] and integration by parts.) ■

If we combine Eqs. [51] and [52], we find that

$$\int T^{kl} \, d^3x = \frac{1}{2} \frac{\partial^2}{\partial t^2} \int T^{00} x^k \, x^l \, d^3x \qquad [53]$$

which shows that most of the integrals appearing on the right side of [47] can be expressed in terms of T^{00}. In essence, the reason why we can express the integral over T^{kl} in terms of derivatives over T^{00} is that whenever the momentum flux density T^{kl} is not zero, the momentum density must accumulate somewhere in the volume, that is, T^{k0} must change. And whenever the energy flux density T^{k0} is nonzero, the energy density must accumulate somewhere in the volume, that is, T^{00} must change.

For nonrelativistic matter, we can approximate

$$T^{00} = [\text{energy density}] \simeq [\text{rest-mass density}] = \rho \qquad [54]$$

and we obtain

$$\phi^{kl}(t, \mathbf{x}) = - \left[\frac{\kappa}{8\pi r} \frac{\partial^2}{\partial t^2} \int \rho(\mathbf{x}') \, x'^k x'^l \, d^3x' \right]_{t-r} \qquad [55]$$

This integral can also be expressed in terms of the quadrupole-moment tensor (see Eq. [1.25])

$$Q^{kl} = \int (3x'^k x'^l - r'^2 \delta_k{}^l) \rho(\mathbf{x'}) \, d^3 x' \qquad [56]$$

in terms of which

$$\phi^{kl}(t, \mathbf{x}) = - \frac{\kappa}{8\pi r} \frac{1}{3} \left[\frac{\partial^2}{\partial t^2} Q^{kl} + \delta_k^l \frac{\partial^2}{\partial t^2} \int r'^2 \rho(\mathbf{x'}) \, d^3 x' \right]_{t-r} \qquad [57]$$

For the calculation of the energy flux, we can omit the term proportional to δ_k^l in the brackets of Eq. [57], because this part of the wave carries away no energy. To see this, note that if the distance r is large enough, then ϕ^{kl} can be regarded, to a sufficient approximation, as a plane wave in the vicinity of the point $\mathbf{x}$. According to the preceding section, the only plane-wave polarizations that carry energy are $\epsilon_\oplus{}^{kl}$ and $\epsilon_\otimes{}^{kl}$ as given by Eqs. [18] and [19]; obviously, δ_k^l is not a polarization of this type. If we omit the δ_k^l term, then

$$\phi^{kl}(t, \mathbf{x}) = - \frac{\kappa}{8\pi r} \frac{1}{3} \ddot{Q}^{kl} \qquad [58]$$

where the dots stand for time derivatives and it is understood that the right side is to be evaluated at the retarded time $t - r$.

The energy flux in the radial direction is given by (see Eq. [3.62])

$$t_{(1)}{}^{0s} n^s = \frac{n^s}{4} \left(2\phi^{\alpha\beta,0} \phi_{\alpha\beta}{}^{,s} - \phi^{,0} \phi^{,s} \right) \qquad [59]$$

where $n^s = (n_x, n_y, n_z) = (x/r, y/r, z/r)$ is the unit vector in the radial direction (see Fig. 5.7). Let us separate $\phi^{\alpha\beta}$ into space and time parts,

$$t_{(1)}{}^{0s} n^s = \frac{n^s}{4} \left(\phi^{kl,0} \phi^{kl,s} - 4\phi^{k0,0} \phi^{k0,s} + 2\phi^{00,0} \phi^{00,s} - \phi^{kk,0} \phi^{ll,s} \right.$$
$$\left. + \phi^{kk,0} \phi^{00,s} + \phi^{00,0} \phi^{kk,s} - \phi^{00,0} \phi^{00,s} \right) \qquad [60]$$

where we have taken into account that, according to Eqs. [58] and [56],

$$\phi^{kk} \propto Q^{kk} = 0$$

In the evaluation of $\phi_{\alpha\beta,s}$ we can neglect the derivative of the factor $1/r$ appearing in Eq. [57], since we are assuming that r is very large.* The only other place where a dependence on r occurs is in the retarded time, $t - r$; this means that r and t appear *only* in the combination $t - r$, and therefore

$$\phi^{\alpha\beta}_{,s} = - \phi^{\alpha\beta}_{,0} \frac{\partial r}{\partial x^s} = - \phi^{\alpha\beta}_{,0} n^s \qquad [61]$$

Accordingly, the derivative appearing in the first term in the parentheses in Eq. [60] can be written

$$\phi^{kl,s} = \phi^{kl,0} n^s \qquad [62]$$

We can obtain convenient expressions for the other terms appearing in Eq. [60] by beginning with our familiar gauge condition

$$\phi^{\mu 0}_{,0} = - \phi^{\mu l}_{,l} \qquad [63]$$

This gives

$$\phi^{k0}_{,0} = - \phi^{kl}_{,l} = \phi^{kl}_{,0} n^l = \phi^{kl,0} n^l \qquad [64]$$

and also

$$\phi^{00,0} = - \phi^{0l}_{,l} = \phi^{0l}_{,0} n^l = \phi^{kl,0} n^k n^l \qquad [65]$$

Substituting Eqs. [62], [64], and [65] into Eq. [60], we obtain

$$t_{(1)}{}^{0s} n^s = \frac{n^s}{4} \left(2\phi^{kl,0} \phi^{kl,0} n^s - 4\phi^{kl,0} n^l \phi^{km,0} n^m n^s \right.$$

$$\left. + \phi^{kl,0} n^k n^l \phi^{mr,0} n^m n^r n^s \right) \qquad [66]$$

If we use $n^s n^s = 1$ and then Eq. [58], this becomes

$$t_{(1)}{}^{0s} n^s = \frac{1}{4}\left(2\phi^{kl,0} \phi^{kl,0} - 4\phi^{kl,0} \phi^{km,0} n^l n^m + \phi^{kl,0} \phi^{mr,0} n^k n^l n^m n^r \right)$$

---

* This approximation is justified whenever $r \gg \lambda$, where λ is the wavelength of the emitted radiation.

$$= \frac{1}{9} \left(\frac{\kappa}{8\pi r} \right)^2 \left[\frac{1}{2} \dddot{Q}^{kl} \dddot{Q}^{kl} - \dddot{Q}^{kl} \dddot{Q}^{km} n^l n^m + \frac{1}{4} \dddot{Q}^{kl} \dddot{Q}^{mr} n^k n^l n^m n^r \right]$$

[67]

The energy radiated by the system per unit solid angle and unit time in the direction n^s is

$$- \frac{d^2E}{dt d\Omega} = r^2 t_{(1)}^{0s} n^s$$

[68]

Obviously, the angular distribution is very complicated; however, the total power radiated can be obtained by integrating Eq. [68] over all solid angles, and the final expression turns out to be quite simple. We have

$$- \frac{dE}{dt} = \int r^2 t_{(1)}^{0s} n^s \, d\Omega$$

$$= \frac{1}{9} \left(\frac{\kappa}{8\pi} \right)^2 \left[\frac{1}{2} \dddot{Q}^{kl} \dddot{Q}^{kl} \int d\Omega - \dddot{Q}^{kl} \dddot{Q}^{km} \int n^l n^m \, d\Omega \right.$$

$$\left. + \frac{1}{4} \dddot{Q}^{kl} \dddot{Q}^{mr} \int n^k n^l n^m n^r \, d\Omega \right]$$

[69]

The average values of $n^l n^m$ and $n^k n^l n^m n^r$ over the surface of a sphere are, respectively,

$$\frac{1}{4\pi} \int n^l n^m \, d\Omega = \frac{1}{3} \delta_l^m$$

[70]

$$\frac{1}{4\pi} \int n^k n^l n^m n^r \, d\Omega = \frac{1}{15} (\delta_k^l \delta_m^r + \delta_k^m \delta_l^r + \delta_k^r \delta_l^m)$$

[71]

■ *Exercise 11.* Show this. (Hint: Unless the indices appearing in [70] and [71] are equal in pairs, the integral vanishes because the integrand is odd. This shows that the right and left sides of these equations must be *proportional*; determine the constant of proportionality.) ■

Substituting Eqs. [70] and [71] into Eq. [69], we obtain

$$- \frac{dE}{dt} = \frac{4\pi}{45} \left(\frac{\kappa}{8\pi} \right)^2 \dddot{Q}^{kl} \dddot{Q}^{kl} \qquad [72]$$

■ *Exercise 12.*   Check this. ■

If we insert the value $\kappa^2 = 16\pi G$ and return to cgs units, we obtain a radiated power of

$$- \frac{dE}{dt} = \frac{G}{45c^5} \dddot{Q}^{kl} \dddot{Q}^{kl} \qquad [73]$$

In Section 1.3 we saw that a shift of the origin of coordinates changes the quadrupole moment by an additive constant (see Eq. [1.28]). Since the emission of radiation involves only the time-dependent part of Q^{kl}, it follows that changes in the origin will not affect our results; we can choose any convenient point as origin.

The radiation we have calculated above is often called *quadrupole radiation*. Since our results all involve the quadrupole-moment tensor, this is as good a name as any. However, it should be kept in mind that if we were to make in the electromagnetic case the same kind of approximations that we have made above in the gravitational case, we would obtain electromagnetic dipole radiation. The analog of gravitational "quadrupole" radiation is electromagnetic dipole radiation--these kinds of radiation correspond to the lowest-order approximation we can make in the retarded solution to the field equation (Eq. [47] and the corresponding equation in electromagnetism).

5.3 EMISSION BY A VIBRATING QUADRUPOLE

According to Eq. [73], any system of masses with a time-dependent quadrupole moment and with $\dddot{Q}^{kl} \neq 0$ will radiate gravitational waves. We can classify the sources of gravitational waves into *periodic* and *bursting* sources. The former have a quadrupole moment that varies periodically; they emit harmonic waves. The latter have a quadrupole moment that varies for a short while in a nonperiodic manner; they emit a burst, or a pulse, of gravitational waves. For example, vibrating masses and rotating masses are periodic sources, whereas a mass suffering a brief and sharp acceleration while colliding with another mass is a bursting source.

The simplest periodic source of gravitational waves consists of two equal masses connected by a spring (see Fig. 5.8). This ideal system is called a linear quadrupole. We will treat the masses as particles, but they could equally well be spherical (for a system of spherical masses, the quadrupole moment is the same as for a system of particles). The ideal linear quadrupole is mainly of theoretical interest. Although such a linear quadrupole is easy to construct in a laboratory, the amount of gravitational radiation is insignificant for masses of laboratory size. There are no astrophysical sources of gravitational radiation in the shape of the linear quadrupole. However, a vibrating star, in an elongated mode of vibration, exhibits some of the general features of the vibrating linear quadrupole, and we can extract some crude estimates for the radiation emitted by such a star if we use the simple formulas valid for the linear quadrupole.

■ *Exercise 13.* Show that the quadrupole moment of a system of spherical masses is the same as the quadrupole moment of a corresponding system of point masses located at the centers of the spheres. ■

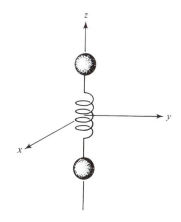

Fig. 5.8 An oscillating quadrupole.

As shown in Fig. 5.8, the masses move along the z-axis; their positions are, respectively,

$$z = \pm(b + a \sin \omega t) \tag{74}$$

For the sake of simplicity, we assume that $a \ll b$. Hence we can approximate

$$z^2 \simeq b^2 + 2ab \sin \omega t \tag{75}$$

Since our system consists of two discrete point masses, the integral [56] for the quadrupole moment reduces to a sum of two terms, which are equal. The retarded value of the quadrupole moment of the system is

$$Q^{kl} \simeq \left[1 + \frac{2a}{b} \sin \omega(t - r) \right] Q(0)^{kl} \tag{76}$$

where

$$Q(0)^{kl} = \begin{pmatrix} -2mb^2 & 0 & 0 \\ 0 & -2mb^2 & 0 \\ 0 & 0 & 4mb^2 \end{pmatrix} \qquad [77]$$

is the static quadrupole moment.

■ **Exercise 14.** Derive Eq. [76]. ■

■ **Exercise 15.** Show that the radiated gravitational field is

$$\phi^{kl} = \frac{\kappa}{8\pi r} \frac{2a}{3b} \, \omega^2 \sin \omega(t - r) \, Q(0)^{kl} \qquad ■ \qquad [78]$$

The angular distribution of radiation given by Eq. [68] is

$$-\frac{d^2E}{dt d\Omega} = \frac{1}{9} \left(\frac{\kappa}{8\pi} \right)^2 \left[\frac{1}{2} (\dddot{Q}^{11})^2 + \frac{1}{2} (\dddot{Q}^{22})^2 + \frac{1}{2} (\dddot{Q}^{33})^2 \right.$$
$$- (\dddot{Q}^{11}n^1)^2 - (\dddot{Q}^{22}n^2)^2 - (\dddot{Q}^{33}n^3)^2$$
$$\left. + \frac{1}{4} (\dddot{Q}^{11}n^1n^1 + \dddot{Q}^{22}n^2n^2 + \dddot{Q}^{33}n^3n^3)^2 \right] \qquad [79]$$

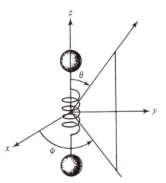

Fig. 5.9 Angular direction for the radiated gravitational wave.

In terms of the usual polar angles, defined in Fig. 5.9, we have

$$n^1 = n_x = \sin \theta \cos \phi \qquad [80]$$
$$n^2 = n_y = \sin \theta \sin \phi \qquad [81]$$
$$n^3 = n_z = \cos \theta \qquad [82]$$

and Eq. [79] reduces to

$$- \frac{d^2E}{dtd\omega} = \left(\frac{\kappa}{8\pi}\right)^2 [2mab\omega^3 \cos \omega(t - r)]^2 \sin^4 \theta \qquad [83]$$

■ **Exercise 16.** Derive [83] from [79]. ■

The radiation pattern is shown in Fig. 5.10. Note that this radiation pattern in no way resembles the electromagnetic quadrupole pattern; in the latter case the flux is zero in at least three perpendicular directions (and their opposite directions). If anything, Fig. 5.10 is vaguely reminiscent of an electromagnetic dipole radiation pattern.

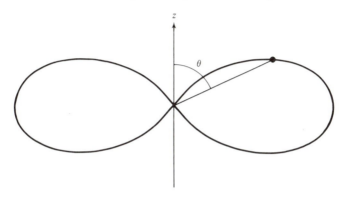

Fig. 5.10 The radiation pattern for the emission of gravitational radiation by a quadrupole oscillating along the z-axis. This is a polar plot of the function $\sin^4 \theta$. The distance from the origin to a point on the curve is proportional to the flux emitted in that direction. The pattern has cylindrical symmetry about the z-axis.

The total emitted power is

$$- \frac{dE}{dt} = \frac{G}{45c^5} \dddot{Q}^{kl} \dddot{Q}^{kl} = \frac{32G}{15c^5} [mab\omega^3 \cos \omega(t - r)]^2 \qquad [84]$$

with a time average

$$- \overline{\frac{dE}{dt}} = \frac{16G}{15c^5} (mab)^2 \omega^6 \qquad [85]$$

Let us define the *damping rate* of the oscillator as the ratio of rate of loss of energy to energy:

$$\gamma_{rad} \equiv - \frac{1}{E} \overline{\frac{dE}{dt}} \qquad [86]$$

Since the energy of each mass is $\frac{1}{2}m\omega^2a^2$, we obtain

$$\gamma_{rad} = \frac{16G}{15c^5} mb^2\omega^4 \qquad\qquad [87]$$

The quantity $1/\gamma_{rad}$ is called the *damping time*; this is the time the oscillator takes to lose a fraction $1/e$ of its energy to radiation.

Although we have derived Eq. [85] for a system of two point masses, or two spherical masses, connected by a spring, it gives an order of magnitude estimate for the gravitational-wave energy emitted in the vibration of any elastic body provided the vibration is, roughly, in one direction.* For example, we can use Eq. [85] to estimate the power radiated by a steel rod in longitudinal vibration; in this case, a is the amplitude of oscillation, m the mass of the rod, b its length, and ω the frequency of its oscillation. A more precise calculation of the radiation emitted by a vibrating elastic body can be performed by replacing the amplitude of oscillation by an effective amplitude (see Eq. [131]).

Radiation emitted by quadrupole vibrations might be important in the case of novas. Nova outbursts occur on white-dwarf stars in binary systems, where accretion from a companion star gradually accumulates nuclear fuel on the dwarf's surface until the fuel reaches a critical mass and explodes. Among other things, the explosion initiates vibrations in the body of the white dwarf, at characteristic frequencies of 0.01 to 1 Hz. The energy released in a nova explosion is typically 10^{45} erg, of which perhaps as much as 10% is deposited in the vibrational motion of the star and is subsequently radiated in the form of gravitational waves.

Higher frequencies and larger amounts of energy are emitted by the vibrations of neutron stars formed in supernova implosions. Since the damping time for the vibrations of a neutron star is quite short (a fraction of a second), this kind of radiation has the form of a burst, and we will deal with it in a later section.

5.4 EMISSION BY A ROTATING QUADRUPOLE

Another simple periodic source of gravitational waves is the rotating quadrupole, consisting of two particles or two spherical masses moving in circular orbits about their common center of mass (see Fig. 5.11). Since the quadrupole moment repeats when the masses move through one-half of their orbit, the frequency of the emitted waves is twice the orbital frequency. The power radiated by this system is

---

* Eq. [85] cannot be applied to the radial pulsations of a sphere; in the case of spherical symmetry, the radiation is exactly zero.

$$- \frac{dE}{dt} = \frac{32G}{5c^5} \left[\frac{m_1 m_2}{m_1 + m_2} \right]^2 r^4 \omega^6 \qquad [88]$$

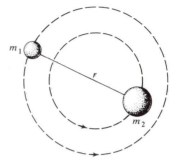

Fig. 5.11 Two spherical masses in circular orbits about their center of mass.

where m_1, m_2 are the masses, r is the distance between them, and ω is orbital frequency.

■ *Exercise 17.* Derive Eq. [88]. ■

■ *Exercise 18.* Show that the radiation emitted in the direction perpendicular to the plane of the orbit is circularly polarized; show that the radiation emitted in the plane of the orbit is "linearly" polarized. ■

We can apply Eq. [88] to a binary star system, consisting of two stars in circular orbits around each other. In this case, the force holding the two masses in their orbits is gravitational, and r and ω are related by Kepler's third law,

$$\omega^2 = \frac{G(m_1 + m_2)}{r^3} \qquad [89]$$

so Eq. [88] becomes

$$- \frac{dE}{dt} = \frac{32G^4}{5c^5 r^5} (m_1 m_2)^2 (m_1 + m_2) \qquad [90]$$

As the gravitating system loses energy by radiation, the distance between the masses decreases at a rate

$$\frac{dr}{dt} = - \frac{64G^3}{5c^5 r^3} m_1 m_2 (m_1 + m_2) \qquad [91]$$

and the orbital frequency increases at a rate

$$\frac{d\omega}{dt} = -\frac{3\omega}{2r}\frac{dr}{dt} = \frac{96}{5}\left[\frac{G(m_1 + m_2)}{c^2 r}\right]^{3/2}\frac{G^2 m_1 m_2}{c^2 r^4}\qquad [92]$$

■ *Exercise 19.* Obtain Eqs. [91] and [92]. (Hint: Use $E = -Gm_1 m_2/2r$, and Eq. [90], and Kepler's third law.) ■

Appreciable amounts of gravitational radiation are emitted by a binary star system provided the orbital radius is small. Table 5.2 gives some examples of known binary systems with small orbital radii (and short orbital periods). The system emitting the most power and producing the largest flux at the Earth is Am CVn, with 3×10^{32} erg/s. This system consists of a blue white dwarf and a low-mass white dwarf (see Fig. 5.12) in an exceptionally small orbit around each other, with a period of only 17.5 min (!). However, the system that produces the largest gravitational-wave amplitude at the Earth is μ Sco, with $\kappa A \simeq 2 \times 10^{-20}$ (although the amplitude of the waves from Am CVn is smaller, their much higher frequency permits them to carry more energy; see Eqs. [28] and [29]). Detectors sensitive to gravitational waves of amplitude $\kappa A \simeq 10^{-20}$ have already been built, but they respond only to waves of much higher frequency than the frequencies radiated by binary systems, and there are no immediate prospects for the direct detection of these gravitational waves.

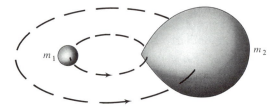

Fig. 5.12 The binary system Am CVn. The compact star has the larger mass (m_1) and is a blue white dwarf; the other star (m_2) is a low-mass white dwarf. There is mass transfer from the latter to the former. (After Faulkner, Flannery, and Warner, 1972.)

However, gravitational waves can be detected indirectly, by the effect that the loss of energy has on the orbital motion of the binary system. In this context, the most interesting system is PSR 1913+16, the binary pulsar discovered by Hulse and Taylor in 1974 during a search for pulsars with the Arecibo radio telescope. This system consists of a pulsar (rotating neutron star that emits radio pulses) in a tight orbit around a companion, either a white dwarf or another neutron star. Hulse and Taylor found that the pulsar frequency exhibits a periodic variation, attributed to the Doppler shift generated by the

TABLE 5.2 BINARY SYSTEMS AS SOURCES OF GRAVITATIONAL RADIATION*

System	Masses	Distance	Wave frequency	Luminosity at Earth	Flux at Earth	Amplitude at Earth
	$(M_\odot)$	(pc)	$(10^{-6}$ Hz)	$(10^{30}$ erg/s)	$(10^{-12}$ erg/cm²·s)	(10^{-22})
Eclipsing binaries						
ι Boo	1.0, 0.5	11.7	86	1.1	68.0	51
μ Sco	12, 12	109	16	51	38.0	210
V Pup	16.5, 9.7	520	16	59	1.9	46
Cataclysmic binaries (novas)						
Am CVn	1.0, 0.041	100	1900	300	240	5
WZ Sge	15, 0.12	75	410	24	37	8
SS Cyg	0.97, 0.83	30	84	2	20	30
Binary X-ray sources (black holes or neutron stars)						
Cyg X-1	30, 6	2500	4.1	1.0	1	4
PSR1913+16	14, 14	5000	70**	0.6	0.2	0.12
			140	2.9	1.1	0.14
			210	5.8	2.1	0.12

* From Douglass and Braginsky (1979).

** This binary emits radiation at harmonic frequencies, because it has an orbit of large eccentricity.

orbital motion of the pulsar around its (undetected) companion. They carefully analyzed the motion by examining the arrival times of individual pulses. Such an analysis of arrival times of individual pulses yields higher accuracy than measurement of the "instantaneous" frequency, because a frequency measurement in a finite time $\Delta \tau$ is subject to the usual uncertainty $\Delta \nu \simeq 1/\Delta \tau$. For instance, if we attempt to measure the "instantaneous" frequency associated with a 1-min segment of the orbit, then $\Delta \nu \simeq 1/60$ Hz, which is a rather large error.

Precise values of the parameters of the binary system have been extracted from a least-squares fit of the arrival times. This analysis not only yielded the orbital period and the projected semi-major axis ($a \sin i$, where i is the angle of inclination of the orbit relative to the line of sight), but also the eccentricity and the rate of precession of the periastron (analogous to the precession of the perihelion of Mercury in the Solar System). Table 5.3 lists the most important parameters of PSR 1913+16. Note that the pulse period and the pulse frequency are known to about 10 significant figures. The analysis also revealed the presence of significant time-dilation effects, which vary in magnitude as the pulsar moves in its elliptical orbit that brings it closer or farther from its companion star. The time-dilation effects arise in part from the gravitational time dilation produced by the gravitational field of the companion, and in part from the "transverse," or special-relativistic, time dilation produced by the orbital speed. The nominal pulsar period listed in Table 5.3 has been corrected for these time-dilation effects (the pulsar period is also subject to a time dilation contributed by the pulsar's own gravitational field; but this is a constant factor, which cannot be determined from the data).

TABLE 5.3 SOME OBSERVED PARAMETERS OF PSR 1913+16

Pulsar period (nominal)	$0.059029995271 \pm 0.000000000002$ s
Projected semi-major axis	2.3418 ± 0.0001 light-seconds
Eccentricity	0.617127 ± 0.000003
Orbital period	27906.98163 ± 0.00002 s
Rate of precession of periastron	4.2263 ± 0.0003^{0} per year
Amplitude of time-dilation factor	0.0044 ± 0.0001
Rate of change of orbital period	$(- 2.40 \pm 0.09) \times 10^{-12}$ s per s
Rate of precession of spin axis	?

As expected for a binary system that loses energy by gravitational radiation, the orbital period of PSR 1913+16 is decreasing, by 2.4×10^{-12} second per second. For a quantitative test of the theoretical prediction for the rate of change of the orbital period, we need to know the masses of the two components (see Eq. [92]). The values of the masses can be extracted from the measured values of the two relativistic effects in the system: the precession of the periastron and the time dilation. The periastron precession depends only on the sum $m_1 + m_2$ of the masses, whereas the time dilation depends on a different combination of m_1 and m_2. Thus, the measured values of the periastron precession and the time dilation (see Table 5.3) determine m_1 and m_2:

$$m_1 = (1.442 \pm 0.003)M_\odot \quad \text{for pulsar}$$

$$m_2 = (1.386 \pm 0.003)M_\odot \quad \text{for companion} \qquad [93]$$

With these values of the masses, the theoretical prediction for the rate of change of the period is -2.38×10^{-12} second per second, that is, 0.99 ± 0.01 times the observed rate of change (Taylor and Weisberg, 1989). Fig. 5.13 shows plots of the observed and the predicted shifts of the time of periastron passage; these plots illustrate the excellent agreement between the theoretical prediction for the decrease of the period and the observed decrease.

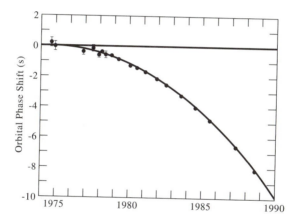

Fig. 5.13 Observed shift of the time of periastron passage (dots) and predicted shift (solid line). Since the period of the orbital motion is decreasing, the system reaches periastron early, which corresponds to a negative time shift. The magnitude of this shift increases from year to year. (From Taylor and Weisberg, 1989.)

Note that for this comparison of theory and observation we cannot use the simple formula [92], because this formula applies only to circular orbits, whereas the orbit of PSR 1913+16 is elliptical, with a substantial eccentricity ($\epsilon = 0.617$). For an elliptical orbit, gravitational radiation is emitted not only at the fundamental frequency, or the orbital frequency,* but also at multiples of this fundamental frequency, because the x and y coordinates of a mass in an elliptical orbit

---

* For a circular orbit, the fundamental (and the only) frequency of gravitational radiation is twice the orbital frequency since the quadrupole moment repeats every half revolution. However, for an elliptical orbit, the quadrupole moment is smaller at periastron than at apastron, and hence the fundamental frequency is the orbital frequency.

are not simple harmonic functions, but harmonic series (Fourier series), which include all multiples of the orbital frequency. In the case of PSR 1913+16, the most power is emitted at 8 times the orbital frequency, but the largest wave amplitude occurs at 4 times the orbital frequency (see Table 5.2).

An additional complication is that the gravitational fields in the system are fairly strong. With $M = 1.44\ M_\odot$ and $r \simeq 10$ km, which is a typical radius for a neutron star, we find $GM/rc^2 \simeq 0.2$ at the surface of the star. It therefore becomes necessary to contemplate corrections to the linear approximation on which the quadrupole formula for gravitational radiation was based. The required corrections have been explored in great detail by Damour (1983a, 1983b, 1983c, 1987) with the conclusion that the quadrupole formula is actually better than expected--it correctly gives the power radiated in the form of gravitational waves even when the background gravitational fields are fairly strong.

The excellent agreement between the observed and the predicted rates of decrease of the orbital period of PSR 1913+16 provides us with good--although circumstantial--evidence for the existence of gravitational radiation. At present, it is the only available observational evidence for gravitational radiation.

The extreme precision with which the parameters of PSR 1913+16 are known permits us to use this pulsar to test for some other relativistic and cosmological effects that might affect the rotational motion or the orbital motion. For instance, as already mentioned in Section 4.7, the relativistic gravitational theory predicts gravimagnetic effects; and the data from the binary pulsar provide us with observational evidence for these gravimagnetic effects, and confirm gravitational spin-orbit coupling. Furthermore, according to some cosmological speculations, the gravitational constant G might gradually decrease, with a characteristic time scale equal to the age of the universe. Such a decrease of the strength of gravitation would result in a gradual expansion of the orbits of all gravitationally bound systems, and a consequent gradual increase of the orbital periods. Since the binary pulsar system displays no residual, unexplained changes in its period, we can set a limit on the rate of change of G. The limit consistent with the observational uncertainties is $|dG/dt|/G < 2 \times 10^{-11}$/yr (Damour et al., 1988).

The orbital motion of a binary system provides us with the simplest example of emission of gravitational waves by a rotating quadrupole. But the rotational motion of almost any system consisting of particles orbiting about a center or an asymmetric body spinning about an axis also will produce a time-dependent quadrupole moment and emit gravitational waves. In this connection, the spinning motion of neutron stars is of considerable astrophysical interest. The spin period of neutrons stars (pulse period of pulsars) is typically in the range from 0.03 to 3 s. If such a spinning neutron star is endowed with some deviation from exact cylindrical symmetry about the axis of rotation, then it constitutes a rotating quadruple, which will emit gravitational radiation. For instance, the Crab pulsar rotates with a period of 0.033 s, and if the dimensions in the equatorial plane differ by 1 part in 1200, then gravitational radiation would be emitted at a rate of 10^{38} erg/s (Ferrari and Ruffini, 1969). This energy loss would account for the observed slowing down of the rate of spin observed for the Crab pulsar. However, since it is known that the Crab emits large amounts of electromagnetic energy, most of the loss of rotational kinetic energy is already accounted for, and the emission rate of gravitational energy (and the deformation in the equatorial plane) had better be smaller. A search for gravitational waves from the Crab pulsar was attempted with a detector tuned to the expected wave frequency, $2 \times (1/0.033)$ Hz = 60 Hz. The results were negative (see Section 5.7).

5.5 EMISSION OF BURSTS OF GRAVITATIONAL RADIATION

We saw in the preceding section that, because of the loss of energy by gravitational radiation, the orbits of the two stars in a binary system gradually shrink--the stars gradually spiral toward each other. This inward motion is slow at first, but as the stars fall into smaller and smaller orbits, of higher and higher frequency, the inward motion accelerates drastically. The stars spiral toward each other faster and faster, emitting a crescendo of gravitational radiation. The swan song of the binary system comes to a sudden end when the stars collide. The collision releases a final blast of gravitational radiation.

The radiation emitted during this coalescence of a binary system is initially periodic, or quasiperiodic, but the radiation emitted during the last few revolutions is burst-like. For example, consider the coalescence of a system of two neutron stars, initially at a distance of, say, $r_0 = 100$ km. If the mass of each star is 1 $M_\odot$, the radiated power is initially 3.3×10^{51} erg/s, and the rate of decrease of the distance is $[- dr/dt]_0 = 2.5 \times 10^6$ cm/s. Integrating Eq. [91], we obtain

$$\frac{1}{4}(r^4 - r_0{}^4) = - \frac{64G^3}{5c^5}m_1 m_2(m_1 + m_2)(t - t_0) \qquad [94]$$

which we can also write as

$$t - t_0 = \left(1 - \frac{r^4}{r_0{}^4}\right)\frac{r_0}{4}\frac{1}{[- dr/dt]_0} \qquad [95]$$

where the subscripts $_0$ indicate quantities evaluated at the initial time t_0. The time interval needed for a decrease in the distance from r_0 to r

$<< r_0$ is therefore

$$\Delta t \simeq \frac{r_0}{4} \frac{1}{[- dr/dt]_0}$$ [96]

This result is (approximately) independent of the final distance r, because the system spends most of its time at the larger distances, and passes through the smaller distances very quickly. If we insert the values of r_0 and $[dr/dt]_0$ given above, we obtain

$$\Delta t \simeq 1.0 \text{ s}$$ [97]

A simple calculation shows that in this time interval the binary system completes $\simeq 130$ revolutions; the shape of the orbit is therefore that of a tightly wound spiral. The frequency of the radiation increases with time; initially, the frequency is $\simeq 160$ Hz, but toward the end it is $\simeq 10$ times as large. The radiated power increases even more drastically; toward the end it is $\simeq 3000$ times as large as initially.

The total amount of energy emitted during the coalescence is equal to the change in the total energy of the system,

$$\Delta E = \frac{G}{2} m_1 m_2 \left(\frac{1}{r_0} - \frac{1}{r} \right)$$ [98]

With $r_0 = 100$ km and $r = 20$ km, the emitted energy amounts to

$$- \Delta E = 5.3 \times 10^{52} \text{ erg}$$ [99]

which is 1% of the rest-mass energy of the system! Most of this energy is emitted during the last stages of the coalescence of the binary system. An extra pulse will be emitted when the neutron stars collide; furthermore, the system that forms after the collision (a larger neutron star or a black hole) may emit more gravitational radiation while it settles into a stationary configuration. The pulse of radiation emitted by a coalescing binary neutron-star system somewhere in our galaxy or in a neighboring galaxy would be of sufficient intensity to be detected with presently available equipment; however, such events are expected to be rare.

Note that the preceding calculation is only approximate; relativistic corrections have been ignored. Even at $r = 100$ km, the orbital speed is $v/c \simeq 1/10$, and by the time the distance reaches $r = 20$ km, the speed is even larger, $v/c \simeq 2/10$. It is also necessary to take into account corrections arising from the nonlinear behavior of the gravitational fields. Our calculation must be regarded merely as an order of magnitude estimate.

The sources of gravitational radiation we have mentioned so far have been periodic or quasiperiodic, and they were based on vibrating or rotating quadrupoles. Since any quadrupole with a nonzero third time derivative will radiate, there are many other possible sources. For example, consider a particle that is accelerated by falling directly toward a star along a radial line (orbit of zero angular momentum). For simplicity, assume that the mass M of the star is large compared with the mass m of the particle. This implies that the star remains at rest and only the particle contributes a time-dependent quadrupole moment. If we take the z-axis along the direction of incidence and the origin at the center of the star (see Fig. 5.14), then Eq. [73] gives

$$- \frac{dE}{dt} = \frac{2Gm^2}{15c^5} (6\dot{z}\dddot{z} + 2z\dddot{z})^2 \qquad [100]$$

■ *Exercise 20.* Derive Eq. [100]. ■

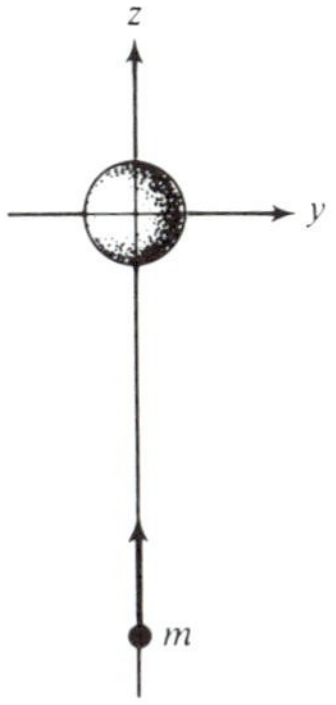

Fig. 5.14 Particle falling radially toward a mass.

If the particle starts out at infinity with zero velocity (orbit of zero energy), then

$$\frac{1}{2} m\dot{z}^2 = \frac{GmM}{|z|} \qquad [101]$$

and

$$\dot{z} = \frac{1}{|z|^{1/2}}(2GM)^{1/2}$$

$$\ddot{z} = \frac{GM}{z^2}$$

$$\dddot{z} = \frac{2GM}{|z|^{7/2}}(2GM)^{1/2} \qquad [102]$$

Eqs. [100] and [102] give

$$- dE = \frac{1}{|z|^{9/2}} \frac{2Gm^2}{15c^5} (2GM)^{5/2} \, dz \qquad [103]$$

The energy radiated during the plunge from $z = -\infty$ to $z = -R$ is obtained by integrating [103]:

$$- \Delta E = \frac{1}{R^{7/2}} \frac{4Gm^2}{105c^5} (2GM)^{5/2} \qquad [104]$$

Obviously, the energy radiated is large when the lower limit R is small. It is therefore advantageous to consider the case of a very compact star or other object. The most compact object that we can think of is a black hole. We will discuss these objects later on; for now, we need to know only that a black hole of mass M has a radius

$$R = \frac{2GM}{c^2} \qquad [105]$$

The gravitational fields at this radius are so intense that any particle (or light, or whatever) that approaches the object nearer than $2GM/c^2$ is unavoidably pulled into the black hole and can nevermore escape.

For the case of a particle falling directly into such a black hole, the lower limit on R is that given by Eq. [105], since any gravitational radiation emitted afterward will remain within the black hole. Substitution into Eq. [104] yields

$$- \Delta E = \frac{2}{105} mc^2 \left(\frac{m}{M} \right)$$

$$= 0.019 mc^2 \left(\frac{m}{M} \right) \qquad [106]$$

Our simple calculation is defective in that we have failed to take into account that the motion becomes relativistic as the particle approaches $R = 2GM/c^2$. Also, nonlinear gravitational effects must be considered. A precise calculation by Davis et al. (1971) shows that

$$- \Delta E = 0.0104 mc^2 \left(\frac{m}{M} \right) \qquad [107]$$

which is fairly close to our simple approximate result [106].

Note that the amount of gravitational-wave energy radiated decreases with M. To obtain a large pulse of radiation we must choose m large and M as small as we dare (remember that we assumed $M \gg m$ in our calculation). If we take a "particle" of mass $M_\odot$ and a black hole of mass $M = 10\ M_\odot$, radius $R = 2GM/c^2 \simeq 30$ km, we obtain

$$- \Delta E = 2 \times 10^{51}\ \text{erg} \qquad [108]$$

Most of this radiation is emitted during the final stages of the plunge, say, when the particle falls from $2R$ to R. The radiation therefore appears as a pulse of length

$$\Delta t \simeq R/[\text{particle velocity}] \simeq R/c$$

$$\simeq 30\ \text{km}/c \simeq 10^{-4}\ \text{s} \qquad [109]$$

Such a pulse will contain frequencies of up to 10^4 Hz. The results of a calculation of the frequency spectrum are shown in Fig. 5.15.

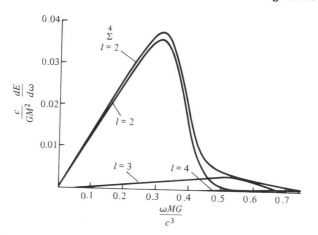

Fig. 5.15 Spectrum of the gravitational radiation emitted by a particle of mass m falling radially into a black hole of mass M. The quantity $dE/d\omega$ gives the amount of energy radiated per unit frequency interval. The curve marked $l = 2$ corresponds to quadrupole radiation; the other curves ($l = 3$, $l = 4$) correspond to multipole radiation of higher order. Note that most of the radiation is emitted with frequencies below $\omega \simeq 0.5c^3/GM$. (From Davis, Ruffini, Press, and Price, 1971.)

As a final example of a source of gravitational radiation we mention gravitational collapse of supernovas (type II supernovas). When a star of several solar masses evolves into a state of high density, the gravitational field can become so intense that the interior pressure is unable to support the weight of the outer layers of the star. The core of the star crushes itself in its own gravitational field and collapses on itself, until it reaches nuclear densities, where the collapse suddenly slows or stops. This sudden stop generates a shock wave that travels outward and rips off the outer layers of the star in a tremendous explosion, which we see as a supernova. Meanwhile, either the core attains a stable configuration as a neutron star or, if the core is too massive, it continues to collapse and ultimately forms a black hole.

Whether gravitational radiation is emitted during the gravitational collapse depends on the presence of an asymmetry. If the collapsing star remains spherically symmetric, it has no quadrupole moment, and will not emit gravitational radiation. However, most stars rotate, and their symmetry is cylindrical, not spherical. During the gravitational collapse, the pole and the equator of such a rotating star collapse at different speeds, and they may even oscillate in and out when the core reaches nuclear densities. This produces a time-dependent, possibly even an oscillating, quadrupole moment. Furthermore, the neutron star formed at the end of the collapse is likely to be in some excited vibrational state. The amount of gravitational energy radiated by the gravitational collapse of a supernova has been estimated at about 10^{-6} $M_\odot c^2$, although it might be considerably larger in the case of a fast-rotating star, which might break up into several, highly asymmetric fragments during the collapse. The pulse of gravitational radiation contains a range of frequencies, from 10^3 to 10^4 Hz.

Another kind of gravitational collapse might occur in a dense galactic nucleus, where a cluster of stars might merge to form a large black hole, of a mass between 10^3 and 10^7 $M_\odot$. Such a collapse could generate low-frequency gravitational waves, between 10 and 10^{-4} Hz.

TABLE 5.4 TYPICAL ASTROPHYSICAL SOURCES OF GRAVITATIONAL RADIATION*

Source	Frequency	Distance**	Amplitude (κA)
Periodic sources			
Binaries	10^{-4} Hz	10 pc	10^{-20}
Nova	10^{-2} to 1	500 pc	10^{-22}
Spinning neutron star (Crab)	60	2 kpc	$<10^{-24}$
Bursting sources			
Coalescence of binary	10 to 10^3	100 Mpc	10^{-21}
Infall of star into $10M_\odot$ b. h.	10^{-4}	10 Mpc	10^{-21}
Supernova	10^3	10 kpc	10^{-18}
Gravitational collapse of $10^4 M_\odot$ star	10^{-1}	3Gpc	10^{-19}

* Based on Thorne (1987).

** Distances have been selected large enough to yield approximately three events per year.

Table 5.4 is a summary of typical astrophysical sources of gravitational radiation, and the corresponding frequency ranges and amplitudes of the gravitational waves reaching the Earth. For each kind of bursting source, the distance listed in this table has been selected large enough to comprise an estimated three events per year.

5.6 THE QUADRUPOLE DETECTOR AND ITS CROSS SECTION

A simple quadrupole oscillator, consisting of two masses connected by a spring, can be used as a detector of gravitational waves. Since we know that the oscillations of this system radiate gravitational waves, we expect that a wave incident on the system will excite the oscillations. We want to calculate what oscillation amplitude the incident gravitational wave can excite and how much energy is absorbed from the wave. However, there is a difficulty: to the extent that both masses accelerate in the same way, no oscillations of the system will be excited. The gravitational force is not directly observable; only the *difference* between the forces acting on the masses produces an observable effect. Even if the experimenter looks for net translational motion of his apparatus, he will not see any, because the experimenter also is accelerated in the gravitational field. Only a relative motion between the two masses, or between them and the experimenter, is noticeable. This means that the observable driving force on the oscillator is the tidal force.

Suppose that a plane gravitational wave, such as that given by Eq. [24], is incident on a simple quadrupole oscillator consisting of two masses joined by a spring placed along the x-axis, with equilibrium positions $x = \pm b$ (see Fig. 5.16). In complex notation, we can write the wave as

$$h^{\mu\nu} = A_\oplus \epsilon_\oplus{}^{\mu\nu} e^{i\omega t - ikz} \qquad [110]$$

If the displacement of the two masses always remains small ($|x - b| << b$), then the tidal force on each mass is given by Eq. [41] as

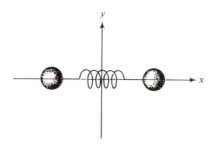

Fig. 5.16 Quadrupole oscillator consisting of two masses joined by a spring along the x-axis.

$$f_x = \left(m \, \tfrac{\kappa}{2} A_\oplus \omega^2 e^{i\omega t} \right) x \simeq \left(m \, \tfrac{\kappa}{2} A_\oplus \omega^2 e^{i\omega t} \right) b \qquad [111]$$

The equation of motion of one of the masses, say, the mass on the positive x-axis, is then

$$m\ddot{x} + m\gamma\dot{x} + m\omega_0^2(x - b) = m \, \tfrac{\kappa}{2} A_\oplus \omega^2 b e^{i\omega t} \qquad [112]$$

Here ω_0^2 is the natural frequency for the free oscillations, and γ is the *damping rate* associated with the frictional forces acting on the oscillator. In terms of the mean (average over one cycle) rate of loss of energy to friction, the damping rate is

$$\gamma = - \frac{1}{E} \left(\overline{\frac{dE}{dt}} \right)_{friction} \qquad [113]$$

Experimenters usually prefer to use the Q, or the "quality factor," of the oscillator, defined by

$$Q = \omega_0/\gamma \qquad [114]$$

The time $\tau_0 = \gamma^{-1}$, or $\tau_0 = \omega_0/Q$, is the *relaxation time*, that is, it is the time interval in which the energy of the free (driving force absent) oscillations decreases by a factor of e.

The gravitational tidal force [111] plays the role of a driving force in Eq. [112]. With this driving force, the steady-state solution of the equation of motion is

$$x - b = \frac{\tfrac{1}{2}\kappa A_\oplus \omega^2 b \; e^{i\omega t}}{-\omega^2 + \omega_0^2 + i\gamma\omega} \qquad [115]$$

■ *Exercise 21.* Check this. ■

If the friction in the oscillator is relatively small, so

$$\gamma \ll \omega_0$$

then the steady-state response of the oscillator has a sharp maximum at the frequency $\omega \simeq \omega_0$. Gravitational waves of this frequency are in resonance with the natural oscillations of the system. For $\omega = \omega_0$, Eq. [115] gives an amplitude of oscillation

$$\tfrac{1}{2}\kappa A_{\oplus}\left(\frac{\omega_0}{\gamma}\right)b \qquad\qquad [116]$$

This is larger than the amplitude of motion for a free particle* by a factor ω/γ. According to Eq. [115], this factor is large compared with 1.

It is customary to characterize the sensitivity of the detector by a parameter called the *cross section*. Let us begin by defining the *scattering cross section*. While the gravitational wave excites oscillations in the quadrupole detector, these oscillations in turn radiate gravitational waves; that is, some of the energy absorbed from the incident gravitational wave is reradiated as a new gravitational wave. The scattering cross section is defined as the ratio of power reradiated to incident flux,

$$\sigma_{scatt} = \frac{[\text{power reradiated}]}{[\text{incident flux}]} \qquad\qquad [117]$$

Here, it is understood that both power and flux are to be averaged over a cycle. The cross section defined by [117] has the dimensions of an area. However, it has nothing to do with any geometrical cross-sectional area of the oscillating masses; the cross section is nothing but a parameter that measures how efficiently the oscillator scatters the radiation.

In our case, the incident wave has a flux $\tfrac{1}{2}A_{\oplus}{}^2\omega^2/c$. To find the radiated power, we use Eq. [85]. The quantity a that appears in this equation is the amplitude of oscillation; according to Eq. [115],

$$a = \frac{\tfrac{1}{2}\kappa A_{\oplus}\omega^2 b}{\left|-\omega^2 + \omega_0{}^2 + i\gamma\omega\right|} \qquad\qquad [118]$$

Hence

$$-\left(\overline{\frac{dE}{dt}}\right)_{rad} = \frac{16G}{15c^5}\frac{(\tfrac{1}{2}m\kappa A_{\oplus}\omega^2 b^2)^2}{(\omega^2 - \omega_0{}^2)^2 + \gamma^2\omega^2}\omega^6 \qquad\qquad [119]$$

and

$$\sigma_{scatt} = -\frac{1}{\tfrac{1}{2}A_{\oplus}{}^2\omega^2/c}\left(\frac{dE}{dt}\right)_{rad}$$

---

* The free-particle result can be obtained from Eq. [115] by taking $\omega_0 = \gamma = 0$.

$$= \frac{128\pi G^2}{15c^8} \frac{(mb^2\omega^4)^2}{(\omega^2 - \omega_0{}^2)^2 + \gamma^2\omega^2} \qquad [120]$$

$$= \frac{15\pi c^2}{2} (\gamma_{rad})^2 \frac{1}{(\omega^2 - \omega_0{}^2)^2 + \gamma^2\omega^2} \qquad [121]$$

where, as in Eq. [87],

$$\gamma_{rad} = \frac{16G}{15c^5} mb^2\omega^4 \qquad [122]$$

Note that σ_{scatt} is independent of the amplitude of the incident wave (amplitudes cancel in Eq. [120]); this is what makes the cross section useful as a measure of how much radiation the quadrupole scatters out of the incident wave.

For the purpose of detecting gravitational radiation, we are more interested in the energy absorbed from the gravitational wave than in the energy scattered. This means we want to know what power the oscillator delivers to the (mechanical) frictional forces in the oscillator. We can define a corresponding absorption cross section as

$$\sigma_{abs} = \frac{[\text{power lost to mechanical friction}]}{[\text{incident flux}]} \qquad [123]$$

We can establish a connection between σ_{scatt} and σ_{abs} if we note that the total damping rate γ, given by Eq. [113], is the sum of two terms

$$\gamma = -\frac{1}{E}\left[\frac{\overline{dE}}{dt}\right]_{total} = -\frac{1}{E}\left[\frac{\overline{dE}}{dt}\right]_{friction} - \frac{1}{E}\left[\frac{\overline{dE}}{dt}\right]_{rad}$$

$$= \gamma_m + \gamma_{rad} \qquad [124]$$

In words: the total "friction" is due to both mechanical friction and radiation damping. By their definitions, σ_{scatt} and σ_{abs} must stand in the ratio γ_{rad}/γ_m, that is,

$$\sigma_{abs} = \frac{\gamma_m}{\gamma_{rad}} \sigma_{scatt} \qquad [125]$$

For any reasonable oscillator we might hope to build, $\gamma_m \gg \gamma_{rad}$, and hence $\gamma_m \simeq \gamma$. Thus

$$\sigma_{abs} \simeq \frac{\gamma}{\gamma_{rad}} \sigma_{scatt} \qquad [126]$$

$$= \frac{15}{2} \pi c^2 \gamma \gamma_{rad} \frac{1}{(\omega^2 - \omega_0{}^2)^2 + \gamma^2 \omega^2} \tag{127}$$

We have calculated the cross sections [121] and [127] for a quadrupole oscillator oriented in the most favorable direction in the tidal field of the wave. The line of vibration of the masses was taken both perpendicular to the direction of incidence and parallel to one of the principal axes of the tidal deformation field. For an oscillator whose masses are constrained to vibrate along a line making an angle θ with the direction of incidence (z-axis) and an angle ϕ with one of the principal axes of the tidal field (see Fig. 5.17), the component of the tidal force along this line is reduced by a factor

$$\sin^2 \theta \cos 2\phi \tag{128}$$

as compared with the most favorable case. This factor is easy to understand: the magnitude of the tidal force is proportional to the transverse dimension of the system ($\simeq \sin \theta$); taking the component of this force along the line of vibration results in another factor of $\sin \theta$; finally, the factor $\cos 2\phi$ simply represents the angular dependence of the (radial) tidal field strength in the transverse plane (see Eq. [44]).

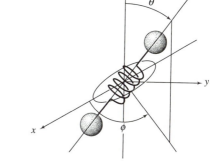

Fig. 5.17 Orientation of the quadrupole detector relative to an incident gravitational wave. The wave propagates along the z-axis. The ellipse indicates the polarization of the wave.

The cross section depends on the square of the component of the tidal force along the line of vibration; hence Eqs. [121] and [127] must be reduced by a factor $\sin^4 \theta \cos^2 2\phi$. This shows that the cross section of the detector is quite sensitive to the orientation and varies between the upper limit given by Eq. [127] and zero. As a compromise, it is convenient to introduce a cross section for random relative orientation between wave and detector. For this, we simply replace the factor $\sin^4 \theta \cos^2 2\phi$ by its average value over all angles,

$$\frac{1}{4\pi} \int \sin^4\theta \, \cos^2 2\phi \, d\Omega = \frac{4}{15} \qquad [129]$$

which gives

$$\bar{\sigma}_{abs} = 2\pi c^2 \gamma \gamma_{rad} \frac{1}{(\omega^2 - \omega_0{}^2)^2 + \gamma^2 \omega^2} \qquad [130]$$

Although our calculations were based on the special case of an ideal linear quadrupole, our results are also valid for an arbitrary mass system vibrating in a mode of cylindrical symmetry. More precisely, since the amplitude of the time-dependent part of the quadrupole tensor is some symmetric matrix Q'^{kl},* it can be diagonalized by a transformation to principal axes. By cylindrical symmetry we mean that two of the diagonal elements of the diagonalized matrix are equal. If we take the axis in the z direction, then $Q'^{11} = Q'^{22}$. Furthermore, since the trace must be zero, $Q'^{33} = -2Q'^{11}$. Let us now write these components as

$$Q'^{11} = Q'^{22} = -\tfrac{1}{2}Q'^{33} = -2mba' \qquad [131]$$

where m is one-half the mass of the oscillating system, b is one-half the length of the system, and a' is *defined* by Eq. [131]. Obviously, a' has the dimensions of a length; it can be regarded as an "effective oscillation amplitude" for the masses in the quadrupole. The diagonalized quadrupole matrix is then

$$Q'^{kl} = \begin{pmatrix} -2mba' & 0 & 0 \\ 0 & -2mba' & 0 \\ 0 & 0 & 4mba' \end{pmatrix} \qquad [132]$$

Note that this has essentially the same form as the time-dependent part of the quadrupole moment for a linear quadrupole (see Eq. [77]); thus an axially symmetric vibrating system is equivalent to a linear quadrupole. The present-day detectors are cylinders operated in a longitudinal oscillation mode; these detectors satisfy our assumptions. The values of a' are usually of the same order of magnitude as a (Eq. [118]); for example, a long cylinder vibrating in the fundamental longitudinal mode has $a'/a = 2/\pi$.

---

* The definition of Q'^{kl} is as follows: $Q^{kl} = [\text{constant}] + Q'^{kl} \sin(\omega t + \delta)$, where Q^{kl} is the total quadrupole moment of the system.

5.7 *EXPERIMENTS WITH DETECTORS OF GRAVITATIONAL RADIATION*

Detectors of gravitational radiation were first built by Weber at the University of Maryland, starting in 1966 (Weber, 1973; Trimble and Weber, 1973). These detectors consist of massive cylinders of aluminum--typically about 1000 or 2000 kg--in which gravitational waves would excite quadrupole oscillations. The resonant frequency for the longitudinal mode of oscillation of the cylinders is typically somewhere around 1000 Hz. The cylinder is suspended inside a vacuum tank by a wire wrapped around its middle. The beam supporting the wire rests on stacks of rubber and metal blocks, which provide good isolation against external mechanical vibrations. The oscillations of the cylinder in its fundamental longitudinal mode are measured by piezo-electric strain transducers bonded to the surface of the cylinder around its middle. Table 5.5 summarizes the characteristics of one of Weber's first detectors, and Fig. 5.18 shows a photograph.

Fig. 5.18 One of Weber's early detectors. A large number of piezoelectric transducers can be seen attached to the waist of the cylinder. When in operation, the detector is placed inside a vacuum tank. (Courtesy J. W. Weber, University of Maryland.)

TABLE 5.5 CHARACTERISTICS OF A WEBER DETECTOR

Mass	1410 kg
Length	153 cm
Diameter	66 cm
Resonant frequency	1660 Hz
Damping rate, γ	$\simeq 0.05$ s
$Q = \omega_0/\gamma$	$\simeq 2 \times 10^5$

According to Eq. [130], the cross section at resonance for a cylinder with the mass, length, and resonant frequency listed in Table 5.5 is*

$$\bar{\sigma}(\omega_0)_{abs} = \frac{32\pi G}{15c^3}\left[\frac{\omega_0}{\gamma}\right]\omega_0 mb^2 \qquad [133]$$

$$\simeq 10^{-19} \text{ cm}^2$$

This is quite small. And what is worse is that the resonance peak in the cross section is quite narrow (only about 0.1 Hz; see Fig. 5.19); off resonance, the cross section is much smaller. Thus, the quadrupole detector will be sensitive to a periodic (or continuous) gravitational wave only if the frequency of the wave is in tune with the resonant frequency of the detector; and it will be sensitive to a wave pulse (or burst) only if the Fourier transform of the wave pulse has a substantial component at the resonant frequency.

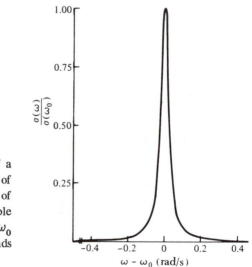

Fig. 5.19 The cross section of a Weber detector as a function of frequency. The characteristics of this dectector are given in Table 5.5. The resonant frequency is ω_0 = 10,430 rad/s (which corresponds to ν_0 = 1660 Hz).

---

* m and b are one-half the mass and length given in Table 5.5.

In the case of a periodic wave, lasting at least as long as the damping time γ^{-1}, the oscillations of the detector reach steady state. In this state, the frictional forces absorb energy from the gravitational wave at a steady rate

$$- \left(\frac{\overline{dE}}{dt} \right)_m = t_{(1)}{}^{03} \bar{\sigma}(\omega_0)_{abs} \qquad [134]$$

and the energy of the oscillator attains a steady value

$$E = - \frac{1}{\gamma} \left(\frac{\overline{dE}}{dt} \right)_{total} \simeq - \frac{1}{\gamma} \left(\frac{\overline{dE}}{dt} \right)_m$$

$$= \frac{1}{\gamma} t_{(1)}{}^{03} \bar{\sigma}(\omega_0)_{abs} \qquad [135]$$

What limits our ability to detect this energy is the thermal noise in the oscillations of the cylinder. For an oscillator kept at temperature T, the thermal fluctuations in energy in each mode of oscillation will be approximately kT, and in order to detect the presence of a gravitational wave, the excitation energy E given by Eq. [135] must be at least as large as kT. Therefore the minimum detectable flux is

$$t_{(1)}{}^{03} \geq \frac{kT\gamma}{\bar{\sigma}(\omega_0)_{abs}} \qquad [136]$$

With the resonant cross section [133] and with $\gamma = \omega_0/Q$, this takes the alternative form

$$t_{(1)}{}^{03} \geq \frac{15c^3}{32\pi G} \frac{kT}{mb^2Q^2} \qquad [137]$$

In terms of the wave amplitude, the flux is (see Eq. [28])

$$t_{(1)}{}^{03} = \frac{1}{2c} A^2 \omega_0{}^2 \qquad [138]$$

Hence the dimensionless wave amplitude κA, which tells us the deviation of the metric from the flat spacetime metric, is

$$\kappa A = \kappa \sqrt{\frac{2ct_{(1)}{}^{03}}{\omega_0{}^2}} = \sqrt{\frac{32\pi G t_{(1)}{}^{03}}{c^3 \omega_0{}^2}} \qquad [139]$$

Corresponding to the minimum detectable flux [137], the minimum detectable value of κA is

$$\kappa A \geq \sqrt{\frac{15kT}{m}} \; \frac{1}{\omega_0 bQ} \qquad [140]$$

Weber's early detectors operated at room temperature, $T \simeq 300$ K. From Eqs. [137] and [140], we find a minimum detectable flux and amplitude

$$t_{(1)}{}^{03} \geq 2 \times 10^4 \; \text{erg/cm}^2 \cdot \text{s}$$

$$\kappa A \geq 6 \times 10^{-21} \qquad [141]$$

However, this estimate of the sensitivity applies only to a wave whose frequency coincides with the resonant frequency of the detector. There are no known astrophysical sources of periodic waves in the kilohertz frequency range, and therefore the estimate [141] is irrelevant. However, as mentioned in Section 5.4, spinning neutron stars might emit some gravitational radiation, with a frequency of 10 to 100 Hz (twice the rotation frequency), and it is feasible to build Weber detectors deliberately tuned to the wave frequency. For instance, a detector has been built to match the frequency of (possible) gravitational waves emitted by the Crab pulsar. This detector has a higher Q and a lower temperature than Weber's early detectors, and it attained a sensitivity of $\kappa A \simeq 7 \times 10^{-23}$ (Owa et al., 1986).

Apart from such exceptional, contrived instances, we cannot rely on a coincidence between the wave frequency and the resonant frequency of the detector. The best we can hope for is that an astrophysical source produces a short pulse of radiation that contains a spread of frequencies that overlaps the resonant frequency. Such a short pulse can be conveniently described by the total energy per unit frequency interval (and unit area) that it carries. If we designate this energy per unit frequency interval and unit area by $t_{(1)}{}^{03}(\omega)$, then the energy absorbed by the detector from the pulse is

$$\Delta E = \int_0^\infty t_{(1)}{}^{03}(\omega) \, \sigma_{abs}(\omega) \, d\omega \qquad [142]$$

We will not bother with a formal derivation because Eq. [142] is rather obvious: by Fourier analysis we can express the pulse, of finite duration in time, as a superposition of harmonic waves $e^{i\omega t}$ of many different frequencies. In essence, $t_{(1)}{}^{03}(\omega)$ gives the total energy, per unit area, that is associated with the wave of frequency ω. Since this wave is of infinite duration, the oscillator reaches steady state with it. The cross section $\sigma(\omega)$, by its definition, tells us what fraction of the total energy, per unit area, is absorbed from the wave of frequency ω. The integral over all the participating frequencies will then give us the desired absorption for the complete pulse.

If $t_{(1)}{}^{03}(\omega)$ is a reasonably smooth function, then most of the contribution to the integral [142] comes from the vicinity of the frequency ω_0, where $\sigma(\omega)$ has a very sharp peak (see Fig. 5.19). A pulse of radiation that lasts a short Δt contains a wide range of frequencies; this means that the function $t_{(1)}{}^{03}(\omega)$ has a broad maximum of width

$$\Delta\omega \simeq \frac{1}{\Delta t} \qquad [143]$$

We will assume that $\Delta t \ll \gamma^{-1}$, so the broad maximum in $t_{(1)}{}^{03}(\omega)$ is much wider than the resonance peak in the cross section of the quadrupole detector. Thus, everywhere in the region of the peak of $\sigma(\omega)$, the function $t_{(1)}{}^{03}(\omega)$ has approximately the same value, $t_{(1)}{}^{03}(\omega) \simeq t_{(1)}{}^{03}(\omega_0)$. We can then factor this function out of the integral, and we obtain

$$\Delta E \simeq t_{(1)}{}^{03}(\omega_0) \int \sigma_{abs}(\omega)\, d\omega \qquad [144]$$

To evaluate this integral, let us use the cross section $\bar{\sigma}_{abs}(\omega_0)$ for random directions. It is convenient to replace the expression [130]

$$\bar{\sigma}_{abs}(\omega) = 2\pi c^2 \gamma \gamma_{rad}(\omega)\; \frac{1}{(\omega^2 - \omega_0{}^2)^2 + \gamma^2 \omega^2} \qquad [145]$$

by

$$\bar{\sigma}_{abs}(\omega) = \frac{2\pi c^2 \gamma \gamma_{rad}(\omega_0)}{(2\omega_0)^2}\; \frac{1}{(\omega - \omega_0)^2 + (\gamma/2)^2} \qquad [146]$$

These two expressions are essentially the same in the vicinity of $\omega = \omega_0$; the fact that they differ elsewhere is of no consequence, since these other regions contribute little to the integral.

■ **Exercise 22.** Check that Eqs. [146] and [145] agree near ω_0, and perform the integral [144]. ■

The final result is

$$\Delta E = \tfrac{1}{2}\pi\gamma l_{(1)}{}^{03}(\omega_0)\bar{\sigma}_{abs}(\omega_0) \qquad [147]$$

If we demand that the energy deposited by the pulse be at least as large as kT, we are lead to a minimum detectable energy per unit area and unit frequency

$$l_{(1)}{}^{03}(\omega_0) \geq \frac{2kT}{\pi\gamma\bar{\sigma}_{abs}(\omega_0)} \qquad [148]$$

or

$$l_{(1)}{}^{03}(\omega_0) \geq \frac{15c^3}{16\pi^2 G}\frac{kT}{mb^2\omega_0{}^2} \qquad [149]$$

So far, in our calculation of the minimum detectable flux, we have ignored the noise contributed by the thermal fluctuations in the amplifier that converts the oscillations of the aluminum cylinder into observable electric signals. In the case of steady-state oscillations induced by a periodic gravitational wave, we have available a long sampling time, and we can therefore operate the amplifier at a very small bandwidth (small $\Delta\nu$). The amplifier noise is proportional to the bandwidth, and it can be neglected if the bandwidth is small. But in the case of a short pulse, we cannot use a long sampling time and a small bandwidth, and we must take the amplifier noise into account.

For the aluminum cylinder, a small sampling time is actually advantageous, because it reduces the thermal fluctuations. Typically, the thermal fluctuations take a time $\tau_0 = \gamma^{-1}$ to change the oscillation energy by kT (this can be readily established by means of the Langevin equation for the oscillator). In a time Δt, short compared with γ^{-1}, they change the thermal energy by only $\simeq kT(\gamma\Delta t)^{1/2}$. Hence, if we select a short sampling time Δt, we reduce the thermal fluctuations. However, a short sampling time implies a large bandwidth for the amplifier, and a large amount of amplifier noise. For the best compromise, we must select the sampling time such that the thermal fluctuations of the cylinder and the amplifier noise are about equal. It can be shown that this implies an effective detector temperature proportional to the geometric mean of the actual temperature T and the noise temperature T_a of the amplifier:

$$T_{eff} = \sqrt{\frac{\pi T T_a}{Q\beta}} \qquad [150]$$

where β is a constant that characterizes the strength of the coupling between the cylinder and the electronic sensor attached to it. With this effective temperature, the expression [149] becomes

$$t_{(1)}{}^{03}(\omega_0) \geq \frac{15c^3}{16\pi^2 G} \frac{kT_{eff}}{mb^2\omega_0{}^2} \tag{151}$$

For the Weber detector, T_{eff} is approximately 20 K, which gives

$$t_{(1)}{}^{03}(\omega_0) \geq 3 \times 10^5 \ (\text{erg/cm}^2)/(\text{rad/s}) \tag{152}$$

This is the minimum detectable energy per unit area and unit frequency interval at the frequency ω_0. To obtain the total energy of the pulse, we need to know the shape of the energy spectrum. The most optimistic assumption, which implies the least total energy, is that the spectrum extends in a broad plateau from zero frequency to some maximum frequency not much larger than the resonant frequency (if the maximum frequency of the spectrum were *smaller* than the resonant frequency, we could not detect the pulse). This means that the bandwidth of the spectrum is roughly equal to the resonant frequency, $\Delta\omega \simeq \omega_0$. For the minimum detectable value given by Eq. [152], the total pulse energy per unit area is then roughly

$$t_{(1)}{}^{03}(\omega_0)\Delta\omega \simeq t_{(1)}{}^{03}(\omega_0)\omega_0$$

$$\simeq 3 \times 10^5 \ (\text{erg/cm}^2)/(\text{rad/s}) \times 2\pi \times 1600 \ \text{rad/s}$$

$$\simeq 10^9 \ \text{erg/cm}^2 \tag{153}$$

From Eq. [151] we can find the minimum detectable Fourier amplitude $A(\omega_0)$, or rather, the absolute value $|A(\omega_0)|$:

$$|A(\omega_0)| = \frac{1}{\omega_0} \sqrt{2ct_{(1)}{}^{03}(\omega_0)} \tag{154}$$

But this Fourier amplitude at one frequency contains insufficient information to determine the amplitude $A(t)$ of the gravitational wave. For a crude estimate, we again assume that the Fourier spectrum extends in a broad plateau from zero frequency to the resonant frequency. Then the integral of the Fourier amplitude is of the order of $A(\omega_0)\Delta\omega \simeq A(\omega_0)\omega_0$, and we obtain the order of magnitude estimate

$$\kappa A \geq \kappa \sqrt{2ct_{(1)}{}^{03}(\omega_0)} \simeq \sqrt{\frac{30}{\pi} \frac{kT_{eff}}{m}} \frac{1}{\omega_0 b} \tag{155}$$

For the Weber detector, this is

$$\kappa A \geq 3 \times 10^{-16} \qquad [156]$$

Even if our detector registers a signal of energy kT, we cannot be absolutely sure that the cause is a gravitational wave, since thermal fluctuations will occasionally rise above this level. The thermal fluctuations obey the Boltzmann distribution; the probability for a thermal energy E is proportional to $e^{-E/kT}$. If we want strong evidence for the arrival of gravitational waves, we must require that the energy deposited in our detector is several times kT. Alternatively, we might operate two, or more, detectors in coincidence; if a signal is observed simultaneously in all of these detectors, then we can be reasonably certain that an external influence is at work, even if these signals are only of a magnitude comparable with the random thermal noise.

Weber operated several detectors in coincidence, and in 1968 he reported signals, with an energy in excess of kT, in his detectors (Weber et al., 1969, 1970a, 1973). He claimed that these signals were arriving at the rate of a few per day, and he claimed that there was some indication that their direction of arrival correlates with the direction of the galactic center (Weber et al., 1970b). The bandwidth of the pulses is at least $\Delta\nu = 630$ Hz, because coincident pulses were observed in cylindrical detectors of a resonant frequency $\nu_0 = 1660$ Hz and also in a disk-shaped detector of a resonant frequency $\nu_0 = 1030$ Hz.

According to the estimate given above, a gravitational pulse would need an energy of about 10^9 erg/cm^2 to make such a signal. If these pulses originate at the galactic center, at a distance of $r = 10^4$ pc = 3×10^{22} cm, the total energy per pulse would be (Gibbons and Hawking, 1971)

$$4\pi r^2 \times 10^9 \text{ erg/cm}^2 = 10^{55} \text{ erg} \simeq 10 \, M_\odot c^2 \qquad [157]$$

With several such pulses per day, the amount of mass converted into gravitational radiation would add up to $\simeq 10^4 \, M_\odot c^2$ per year. Such a large rate of loss of mass, if continuous, is in contradiction with astronomical evidence, since the reduction of the mass would imply a reduction of the force that holds the Galaxy together, and the Galaxy would expand (Sciama, Field, and Rees, 1969).

The sensational claims of Weber stimulated several other experimenters to seek confirmation of his discoveries. Detectors of the Weber type, but of somewhat higher sensitivity, were built in quick succession by Braginsky (University of Moscow), Tyson (Bell Laboratories), Douglass (University of Rochester), Drever (University of Glasgow), Maischberger (Frascati), Kafka (Munich), and Levine and Garwin (IBM Research Center). None of these experimenters found any pulses of gravitational radiation. The weight of the evidence suggests that Weber's apparent coincidences were the result of faulty data analysis, which permitted chance coincidences ("accidentals") to masquerade as real coincidences (for a critical assessment, see Tyson and Giffard, 1978; Garwin, 1974, 1975).

By now, several groups of experimenters have built larger, much more sensitive quadrupole detectors (see Table 5.6). These new detectors attain a sensitivity of up to $\kappa A \simeq 10^{-18}$ by a higher value of Q (resulting from the use of a new aluminum alloy in most cases; although the highest Q has been attained with a niobium cylinder), better electronic sensors, low-noise amplifiers, and low-temperature operation. All the detectors listed in Table 5.6 operate at 4.2 K, the temperature of liquid helium; but their effective temperatures are even lower, as low as a few mK. Their resonant frequencies are around 1000 to 2000 Hz, except for the Tokyo detector which has an unsual shape giving it a much lower resonant frequency of 60 Hz, the frequency of the gravitational waves (if any) emitted by the rotational motion of the Crab pulsar. None of the detectors has found anything so far.

TABLE 5.6 QUADRUPOLE DETECTORS OF GRAVITATIONAL RADIATION

Location	Mass	Q	Frequency	T	T_{eff}
Rome	2270 kg	10^7	916 Hz	4.2 K	12 mK
Stanford	4800	5×10^6	840	4.2	15
UWA	1500	2×10^8	703	4.2	20
LSU	2270	10^7	915	4.2	96
Maryland	1200	3×10^6	1700	4.2	5
Frascati	390	2×10^6	1800	4.2	100
Tokyo	1200	2×10^7	60	4.2	–*

* Used for resonant detection of periodic waves of 60 Hz; minimum detectable amplitude of such waves is $\kappa A \simeq 10^{-22}$.

Other groups of experimenters have built interferometric detectors, which measure the displacement that an incident gravitational wave produces between a pair of free masses, rather than masses joined by springs, as in the quadrupole detectors. For a pair of free masses, the gravitational wave produces the displacement given by Eqs. [38] and [39]. In practice, free masses can be simulated by suspending the masses from wires, like pendulums. If the frequency of the pendulum is much lower than the frequency of the gravitational wave, then the suspended mass will behave like a free mass (as is well known, a harmonic oscillator behaves like a free particle if the resonant frequency is much lower than the frequency of the driving force). The pendulum suspension typically has a frequency of 1 Hz, and the gravitational-wave frequency must therefore be substantially larger than 1 Hz.

The displacement of the suspended masses is detected by interferometric methods, based on the principle of either a Michelson interferometer or a Fabry-Perot interferometer with resonant cavities. Fig. 5.20 shows an experimental arrangement based on the Michelson interferometer. The free masses carry mirrors, and a beam of light, supplied by a laser, is reflected along the arms and then allowed to interfere with itself. An incident gravitational wave displaces the end mirrors in the two arms by different amounts, and thereby generates a phase shift between the interfering light beams. The resulting change of intensity of the combined beams is measured by a photodetector. The measurement must be completed within half a cycle of the gravitational wave, since after half a cycle the wave will displace the end mirrors in opposite directions, and the phase shift between the interfering light beams will return to zero.

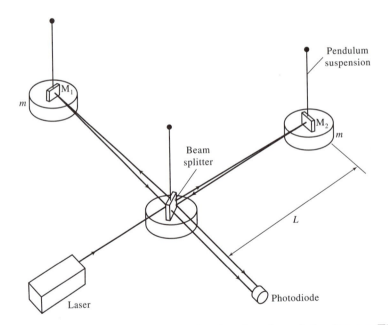

Fig. 5.20 A Michelson interferometer for the detection of gravitational waves. The mirrors M_1 and M_2 at the ends of the arms have a pendulum suspension.

For high sensitivity, the arms of the interferometer must be made long, either in actual length or in effective length, by multiple back-and-forth reflections in each arm. Also, an intense laser beam is needed, so that even a quite small (and faint) phase shift becomes detectable.

It turns out that, to achieve maximum sensitivity, it is desirable to adjust the interferometer in such a way that in the absence of a gravitational wave, the light beams emerging from the two arms are out of phase, $\phi = \pi$, and they interfere destructively. When a gravitational wave then changes the length of one arm relative to the other by an amount x, the phase between emerging light beams changes by an additional amount $\Delta\phi = 4\pi nx/\lambda$, where n is the number of back-and-forth reflections along each arm and λ is the wavelength of the light. The resulting amplitude of the light signal is then proportional to

$$1 + e^{i\pi + 4\pi nx/\lambda} \qquad [158]$$

and the intensity is proportional to

$$\sin^2\left(\frac{2\pi nx}{\lambda}\right) \qquad [159]$$

The number of photons that reach the detector is proportional to this intensity. If the number of photons supplied by the laser during the measurement time is N, the number of photons that are detected in the emerging light beam is

$$N_{out} = N \sin^2\left(\frac{2\pi nx}{\lambda}\right) \qquad [160]$$

This equation permits us to calculate x from a measurement of the number N_{out} of emerging photons. But the number of photons is subject to fluctuations, due to statistics. The magnitude of the fluctuation is

$$\delta N_{out} = \sqrt{N_{out}} = N^{1/2} \sin\left(\frac{2\pi nx}{\lambda}\right) \qquad [161]$$

Corresponding to this fluctuation, there will be an uncertainty in the value of x calculated from Eq. [160]:

$$\delta x = \frac{\delta N_{out}}{4\pi nN/\lambda \, \sin(2\pi nx/\lambda) \, \cos(2\pi nx/\lambda)}$$

$$= \frac{\lambda}{4\pi nN^{1/2} \cos(2\pi nx/\lambda)} \qquad [162]$$

The minimum gravitational-wave amplitude κA detectable by the interferometer is $\kappa A = \delta x/L$, or

$$\kappa A = \frac{\lambda}{4\pi n L N^{1/2} \cos(2\pi n x/\lambda)} \qquad [163]$$

From this equation, we see that it is indeed advantageous to operate the interferometer at an interference minimum, where $\cos(2\pi n x/\lambda) = 1$. Furthermore, we see that a long arm length L, a large number n of reflections, and a large number of photons (a high laser power) are desirable.

According to Eq. [163] it would seem that merely by increasing the number of photons, we can make the sensitivity κA as small as we please. However, there are other limitations on the measurement: a large laser power implies a large radiation pressure on the mirrors, and a large fluctuation in this pressure. During n reflections by the mirror, a photon deposits a momentum $2n \times 2\pi\hbar/\lambda$ on the mirror. When N photons strike the mirror, the fluctuation in their number is $\sqrt{N}$, and the uncertainty in the momentum deposited on the mirror is

$$\delta p = \frac{4\pi n \hbar}{\lambda} \sqrt{N} \qquad [164]$$

If the measurement takes a time τ, the momentum uncertainty leads to a position uncertainty

$$\delta x = \frac{\tau \delta p}{m} = \frac{\tau}{m} \frac{4\pi n \hbar \sqrt{N}}{\lambda} \qquad [165]$$

where m is the mass of the mirror. This uncertainty increases with the number of photons, that is, it increases with the laser power, and it prevents us from reducing δx by increasing N. The best compromise between Eqs. [162] and [165] is achieved by making these two uncertainties equal, which tells us that the optimum choice for the number of photons is $N = (1/4\pi)^2(m\lambda^2/\tau n^2 \hbar)$, and that the minimum uncertainty in x takes the simple form

$$\delta x = \sqrt{\frac{\hbar \tau}{m}} \qquad [166]$$

It is easy to check that this sensitivity corresponds to the quantum limit on measurements set by the Heisenberg principle, $\delta x \delta p \simeq \hbar$. From Eq. [166] we see that the minimum detectable gravitational-wave amplitude is

$$\kappa A = \sqrt{\frac{\hbar \tau}{mL^2}} \qquad [167]$$

Large interferometers, now under development (see below), will have mirror masses of 100 kg, arm lengths of 3 km, and integration times of $\tau = 0.001$ s. For such an interferometer, Eq. [167] gives us a sensitivity $\kappa A \simeq 10^{-23}$. However, we can only attain the required number N of photons within the time $\tau = 0.001$ s if the laser power is quite large, about 500 kW, which is much larger than the power delivered by the best available lasers (pulsed lasers can easily deliver more power in short spurts, but for our interferometer we need a stabilized laser with steady power). Note that Eq. [167] is independent of the number n of reflections. However, this independence is valid only if the laser power is at the optimum level. The interferometers built so far are "underpowered," and their sensitivity improves if n is large (see Eq. [162]).

■ *Exercise 23.* Show that the values of δx and δp in Eqs. [166] and [164] satisfy the uncertainty principle $\delta x \delta p \simeq \hbar$. ■

The first interferometric detector, of modest sensitivity, was built by Forward and Moss in 1972. More sensitive interferometric detectors were built by Billing (Munich), Drever (first at Glasgow, later at Caltech), Hough (Glasgow), and Weiss (MIT). Most of the interferometric detectors now in operation are listed in Table 5.7. All these interferometers have arm lengths of a few meters or a few tens of meters (see Fig. 5.21), and they attain sensitivities of the order of $\kappa A \simeq 10^{-19}$ for gravitational waves of a frequency of about 1 kHz. Thus, these interferometric detectors are somewhat more sensitive than the available quadrupole detectors.

Larger interferometric detectors, with arm lengths of several kilometers, are under development. These detectors, called LIGOs (*L*aser *I*nterferometer *G*ravity *W*ave *O*bservatories), are expected to attain sensitivities of the order of $\kappa A \simeq 10^{-22}$. The sensitivities of these LIGOs will be much higher than those of quadrupole detectors, and we hope that LIGOs will detect gravitational wave bursts emitted by supernovas.

TABLE 5.7 INTERFEROMETRIC DETECTORS OF GRAVITA-
TIONAL RADIATION*

Location	Arm length	n	Sensitivity (κA)
MIT	6 m	130	10^{-20}
Munich	3	100	10^{-19}
Glasgow	1.5	3000	10^{-19}
Caltech	40	2000	10^{-19}
Orsay	5		
Pisa	2		
Caltech LIGO	4000 ?	?	10^{-22} ?

* Based on Blair (1991).

Fig. 5.21 Laser interferometric detector at the Massachusetts Institute of Technology. This detector is a prototype, intended to test the optics for a full-scale LIGO.
(Courtesy D. Shoemaker, MIT.)

FURTHER READING

A simple, elementary introduction to gravitational waves is given by Schutz, B. F., Am. J. Phys. **52**, 412 (1984).

Selected information about sources of gravitational radiation is contained in Rees, M., Ruffini, R., and Wheeler, J. A. W., *Black Holes, Gravitational Waves, and Cosmology* (Gordon and Breach, New York, 1974). This book also presents a neat derivation of the cross section for absorption (Eq. [130]) by a method that relies on detailed balance. Yet another derivation is given in Weinberg, S., *Gravitation and Cosmology* (Wiley, New York, 1972); this derivation relies simply on energy conservation (or the optical theorem).

The article "Gravitational-Wave Astronomy" by Press, W. H., and Thorne, K. S., Ann. Rev. Astron. Astrophys. **10**, 335 (1972) speculates about the future use of gravitational antennas to carry out astronomical observations. Table 5.1 is based on this article, which also contains an analysis of detector sensitivity and a warning against the indiscriminate use of the cross section as a measure of this sensitivity.

An account of the early experiments with quadrupole detectors and of the controversy over Weber's claims will be found in Tyson, J. A., and Giffard, R. P., Ann. Rev. Astron. Astrophys. **16**, 521 (1978), reprinted in Worden, P. W., and Everitt, C. W. F., eds., *Gravity and Inertia* (American Assoc. of Physics Teachers, Stony Brook, 1983). This article also includes a concise review of sources of radiation and a discussion of detector sensitivity and the role of coincidence schemes.

The following references supply a wealth of information about possible astrophysical sources of gravitational radiation, quadrupole and interferometric detectors, and improved detection schemes that might be implemented in the next generation of detectors:

Smarr, L., ed., *Sources of Gravitational Radiation* (Cambridge University Press, Cambridge, 1979). (Strengths of the radiation that different kinds of sources deliver to detectors on the Earth are displayed on convenient diagrams. Warning: two of these diagrams, Figs. 3 and 4, pp. 484-85, are accidentally interchanged.)

Douglass, D. H., and Braginsky, V. B., "Gravitational-radiation Experiments," in Hawking, S. W., and Israel, W., eds., *General Relativity* (Cambridge University Press, Cambridge, 1979).

Thorne, K. S., "Gravitational Radiation," in Hawking, S. W., and Israel, W., eds., *Three Hundred Years of Gravitation* (Cambridge University Press, Cambridge, 1987). (More diagrams of source strengths.)

Schutz, B. F., ed., *Gravitational Wave Data Analysis* (Kluwer, Dordrecht, 1989).

Blair, D. G., *The Detection of Gravitational Waves* (Cambridge University Press, Cambridge, 1991).

The last of these references provides a good review of the state of the art in detectors; it includes a complete survey of quadrupole and interferometric detectors now in operation or under construction. The current trend in detectors is toward laser interferometers of large size (LIGOs). Abramovici, A., et al., Science **256**, 325 (1992) and Vogt, R. E., "The U. S. LIGO Project," in Humitaka, S., Nakamura, T., and Ruffini, R., eds., *Proceedings of the Sixth Marcel Grossmann Meeting on General Relativity* (World Scientific, Singapore, 1992) give good surveys of LIGOs now under construction or development.

REFERENCES

Damour, T., "Radiation Damping in General Relativity," in Ning, H., eds., *Proceedings of the Third Marcel Grossmann Meeting on General Relativity* (Science Press and North-Holland, Amsterdam, 1983a).

Damour, T., Phys. Rev. Lett. **51**, 1019 (1983b).

Damour, T., "Gravitational Radiation and the Motion of Compact Bodies," in Deruelle, N. and Piran, T., eds., *Gravitational Radiation* (North-Holland, Amsterdam, 1983).

Damour, T., "The Problem of Motion in Newtonian and Einsteinian Gravity," in Hawking, S. W., and Israel, W., eds., *Three Hundred Years of Gravitation* (Cambridge University Press, Cambridge, 1987).

Damour, T., Gibbons, G. W., and Taylor, J. H., Phys. Rev. Lett. **61**, 1151 (1988).

Davis, M., Ruffini, R., Press, W. H., and Price, R. H., Phys. Rev. Lett. **27**, 1466 (1971).

Deruelle, N., and Piran, T., eds., *Gravitational Radiation* (North-Holland, Amsterdam, 1983).

Douglass, D. H., and Braginsky, V. B., "Gravitational-radiation Experiments," in Hawking, S. W., and Israel, W., *General Relativity* (Cambridge University Press, Cambridge, 1979).

Faulkner, J., Flannery, B. P., and Warner, B., Ap. J. **175**, L79 (1972).

Ferrari, A., and Ruffini, R., Ap. J. **158**, L71 (1969).

Garwin, R. L., *Phys. Today*, December 1974, p.9.; ibid., November 1975, p. 13.

Gibbons, G. W., and Hawking, S. W., Phys. Rev. D **4**, 2191 (1971).

Levine, J. L., and Garwin, R. L., Phys. Rev. Lett. **33**, 794 (1974).

Moss, G., Miller, L., and Forward, R., Applied Optics **10**, 2495 (1971).

Owa, S., Fujimoto, M-K., Hirakawa, H., Morimoto, K., Suzuki, T., and Tsubono, K., in Ruffini, R., ed., *Proceedings of the Fourth Marcel Grossmann Meeting on General Relativity* (Elsevier, Amsterdam, 1986).

Press, W. H., and Thorne, K. S., Ann. Rev. Astron. Astrophys. **10**, 335 (1972).

Rees, M., Ruffini, R., and Wheeler, J. A., *Black Holes, Gravitational Waves, and Cosmology* (Gordon and Breach, New York, 1974).

Sciama, D. W., Field, G. F., and Rees, M. J., Phys. Rev. Lett. **23**, 1514 (1969).

Taylor, J. H., and Weisberg, J. M., Astrophys. J. **345**, 434 (1989).

Thorne, K. S., "Gravitational Radiation," in Hawking, S. W., and Israel, W., eds., *Three Hundred Years of Gravitation* (Cambridge University Press, Cambridge, 1987).

Trimble, V., and Weber, J., Ann. N.Y. Acad. Sci. **224**, 93 (1973).

Tyson, J. A., Ann. N. Y. Acad. Sci. **224**, 74 (1973); Phys. Rev. Lett. **31**, 326 (1973).

Tyson, J. A., and Giffard, R. P., Ann. Rev. Astr. and Astrophys., **16**, 521 (1978).

Weber, J., in Bertotti, B., ed., *Proceedings of Course 56 of the International School of Physics "Enrico Fermi"* (Academic Press, New York, 1973).

Weber, J., et al., Phys. Rev. Lett **22**, 1320 (1969)

Weber, J., et al., Phys. Rev. Lett. **24**, 276 (1970a).

Weber, J., et al., Phys. Rev Lett. **25**, 180 (1970b).

Weber, J., et al., Phys. Rev. Lett. **31**, 779 (1973).

PROBLEMS

1. Show that the gravitational waves $\phi_\oplus{}^{\mu\nu}$ and $\phi_\otimes{}^{\mu\nu}$ (see Eqs. [24] and [25]) carry a momentum flux equal to their energy flux.

2. For each of the gauge waves [11] - [14], prove that the energy-momentum tensor is zero. Furthermore, prove that if a gravitational wave consists of a superposition of a transverse wave and a gauge wave, the contribution of the gauge wave to the net energy-momentum tensor is zero.

3. For each of the gauge waves [11] - [14], find a gauge transformation analogous to Eq. [22] that cancels the wave.

4. A plane gravitational wave of amplitude A and frequency ω traveling in the z direction has an energy-momentum tensor

$$t_{(1)}{}^{\mu\nu} = A^2\, k^\mu k^\nu\, \sin^2 k_\alpha x^\alpha$$

(a) Write down all components of $t_{(1)}{}^{\mu\nu}$ as a matrix. Your answer should be expressed in terms of A and ω.
(b) Suppose that you measure the energy-momentum tensor in a new reference frame moving with velocity v in the x direction of the old reference frame. Find all components of the new energy-momentum tensor $t'_{(1)}{}^{\mu\nu}$.
(c) What is the direction (relative to the z'-axis) of the energy flow in the new reference frame?

5. Suppose that the wave $\phi_\otimes{}^{\mu\nu}$ (see Eq. [25]) is observed from a new reference frame moving at velocity v along the z-axis of the old reference frame.
(a) Find the polarization tensor of the wave in the new frame and express it as a superposition of the standard polarizations given by Eqs. [9]-[14].
(b) Find the amplitude $A'_\otimes$ of the wave in the new frame.
(c) Find the energy flux in the new frame and compare it with the flux in the old frame.

6. An extremely low frequency gravitational wave is incident on the Earth. Assuming that the frequency is so low that quasistatic conditions prevail, use the methods of Section 1.8 to find an expression for the height of the tide raised by this wave on the Earth. Describe how the shape of the tidal surface differs from that generated by the tidal forces of the Moon.

7. Suppose that a plane gravitational wave of frequency $\omega = 10^{-5}$ radian/s with an energy flux of 10^7 erg/cm^2 is incident on the Earth. The tidal torques will change the direction of the spin of the Earth. Estimate the maximum rate of change of direction that this wave can generate.

8. Use Eq. [85] to estimate the radiation rate for a "rod" of steel of radius 1 m, length 40 m, oscillating longitudinally with the maximum amplitude permitted by

the tensile strength of steel, about 10 cm. Assume that the rod is oscillating in its lowest acoustic mode, that is, $\frac{1}{2}\lambda_{sound}$ = 40 m.

9. Prove that the spherically symmetric oscillations ("breathing mode") of a spherical star cannot emit any gravitational radiation.

10. Calculate the rate at which the Earth radiates gravitational energy as it orbits around the Sun. What is the corresponding rate of change of the orbital radius?

11. The system W UMa consists of two stars in a very tight orbit with a period of only 8 hr. The relevant parameters of the system are

$$m_1 = 0.76 \quad m_2 = 0.57 \quad r = 1.5 \times 10^{11} \text{ cm} \quad \text{period 8.0 hr}$$

and the distance is 110 pc.
(a) Calculate the power radiated in the form of gravitational radiation.
(b) Calculate the flux produced at the Earth. Assume an average orientation of the system relative to the Earth.
(c) Calculate the rate of decrease of the separation between the two stars.

12. Calculate the power radiated in the form of gravitational waves by a stone of mass 3 kg being whirled around a circle of radius 1 m at a frequency of 10 radians per second.

13. A simple harmonic oscillator consists of a mass m that oscillates along the z-axis at frequency ω according to $z = A \sin \omega t$. What is the power radiated by this oscillator in the form of gravitational waves?

14. A grain of dust of mass m carries a single excess electron so its net electric charge is $-e$. The grain orbits in interstellar space in a circular orbit under the influence of a uniform magnetic field $\mathbf{B}$. The speed of the grain is v; the speed is non-relativistic ($v \ll c$).
(a) Use the quadrupole formula to calculate the power radiated by the grain in the form of gravitational waves.
(b) Compare with the power radiated in the form of electromagnetic waves, $dE/dt = 2(e^2/3c^3) a^2$. Show that the gravitational power exceeds the electromagnetic power if the momentum mv of the grain exceeds a certain critical value. What is this critical value?
(c) For a grain of $m = 10^{-3}$ g, $v = 10^9$ cm/s, the electromagnetic power is small (negligible) compared with the gravitational power. How long does it take this grain to lose 1% of its energy in a magnetic field of 10^{-4} gauss?

15. For each of the first three binary systems listed in Table 5.2, calculate the rate of change of the orbital period. Assume the orbits are circular. Are the rates of change of the periods observable?

16. A symmetric dumbbell consists of two equal masses m attached to the two ends of a massless rod of length l (see Fig. 5.22). The dumbbell rotates in the x-y plane with angular velocity ω about its center, so the position of the first mass is

$$x = \frac{l}{2}\cos\omega t \qquad y = \frac{l}{2}\sin\omega t \qquad z = 0$$

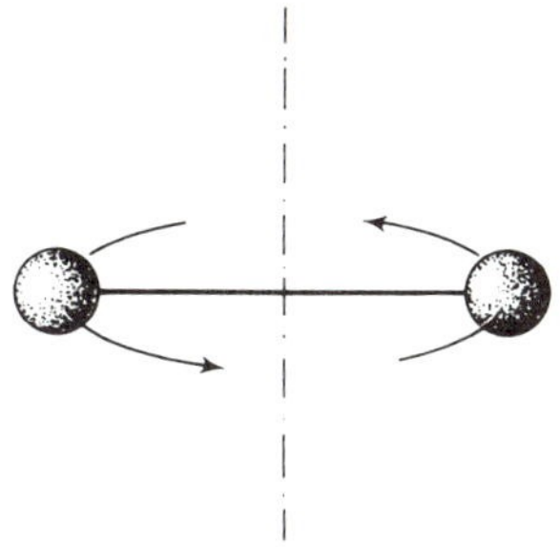

Fig. 5.22 A rotating symmetric dumbbell.

(a) What is the quadrupole moment tensor of this system as a function of time?
(b) What is the rate dE/dt at which gravitational energy is radiated?
(c) Use the result of Part (b) for a rough estimate of the gravitational energy radiated by a uniform rod of steel of mass 50 kg, length 1 m rotating at 2×10^3 radians per second about its center.

17. In a desperate attempt to generate gravitational radiation artificially, we take two large battleships of 70,000 tons each, and we make them collide head-on at 40 km/h. Assume that during the collision the battleships decelerate at a constant rate and come to rest in 2.0 s.
(a) Estimate the gravitational energy radiated during the collision. Treat the battleships as point masses.
(b) Assume that the radiated gravitational energy is uniformly distributed over all directions in space. Could a detector of the Weber type (with the sensitivity given by Eqs. [153] and [156]), placed at a distance of 200 m from the impact point, detect the radiation?

18. A meteoroid of a mass of 2×10^7 kg strikes the surface of the Earth with a vertical velocity of 11 km/s. The meteoroid comes to rest after penetrating the Earth to a depth of 200 m. Use Eq. [100] to estimate the total gravitational energy radiated during the impact; assume constant deceleration. Describe the polarization of the pulse of gravitational radiation that reaches an observer on the surface of the Earth at some short distance from the impact point. Can an antenna of the Weber type (with the sensitivity given by Eqs. [153] and [156]) detect this gravitational radiation if placed near the impact point?

19. Two stars, similar to the Sun, collide head-on with a relative velocity of $v/c =$

10^{-3}. Roughly estimate the gravitational energy radiated during the impact. Treat the stars as mass points that come to rest after moving through a distance of about one solar radius; pretend that the deceleration is constant. If such a collision takes place at the center of our Galaxy, can we detect the gravitational radiation with an available Weber detector (see the sensitivities given by Eqs. [153] and [156])?

20. A cannon placed at the origin of coordinates fires a shot of mass 50 kg in the horizontal direction (see Fig. 5.23). The barrel of the cannon has a length of 2.0 m; the shot has a uniform acceleration while in the barrel and emerges with a muzzle velocity of 300 m/s. Calculate the gravitational radiation field $\kappa h_{\mu\nu}$ generated by the shot at the point P on the z-axis at a vertical distance of 20 m above the cannon. What is the maximum value of κh_{11}? Ignore the gravitational field of the Earth.

Fig. 5.23 Shot fired from a cannon.

21. A neutron star can be regarded as a sphere of nearly incompressible fluid held together by gravitation.
(a) Show that the fundamental mode of vibration (elongation) has a frequency of the order of $\omega \simeq \sqrt{G\rho}$, where ρ is the density. Evaluate this frequency for a neutron star of $M \simeq 0.7M_\odot$, $R \simeq 10$ km. (Hint: The speed of sound waves in a fluid is $v \simeq \sqrt{(p/\rho)}$, where p is the pressure. Estimate the typical pressure required to support a sphere against its own gravity.)
(b) If the neutron star is elongated in some direction by an amount δR during its formation in the supernova implosion, the corresponding vibrational energy is of the order of $\frac{1}{2}m\omega^2(\delta R)^2$. Evaluate this energy for $\delta R/R \simeq 10^{-1}$.
(c) If all of this energy is radiated in the form of gravitational waves in a pulse lasting $\simeq 1$ s, what would be the radiated power? What would be the energy flux at the Earth at a distance of, say, 10^4 light-years?

22. A long thin rod of iron lies along the z-axis. When in equilibrium, the rod occupies the interval $-b \le z \le b$. The rod oscillates in a longitudinal acoustic mode with

$$\psi(z, t) = a \sin \frac{\pi z}{2b} \cos \frac{\pi vt}{2b}$$

where $\psi(z, t)$ is the longitudinal displacement suffered by a point whose equilibrium position is z, and where v is the velocity of sound waves along the rod. Find the "effective oscillation amplitude" a' defined by Eq. [131]. Repeat the calculation for the mode

$$\psi(z, t) = a \sin \frac{3\pi z}{2b} \cos \frac{3\pi v t}{2b}$$

Which of these modes has the shorter (gravitational) damping rate?

6. *Riemannian Geometry*

I KNOW WHY THERE ARE SO MANY PEOPLE WHO LOVE CHOPPING WOOD. IN THIS ACTIVITY ONE IMMEDIATELY SEES THE RESULTS.

Albert Einstein

The linear tensor theory of gravitation that we developed by analogy with electrodynamics started out as the theory of a tensor field in a flat spacetime background. The geometric interpretation of this tensor field emerged only as an afterthought. However, the analysis of space-time measurements (clocks for time measurements and also for distance, by means of the radar-ranging procedure) has shown us that the flat spacetime background is purely fictitious--in a gravitational field, the real geometry measured by our instruments is the geometry of a curved space, that is, a Riemannian space.

Mathematically, a Riemannian space is a differentiable manifold endowed with a topological structure and a geometric structure. In the discussion of the geometric structure of a curved space we can make a distinction between the *affine* geometry and the *metric* geometry. These two kinds of geometries correspond to two different ways in which we can detect the curvature of a space. One way is by examination of the behavior of parallel line segments, or parallel vectors. For example, on the surface of a sphere, we can readily detect the curvature by transporting a vector around a closed path, always keeping the vector as parallel to itself as possible. Fig. 6.1 shows what happens if we parallel transport a vector around a "triangular" path on the sphere. The final vector differs in direction from the initial vector, whereas on a flat surface the final vector would not differ. Such changes in a vector produced by parallel transport characterize the affine geometry (the word *affine* means *connected*, and refers to how parallels at different places are connected, or related).

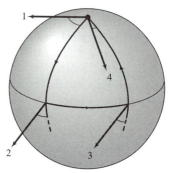

Fig. 6.1 Parallel transport of a vector. We start at the north pole, and parallel transport the vector along a meridian to the equator, then some distance along the equator, and then return to the north pole along another meridian.

Another way in which we can detect curvature is by measurements of lengths and areas. For example, we can draw a circle on the sphere, and check the radius vs. the circumference, or the radius vs. the area. Both the circumference and the area of such a circle are smaller than for a circle on a flat surface (see Fig. 6.2). Such deviations from what we expect on a flat surface characterize the metric geometry.

Fig. 6.2 A circle on the sphere has a circumference smaller than 2π times its radius. If the radius exceeds the distance from the pole to the equator, the circumference *decreases* with increasing radius.

Note that the lateral surface of a cylinder is a flat space, according to both of our criteria for curvature: parallel transport of a vector along this surface does not alter its direction, and measurements of lengths and areas reveal no deviations from what we find on a flat surface (this can be easily seen by slicing the lateral surface of the cylinder open and spreading it out on a table). The lateral surface of a cylinder is said to have zero intrinsic curvature, but nonzero extrinsic curvature. By the latter is meant how the surface curves into dimensions that lie outside the surface. When we perform parallel transport or when we measure lengths and areas within some given curved surface, we can detect only the intrinsic curvature, not the extrinsic curvature. Since we are confined to the four dimensions of spacetime, our measurements permit us to detect only the intrinsic curvature of spacetime, and this is the only curvature we will deal with in our study of Riemannian geometry.

Before we can proceed with the study of "gravitation as geometry," we need to develop some mathematical concepts, in particular the concepts of tensors and tensor fields in a Riemannian spacetime. We begin with the general definitions of tensors. These definitions require no more than the differential structure of spacetime--we do not need the affine or the metric structure. As a next step, we will attempt to construct tensors by differentiation; we then need to introduce the affine structure, since for differentiation we have to calculate the change experienced by a vector during a displacement, and we need to take into account how much of this change is caused by parallel displacement. And as a final step, we will want to calculate the lengths of vectors, for which we need to introduce the metric structure.

Throughout most of this chapter, we will describe tensors by the formalism of the absolute differential calculus of Ricci and Levi-Civita (Ricci and Levi-Civita, 1901; Levi-Civita, 1929). In this formalism, tensors are represented by their components relative to a general set of coordinates, such as spherical, cylindrical, or any other kind of curvilinear coordinates. The tensor calculus is said to be *absolute* because it is independent of the details of the choice of coordinates; that is, the equations have the same form in all coordinate systems. Thus, the details of the coordinate system need not be specified for the purpose of developing the mathematics of tensor calculus (of course, when we want to do *numerical* computations, we must specify the details of our choice of coordinates). The tensor calculus for Lorentz tensors developed in Chapter 2 is a special case of the general tensor calculus.

Tensors can also be described by the formalism of differential forms of Cartan. Vectors, differential forms, and tensors are then operators, represented by abstract symbols, as in Section 2.6. In many instances, the proofs of theorems of tensor calculus take a very elegant and concise form when expressed in the language of differential forms. But in other instances this language results in excessive "orgies of formalism" (Weyl, 1922), and then component language is our salvation. Furthermore, we must not forget that the physicist who wishes to measure a tensor has no choice but to set up a coordinate system, and then measure the numerical values of the components. Thus, to carry out the comparison of theory and experiment, the physicist cannot ultimately avoid the language of components; only a pure mathematician can adhere exclusively to the abstract, coordinate-free language of differential forms. We will develop Riemannian geometry in the language of differential forms in Section 6.7, and in later chapters we will occasionally use this language, when convenient.

6.1 GENERAL COORDINATES AND TENSORS

Unless the spacetime is flat, there is no good reason to use rectangular coordinates. In flat spacetime, rectangular coordinates are preferred because both the metric ($g_{\mu\nu} = \eta_{\mu\nu}$) and the Lorentz transformations take their simplest form when expressed in these coordinates. In a curved spacetime, the use of rectangular coordinates, or some modified version of rectangular coordinates, does not necessarily result in any simplification of the metric, except in those regions that are far away from all gravitational fields, where spacetime becomes asymptotically flat. We will use *general coordinates* to describe points in spacetime. By this is meant that to each point of spacetime we attach a set of four numbers that serve to identify the point. These numbers can be assigned according to any arbitrary, but well-defined, rule. The coordinates should be assigned in such a way that they vary more or less continuously as we move through spacetime; a finite number of discontinuities, such as found in polar coordinates at $\phi = 2\pi$, will of course be acceptable.

On a sheet of paper it is easy to label a point by coordinates: we simply mark the point with an X and write the coordinate values next to the X. In empty spacetime this will not do. One way to label spacetime points is to imagine all of space to be filled with small clocks, each of which carries a dog tag with the values of the spatial coordinates printed on it. There is no need that these clocks be at "rest"; a continuous streaming will keep the coordinate values smooth and acceptable. The coordinates of spacetime points in the empty space between adjacent clocks can be obtained by interpolation, using meter sticks or a radar-ranging procedure (the interpolation procedure can be performed as in flat spacetime, since locally any curved spacetime is approximately flat). A somewhat more elegant proposal for labeling spacetime points is the following: Suppose that at some distance from the region where we want to introduce coordinates, there are three "fixed" stars (red, white, and blue), which are not collinear. For the spatial coordinates of a given point in space we can then adopt the three angles between the stars as seen from the given point (see Fig. 6.3).* For the time coordinate we can adopt the angle between, say, the red star and a fourth star (yellow) which is in motion with respect to red, white, and blue. Note that it is not really necessary that the red, white, and blue stars be fixed; even if they are in motion, the coordinates will be well defined. Also note that the bending of light by gravitational fields does no harm to these coordinates, since we are merely trying to find some well-defined procedure for associating

---

* This construction of coordinates can fail for some exceptional points; for instance, in flat spacetime, all points on the circle that passes through the red, white, and blue stars yield the same angles—our construction of coordinates is unable to discriminate among such points.

four numbers with each given point of spacetime. We might call the numbers obtained by this construction *astrogator coordinates*, since an astro-navigator would probably use coordinates of this kind.

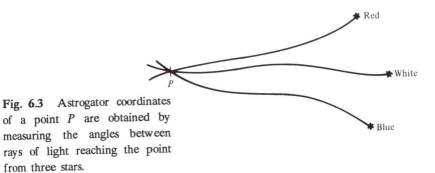

Fig. 6.3   Astrogator coordinates of a point *P* are obtained by measuring the angles between rays of light reaching the point from three stars.

   These astrogator coordinates are intended only to serve as a demonstration that an operational definition of spacetime coordinates is possible. In practice, our choice of coordinates will be governed by our desire to keep the solution of the field equations as simple as possible. In the theory of the (electrostatic) potential, the advantages of a clever choice of coordinates are well known. Unfortunately, in a curved spacetime the coordinates that lead to the simplest mathematics often lack any obvious physical interpretation, and it will often be possible to give an operational definition of the coordinate points only *after* we have solved the field equation for the spacetime metric, and after we have calculated just how much the light rays used to locate coordinate points are bent. We already discussed one instance of this difficulty at the end of Section 4.4.
   We will use the notation $x^\mu = (x^0, x^1, x^2, x^3)$ for general coordinates. It is important to keep in mind that the index on the coordinates is always an *upper* (contravariant) index. Although there exists a general procedure for raising and lowering indices (it involves the *metric* tensor of the space, see below), this procedure gives nothing useful when applied to x^μ. Thus, we will avoid using x_μ, except when dealing, as we were in earlier chapters, with simple rectangular coordinates in flat spacetime.
   Let us see what happens if we transform from one set of general coordinates to another. Suppose that the old coordinates are x^μ and the new x'^μ. The new coordinates are then functions of the old,

$$x'^\mu = x'^\mu(x) \qquad [1]$$

and the old are functions of the new,

$$x^\mu = x^\mu(x') \qquad [2]$$

Accordingly, the differentials of the coordinates obey the transformation law

$$dx'^{\mu} = \frac{\partial x'^{\mu}}{\partial x^{\nu}} \, dx^{\nu} \tag{3}$$

$$dx^{\mu} = \frac{\partial x^{\mu}}{\partial x'^{\nu}} \, dx'^{\nu} \tag{4}$$

In the case of the Lorentz transformations, the partial derivatives $\partial x^{\mu}/\partial x'^{\nu}$ were constants. In the present case of general coordinate transformations, these partial derivatives are functions of x^{μ} (or x'^{μ}). Note that the transformations [3] and [4] are the inverses of one another, and therefore

$$\frac{\partial x'^{\mu}}{\partial x^{\beta}} \frac{\partial x^{\beta}}{\partial x'^{\nu}} = \delta^{\mu}_{\nu} \tag{5}$$

This also is obvious from the chain rule for derivatives.

The transformation law for the differential operator ∂_{μ} is

$$\frac{\partial}{\partial x'^{\mu}} = \frac{\partial x^{\nu}}{\partial x'^{\mu}} \frac{\partial}{\partial x^{\nu}} \tag{6}$$

$$\frac{\partial}{\partial x^{\mu}} = \frac{\partial x'^{\nu}}{\partial x^{\mu}} \frac{\partial}{\partial x'^{\nu}} \tag{7}$$

In Chapter 2 we defined vectors and tensors by their behavior under Lorentz transformations. Now we will define them by their behavior under general coordinate transformations. These new definitions are natural generalizations of our earlier definitions. The transformation equations [3] and [6] for differentials and for differential operators serve as prototypes for the transformation equations of vectors. Thus, contravariant vectors have the transformation law of dx^{μ}, covariant vectors that of ∂_{μ}:

A four-component object A^{μ} is a contravariant vector under general coordinate transformations if it transforms according to

$$A'^{\mu} = \frac{\partial x'^{\mu}}{\partial x^{\nu}} A^{\nu} \tag{8}$$

A four-component object B_{μ} is a covariant vector if

$$B'_{\mu} = \frac{\partial x^{\nu}}{\partial x'^{\mu}} B_{\nu} \tag{9}$$

Note that x^{μ} *is not a vector* with respect to general coordinate transformations: the transformation law for x^{μ} is given by Eq. [1], and this does not have the form of Eq. [8]. In the exceptional case of a *linear* (and homogeneous) transformation, Eq. [1] becomes

$$x'^{\mu} = b^{\mu}_{\nu} x^{\nu} \qquad [10]$$

where b^{μ}_{ν} is a constant matrix. In this case, x^{μ} does have the transformation law [8], since Eq. [10] can then also be written as

$$x'^{\mu} = \frac{\partial x'^{\mu}}{\partial x^{\nu}} x^{\nu} \qquad [11]$$

Thus, x^{μ} is a vector with respect to such linear transformations. This is why, in Chapter 2, we were able to treat the rectangular coordinate x^{μ} as a vector with respect to Lorentz transformations.

Tensors of higher rank have the transformation laws

$$A'^{\alpha\beta...\lambda} = \frac{\partial x'^{\alpha}}{\partial x^{\mu}} \frac{\partial x'^{\beta}}{\partial x^{\nu}} \cdots \frac{\partial x'^{\lambda}}{\partial x^{\kappa}} A^{\mu\nu...\kappa} \qquad [12]$$

and

$$B'_{\alpha\beta...\lambda} = \frac{\partial x^{\mu}}{\partial x'^{\alpha}} \frac{\partial x^{\nu}}{\partial x'^{\beta}} \cdots \frac{\partial x^{\kappa}}{\partial x'^{\lambda}} B_{\mu\nu...\kappa} \qquad [13]$$

if they are contravariant and covariant, respectively.

■ *Exercise 1.* Show that δ^{ν}_{μ} transforms as a tensor covariant in μ, contravariant in ν. ■

We must keep in mind that a multicomponent object is (or is not) a tensor with respect to a given class of transformations. Thus, it is quite possible for a quantity to be a tensor with respect to Lorentz transformations, but not a tensor with respect to general coordinate transformations; the converse is of course not possible. By the word *tensor*, without qualifiers, we will hereafter mean *tensor with respect to general coordinate transformations.*

Using Eq. [5], we can now easily show that

$$A^{\mu} B_{\mu} \qquad [14]$$

is a scalar if A^{μ} and B_{ν} are vectors, and that

$$A^{\mu\nu\alpha}_{\alpha} \qquad [15]$$

is a second-rank contravariant tensor if $A^{\mu\nu\alpha}_{\beta}$ is a tensor contravariant in $\mu\nu\alpha$, covariant in β, and so on.

■ *Exercise 2.* Show this. ■

We can also easily show that the derivative $\partial_\mu\phi$ of a scalar function is a vector. However, it does not follow from the transformation laws that the derivative $\partial_\mu A_\nu$ of a vector is a tensor. As we will see in the next section, to construct tensors by differentiation requires a somewhat more complicated procedure.

6.2 PARALLEL TRANSPORT; THE COVARIANT DERIVATIVE

In Chapter 2 we saw that in flat spacetime, with rectangular coordinates x^μ, it is easy to construct higher-rank tensors by differentiation. For example, if A_μ is a tensor of first rank, then $\partial_\nu A_\mu$ is a tensor of second rank. In a curved spacetime, and even in flat spacetime with curvilinear coordinates, simple differentiation does not yield a tensor quantity. For example, suppose that A_μ is a tensor of first rank in flat spacetime in, say, spherical coordinates, with $x^0 = t$, $x^1 = r$, $x^2 = \theta$, $x^3 = \phi$. Then simple differentiation yields $\partial_\nu A_\mu = (\partial A_\mu/\partial t, \partial A_\mu/\partial r, \partial A_\mu/\partial\theta, \partial A_\mu/\partial\phi)$, which does not have the correct transformation properties for a second-rank tensor. We can see this by examining what happens if we attempt to transform to some other coordinates, say, from the spherical coordinates x^μ to rectangular coordinates x'^μ. The transformation law for A_μ is

$$A'_\alpha = \frac{\partial x^\mu}{\partial x'^\alpha} A_\mu \qquad [16]$$

and hence

$$\frac{\partial A'_\alpha}{\partial x'^\beta} = \frac{\partial x^\mu}{\partial x'^\alpha}\frac{\partial A_\mu}{\partial x'^\beta} + \frac{\partial^2 x^\mu}{\partial x'^\alpha \partial x'^\beta} A_\mu$$

$$= \frac{\partial x^\mu}{\partial x'^\alpha}\frac{\partial x^\nu}{\partial x'^\beta}\frac{\partial A_\mu}{\partial x^\nu} + \frac{\partial^2 x^\mu}{\partial x'^\alpha \partial x'^\beta} A_\mu \qquad [17]$$

The right side of Eq. [17] differs from the transformation law for a second-rank tensor by the extra term $\partial^2 x^\mu/(\partial x'^\alpha \partial x'^\beta)\, A_\mu$, which contains second derivatives of the spherical coordinates x^μ relative to the rectangular coordinates x'^α. Since the relation between the spherical and the rectangular coordinates is nonlinear, these derivatives are nonzero. Their presence "spoils" the transformation law for $\partial A_\mu/\partial x^\nu$, and we see that this quantity is not a second-rank tensor with respect to general coordinate transformations.

We can gain some further insight into the reason for the failure of the tensor transformation law for the derivative $\partial A_\mu / \partial x^\nu$ by examining the definition of this derivative,

$$\frac{\partial A_\mu}{\partial x^\nu} = \lim_{dx \to 0} \frac{A_\mu(x + dx) - A_\mu(x)}{dx^\nu} \qquad [18]$$

The trouble with this expression is that the numerator is not a vector. This numerator is the difference between two vectors located at two separate points. Although the difference between two vectors located at the *same* point is a vector, the difference between two vectors located at *separate* points is not a vector, since the transformation law [9] depends on position. This suggests that to obtain a numerator that is a vector, we should parallel transport $A_\mu(x)$ from x to $x + dx$ before performing the subtraction, so the two vectors in the numerator are located at the same point. Thus, the construction of tensors by differentiation requires that we first construct a definition of parallel transport of a vector.

We will denote by δA_μ the change produced in the components of the vector A_μ by its parallel transport during a small displacement dx^β. This change δA_μ must be linear in dx^β; and it must also be linear in A_μ, since the change produced by parallel transport of a vector sum $A_\mu + B_\mu$ must equal the sum of the changes of the individual vectors. Therefore δA_μ must be of the general form

$$\delta A_\mu = \Gamma^\nu{}_{\mu\beta} A_\nu \, dx^\beta \qquad [19]$$

where the $4 \times 4 \times 4$-component object $\Gamma^\nu{}_{\mu\beta}$ is some function of position. As we will see later, $\Gamma^\nu{}_{\mu\beta}$ is related to the derivatives of the metric tensor. However, since we have not yet introduced a metric tensor in our spacetime, we will for now regard $\Gamma^\nu{}_{\mu\beta}$ as some arbitrarily specifiable function that characterizes parallel transport in our spacetime, in other words, it characterizes the affine geometry of spacetime. The objects $\Gamma^\nu{}_{\mu\beta}$ are called the *Christoffel symbols*, or the *connection coefficients.**

Eq. [19] defines the parallel transport of a covariant vector. The parallel transport of a contravariant vector B^μ is determined by the requirement that $A_\mu B^\mu$ is a scalar, which cannot change under parallel transport. Thus,

---

* The name *Christoffel symbol* is sometimes reserved for the special values that the coefficients $\Gamma^\nu{}_{\mu\beta}$ take when the metric geometry is specified (see Eq. [63]). However, we will not bother with this fine distinction of terminology.

$$0 = \delta(A_\mu B^\mu) = \delta A_\mu B^\mu + A_\mu \delta B^\mu$$

$$= (\Gamma^\nu_{\mu\beta} A_\nu dx^\beta B^\mu + A_\mu \delta B^\mu)$$

$$= A_\mu (\Gamma^\mu_{\nu\beta} dx^\beta B^\nu + \delta B^\mu)$$

Since this is true for any arbitrary A_μ, the coefficient multiplying A_μ in this equation must vanish, which tells us that

$$\delta B^\mu = - \Gamma^\mu_{\nu\beta} B^\nu dx^\beta \qquad [20]$$

We will assume that $\Gamma^\nu_{\mu\beta}$ is symmetric in its lower indices, that is,

$$\Gamma^\nu_{\mu\beta} = \Gamma^\nu_{\beta\mu} \qquad [21]$$

To understand the significance of this symmetry condition, let us explore what would happen if it were *not* satisfied. Consider two small displacements $d\xi^\beta$ and $d\zeta^\beta$, which start at the same spacetime point, and attempt to construct the parallelogram that has these vectors for two of its sides. To perform this construction, we can parallel transport $d\xi^\beta$ along $d\zeta^\beta$ (see Fig. 6.4); this gives us the displacement vector for the "opposite" corner of the parallelogram:

$$d\zeta^\beta + (d\xi^\beta + \delta d\xi^\beta) = d\zeta^\beta + d\xi^\beta - \Gamma^\beta_{\nu\mu} d\xi^\nu d\zeta^\mu$$

Fig. 6.4 Construction of a parallelogram. (a) Parallel transport of $d\xi^\beta$ along $d\zeta^\beta$. (b) Parallel transport of $d\zeta^\beta$ along $d\xi^\beta$.

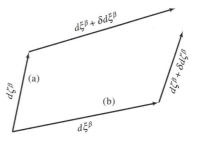

Alternatively, we can parallel transport $d\zeta^\beta$ along $d\xi^\beta$; this gives us a different expression for the opposite corner:

$$d\xi^\beta + (d\zeta^\beta + \delta d\zeta^\beta) = d\xi^\beta + d\zeta^\beta - \Gamma^\beta_{\nu\mu} d\zeta^\nu d\xi^\mu$$

Comparing these expressions, we see that the two alternative constructions agree if and only if $\Gamma^\beta_{\nu\mu}$ is symmetric in μ and ν. If $\Gamma^\beta_{\nu\mu}$ were not symmetric, the parallelogram would fail to close. This would mean that the geometry of the curved spacetime differs from a flat geometry even on a small scale--the curved spacetime could not be approximated locally by a flat spacetime. A geometry with an asymmetric Christoffel symbol is said to have *torsion*. Although an attempt has been made to incorporate torsion into gravitational theory ("Einstein-Cartan theory"), the physical foundations of this attempt are shaky, and we will ignore it.*

That parallel transport produces changes in the components of a vector should not cause surprise. It happens even in flat spacetime, if we use spherical coordinates, or other curvilinear coordinates. When we parallel transport a constant vector along some curve, its components with respect to rectangular coordinates do not change, but its components with respect to spherical coordinates obviously do change (see Fig. 6.5).

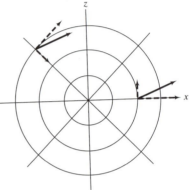

Fig. 6.5 Parallel transport of a constant vector in flat spacetime. The vector is in the meridional plane, ϕ = 0. The magnitude and the direction of the vector do not change during parallel transport, but its radial and tangential components (indicated by broken lines) do change.

Note that in flat spacetime, we can exploit the known behavior of a vector under parallel transport to calculate the Christoffel symbols in any given curvilinear coordinate system. For instance, suppose that a vector in spherical coordinates is initially in the radial direction, $A^\mu = (0, A^1, 0, 0)$, and we parallel displace it from (t, r, θ, ϕ) to $(t, r, \theta + d\theta,$

---

* Cartan's modification of Einstein's theory attempts to take the spin density of elementary particles as the source of torsion. But this introduces ambiguities because we do not know the "genuine" spin content of elementary particles—for instance, we do not know the internal structure of quarks, and we are therefore unable to say whether their ostensible spin $\hbar/2$ is genuine spin, or a combination of spin and orbital angular momentum of the constituents (if any) within the quarks. The proposal for spin as source of a gravitational torsion field is just as ambiguous as a proposal for, say, rest-mass current density as the source of a gravitational (vector) field. Mathematically, the rest-mass current density ρu^μ of elementary particles is a well-defined quantity, but physically its value is ambiguous because we do not know how much of the ostensible rest mass of the known elementary particles is genuine rest mass, and how much is kinetic and binding energy of their constituents. The ambiguities in the source of torsion make any calculations in the Einstein-Cartan theory futile.

ϕ). Then, as we can see from Fig. 6.6, the change in this vector is δA^μ = $(0, 0, -A^1 d\theta/r, 0)$, and hence Eq. [20] tells us that $\Gamma^2{}_{12} = 1/r$. All the other Christoffel symbols in spherical coordinates can be calculated in a similar way, by simple geometrical considerations.

Fig. 6.6 Parallel displacement of a constant vector in flat space-time. During a small displacement $d\theta$ the vector acquires a tangential component of length $-A^1 d\theta$. However, in spherical coordinates, the corresponding θ component is this length divided by r, that is, $-A^1 d\theta/r$ (just as the θ component of an arbitrary displacement dx^μ is the length $r d\theta$ divided by r, that is, $r d\theta/r = d\theta$).

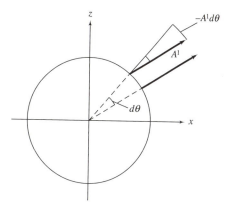

■ *Exercise 3.* By similar arguments show that the nonzero Christoffel symbols for flat spacetime in spherical coordinates are

$$\Gamma^1{}_{22} = -r$$
$$\Gamma^1{}_{33} = -r \sin^2\theta$$
$$\Gamma^2{}_{12} = \Gamma^2{}_{21} = 1/r$$
$$\Gamma^2{}_{33} = -\sin\theta \cos\theta$$
$$\Gamma^3{}_{13} = \Gamma^3{}_{31} = 1/r$$
$$\Gamma^3{}_{23} = \Gamma^3{}_{32} = \cot\theta \quad ■$$

[22]

■ *Exercise 4.* A more interesting example is parallel transport on the surface of a sphere. This surface is a curved two-dimensional space, usually called a two-sphere. Although we are interested mainly in four-dimensional spacetime, the definition [19] for parallel transport applies to a Riemannian space of any number of dimensions; but in an n-dimensional space, the indices μ and ν take the values 1, 2, 3,..., n. Consider the surface of a sphere, and use coordinates $x^1 = \theta$ and $x^2 = \phi$, where θ and ϕ are the usual polar and azimuthal angles on this surface. The Christoffel symbols are

$$\Gamma^1{}_{11} = 0 \quad \Gamma^1{}_{12} = 0 \quad \Gamma^1{}_{22} = -\sin\theta \cos\theta$$
$$\Gamma^2{}_{11} = 0 \quad \Gamma^2{}_{12} = \cot\theta \quad \Gamma^2{}_{22} = 0$$

[23]

Convince yourself that the results for the parallel transport of a vector along a meridian (ϕ = [constant]; see Fig. 6.7) and along the equator ($\theta = \pi/2$) are intuitively reasonable. ■

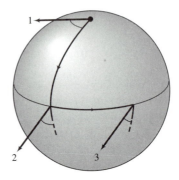

Fig. 6.7 Parallel transport of a vector along a meridian and along the equator of a sphere.

Now that we have available a definition of parallel transport, we can construct a genuine tensor by differentiation. To make the numerator in the derivative [18] into a vector, we parallel transport $A_\mu(x)$ from x to $x + dx$ before performing the subtraction. We then obtain

$$\lim_{dx \to 0} \frac{A_\mu(x + dx) - A_\mu(x) - \delta A_\mu}{dx^\nu}$$

$$= \lim_{dx \to 0} \frac{A_\mu(x + dx) - A_\mu(x)}{dx^\nu} - \Gamma^\alpha{}_{\mu\beta} A_\alpha \frac{dx^\beta}{dx^\nu}$$

$$= \frac{\partial A_\mu}{\partial x^\nu} - \Gamma^\alpha{}_{\mu\nu} A_\alpha \qquad [24]$$

This is called the *covariant derivative* of the vector A_μ. It is customary to indicate a covariant derivative by a semicolon, in distinction to the ordinary derivative, indicated by a comma:

$$A_{\mu;\nu} = A_{\mu,\nu} - \Gamma^\alpha{}_{\mu\nu} A_\alpha \qquad [25]$$

By construction, the covariant derivative is a second-rank tensor. The word *covariant* as used in this context merely serves to distinguish $A_{\mu;\nu}$ from the ordinary derivative $A_{\mu,\nu}$. Since we already have been using *covariant* to describe lower indices on a tensor, it might be better to call [25] the *tensor derivative*; but *covariant* is common usage, and what meaning is intended must be decided from the context. For example, [25] is sometimes called the covariant derivative of a covariant vector.

A similar construction based on the parallel-transport formula for a contravariant vector B^μ tells us that the covariant derivative of such a vector is

$$B^\mu_{\ ;\nu} = \frac{\partial B^\mu}{\partial x^\nu} + \Gamma^\mu_{\ \alpha\nu} B^\alpha = B^\mu_{\ ,\nu} + \Gamma^\mu_{\ \alpha\nu} B^\alpha \qquad [26]$$

Furthermore, we can proceed to the construction of covariant derivatives for tensors of rank higher than one. For such higher-rank tensors, the covariant derivative differs from the ordinary derivative by as many extra terms as there are indices. For example, the covariant derivatives of the tensors $B^{\mu\nu}$ and $A^{\mu\nu}_{\ \ \sigma}$ are, respectively,

$$B^{\mu\nu}_{\ \ ;\beta} = \frac{\partial B^{\mu\nu}}{\partial x^\beta} + \Gamma^\mu_{\ \alpha\beta} B^{\alpha\nu} + \Gamma^\nu_{\ \alpha\beta} B^{\mu\alpha} \qquad [27]$$

and

$$A^{\mu\nu}_{\ \ \sigma;\beta} = \frac{\partial A^{\mu\nu}_{\ \ \sigma}}{\partial x^\beta} + \Gamma^\mu_{\ \alpha\beta} A^{\alpha\nu}_{\ \ \sigma} + \Gamma^\nu_{\ \alpha\beta} A^{\mu\alpha}_{\ \ \sigma} - \Gamma^\alpha_{\ \sigma\beta} A^{\mu\nu}_{\ \ \alpha} \qquad [28]$$

Note that in these covariant derivatives, each tensor index receives the same treatment as the vector index in Eqs. [25] and [26]. The construction of these covariant derivatives is based on the formulas for the parallel transport of higher-rank tensors, which can be obtained from the requirement that scalars, such as $A_\mu C_\nu B^{\mu\nu}$, must not change under parallel transport.

■ *Exercise 5.* Derive the formula for the parallel transport of the tensor $B^{\mu\nu}$.
■

By convention, the covariant derivative of a scalar field is simply taken to be the ordinary derivative, $\phi_{;\alpha} \equiv \phi_{,\alpha}$.

As a consequence of Eq. [27], the covariant derivative of the second-rank tensor $A^\mu B^\nu$ can be written

$$(A^\mu B^\nu)_{;\beta} = A^\mu_{\ ;\beta} B^\nu + A^\mu B^\nu_{\ ;\beta} \qquad [29]$$

This shows that covariant differentiation obeys the familiar rule for the differentiation of a product. This rule is also valid for products of tensors of higher rank that those in Eq. [29].

■ *Exercise 6.* Check Eq. [29]. ■

Although we have written the Christoffel in a notation usually reserved for third-rank tensors, the Christoffel symbol is not a tensor-- it does not obey the transformation law for a third-rank tensor. The transformation law for the Christoffel symbol must be such that the sum $\partial A_\mu / \partial x^\nu - \Gamma^\alpha{}_{\mu\nu} A_\alpha$ is a tensor; we know that in this sum the transformation of the first term deviates from a tensor (see Eq. [17]), and therefore the transformation of the second term must deviate by an opposite amount. This requirement leads to the following transformation law for $\Gamma^\mu{}_{\alpha\beta}$:

$$\Gamma'^\mu{}_{\alpha\beta} = \frac{\partial x'^\mu}{\partial x^\nu} \frac{\partial x^\sigma}{\partial x'^\alpha} \frac{\partial x^\tau}{\partial x'^\beta} \Gamma^\nu{}_{\sigma\tau} + \frac{\partial x'^\mu}{\partial x^\nu} \frac{\partial^2 x^\nu}{\partial x'^\alpha \partial x'^\beta} \qquad [30]$$

Eq. [30] can also be written as

$$\Gamma'^\mu{}_{\alpha\beta} = \frac{\partial x'^\mu}{\partial x^\nu} \frac{\partial x^\sigma}{\partial x'^\alpha} \frac{\partial x^\tau}{\partial x'^\beta} \Gamma^\nu{}_{\sigma\tau} - \frac{\partial x^\nu}{\partial x'^\alpha} \frac{\partial x^\sigma}{\partial x'^\beta} \frac{\partial^2 x'^\mu}{\partial x^\sigma \partial x^\nu} \qquad [31]$$

■ *Exercise 7.* Derive Eq. [30]. ■

■ *Exercise 8.* Derive Eq. [31]. (Hint: Differentiate

$$\frac{\partial x'^\mu}{\partial x^\nu} \frac{\partial x^\nu}{\partial x'^\alpha} = \delta^\mu_\alpha$$

with respect to x'^β, and obtain the identity

$$\frac{\partial x'^\mu}{\partial x^\nu} \frac{\partial^2 x^\nu}{\partial x'^\alpha \partial x'^\beta} + \frac{\partial x^\nu}{\partial x'^\alpha} \frac{\partial x^\sigma}{\partial x'^\beta} \frac{\partial^2 x'^\mu}{\partial x^\sigma \partial x^\nu} = 0 \,)$$ ■

Note that the difference between Eq. [30] and the transformation law of a third-rank tensor is due to the presence of the second term on the right side. In the special case of a linear transformation, say,

$$x'^\mu = b^\mu_\nu x^\nu$$

where b^μ_ν is a constant, this second term is absent, and then $\Gamma^\mu{}_{\alpha\beta}$ does behave like a tensor. $\Gamma^\mu{}_{\alpha\beta}$ is an example of an object that is a tensor with respect to linear transformations (such as Lorentz transformations), but is not a tensor with respect to general nonlinear coordinate transformations. To emphasize the nontensor character of the Christoffel symbol, the notation $\{^\mu_{\alpha\beta}\}$ is sometimes used instead of $\Gamma^\mu{}_{\alpha\beta}$.

6.3 THE GEODESIC EQUATION

In a curved space, we can define geodesic curves in two different ways: as straightest curves or as curves of extremal length. To construct a straightest possible curve, we take a short segment of the curve and move it forward, step by step, always parallel to itself (see Fig. 6.8). To construct a curve of extremal length, we stretch a string (or, in spacetime, we measure proper time with a clock), and we find the path that yields the shortest length (or the extremal proper time). The first method for determining a geodesic relies on parallel transport, that is, it relies on the affine geometry of the curved space. The second method relies on measurements of length (or proper time), that is, it relies on the metric geometry. In this section we will write the geodesic equation according to the first method, exploiting our definition of parallel transport. In the next section we will introduce the metric tensor, and we will then reexamine the geodesic equation. We will see that the geodesics determined by the metric coincide with those determined by parallel transport.

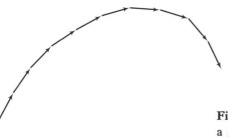

Fig. 6.8 Step-by-step construction of a geodesic curve by parallel displacement of a small segment of the curve.

Consider a small segment dx^α of a geodesic curve (see Fig. 6.8). If we displace this segment forward, parallel to itself, through a displacement dx^α equal to itself, the segment suffers a change

$$\delta dx^\alpha = - \Gamma^\alpha_{\mu\beta} dx^\mu dx^\beta \qquad [32]$$

To convert this into a differential equation, we introduce a parametrization of the curve, with a parameter s that increases monotonically along the curve. If ds is the increment that corresponds to the displacement dx^α, we can divide Eq. [32] by $(ds)^2$, and rewrite it as

$$\frac{d^2x^\alpha}{ds^2} + \Gamma^\alpha_{\mu\beta} \frac{dx^\mu}{ds} \frac{dx^\beta}{ds} = 0 \qquad [33]$$

This is the differential equation for the geodesic. The parameter s in this equation has, as of yet, no physical interpretation; it is merely a mathematical parameter. However, after we introduce the metric in the next section, we will be able to identify this parameter with the proper time.

Note that if we change the parametrization of the curve, by defining a new parameter $dq = f(q)ds$, the geodesic equation becomes

$$\frac{d}{dq}\left[f(q)\frac{dx^\alpha}{dq}\right] + f(q)\Gamma^\alpha{}_{\mu\beta}\frac{dx^\mu}{dq}\frac{dx^\beta}{dq} = 0 \qquad [34]$$

This is an alternative form of the geodesic equation. It looks more complicated than Eq. [33], but, of course, it yields exactly the same curve in spacetime--only the parametrization has changed, the curve has not changed. If we multiply Eq. [34] by $(dq)^2$, we recognize that it expresses the parallel transport of $f(q)dx^\alpha$ along the curve. Since the direction of the vector $f(q)dx^\alpha$ is the same as the direction of dx^α, we could have obtained Eq. [34] by starting our derivation of the geodesic equation with the parallel displacement of $f(q)dx^\alpha$. Eq. [33] is called the *normal form* of the geodesic equation, and we will give it preference over the more complicated form displayed in Eq. [34].

■ *Exercise 9.* The Christoffel symbols for the two-dimensional curved space consisting of the surface of a sphere are given by Eq. [22]. Show that the equations for the geodesic curves in this space are

$$\frac{d^2\theta}{ds^2} - \sin\theta\,\cos\theta\,\left(\frac{d\phi}{ds}\right)^2 = 0$$

$$\frac{d^2\phi}{ds^2} + \cot\theta\,\frac{d\theta}{ds}\frac{d\phi}{ds} = 0$$

Show that $\theta = \pi/2$, $\phi = $ [constant] $\times$ s is a solution of these equations; show that $\theta = $ [constant] $\times$ s, $\phi = $ [constant] is another solution. Thus, the equator is a geodesic and so is any meridian. More generally, any great circle is a geodesic for the sphere, since rotational symmetry permits us to rotate the coordinates so as to make the equator coincide with any great circle. ■

At any one point P of spacetime we can simplify the geodesic equation by introducing a new set of coordinates x'^μ such that the Christoffel symbols are zero at this one point. To accomplish this trick, we use the quadratic coordinate transformation

$$x'^\mu = x^\mu - x^\mu(P) + \tfrac{1}{2}\Gamma^\mu{}_{\alpha\beta}(P)[x^\alpha - x^\alpha(P)][x^\beta - x^\beta(P)] \qquad [35]$$

where $x^\mu(P)$ are the old coordinates of the point P. Note that Eq. [35] takes P as the origin of the new coordinates $[x'(P) = 0]$. From Eq. [35] we obtain

$$\left.\frac{\partial x'^\mu}{\partial x^\nu}\right|_P = \delta^\mu_\nu \quad \text{and} \quad \left.\frac{\partial^2 x'^\mu}{\partial x^\sigma \partial x^\nu}\right|_P = \Gamma^\mu_{\sigma\nu}(P) \qquad [36]$$

which, when inserted into Eq. [31], gives us $\Gamma'^\mu_{\alpha\beta}(P) = 0$. Note that this result hinges on the symmetry of $\Gamma^\mu_{\alpha\beta}$ in α and β. If $\Gamma^\mu_{\alpha\beta}$ had an antisymmetric part, this would be left unchanged by the transformation [35], since only the symmetric part of $\Gamma^\mu_{\alpha\beta}$ contributes to Eqs. [35] and [36].

■ *Exercise 10.* Carry out the calculations that lead from Eq. [36] to the result $\Gamma'^\mu_{\alpha\beta} = 0$. ■

With $\Gamma'^\mu_{\alpha\beta} = 0$, the geodesic equation at this point takes the simple form

$$\frac{d^2 x'^\mu}{ds^2} = 0 \qquad [37]$$

which says that locally, in a small neighborhood of the point P, the geodesic is a straight line.

The coordinates x'^μ that, at a given point, make the Christoffel symbols zero and give the geodesic equation its simple form [37] are called *local geodesic coordinates* (at that point). It must be emphasized that the coordinates are geodesic only at one spacetime point, that is, at one place and at one time. The Christoffel symbols are zero only at one spacetime point P; if we want geodesic coordinates at a different place or a different time, we must carry out the transformation [35] for this different place or time.

Geometrically, the local geodesic coordinates correspond to replacing the curved space by a small flat patch tangent to the former at the given point P. For instance, in the curved two-dimensional space consisting of the surface of a sphere, we can visualize the construction of local geodesic coordinates by placing a small flat rectangular patch tangent to the sphere at the given point P (see Fig. 6.9). The rectangular coordinates painted on this flat patch constitute local geodesic coordinates. This construction is, of course, valid only over an infinitesimal region. When we move away from the point P, the Christoffel symbols begin to deviate from zero. In fact, as we will see later, the first derivatives of $\Gamma^{\mu}{}_{\alpha\beta}$ cannot all be made zero by choice of coordinates.

Note that in local geodesic coordinates, at a given point, parallel transport of a vector reduces to $\delta A_{\mu} = 0$, that is, parallel transport leaves the components of the vector constant, just as in rectangular coordinates in flat spacetime. Furthermore, in local geodesic coordinates, the covariant derivative of a vector or of a tensor coincides with the ordinary derivative. This can be exploited to simplify some calculations with covariant derivatives. For instance, to verify that some expression containing covariant derivatives is equal to some other covariant derivative or some other tensor, we can introduce local geodesic coordinates and eliminate all the terms involving Christoffel symbols.

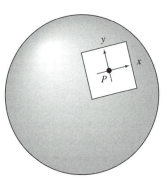

Fig. 6.9 Construction of local geodesic coordinates near a point P on the surface of a sphere. A small flat patch of paper is placed tangent to the sphere at this point. The rectangular coordinates painted on this flat patch give us local geodesic coordinates.

It is also possible to introduce new coordinates, called *Fermi coordinates*, such that the Christoffel symbols are zero at every point on a given geodesic curve. These coordinates are constructed by erecting basis vectors along the coordinate axes at the initial point P. The x^0-axis and the corresponding basis vector are assumed to lie along the geodesic curve; the x^1-, x^2-, and x^3-axes and their correponding basis vectors lie in the surface $x^0 = $ [constant], for the given initial value of x^0. The basis vectors are then carried forward along the geodesic by parallel transport, and the new basis vectors obtained in this way are used to define the coordinate axes at any later point on the geodesic (see Fig. 6.10). Since these coordinate axes are constructed by parallel transport, it is obvious that whenever we parallel transport some vector along the geodesic, its components cannot change; this immediately implies that all the Christoffel symbols concerned with parallel transport along the geodesic must be zero everywhere on the geodesic ($\Gamma^{\alpha}{}_{\beta 0} = 0$). It can be shown that all the other Christoffel symbols are also zero (Misner, Thorne, and Wheeler, 1973). Physically, the Fermi coordinates along a geodesic represent a freely falling reference frame, whose spatial orientation is defined by gyroscopes.

In the curved two-dimensional space consisting of the surface of a sphere, we can visualize the construction of Fermi coordinates by taking a narrow strip or tape of paper and placing it along the geodesic (a great circle; see Fig. 6.11). The rectangular coordinates painted on this strip of paper constitute Fermi coordinates along the geodesic.

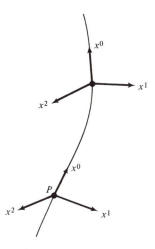

Fig. 6.10 Basis vectors for Fermi coordinates along a geodesic. The basis vectors at different times x^0 are related by parallel transport.

Fermi coordinates can also be introduced along an arbitrary, nongeodesic curve, that is, the worldline of a particle subjected to a nongravitational force. However, along a nongeodesic worldline, the basis vectors cannot be constructed by carrying an initial set of basis vectors forward along the curve by parallel transport; instead, a different kind of transport must be used, called Fermi–Walker transport (Misner, Thorne, and Wheeler, 1973). Furthermore, along a nongeodesic worldline, not all of the Christoffel symbols can be made zero (the nonzero

Fig. 6.11 Construction of Fermi coordinates along and near a geodesic (great circle) on the surface of a sphere. A strip of paper is placed along the geodesic. The rectangular coordinates painted on this strip of paper give us Fermi coordinates. Note that the strip of paper is a flat two-dimensional space; it has no intrinsic curvature, even though it is bent along the great circle (compare the example of the surface of a cylinder discussed in the introduction to this chapter).

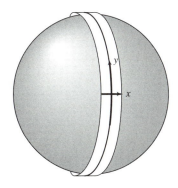

Christoffel symbols reveal how much Fermi-Walker transport differs from parallel transport). The Fermi coordinates along a nongeodesic worldline represent a reference frame accelerated by some nongravitational force, but whose spatial orientation is still defined by gyroscopes (the gyroscopes are accelerated by the nongravitational force, but this force exerts no torque).

6.4 THE METRIC TENSOR

In Chapter 3, we defined the metric tensor $g_{\mu\nu}(x)$ in terms of the spacetime interval ds^2 between two points separated by a coordinate displacement dx^μ (see Eq. [3.125]):

$$ds^2 = g_{\mu\nu}(x)\, dx^\mu dx^\nu \qquad [38]$$

According to our sign convention for $g_{\mu\nu}$, the time component of the metric tensor is positive ($g_{00} > 0$), and the diagonal space components are negative ($g_{kk} < 0$). Depending on the displacement dx^μ, the spacetime interval ds^2 can be positive, negative, or zero. If $ds^2 > 0$ (a timelike interval), then ds^2 represents the proper time accumulated by a clock during the displacement dx^μ. In this case, we will usually introduce the notation $d\tau^2$ for ds^2.

Without any loss of generality, we may assume that the metric tensor is symmetric, $g_{\mu\nu} = g_{\nu\mu}$. Any antisymmetric part in $g_{\mu\nu}$ would simply drop out of Eq. [38], since $dx^\mu dx^\nu$ is symmetric in μ, ν.

To establish that $g_{\mu\nu}$ is indeed a tensor, we rely on the invariance of the spacetime interval ds^2,

$$ds^2 = ds'^2 = g'_{\alpha\beta} dx'^\alpha dx'^\beta \qquad [39]$$

from which it follows that

$$g'_{\alpha\beta} dx'^\alpha dx'^\beta = g_{\mu\nu} dx^\mu dx^\nu \qquad [40]$$

By the definition of the partial derivatives with respect to x'^μ and x'^ν, we then obtain

$$g'_{\alpha\beta} = g_{\mu\nu} \frac{\partial x^\mu}{\partial x'^\alpha} \frac{\partial x^\nu}{\partial x'^\beta} \qquad [41]$$

which is exactly what is required for a second-rank, covariant tensor.

Although our main concern is with curved spacetime, the transformation equation [41] is perfectly general and can be used to obtain the metric tensor $g'_{\mu\nu}$ that describes flat spacetime in curvilinear coordinates. For example, if $x'^0 \equiv t$, $x'^1 \equiv r$, $x'^2 \equiv \theta$, $x'^3 \equiv \phi$ are the usual spherical coordinates, then Eq. [41] with $x^0 = t$, $x^1 = x$, $x^2 = y$, $x^3 = z$, and $g_{\mu\nu} = \eta_{\mu\nu}$ gives

$$
g'_{\alpha\beta} = \begin{bmatrix} 1 & 0 & 0 & 0 \\ 0 & -1 & 0 & 0 \\ 0 & 0 & -r^2 & 0 \\ 0 & 0 & 0 & -r^2 \sin^2\theta \end{bmatrix} \qquad [42]
$$

The corresponding expression for the interval is

$$
ds^2 = g'_{\alpha\beta} dx'^{\alpha} dx'^{\beta}
$$

$$
= dt^2 - dr^2 - r^2 d\theta^2 - r^2 \sin^2\theta \, d\phi^2 \qquad [43]
$$

■ *Exercise 11.* Use Eq. [41] to derive Eq. [42]. ■

Note that the expression [43] can also be obtained directly by expressing the differentials $dt^2 - dx^2 - dy^2 - dz^2$ in terms of dt, dr, $d\theta$, $d\phi$; the metric tensor $g'_{\alpha\beta}$ can then be found by inspection of [43]. This procedure is mathematically equivalent to using the transformation law [41], but perhaps somewhat quicker.

By means of a coordinate transformation it is always possible to make the tensor $g'_{\mu\nu}$ equal to $\eta_{\mu\nu}$ at any given point P. According to Eq. [41], to accomplish this we need a local reference frame with coordinates x'^{α} such that

$$
\eta_{\alpha\beta} = g_{\mu\nu} \frac{\partial x^{\mu}}{\partial x'^{\alpha}} \frac{\partial x^{\nu}}{\partial x'^{\beta}} \qquad [44]
$$

This can be regarded as a set of 10 equations for the coefficients $\partial x^{\mu}/\partial x'^{\alpha}$; since there are 16 such coefficients, it is always possible to find a solution of these 10 (undetermined) equations. This solution gives us a local coordinate system with $g'_{\mu\nu} = \eta_{\mu\nu}$ at the one point P. However, in curved spacetime it is not possible to find a solution that simultaneously makes $g'_{\mu\nu} = \eta_{\mu\nu}$ at all other points of spacetime; in other words, the solution of Eq. [44] is not integrable for $x^{\mu}(x')$. By contrast, in flat spacetime, the solution of Eq. [44] can be integrated, and the local (rectangular) coordinate system can be extended over all spacetime.

We already discussed the physical interpretation of the components of the metric tensor in Section 3.6. The components g_{00} and g_{kl} determine how measured times and lengths are related to coordinate displacements. A coordinate displacement dx^0 implies a time $\sqrt{g_{00}} \, dt$ as measured by a clock at rest, and a coordinate displacement dx^k (with $dx^0 = 0$) implies a measured distance $dl^2 = - g_{kl} dx^k dx^l$. It is convenient to define a 3×3 matrix $^{(3)}g_{kl} = -g_{kl}$; this matrix has positive diagonal components, and it gives us the geometry of three-dimensional space at one instant of time,

$$dl^2 = {}^{(3)}g_{kl}dx^k dx^l \qquad [45]$$

The physical interpretation of the components g_{0k} is that they determine how much the time coordinate x^0 is desynchronized when we compare it with the time coordinate x'^0 defined by a local set of synchronized clocks, instantaneously at rest, but in free fall. In the local freely falling reference frame of such clocks, light propagates at its standard speed, $c = 1$. Hence, we have $ds^2 = 0$ in this reference frame, and by invariance of the spacetime interval, we then also have $ds^2 = 0$ in any other reference frame, or

$$g_{\mu\nu}dx^\mu dx^\nu = 0 \qquad [46]$$

For two neighboring clocks, synchronization can be checked by sending a light signal from clock 1 to clock 2 and reflecting it back to clock 1. Suppose that the light signal arrives at clock 2 when this clock reads $x'^0 = 0$, and that the light signal departed from clock 1 when $x'^0 = dx'^0(-)$ [where $dx'^0(-) < 0$], and returns to clock 1 when $x'^0 = dx'^0(+)$. Synchronization holds if $dx'^0(-) = -dx'^0(+)$, that is, if the zero reading for clock 1 lies exactly halfway between the readings at the departure and the return of the signal. However, for the time coordinate x^0, this synchronization condition fails. With a given value of dx^k, Eq. [46] can be regarded as a quadratic equation for dx^0. The solutions of this quadratic equation are

$$dx^0(-) = \frac{1}{g_{00}}\left[g_{0k}dx^k - \sqrt{(g_{0k}g_{0l} - g_{kl}g_{00})dx^k dx^l}\right] \qquad [47]$$

$$dx^0(+) = \frac{1}{g_{00}}\left[g_{0k}dx^k + \sqrt{(g_{0k}g_{0l} - g_{kl}g_{00})dx^k dx^l}\right] \qquad [48]$$

These two solutions give us the departure and the return times for the light signal, in terms of the x^0 time coordinate. Since $dx^0(-) \neq -dx^0(+)$, the x^0 time coordinate is desynchronized relative to the freely falling synchronized clocks (see Fig. 6.12). Note that the unequal magnitudes of the two solutions [47] and [48] imply different speeds dl/dx^0 for light signals in the two directions; this can be attributed to the desynchronization of the x^0 coordinate--light seems to propagate differently in the two directions because the zero of the x^0 coordinate at different positions is misadjusted.

According to Fig. 6.12, for a given displacement dx^k, the amount by which the x^0 coordinate deviates from the (synchronized) x'^0 coordinate is

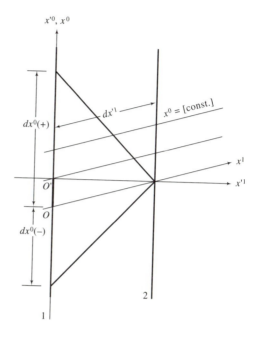

Fig. 6.12 The equal-time surfaces x^0 = [constant] are not orthogonal to the x^0-axis; this permits unequal magnitudes of $dx^0(-)$ and $dx^0(+)$. The slant of the equal-time surfaces x^0 = [constant] relative to the equal-time surfaces x'^0 = [constant] implies a desynchronization between these time coordinates (the zeros of these coordinates coincide at clock 2, but they differ at clock 1). The segment OO' represents the amount of desynchronization associated with the given displacement dx^1.

$$dx^0(+) - \frac{1}{2}\Big[dx^0(+) + |dx^0(-)|\Big] = \frac{1}{2}\Big[dx^0(+) + dx^0(-)\Big]$$

$$= -\frac{g_{0k}\,dx^k}{g_{00}} \qquad [49]$$

This formula tells us that g_{0k}/g_{00} can be interpreted as the amount of desynchronization per unit displacement dx^k.*

Although the time and the space components of $g_{\mu\nu}$ can be interpreted in terms of separate measurements of time and space, this interpretation becomes unnecessary and irrelevant if we perform our measurements with the generalized radar-ranging procedure described in Section 2.1. For any given displacement dx^μ, regardless of whether it is timelike or spacelike, this procedure permits a direct measurement of the corresponding spacetime interval ds^2. This means we can immediately extract the component $g_{\mu\nu}$ from the measured value of ds^2,

---

* One of the consequences of this failure of synchronization when $g_{0k} \neq 0$ is that the metric $^{(3)}g_{kl}$ of the three-dimensional space at one instant of time is not what is measured by a meter stick at rest, but rather what is measured by moving meter sticks. The motion of the stick used for the measurement of dl^2 or a displacement from, say, $x^1 = 0$ to $x^1 = dx^1$ must be adjusted so in the reference frame of the stick its ends pass through the points 0 and dx^1 simultaneously.

$$g_{\mu\nu} = \frac{ds^2}{dx^\mu \, dx^\nu}$$

and we do not need to discriminate between those measurements that involve a time displacement and those that involve a space displacement.

The *inverse* $g^{\mu\nu}$ of the metric tensor $g_{\mu\nu}$ is defined by

$$g_{\mu\alpha} g^{\alpha\nu} = \delta^\nu_\mu \qquad [50]$$

■ *Exercise 12.* Find the tensor inverse to [42]. ■

We will use the tensors $g^{\mu\nu}$ and $g_{\mu\nu}$ to raise and to lower indices, respectively, as in the following examples*:

$$dx_\mu \equiv g_{\mu\nu} dx^\nu \qquad [51]$$

$$\partial^\mu \equiv g^{\mu\nu} \partial_\nu \qquad [52]$$

$$B^{\mu\alpha} \equiv g^{\alpha\beta} B^\mu{}_\beta \qquad [53]$$

$$A^{\nu\alpha\beta\ldots} \equiv g^{\mu\nu} A_\mu{}^{\alpha\beta\ldots} \qquad [54]$$

Note that a tensor index belonging to a covariant derivative can be raised in the same way:

$$A^{\mu;\beta} = g^{\beta\nu} A^\mu{}_{;\nu} \qquad [55]$$

■ *Exercise 13.* Show that the quantities appearing on the left sides of Eqs. [51] - [55] have the transformation properties indicated by the position of their indices if it is assumed that the quantities on the right side transform as indicated by their indices. ■

Next, we establish the relation between the affine geometry and the metric geometry, that is, the relation between the Christoffel symbol $\Gamma^\mu{}_{\alpha\beta}$ and the metric tensor $g_{\mu\nu}$. The relation emerges from the requirement that if we parallel transport a vector A^μ, its "length" $A^\mu A^\nu g_{\mu\nu}$ must remain unchanged. More generally, if we parallel transport two vectors A^μ and B^μ, their scalar product $A^\mu B^\mu g_{\mu\nu}$ must remain unchanged,

---

* Although we will not use general coordinates with lower indices, we will sometimes use differentials with lower indices, such as $dx_\mu \equiv g_{\mu\nu} dx^\nu$.

$$(A^\mu B^\nu g_{\mu\nu})_{,\beta} = 0 \qquad [56]$$

Since $A^\mu B^\nu g_{\mu\nu}$ is a scalar, we can replace the ordinary derivative by a covariant derivative, and if we then differentiate term by term, we find

$$(A^\mu B^\nu g_{\mu\nu})_{;\beta} = A^\mu_{\;;\beta} B^\nu g_{\mu\nu} + A^\mu B^\nu_{\;;\beta} g_{\mu\nu} + A^\mu B^\nu g_{\mu\nu;\beta} = 0 \quad [57]$$

Since the only changes in the vectors A^μ and B^ν are those contributed by parallel transport, the covariant derivatives of these vectors are zero, and Eq. [57] reduces to

$$A^\mu B^\nu g_{\mu\nu;\beta} = 0 \qquad [58]$$

In this equation, the vectors A^μ and B^ν are arbitrary; hence the coefficient of $A^\mu B^\nu$ must vanish:

$$g_{\mu\nu;\beta} = 0 \qquad [59]$$

Thus, the consistency between the affine geometry and the metric geometry demands that *the covariant derivative of the metric tensor is zero*. This means that the metric tensor at any place can be obtained from the metric tensor at some other place by parallel transport. However, as we will see below, it is best to view Eq. [59] as a restriction on the Christoffel symbols, rather than a restriction on the metric tensor.

If we use the standard expression for the covariant derivative of a second-rank tensor, Eq. [59] becomes

$$\frac{\partial g_{\mu\nu}}{\partial x^\beta} - g_{\alpha\nu}\Gamma^\alpha_{\;\beta\mu} - g_{\mu\alpha}\Gamma^\alpha_{\;\beta\nu} = 0 \qquad [60]$$

This permits us to express the derivatives of the metric tensor in terms of the Christoffel symbols:

$$\frac{\partial g_{\mu\nu}}{\partial x^\beta} = g_{\alpha\nu}\Gamma^\alpha_{\;\beta\mu} + g_{\mu\alpha}\Gamma^\alpha_{\;\beta\nu} \qquad [61]$$

Conversely, we can express the Christoffel symbols in terms of the derivatives of the metric tensor. For this, we permute the indices μ, ν, β in Eq. [61] cyclically, to obtain

$$\frac{\partial g_{\beta\mu}}{\partial x^\nu} = g_{\alpha\mu}\Gamma^\alpha_{\;\nu\beta} + g_{\beta\alpha}\Gamma^\alpha_{\;\nu\mu} \qquad [62]$$

$$\frac{\partial g_{\nu\beta}}{\partial x^{\mu}} = g_{\alpha\beta}\Gamma^{\alpha}_{\mu\nu} + g_{\nu\alpha}\Gamma^{\alpha}_{\mu\beta} \tag{63}$$

Now, add Eqs. [62] and [63] and subtract Eq. [61]; the result is

$$\frac{\partial g_{\beta\mu}}{\partial x^{\nu}} + \frac{\partial g_{\nu\beta}}{\partial x^{\mu}} - \frac{\partial g_{\mu\nu}}{\partial x^{\beta}} = 2g_{\alpha\beta}\Gamma^{\alpha}_{\mu\nu}$$

which yields

$$\Gamma^{\alpha}_{\mu\nu} = \frac{1}{2}g^{\alpha\beta}\left(\frac{\partial g_{\beta\mu}}{\partial x^{\nu}} + \frac{\partial g_{\nu\beta}}{\partial x^{\mu}} - \frac{\partial g_{\mu\nu}}{\partial x^{\beta}}\right) \tag{64}$$

or

$$\Gamma^{\alpha}_{\mu\nu} = \frac{1}{2}g^{\alpha\beta}(g_{\beta\mu,\nu} + g_{\nu\beta,\mu} - g_{\mu\nu,\beta}) \tag{65}$$

According to this equation, the metric tensor completely determines the Christoffel symbols, and therefore the metric geometry completely determines the affine geometry. Note that the converse is not true--if we arbitrarily specify the Christoffel symbols, we will not always be able to find a metric tensor consistent with these Christoffel symbols (Eq. [61] is not necessarily integrable). Hereafter, we will always regard the metric as primary, and we will assume that the Christoffel symbols are calculated from the metric tensor.

■ *Exercise 14.* As described in Exercise 4, the surface of a sphere is a two-dimensional curved space. Suppose that the radius of the sphere is b, and use the coordinates $x^1 = \theta$ and $x^2 = \phi$. Then distances along the sphere are given by

$$dl^2 = b^2 d\theta^2 + b^2 \sin^2\theta \, d\phi^2 \tag{66}$$

and therefore the metric tensor $^{(2)}g_{\mu\nu}$ for this two-dimensional space is

$$^{(2)}g_{\mu\nu} = \begin{pmatrix} b^2 & 0 \\ 0 & b^2 \sin^2\theta \end{pmatrix} \tag{67}$$

(In this example we deal only with distances rather than spacetime intervals; we have inserted a superscript $^{(2)}$ to prevent any confusion between $^{(2)}g_{\mu\nu}$ and the metric of spacetime.) For this metric tensor, calculate the Christoffel symbols $\Gamma^k{}_{mn}$ according to Eq. [65], with $k, m, n = 1, 2$, and verify that they are as stated in Eq. [23].

■

■ *Exercise 15.* In a later chapter (see Section 7.4), the following spacetime interval for a curved spacetime will play an important role:

$$ds^2 = e^{N(r)} dt^2 - e^{L(r)} dr^2 - r^2 d\theta^2 - r^2 \sin^2\theta \, d\phi^2 \qquad [68]$$

where N and L are some functions of r, but not functions of t, θ, or ϕ (we write these functions as exponentials e^N and e^L because this simplifies some expressions in the calculations). With $x^0 \equiv t, x^1 \equiv r, x^2 \equiv \theta, x^3 \equiv \phi$, the metric tensor is

$$g_{\mu\nu} = \begin{pmatrix} e^N & 0 & 0 & 0 \\ 0 & -e^L & 0 & 0 \\ 0 & 0 & -r^2 & 0 \\ 0 & 0 & 0 & -r^2 \sin^2\theta \end{pmatrix} \qquad [69]$$

Derive the following expressions for the Christoffel symbols (the primes indicate derivatives, $L' = \partial L/\partial r$, $N' = \partial N/\partial r$):

$$\Gamma^0{}_{01} = \Gamma^0{}_{10} = \tfrac{1}{2} N'$$

$$\Gamma^1{}_{00} = \tfrac{1}{2} N' e^{N-L}$$

$$\Gamma^1{}_{11} = \tfrac{1}{2} L'$$

$$\Gamma^1{}_{22} = - r e^{-L}$$

$$\Gamma^1{}_{33} = - r \sin^2\theta \, e^{-L}$$

$$\Gamma^2{}_{12} = \Gamma^2{}_{21} = 1/r$$

$$\Gamma^2{}_{33} = - \sin\theta \cos\theta$$

$$\Gamma^3{}_{13} = \Gamma^3{}_{31} = 1/r$$

$$\Gamma^3{}_{23} = \Gamma^3{}_{32} = \cot\theta \qquad [70]$$

All the other Christoffel symbols are zero. ■

It is easy to verify that the covariant derivative of the inverse of the metric tensor is also identically zero:

$$g^{\mu\nu}{}_{;\alpha} = 0 \qquad [71]$$

■ *Exercise 16.* Check this. ■

It follows that the operation of raising or lowering of indices commutes with covariant differentiation. For example,

$$(g^{\mu\nu} A_\nu)_{;\alpha} = g^{\mu\nu} A_{\nu;\alpha} \qquad [72]$$

Let us now consider geodesics in the metric geometry. Given the metric $g_{\mu\nu}$, we define the geodesic curves by the requirement of extremal proper time,

$$\delta \int_{P_1}^{P_2} ds = 0 \qquad [73]$$

or

$$\delta \int_{P_1}^{P_2} \sqrt{g_{\mu\nu} \frac{dx^\mu}{d\tau} \frac{dx^\nu}{d\tau}} \, d\tau = 0 \qquad [74]$$

As we already showed in Section 3.6, the Euler-Lagrange equation that results from this variational principle is

$$\frac{d}{d\tau}\left[g_{\mu\nu} \frac{dx^\nu}{d\tau} \right] - \frac{1}{2} g_{\alpha\beta,\mu} \frac{dx^\alpha}{d\tau} \frac{dx^\beta}{d\tau} = 0 \qquad [75]$$

Since

$$\frac{dg_{\mu\nu}}{d\tau} = g_{\mu\nu,\alpha} \frac{dx^\alpha}{d\tau} \qquad [76]$$

we can write Eq. [75] as

$$g_{\mu\nu} \frac{d^2 x^\nu}{d\tau^2} + g_{\mu\nu,\alpha} \frac{dx^\alpha}{d\tau} \frac{dx^\nu}{d\tau} - \frac{1}{2} g_{\alpha\beta,\mu} \frac{dx^\alpha}{d\tau} \frac{dx^\beta}{d\tau} = 0 \qquad [77]$$

Multiplying this by $g^{\sigma\mu}$, we obtain

$$\frac{d^2 x^\sigma}{d\tau^2} + \frac{1}{2} g^{\sigma\mu}(2g_{\mu\beta,\alpha} - g_{\alpha\beta,\mu}) \frac{dx^\alpha}{d\tau} \frac{dx^\beta}{d\tau} = 0 \qquad [78]$$

In view of the symmetry of $dx^\alpha dx^\beta$ in α and β, the term $2g_{\mu\beta,\alpha}(dx^\alpha/d\tau)(dx^\beta/d\tau)$ equals $(g_{\mu\beta,\alpha} + g_{\mu\alpha,\beta})(dx^\alpha/d\tau)(dx^\beta/d\tau)$, and we then recognize that the combination of derivatives of the metric tensor appearing in Eq. [78] is just the Christoffel symbol $\Gamma^\sigma_{\alpha\beta}$. Thus, Eq. [78] becomes

$$\frac{d^2x^\sigma}{d\tau^2} + \Gamma^\sigma{}_{\alpha\beta}\, \frac{dx^\alpha}{d\tau}\, \frac{dx^\beta}{d\tau} = 0 \qquad [79]$$

This geodesic equation obtained from the metric coincides with the geodesic equation Eq. [33] obtained in the preceding section, from parallel transport. However, the purely mathematical parameter s we introduced in Eq. [33] is now seen to play the role of proper time.

Note that we can rewrite the geodesic equation in the alternative forms

$$\frac{d}{d\tau}\left(\frac{dx^\sigma}{d\tau}\right) + \Gamma^\sigma{}_{\alpha\beta}\, \frac{dx^\alpha}{d\tau}\, \frac{dx^\beta}{d\tau} = 0$$

and

$$\frac{dP^\sigma}{d\tau} + \Gamma^\sigma{}_{\alpha\beta} P^\alpha\, \frac{dx^\beta}{d\tau} = 0 \qquad [80]$$

where $P^\sigma = m\, dx^\sigma/d\tau$ is the momentum of the particle and m the rest mass. These equations say that the four-velocity and the momentum are parallel transported along the geodesic.

We know from the preceding section that it is always possible to find coordinates (local geodesic coordinates) such that at some given point P of spacetime the Christoffel symbols are zero. According to Eq. [61], the first derivatives of the metric tensor are then also zero at that point. Furthermore, by an extra coordinate transformation, it is possible to make the metric tensor at that point equal to the metric tensor of flat spacetime, $g''_{\mu\nu} = \eta_{\mu\nu}$. To see this, suppose that in the geodesic coordinates x'^μ defined by Eq. [35], the metric at the point P is $g'_{\mu\nu}$. A transformation from the coordinates x'^μ to new coordinates x''^μ will give us $g''_{\mu\nu} = \eta_{\mu\nu}$ (and $g''^{\mu\nu} = \eta^{\mu\nu}$) at the point P provided

$$\eta^{\mu\nu} = \frac{\partial x''^\mu}{\partial x'^\alpha}\, \frac{\partial x''^\nu}{\partial x'^\beta}\, g'^{\alpha\beta} \qquad [81]$$

As we already know from Eq. [44], this can be regarded as a set of 10 equations for the unknown coefficients $\partial x''^\mu/\partial x'^\nu$, evaluated at the point P, and this set always has a solution. But besides making $g''_{\mu\nu} = \eta_{\mu\nu}$, we also want to keep the Christoffel symbols zero. We can achieve this by demanding a linear relationship between the coordinates x''^μ and x'^μ in the vicinity of the point P:

$$x''^\mu = \frac{\partial x''^\mu}{\partial x'^\nu}\bigg|_P x'^\nu \qquad\qquad [82]$$

Here the coefficients have been labeled with the subscript p to emphasize that they are evaluated at the point P, that is, they are constants. Such a linear transformation does not change the Christoffel symbols--according to Eq. [31], they remain zero.

So far, we have treated the geodesic equation purely mathematically, without establishing any connection to physics. But our discussion of the linear approximation in Chapter 3 leads us to expect that the geodesics of the curved spacetime are the worldlines of freely falling particles. We can establish this result in several ways, by appeal to the equivalence principle. A simple argument is the following. Consider a local freely falling reference frame, such as the rest frame of a freely falling clock. We assume that the coordinate grid of this freely falling reference frame, at one instant of time, is rectangular, with axes and grid lines constructed by parallel transport. Then the Christoffel symbols are zero in these coordinates, that is, the coordinates are geodesic, and we can assume that the metric of spacetime is locally that of flat spacetime, $g'_{\mu\nu} = \eta_{\mu\nu}$. Consider a particle also in free fall, near the origin of the reference frame; this particle has zero acceleration relative to this reference frame, since, according to the principle of equivalence, both the particle and the reference frame accelerate at the same rate in the gravitational field. Hence the equation of motion of the particle is

$$\frac{d^2 x'^\mu}{dt^2} = 0 \qquad\qquad [83]$$

But this equation is exactly the geodesic equation in local geodesic coordinates (see Eq. [37]). Thus, the worldlines of freely falling particles coincide with the geodesics. We can now perform the transformation from general curvilinear coordinates to geodesic coordinates in reverse; it then leads us from Eq. [83] to Eq. [33], the geodesic equation in general coordinates.

6.5 THE RIEMANN CURVATURE TENSOR

When we take consecutive covariant derivatives of a tensor field, we must pay careful attention to the order in which we carry out the differentiations. For example, suppose we take two consecutive covariant derivatives of a vector field A_β, say, with respect to x^μ and x^ν. Depending on which derivative we take first, we obtain

$$A_{\beta;\mu;\nu} \tag{84}$$

or

$$A_{\beta;\nu;\mu} \tag{85}$$

In general, these covariant derivatives are not equal; that is, *covariant derivatives do not commute.*

■ *Exercise 17.* Show that in the exceptional case of covariant differentiation of a scalar field, the derivatives commute:

$$\phi_{;\mu;\nu} = \phi_{;\nu;\mu} \tag{86}$$

(Hint: The first covariant derivative $\phi_{;\nu}$ is simply $\phi_{,\nu}$.) ■

Let us calculate the difference between [84] and [85]. Since $A_{\beta;\mu}$ is a second-rank tensor, we have

$$A_{\beta;\mu;\nu} = A_{\beta;\mu,\nu} - \Gamma^\sigma_{\beta\nu} A_{\sigma;\mu} - \Gamma^\sigma_{\mu\nu} A_{\beta;\sigma} \tag{87}$$

and since A_β is a first-rank tensor,

$$A_{\beta;\mu;\nu} = (A_{\beta,\mu} - \Gamma^\alpha_{\beta\mu} A_\alpha)_{,\nu} - \Gamma^\sigma_{\beta\nu}(A_{\sigma,\mu} - \Gamma^\alpha_{\sigma\mu} A_\alpha) - \Gamma^\sigma_{\mu\nu}(A_{\beta,\sigma} - \Gamma^\alpha_{\beta\sigma} A_\alpha)$$

$$= [A_{\beta,\mu,\nu} - \Gamma^\alpha_{\beta\mu} A_{\alpha,\nu} - \Gamma^\sigma_{\beta\nu} A_{\sigma,\mu} - \Gamma^\sigma_{\mu\nu}(A_{\beta,\sigma} - \Gamma^\alpha_{\beta\sigma} A_\alpha)]$$
$$- \Gamma^\alpha_{\beta\mu,\nu} A_\alpha + \Gamma^\sigma_{\beta\nu}\Gamma^\alpha_{\sigma\mu} A_\alpha \tag{88}$$

The corresponding expression for $A_{\beta;\nu;\mu}$ can be obtained from [88] by simply exchanging μ and ν. Obviously the quantity contained in the brackets in Eq. [88] is symmetric in μ and ν, and hence will not contribute to $A_{\beta;\mu;\nu} - A_{\beta;\nu;\mu}$. The remaining terms give

$$A_{\beta;\mu;\nu} - A_{\beta;\nu;\mu} = (-\Gamma^\alpha_{\beta\mu,\nu} + \Gamma^\alpha_{\beta\nu,\mu})A_\alpha + (\Gamma^\sigma_{\beta\nu}\Gamma^\alpha_{\sigma\mu} - \Gamma^\sigma_{\beta\mu}\Gamma^\alpha_{\sigma\nu})A_\alpha \tag{89}$$

We can rewrite this as

$$A_{\beta;\mu;\nu} - A_{\beta;\nu;\mu} = R^\alpha_{\beta\mu\nu} A_\alpha \tag{90}$$

where

$$R^\alpha_{\beta\mu\nu} = -\Gamma^\alpha_{\beta\mu,\nu} + \Gamma^\alpha_{\beta\nu,\mu} + \Gamma^\sigma_{\beta\nu}\Gamma^\alpha_{\sigma\mu} - \Gamma^\sigma_{\beta\mu}\Gamma^\alpha_{\sigma\nu} \tag{91}$$

The left side of Eq. [90] is a tensor; it is the difference between two tensors. Hence the right side must also be a tensor. Since A_α is an arbitrary vector, the $4 \times 4 \times 4 \times 4$-component object $R^\alpha{}_{\beta\mu\nu}$ must be a tensor of rank four.* It is called the *Riemann curvature tensor*.

We can lower the first index in $R^\alpha{}_{\beta\mu\nu}$ to obtain a purely covariant tensor,

$$R_{\alpha\beta\mu\nu} = g_{\alpha\sigma} R^\sigma{}_{\beta\mu\nu} \qquad [92]$$

This tensor satisfies several important identities:

$$R_{\alpha\beta\mu\nu} = - R_{\beta\alpha\mu\nu} \qquad [93]$$

$$R_{\alpha\beta\mu\nu} = - R_{\alpha\beta\nu\mu} \qquad [94]$$

$$R_{\alpha\beta\mu\nu} = R_{\mu\nu\alpha\beta} \qquad [95]$$

$$R_{\alpha\beta\mu\nu} + R_{\alpha\mu\nu\beta} + R_{\alpha\nu\beta\mu} = 0 \qquad [96]$$

■ *Exercise 18.* Derive Eqs. [93]-[96]. ■

Although the tensor $R_{\alpha\beta\mu\nu}$ has $4 \times 4 \times 4 \times 4 = 256$ components, only 20 of these components are independent. Taking into account the identities [93] and [94], we see that $R_{\alpha\beta\mu\nu}$ can be regarded as an anti-symmetric second-rank tensor in α, β and as an antisymmetric second-rank tensor in μ, ν. Since a second-rank antisymmetric tensor has 6 independent components, we are then left with no more than 6×6 components for $R_{\alpha\beta\mu\nu}$. We can write these 6×6 components in the form of a 6×6 matrix R_{AB}, where A stands for the pair $\alpha\beta$ and B for the pair $\mu\nu$; the values of A and B may be taken as 01, 02, 03, 12, 13, 23. The identity [95] says that R_{AB} is a symmetric matrix; and such a 6×6 symmetric matrix has 21 independent components. Finally, the identity [96] imposes *one* constraint $R_{0123} - R_{0231} + R_{0312} = 0$ among these 21 components (it is easy to see that unless all the indices in Eq. [96] are different, this equation coincides with one of Eqs. [93]-[95]). Hence we are left with only 20 independent components for the Riemann tensor.

The Riemann tensor is the only tensor that can be formed by taking linear combinations of the second derivatives of the metric. To prove this, we can use a simple counting argument. We begin by taking geodesic coordinates x'^μ with origin at some point P and perform a coordinate transformation to a new set of coordinates

---

* Of course, you can show explicitly that $R^\alpha{}_{\beta\mu\nu}$ is a tensor by using Eq. [91] and the transformation law for $\Gamma^\alpha{}_{\beta\mu}$. Try it; you won't like it.

$$x''^\mu = x'^\mu + \tfrac{1}{6} a^\mu_{\alpha\beta\gamma} x'^\alpha x'^\beta x'^\gamma \qquad\qquad [97]$$

where $a^\mu_{\alpha\beta\gamma}$ is a constant matrix. At the point $x'^\mu = 0$, this gives

$$\frac{\partial x''^\mu}{\partial x'^\nu} = \delta^\mu_\nu \qquad \frac{\partial^2 x''^\mu}{\partial x'^\alpha \partial x'^\beta} = 0 \qquad \frac{\partial^3 x''^\mu}{\partial x'^\alpha \partial x'^\beta \partial x'^\gamma} = a^\mu_{\alpha\beta\gamma}$$

Since the transformation laws for $g_{\mu\nu}$ and $g_{\mu\nu,\alpha}$ involve only the first and second derivatives of x''^μ, it follows that $g''_{\mu\nu} = g'_{\mu\nu}$ and $g''_{\mu\nu,\alpha} = g'_{\mu\nu,\beta}$. The transformation law for $g_{\mu\nu,\alpha,\beta}$, obtained from Eq. [41] by differentiating twice, involves the third derivative of x''^μ, and hence $g''_{\mu\nu,\alpha,\beta}$ is not equal to $g'_{\mu\nu,\alpha,\beta}$. In fact, the values of $a^\mu_{\alpha\beta\gamma}$ are at our disposal, and we can adjust the values of the second derivatives $g''_{\mu\nu,\alpha,\beta}$ to some extent. There are 100 derivatives $g''_{\mu\nu,\alpha,\beta}$. Since there are 80 independent coefficients $a^\mu_{\alpha\beta\gamma}$, we can impose 80 conditions on these 100 derivatives.

■ **Exercise 19.** Show that only 80 of the coefficients $a^\mu_{\alpha\beta\gamma}$ are independent. (Hint: $x'^\alpha x'^\beta x'^\gamma$ in Eq. [97] is symmetric under exchanges of α, β, and γ; hence $a^\mu_{\alpha\beta\gamma}$ must also have this symmetry.) ■

We therefore conclude that there exist only 20 independent linear combinations of second derivatives that are unaffected by the choice of $a^\mu_{\alpha\beta\gamma}$. If we now want to construct *tensors* out of the second derivatives, we must use these 20 linear combinations exclusively; the reason for this is that the tensor transformation law involves only $\partial x''^\mu / \partial x'^\nu = \delta^\mu_\nu$, and therefore $a^\mu_{\alpha\beta\gamma}$ may not enter into it. What tensors can be constructed with the available 20 linear combinations? There is one and only one: the tensor $R''_{\alpha\beta\mu\nu}$. This tensor is constructed of second derivatives and it has exactly 20 independent components; it therefore exhausts all the viable linear combinations of second derivatives. The fact that the Riemann tensor is the only tensor that contains the second derivatives of the metric linearly played an important role in the discovery of the differential equations for the gravitational field (see Section 7.3).

Besides the purely algebraic identities [93]-[96], the Riemann tensor also satisfies some differential identities. The covariant derivative $R^\alpha_{\beta\mu\nu;\sigma}$ is a tensor of fifth rank. The following identity holds:

$$R^\alpha_{\beta\mu\nu;\sigma} + R^\alpha_{\beta\nu\sigma;\mu} + R^\alpha_{\beta\sigma\mu;\nu} = 0 \qquad\qquad [98]$$

To derive Eq. [98], we can use a trick. If we introduce geodesic coordinates, then $\Gamma^{\mu}{}_{\alpha\beta} = 0$, and

$$R^{\alpha}{}_{\beta\mu\nu} = -\Gamma^{\alpha}{}_{\beta\mu,\nu} + \Gamma^{\alpha}{}_{\beta\nu,\mu} \qquad [99]$$

Therefore Eq. [98] reduces to

$$-\Gamma^{\alpha}{}_{\beta\mu,\nu,\sigma} + \Gamma^{\alpha}{}_{\beta\nu,\mu,\sigma} - \Gamma^{\alpha}{}_{\beta\nu,\sigma,\mu} + \Gamma^{\alpha}{}_{\beta\sigma,\nu,\mu} - \Gamma^{\alpha}{}_{\beta\sigma,\mu,\nu} + \Gamma^{\alpha}{}_{\beta\mu,\sigma,\nu} = 0$$

which is a trivial identity. We have therefore established the equation [98] in geodesic coordinates. Since this equation is a *tensor* equation, its validity in one coordinate system implies its validity in all.

By contraction on the first and last index of $R^{\alpha}{}_{\beta\mu\nu}$ we obtain a second-rank tensor $R_{\beta\mu}$:

$$R_{\beta\mu} = R^{\alpha}{}_{\beta\mu\alpha} \qquad [100]$$

This tensor is called the *Ricci tensor*; it is symmetric in β, μ:

$$R_{\beta\mu} = R_{\mu\beta} \qquad [101]$$

■ *Exercise 20.* Show this. ■

By a further contraction, we obtain a scalar

$$R = R^{\beta}{}_{\beta} = R^{\alpha\beta}{}_{\beta\alpha} \qquad [102]$$

This one-component object is called the *curvature scalar*.
The following differential equation holds identically:

$$\left(R^{\mu}{}_{\nu} - \tfrac{1}{2}\delta^{\mu}_{\nu}R\right)_{;\mu} = 0 \qquad [103]$$

The identities [98] and [103] are known as *Bianchi identities*.

■ *Exercise 21.* Derive Eq. [103]. (Hint: Use the same method as for obtaining Eq. [98]. You can also derive Eq. [103] directly from [98].) ■

■ *Exercise 22.* Show that the metric [69] and the Christoffel symbols [70] lead to the following components for the Riemann tensor:

$$R^{0}{}_{101} = -\tfrac{1}{2}N'' + \tfrac{1}{4}L'N' - \tfrac{1}{4}(N')^2$$

$$R^{0}{}_{303} = -\tfrac{1}{2}rN'e^{-L}\sin^2\theta$$

$$R^1_{313} = \tfrac{1}{2}rL'e^{-L}\sin^2\theta$$

$$R^0_{202} = -\tfrac{1}{2}rN'e^{-L}$$

$$R^1_{212} = \tfrac{1}{2}rL'e^{-L}$$

$$R^2_{323} = (1 - e^{-L})\sin^2\theta \qquad [104]$$

All components not related to the above by the identities [93]-[96] are zero. ∎

■ *Exercise 23.* Show that the Ricci tensor obtained from Eq. [104] has the components

$$R_{00} = e^{N-L}\left[-\tfrac{1}{2}N'' + \tfrac{1}{4}L'N' - \tfrac{1}{4}(N')^2 - \frac{N'}{r_0}\right]$$

$$R_{11} = \tfrac{1}{2}N'' - \tfrac{1}{4}L'N' + \tfrac{1}{4}(N')^2 - \frac{L'}{r}$$

$$R_{22} = e^{-L}\left[1 + \tfrac{1}{2}r(N' - L')\right] - 1$$

$$R_{33} = \sin^2\theta\, e^{-L}\left[1 + \tfrac{1}{2}r(N' - L')\right] - \sin^2\theta \qquad [105]$$

All other components are zero. Show that the curvature scalar has the value

$$R = e^{-L}\left[-N'' + \tfrac{1}{2}L'N' - \tfrac{1}{2}(N')^2 + \frac{2}{r}(L' - N') - \frac{2}{r^2}\right] + \frac{2}{r^2} \quad ∎ \qquad [106]$$

We can see what $R^\alpha_{\beta\mu\nu}$ has to do with the curvature of spacetime if we examine what happens to a *constant* vector a^α when we parallel transport it from a point P to a point P' along two *different* paths. Fig. 6.13 shows the points P and P' connected by a path PP_1P' consisting of a displacement $d\xi^\mu$ followed by a displacement $d\zeta^\mu$; path PP_2P' involves the same displacements, but carried out in the opposite

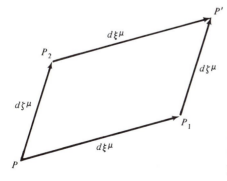

Fig. 6.13 Spacetime points P and P' connected by the two alternative paths PP_1P' and PP_2P'.

order. All the displacements are infinitesimal.

The parallel transport of a^α from P to P_1 gives a vector

$$a^\alpha - \Gamma^\alpha_{\beta\mu}(P)a^\beta \, d\xi^\mu$$

and the further transport of this vector from P_1 to P' gives

$$a^\alpha - \Gamma^\alpha_{\beta\mu}(P)a^\beta \, d\xi^\mu - \Gamma^\alpha_{\sigma\nu}(P_1)[a^\sigma - \Gamma^\sigma_{\beta\mu}(P)a^\beta \, d\xi^\mu]d\varsigma^\nu \qquad [107]$$

Since

$$\Gamma^\alpha_{\sigma\nu}(P_1) \simeq \Gamma^\alpha_{\sigma\nu}(P) + \Gamma^\alpha_{\sigma\nu,\mu}(P)d\xi^\mu \qquad [108]$$

we can write [107] as

$$a^\alpha - \Gamma^\alpha_{\beta\mu}a^\beta \, d\xi^\mu - \Gamma^\alpha_{\sigma\nu}a^\sigma \, d\varsigma^\nu + \Gamma^\alpha_{\sigma\nu}\Gamma^\sigma_{\beta\mu}a^\beta \, d\xi^\mu \, d\varsigma^\nu$$
$$- \Gamma^\alpha_{\sigma\nu,\mu}a^\sigma \, d\xi^\mu \, d\varsigma^\nu \qquad [109]$$

where a term of third order in the (small) displacements has been omitted, and where all Γ's are evaluated at the point P.

The parallel transport of a^α along the path PP_2P' gives a result similar to [109], but with $d\xi$ and $d\varsigma$ interchanged:

$$a^\alpha - \Gamma^\alpha_{\beta\mu}a^\beta \, d\varsigma^\mu - \Gamma^\alpha_{\sigma\nu}a^\sigma \, d\xi^\nu + \Gamma^\alpha_{\sigma\nu}\Gamma^\sigma_{\beta\mu}a^\beta \, d\varsigma^\mu \, d\xi^\nu$$
$$- \Gamma^\alpha_{\sigma\nu,\mu}a^\sigma \, d\varsigma^\mu \, d\xi^\nu \qquad [110]$$

If we exchange the (dummy) indices ν and μ in the last two terms of [109], we find that the difference between the expressions [109] and [110] is

$$(\Gamma^\alpha_{\beta\nu,\mu} - \Gamma^\alpha_{\beta\mu,\nu} + \Gamma^\alpha_{\sigma\mu}\Gamma^\sigma_{\beta\nu} - \Gamma^\alpha_{\sigma\nu}\Gamma^\sigma_{\beta\mu})a^\beta \, d\xi^\nu \, d\varsigma^\mu \qquad [111]$$

This shows that the result of parallel transport is *path dependent*. The combination of Christoffel symbols and their derivatives appearing in the expression [111] is the Riemann curvature tensor (compare Eq. [91]). Thus, the results obtained along two paths differ by a term involving the curvature tensor,

$$\Delta a^\alpha = R^\alpha_{\beta\mu\nu}a^\beta \, d\xi^\nu \, d\varsigma^\mu \qquad [112]$$

Since the change produced by parallel transport along the path PP_2P' is opposite to that produced along the reversed path $P'P_2P$, it follows that Δa^α may also be regarded as the change in the vector a^α due to parallel transport around the closed loop $PP_1P'P_2P$.

■ *Exercise 24.* Show that the corresponding result for a covariant vector is

$$\Delta b_\alpha = - R_\alpha{}^\beta{}_{\mu\nu} b_\beta \, d\xi^\nu \, d\zeta^\mu \qquad [113]$$

[Hint: Rather than go through the long calculation, use the condition $\Delta(b_\alpha a^\alpha) = 0$ for arbitrary a^α.] ■

Eqs. [112] and [113] show that successive parallel displacements do not commute. Since, as we have seen in Section 6.2, the covariant derivatives can be defined in terms of parallel displacements, it is not at all surprising that successive covariant derivatives also fail to commute.

In a flat spacetime, the result of parallel transport does not depend on the path taken. Hence, in flat spacetime we necessarily have $R^\alpha{}_{\beta\mu\nu}$ = 0. In a curved spacetime, the result of parallel transport depends on the path taken. The tensor $R^\alpha{}_{\beta\mu\nu}$ characterizes the path dependence in parallel transport, and serves as a quantitative measure of the curvature of spacetime. If $R^\alpha{}_{\beta\mu\nu} \neq 0$, the spacetime is curved. We can also show the converse: if $R^\alpha{}_{\beta\mu\nu} = 0$ throughout spacetime, then the spacetime is flat. In other words, if the curvature tensor is zero, then the metric can differ from $\eta_{\mu\nu}$ by only a coordinate transformation. A simple argument is the following: Choose a point P and introduce local geodesic coordinates at that point. Then parallel transport basis vectors from the point P to all other points, and construct coordinate axes at all other points. If $R^\alpha{}_{\beta\mu\nu} = 0$, the result is path independent and therefore uniquely defined. Since the coordinate axes have been constructed by parallel transport, any vector transported with respect to the axes will suffer no change. Hence, $\Gamma^\mu{}_{\alpha\beta} = 0$ everywhere, and therefore $g_{\mu\nu}$ is constant. Since $g_{\mu\nu} = \eta_{\mu\nu}$ at the initial point P, the metric must be $\eta_{\mu\nu}$ everywhere. This shows that the spacetime is flat.

Given a metric tensor $g_{\mu\nu}$, we can calculate the Riemann curvature tensor and decide unambiguously whether this metric tensor describes flat spacetime in some complicated curvilinear coordinates or whether it describes curved spacetime by calculating the curvature tensor. We have seen that, in gravitational theory, $g_{\mu\nu}$ (or $h_{\mu\nu}$) plays a role analogous to that of the vector potential of electrodynamics. The curvature tensor $R^\alpha{}_{\beta\mu\nu}$ plays a role analogous to that of the **E** and **B** fields of electrodynamics. A gravitational field is said to be present if, and only if, $R^\alpha{}_{\beta\mu\nu}$ has a nonzero component. As we will see in the next section, the tensor $R^\alpha{}_{\beta\mu\nu}$ is the relativistic generalization of the tidal-force tensor introduced in Eq. [1.51].

Table 6.1 summarizes the most important formulas of tensor analysis in Riemannian geometry. The formulas are given in terms of a first-rank tensor (a vector); the generalization to higher-rank tensors is obvious. The concepts of derivative along a curve and of geodesic deviation will be discussed in the next section; the corresponding formulas have been included in Table 6.1 for the sake of completeness.

TABLE 6.1   TENSOR-ANALYSIS FORMULAS

Transformation of contravariant vector $A'^{\mu} = \dfrac{\partial x'^{\mu}}{\partial x^{\nu}} A^{\nu}$

Transformation of covariant vector $A'_{\mu} = \dfrac{\partial x^{\nu}}{\partial x'^{\mu}} A_{\nu}$

Parallel transport $\delta A^{\mu} = - \Gamma^{\mu}_{\alpha\beta} A^{\alpha} \delta x^{\beta}$

$\delta A_{\mu} = \Gamma^{\alpha}_{\mu\beta} A_{\alpha} \delta x^{\beta}$

Covariant derivative of
  contravariant vector $A^{\mu}_{;\beta} = A^{\mu}_{,\beta} + \Gamma^{\mu}_{\alpha\beta} A^{\alpha}$

Covariant derivative of
  covariant vector $A_{\mu;\beta} = A_{\mu,\beta} - \Gamma^{\alpha}_{\mu\beta} A_{\alpha}$

Spacetime interval $ds^2 = g_{\mu\nu} dx^{\mu} dx^{\nu}$

Inverse metric tensor $g_{\mu\alpha} g^{\alpha\nu} = \delta^{\nu}_{\mu}$

Christoffel symbol $\Gamma^{\mu}_{\alpha\beta} = \frac{1}{2} g^{\mu\nu} (g_{\nu\alpha,\beta} + g_{\beta\nu,\alpha} - g_{\alpha\beta,\nu})$

Geodesic coordinates
  (at one spacetime point) $g'_{\mu\nu,\alpha} = 0, \; g'_{\mu\nu} = \eta_{\mu\nu}$

Geodesic equation $\dfrac{d^2 x^{\mu}}{d\tau^2} + \Gamma^{\mu}_{\alpha\beta} \dfrac{dx^{\alpha}}{d\tau} \dfrac{dx^{\beta}}{d\tau} = 0$

Riemann tensor $R^{\alpha}_{\beta\mu\nu} = - \Gamma^{\alpha}_{\beta\mu,\nu} + \Gamma^{\alpha}_{\beta\nu,\mu} + \Gamma^{\sigma}_{\beta\nu} \Gamma^{\alpha}_{\sigma\mu} - \Gamma^{\sigma}_{\beta\mu} \Gamma^{\alpha}_{\sigma\nu}$

Ricci tensor $R_{\mu\nu} = R^{\alpha}_{\mu\nu\alpha}$

Curvature scalar $R = R^{\mu}_{\mu}$

Commutation relation for
  derivatives $A_{\beta;\mu;\nu} - A_{\beta;\nu;\mu} = R^{\alpha}_{\beta\mu\nu} A_{\alpha}$

Change in vector produced by
  parallel transport around closed
  path $(0 \to d\xi \to d\zeta \to d\zeta \to 0)$ $\Delta A^{\alpha} = R^{\alpha}_{\beta\mu\nu} A^{\beta} d\xi^{\nu} d\zeta^{\mu}$

Derivative along a curve $\dfrac{DA^{\mu}}{D\tau} = A^{\mu}_{;\nu} \dfrac{dx^{\nu}}{d\tau} = \dfrac{dA^{\mu}}{d\tau} + \Gamma^{\mu}_{\alpha\beta} A^{\alpha} \dfrac{dx^{\beta}}{d\tau}$

Geodesic deviation $\dfrac{D^2 s^{\mu}}{D\tau^2} = - R^{\mu}_{\alpha\sigma\beta} s^{\sigma} \dfrac{dx^{\alpha}}{d\tau} \dfrac{dx^{\beta}}{d\tau}$

6.6 GEODESIC DEVIATION AND TIDAL FORCES

We defined the Riemann curvature tensor in terms of entirely geo-metrical concepts. But this tensor also has a simple physical interpretation: $R^{\alpha}{}_{\beta\mu\nu}$ is the tidal force field, that is, it gives the relative acceleration between two particles in free fall. Before we can calculate this relative acceleration, we need to introduce the concept of *differentiation along a curve*. What is involved here is the following. Consider a curve in spacetime, possibly, but not necessarily, a geodesic. We can specify displacements along the curve by giving the corresponding increment in the "length" or the proper time τ. Suppose that the curve is immersed in a vector field A^{μ}. If we want to know the rate of change of A^{μ} along the curve, we must calculate

$$\frac{dA^{\mu}}{d\tau}$$

However, this is not a good measure of how much A^{μ} is "really" changing, since part, or maybe all, of the contribution to $dA^{\mu}/d\tau$ could be due to the curvilinear coordinates used to define the components A^{μ}. Also, $dA^{\mu}/d\tau$ is *not* a vector (for the same reason that $\partial A_{\mu}/\partial x^{\nu}$ is not a tensor). A better measure of the rate of change of A^{μ} along the curve is the *vector* quantity

$$\frac{DA^{\mu}}{D\tau} \equiv A^{\mu}{}_{;\beta}\frac{dx^{\beta}}{d\tau} = \frac{dA^{\mu}}{d\tau} + \Gamma^{\mu}{}_{\alpha\beta}A^{\alpha}\frac{dx^{\beta}}{d\tau} \qquad [114]$$

This is a vector, because it is the product of the tensor $A^{\mu}{}_{;\nu}$ and the vector $dx^{\nu}/d\tau$. The quantity $DA^{\mu}/D\tau$ is called the *derivative along the curve*. In local geodesic coordinates (in which $\Gamma^{\mu}{}_{\alpha\beta} = 0$), the derivative $DA^{\mu}/D\tau$ along a curve reduces to the ordinary derivative $dA^{\mu}/d\tau$.

■ *Exercise 25.* $dA^{\mu}/d\tau$ is not a vector, but $dx^{\nu}/d\tau$ is a vector. Explain this. [Hint: x^{ν} is not a vector (see Eq. [1]), but dx^{ν} is a vector (see Eq. [3]).] ■

In terms of parallel transport, we can understand Eq. [114] as follows: $DA_{\mu}/D\tau$ is obtained from $[A_{\mu}(\tau + d\tau) - A^{\mu}(\tau)]/d\tau$ by subtracting from the latter expression that part that is entirely due to parallel transport of the vector $A^{\mu}(\tau)$ from the point τ to the point $\tau + d\tau$ on the curve (compare Eq. [20]), and by then taking the limit $d\tau \to 0$. Note that if the vector field $A^{\mu}(x)$ is a "constant" vector field, obtained by parallel transporting some initial vector from an initial point to all regions of spacetime, then the derivative $DA^{\mu}/D\tau$ is zero.

The expression [114] is valid for any curve. Let us now concentrate on geodesics and find the *second* derivative along a geodesic curve:

$$\frac{D^2A^\mu}{D\tau^2} = \frac{D}{D\tau}\left(\frac{DA^\mu}{D\tau}\right) = \frac{d}{d\tau}\left(\frac{DA^\mu}{D\tau}\right) + \Gamma^\mu{}_{\alpha\beta}\frac{DA^\alpha}{D\tau}\frac{dx^\beta}{d\tau}$$

$$= \frac{d}{d\tau}\left[\frac{dA^\mu}{d\tau} + \Gamma^\mu{}_{\alpha\beta}A^\alpha\frac{dx^\beta}{d\tau}\right] + \Gamma^\mu{}_{\alpha\beta}\left[\frac{dA^\alpha}{d\tau} + \Gamma^\alpha{}_{\kappa\lambda}A^\kappa\frac{dx^\lambda}{d\tau}\right]\frac{dx^\beta}{d\tau}$$

$$= \frac{d^2A^\mu}{d\tau^2} + \Gamma^\mu{}_{\alpha\beta,\nu}\frac{dx^\nu}{d\tau}A^\alpha\frac{dx^\beta}{d\tau} + 2\Gamma^\mu{}_{\alpha\beta}\frac{dA^\alpha}{d\tau}\frac{dx^\beta}{d\tau}$$

$$- \Gamma^\mu{}_{\alpha\beta}A^\alpha\Gamma^\beta{}_{\kappa\lambda}\frac{dx^\kappa}{d\tau}\frac{dx^\lambda}{d\tau} + \Gamma^\mu{}_{\alpha\beta}\Gamma^\alpha{}_{\kappa\lambda}A^\kappa\frac{dx^\lambda}{d\tau}\frac{dx^\beta}{d\tau} \qquad [115]$$

The next to last term comes from evaluating $d^2x^\beta/d\tau^2$ with the geodesic equation.

We can now begin our calculation of the relative acceleration of two particles moving on neighboring geodesics by writing down the equation of motion of each. Suppose that for a given value of τ along each geodesic, the coordinates of the particles are $x^\mu(\tau)$ and $x^\mu(\tau) + s^\mu(\tau)$ (see Fig. 6.14). Then the geodesic equations for the two particles are

$$\frac{d^2x^\mu}{d\tau^2} + \Gamma^\mu{}_{\alpha\beta}(x)\frac{dx^\alpha}{d\tau}\frac{dx^\beta}{d\tau} = 0 \qquad [116]$$

and

$$\frac{d^2(x^\mu + s^\mu)}{d\tau^2} + \Gamma^\mu{}_{\alpha\beta}(x + s)\left(\frac{dx^\alpha}{d\tau} + \frac{ds^\alpha}{d\tau}\right)\left(\frac{dx^\beta}{d\tau} + \frac{ds^\beta}{d\tau}\right) = 0 \quad [117]$$

We will assume that s^μ and $ds^\mu/d\tau$ are infinitesimal; this means that the particles are near to each other and remain near for a fairly long time. If we approximate

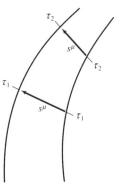

Fig. 6.14 Each of two adjacent geodesics is parametrized by its proper time τ. The displacement vector s^μ connects pairs of points with the same value of τ on the two geodesics.

$$\Gamma^{\mu}{}_{\alpha\beta}(x + s) \simeq \Gamma^{\mu}{}_{\alpha\beta}(x) + \Gamma^{\mu}{}_{\alpha\beta,\sigma}s^{\sigma} \qquad [118]$$

and keep only first-order terms in s^{σ}, the difference between Eqs. [116] and [117] yields

$$\frac{d^2 s^{\mu}}{d\tau^2} = - \Gamma^{\mu}{}_{\alpha\beta,\sigma}s^{\sigma} \frac{dx^{\alpha}}{d\tau} \frac{dx^{\beta}}{d\tau} - 2 \Gamma^{\mu}{}_{\alpha\beta} \frac{ds^{\alpha}}{d\tau} \frac{dx^{\beta}}{d\tau} \qquad [119]$$

The second derivative of s^{μ} along the geodesic can then be calculated by substituting Eq. [119] into [115], with the following result:

$$\frac{D^2 s^{\mu}}{D\tau^2} = - R^{\mu}{}_{\alpha\sigma\beta}s^{\sigma} \frac{dx^{\alpha}}{d\tau} \frac{dx^{\beta}}{d\tau} \qquad [120]$$

■ **Exercise 26.** Carry out the calculation leading to Eq. [120]. ■

Eq. [120] is called the equation of *geodesic deviation*. This equation gives us the relativistic generalization of our Newtonian result for the tidal force.

For a comparison of the relativistic and Newtonian equations for the tidal force, assume that the particles under consideration are moving slowly, with

$$\frac{dx'^{\alpha}}{d\tau} \simeq (1, 0, 0, 0) \qquad [121]$$

Furthermore, assume that $s^0 = 0$; this simply means that the particle accelerations are compared at equal times. Then Eq. [120] reduces to

$$\frac{d^2 s'^{\mu}}{dt'^2} \simeq - R'^{k}{}_{0l0}s'^{l} \qquad [122]$$

The tidal force is therefore

$$f^k \simeq - mR'^{k}{}_{0l0}s'^{l} \qquad [123]$$

where m is the mass of the particle and s'^{l} its displacement from the origin. Note that this equation is valid only if the displacement and the velocity are small ($s'^{l} \to 0$, $ds'^{l}/d\tau \to 0$).

To establish that this equation for the tidal force is in agreement with the Newtonian expression (Eq. [1.52]), we must check that the old (Eq. [1.51]) and new (Eq. [91]) definitions of $R^{k}{}_{0l0}$ coincide in the case of weak static gravitational fields. In the linear approximation, the two terms quadratic in $\Gamma^{\alpha}{}_{\beta\mu}$ can be omitted from Eq. [91]. Furthermore, in this approximation,

$$\Gamma^\alpha{}_{\beta\mu} = \frac{\kappa}{2} \eta^{\alpha\sigma}(h_{\sigma\beta,\mu} + h_{\mu\sigma,\beta} - h_{\beta\mu,\sigma}) \qquad [124]$$

This results in

$$R^\alpha{}_{\beta\mu\nu} \simeq -\frac{\kappa}{2} \eta^{\alpha\sigma}(h_{\mu\sigma,\beta,\nu} - h_{\beta\mu,\sigma,\nu} - h_{\nu\sigma,\beta,\mu} + h_{\beta\nu,\sigma,\mu}) \qquad [125]$$

Therefore

$$R^k{}_{0l0} \simeq \frac{\kappa}{2} (h_{lk,0,0} - h_{0l,k,0} - h_{0k,0,l} + h_{00,k,l}) \qquad [126]$$

In the Newtonian limit, all the terms containing time derivatives can be omitted. Using Eq. [3.102],

$$\tfrac{1}{2}\kappa h_{00} = \Phi \qquad [127]$$

we then find

$$R^k{}_{0l0} = \frac{\partial^2\Phi}{\partial x^k \partial x^l} \qquad [128]$$

which is in agreement with Eq. [1.51].

The tidal-force equation [123] can be used for measuring the components $R^k{}_{0l0}$ of the Riemann tensor. We already discussed in Section 1.8 how such measurements of the tidal force can be carried out; we emphasize once more that these measurements can be performed *locally*.

What about the other components of the Riemann tensor? These components can be obtained by measuring the components $R^k{}_{0l0}$ in a sufficiently large number of other reference frames that have some velocity relative to the original reference frame. Lorentz transformations between these reference frames relate the components $R^k{}_{0l0}$ in one frame to the general components $R^\alpha{}_{\beta\mu\nu}$ in another and therefore can be used to find all the components. The Riemann tensor is therefore entirely determined by tidal-force measurements.

To conclude this section, we will show that the tidal force [123] satisfies the condition

$$\frac{\partial f^k}{\partial s'^k} = 0 \qquad [129]$$

This means that the tidal force field can be represented graphically by field lines. Of course, we already came across the condition [129] in Newton's theory (Eq. [1.55]) and in the linear tensor theory (Eq. [5.43]), but for a general proof we need to use the exact field equations.

According to Eq. [123],

$$\frac{\partial f'^k}{\partial s'^k} = -mR'^k{}_{0l0}\,\frac{\partial s'^l}{\partial s'^k} = -mR'^k{}_{0k0} \tag{130}$$

Since $R^\alpha{}_{\beta\mu\nu}$ is antisymmetric in μ and ν, we have $R'^0{}_{000} = 0$, and Eq. [130] is therefore equivalent to

$$\frac{\partial f'^k}{\partial s'^k} = -mR'^\mu{}_{0\mu0} = mR'_{00} \tag{131}$$

where $R'_{00} \equiv R'^\mu{}_{00\mu}$ is a component of the Ricci tensor. As we will see in the next chapter, Einstein's field equations *in vacuo* tells us that $R'_{00} = 0$, and hence Eq. [129] is valid as long as we remain outside of the mass distribution that generates the gravitational field.

6.7 DIFFERENTIAL FORMS IN CURVED SPACE

In the preceding sections we have developed the mathematics of Riemannian geometry in component language. In this section we will redevelop the mathematics in the language of tangent vectors and differential forms. This gives us more concise equations, and also permits us to derive some identities in a more elegant manner.

The notions of tangent vector and of differential forms are equally valid in curved spacetime and in flat spacetime. As in Section 2.6, the tangent vectors are linear differential operators defined on test functions, and 1-forms are duals of tangent vectors. Given an arbitrary system of coordinates, we can introduce the basis vectors $\partial/\partial x^\mu$ and the basis 1-forms $\mathbf{d}x^\mu$, so any tangent vector $\mathbf{u}$ and any 1-form α can be expressed as a superposition of basis vectors or basis 1-forms, respectively:

$$\mathbf{u} = u^\mu\,\frac{\partial}{\partial x^\mu} \tag{132}$$

$$\alpha = \alpha_\mu\,\mathbf{d}x^\mu \tag{133}$$

When the basis 1-forms act on the basis vectors, the result is the same as in flat spacetime,

$$dx^\mu \left(\frac{\partial}{\partial x^\mu} \right) = \frac{\partial x^\mu}{\partial x^\nu} = \delta^\mu_\nu \qquad [134]$$

Under a transformation of coordinates, the components u^μ and α_μ introduced in equations [132] and [133] obey the usual transformation laws for contravariant and covariant vector components. As in Section 2.6, these transformation laws follow directly from the transformation laws for the basis vectors and the basis 1-forms,

$$\frac{\partial}{\partial x'^\nu} = \frac{\partial x^\mu}{\partial x'^\nu} \frac{\partial}{\partial x^\mu} \qquad [135]$$

and

$$dx'^\nu = \frac{\partial x'^\nu}{\partial x^\mu} dx^\mu \qquad [136]$$

■ *Exercise 27.* Use Eqs. [135] and [136] to derive the transformation laws for u^μ and α_μ. ■

The notion of tensor can also be taken over unchanged from Section 2.6. For instance, a tensor of type $\binom{1}{2}$ is a linear map that yields a real number when acting on a 1-form and on two tangent vectors. Such a tensor can be written as a superposition of tensor products of the basis vectors and basis 1-forms,

$$\mathbf{T} = T^\sigma_{\mu\nu} \frac{\partial}{\partial x^\sigma} \otimes dx^\mu \otimes dx^\nu \qquad [137]$$

It is easy to show that, under a coordinate transformation, the components $T^\sigma_{\mu\nu}$ of such a tensor obey the usual transformation law, contravariant in the first index, and covariant in the other two indices.

■ *Exercise 28.* Show this. ■

In the language of differential forms, the definition of the covariant derivative requires parallel transport, in much the same way as in component language. However, the definition of the covariant derivative per se is a bit awkward, and it is easier to begin with the covariant derivative along a curve (compare Eq. [114]). Suppose that the curve is parametrized by λ, and that at a given point, the curve has the tangent vector $\mathbf{u} = d/d\lambda$. Along this curve, an arbitrary tensor field $\mathbf{T}$ can be regarded as a function of λ. The covariant derivative of such a tensor field $\mathbf{T}$ along the curve is then defined by

$$\mathbf{\nabla_u T} = \lim_{d\lambda \to 0} \frac{T(\lambda + d\lambda) - T(\lambda) - \delta T}{d\lambda} \qquad [138]$$

where, as in Section 6.2, δT is the change in $\mathbf{T}$ produced by parallel transport. Note that this covariant derivative along a curve is a tensor of the same rank as the original tensor.

The covariant derivative per se can next be constructed in terms of the covariant derivative along a curve. To accomplish this construction, we demand that the covariant derivative $\mathbf{\nabla T}$ be a tensor of rank higher by one than $\mathbf{T}$, such that when $\mathbf{\nabla T}$ acts on an arbitrary tangent vector $\mathbf{u}$ the result is $\mathbf{\nabla_u T}$. However, in most of the calculations hereafter, we will find it sufficient to deal with the covariant derivative along a curve, and we will not need the covariant derivative per se.

The definition [138] is valid for tensors of arbitrary rank. In the special case of a function f (that is, a tensor of rank zero), the covariant derivative reduces to an ordinary derivative,

$$\mathbf{\nabla_u} f = \frac{df}{d\lambda} \qquad [139]$$

For a tensor product, the derivative obeys the usual rules,

$$\mathbf{\nabla_u} (\mathbf{T} \otimes \mathbf{R}) = \mathbf{\nabla_u T} \otimes \mathbf{R} + \mathbf{T} \otimes \mathbf{\nabla_u R} \qquad [140]$$

To discover the components of the covariant derivative, take a curve that coincides with one of the lines of the coordinate grid, so the tangent vector $\mathbf{u}$ coincides with one of the basis vectors, say, $\mathbf{u} = \partial/\partial x^\mu$. For convenience, we write $\mathbf{\nabla_u} = \mathbf{\nabla}_\mu$, since it is rather awkward to write the full subscript $\partial/\partial x^\mu$. Let us evaluate the covariant derivative of some vector $\mathbf{v}$. In terms of basis vectors, such a vector $\mathbf{v}$ can be written

$$\mathbf{v} = v^\nu \frac{\partial}{\partial x^\nu} \qquad [141]$$

If we differentiate this term by term, we find

$$\mathbf{\nabla}_\mu \mathbf{v} = \frac{\partial v^\nu}{\partial x^\mu} \frac{\partial}{\partial x^\nu} + v^\nu \mathbf{\nabla}_\mu \frac{\partial}{\partial x^\nu} \qquad [142]$$

Now suppose that the vector field $\mathbf{v}$ has been constructed by parallel transport. For such a vector field, the covariant derivative is zero,

$$\frac{\partial v^\nu}{\partial x^\mu} \frac{\partial}{\partial x^\nu} + v^\nu \mathbf{\nabla}_\mu \frac{\partial}{\partial x^\nu} = 0 \qquad [143]$$

To extract the components from this equation, we let the dual vector dx^α act on it:

$$\frac{\partial v^\nu}{\partial x^\mu} dx^\alpha \left(\frac{\partial}{\partial x^\nu} \right) + v^\nu dx^\alpha \left(\nabla_\mu \frac{\partial}{\partial x^\nu} \right) = 0 \qquad [144]$$

or

$$\frac{\partial v^\alpha}{\partial x^\mu} + v^\nu dx^\alpha \left(\nabla_\mu \frac{\partial}{\partial x^\nu} \right) = 0 \qquad [145]$$

This tells us how the components of the vector change under parallel transport. Comparing this with the standard expression for parallel transport in component language, $\delta v^\alpha + \Gamma^\alpha{}_{\mu\nu} v^\nu dx^\mu = 0$, or

$$\frac{\partial v^\alpha}{\partial x^\mu} + \Gamma^\alpha{}_{\mu\nu} v^\nu = 0 \qquad [146]$$

we see that

$$dx^\alpha \left(\nabla_\mu \frac{\partial}{\partial x^\nu} \right) = \Gamma^\alpha{}_{\mu\nu} \qquad [147]$$

Thus, the covariant derivatives of basis vectors are directly related to the Christoffel symbols.

■ *Exercise 29.* Show that

$$\nabla_\mu dx^\alpha \left(\frac{\partial}{\partial x^\nu} \right) = -\Gamma^\alpha{}_{\mu\nu} \qquad [148]$$

(Hint: The covariant derivative of Eq. [134] is zero.) ■

In the language of differential forms, the geodesic equation asserts that the tangent vector **u** is parallel transported along the curve,

$$\nabla_u u = 0 \qquad [149]$$

If we write this out in components, it yields the standard form [33] of the geodesic equation.

■ *Exercise 30.* Show this. ■

The metric tensor **g** of the curved spacetime is a second-rank tensor of the type $\binom{0}{2}$. Acting on a pair of vectors, this tensor produces a real number **g(u, w)**. In terms of basis 1-forms, the metric tensor can be expressed as

$$\mathbf{g} = g_{\mu\nu}\, \mathbf{d}x^{\mu} \otimes \mathbf{d}x^{\nu} \qquad [150]$$

The inverse metric tensor $\mathbf{g}^{-1}$ is a second-rank tensor of the type $\binom{2}{0}$, such that

$$\mathbf{g}\mathbf{g}^{-1} = \mathbf{g}^{-1}\mathbf{g} = 1 \qquad [151]$$

In terms of basis vectors,

$$\mathbf{g}^{-1} = g^{\mu\nu}\, \frac{\partial}{\partial x^{\mu}} \otimes \frac{\partial}{\partial x^{\nu}} \qquad [152]$$

It is then easy to verify that Eq. [151] is satisfied, provided that

$$g_{\mu\alpha}g^{\alpha\nu} = \delta^{\mu}_{\nu} \qquad [153]$$

■ *Exercise 31.* Multiply **g** and $\mathbf{g}^{-1}$ term by term and obtain Eq. [153]. ■

The covariant derivative of the metric tensor must be identically zero,

$$\nabla_{\mathbf{u}}\mathbf{g} \equiv 0 \qquad [154]$$

so that **g(v, w)** = [constant] whenever **v** and **w** are parallel-transported vectors. In terms of components, this requirement implies the usual relation between $g_{\mu\nu,\alpha}$ and the Christoffel symbols.

The definition of the Riemann curvature tensor can be introduced by examining the difference between successive covariant derivatives taken in opposite order. Thus, consider a 1-form α, and compare the covariant derivatives $\nabla_{\mu}\nabla_{\nu}\alpha$ and $\nabla_{\nu}\nabla_{\mu}\alpha$, evaluated at some given point. The difference between these derivatives is

$$(\nabla_{\mu}\nabla_{\nu} - \nabla_{\nu}\nabla_{\mu})\alpha \qquad [155]$$

We can show that this difference depends only on the value of the 1-form at the given point, but does not depend on the value of the 1-form at nearby points; that is, the derivative terms in the definition of the second covariant derivatives cancel against each other, leaving only the terms arising from parallel transport, which terms are linear functions of the 1-form. The proof can be most easily given by using components, as we did in Eqs. [87]-[91], but for purists there is a rather more tedious proof that does not use components. In any case, the operator $(\nabla_\mu \nabla_\nu - \nabla_\nu \nabla_\mu)$ generates a linear map from the 1-form α to the 1-form $(\nabla_\mu \nabla_\nu - \nabla_\nu \nabla_\mu)\alpha$. If we let this 1-form act on a vector $\mathbf{w}$, the result is a real number. Since this real number also depends on the basis vectors $\partial/\partial x^\mu$ and $\partial/\partial x^\nu$, as indicated by the abbreviated subscript notation in Eq. [155], we write it as $\mathbf{R}(\alpha, \partial/\partial x^\mu, \partial/\partial x^\nu, \mathbf{w})$, that is,

$$(\nabla_\mu \nabla_\nu - \nabla_\nu \nabla_\mu)\alpha(\mathbf{w}) = \mathbf{R}\left(\alpha, \frac{\partial}{\partial x^\mu}, \frac{\partial}{\partial x^\nu}, \mathbf{w}\right) \qquad [156]$$

Here we see that $\mathbf{R}$ is a map from one 1-form α and three tangent vectors (two of which are basis vectors, $\partial/\partial x^\mu$ and $\partial/\partial x^\nu$) to one 1-form. Thus, $\mathbf{R}$ is a tensor of type $\binom{1}{3}$.

Actually, for the general definition of the tensor $\mathbf{R}$, it is not sufficient to examine its action on the basis vectors $\partial/\partial x^\mu$ and $\partial/\partial x^\nu$, but on general tangent vectors, such as $\mathbf{u} = u^\mu \partial/\partial x^\mu$ and $\mathbf{v} = v^\nu \partial/\partial x^\nu$. Such vectors do not commute, since the derivative in each acts on the coefficients appearing in the other,

$$\mathbf{u}\mathbf{v} - \mathbf{v}\mathbf{u} = u^\mu \frac{\partial}{\partial x^\mu}\left[v^\nu \frac{\partial}{\partial x^\nu}\right] - v^\nu \frac{\partial}{\partial x^\nu}\left[u^\mu \frac{\partial}{\partial x^\mu}\right]$$

$$= u^\mu \frac{\partial v^\nu}{\partial x^\mu}\frac{\partial}{\partial x^\nu} - v^\nu \frac{\partial u^\mu}{\partial x^\nu}\frac{\partial}{\partial x^\mu} \qquad [157]$$

Thus, the commutator of two tangent vectors is, again, a tangent vector. The extra term resulting from the noncommutativity of $\mathbf{u}$ and $\mathbf{v}$ spoils the cancellation of derivatives in the operator on the left side of Eq. [156], and therefore destroys the tensorial character of the operator. But we can restore the cancellation by subtracting $\mathbf{u}\mathbf{v} - \mathbf{v}\mathbf{u}$, which gives us a corrected operator:

$$(\nabla_\mathbf{u} \nabla_\mathbf{v} - \nabla_\mathbf{v} \nabla_\mathbf{u} - \mathbf{u}\mathbf{v} + \mathbf{v}\mathbf{u})\alpha(\mathbf{w}) = \mathbf{R}(\alpha, \mathbf{u}, \mathbf{v}, \mathbf{w}) \qquad [158]$$

This is the general definition of the Riemann tensor $\mathbf{R}$ in the language of differential forms.

The components of this tensor are $R^\sigma_{\beta\mu\nu}$,

$$\mathbf{R} = R^\sigma_{\beta\mu\nu} \frac{\partial}{\partial x^\sigma} \otimes \mathbf{d}x^\beta \otimes \mathbf{d}x^\mu \otimes \mathbf{d}x^\nu \qquad [159]$$

The symmetry properties of $\mathbf{R}$ are somewhat awkward to state in the language of differential forms, and we will not trouble to rederive them.

The second-rank Ricci tensor $\overline{R}$ is a tensor of type $\binom{0}{2}$, obtained in the usual way by contraction on the first and last slots. Acting on two vectors, it produces a real number:

$$\overline{R}(u, v) = R\left(\frac{\partial}{\partial x^\gamma}, u, v, dx^\gamma\right) \qquad [160]$$

From the Ricci tensor we can construct the curvature scalar by first multiplying with g, which gives us a tensor of type $\binom{1}{1}$, and then by contracting,

$$g \otimes \overline{R}\left(\frac{\partial}{\partial x^\gamma}, dx^\lambda, \frac{\partial}{\partial x^\lambda}, dx^\gamma\right) \qquad [161]$$

Finally, let us rewrite the equation of geodesic deviation in the language of differential forms. Instead of rederiving the equation of geodesic deviation from first principles, we will simply translate Eq. [120] from component language into the new language. We designate the tangent vectors along the geodesic by u, and we designate the tangent vector from one geodesic to the adjacent geodesic by s. The components of u are $dx^\alpha/d\tau$, and the components of s are s^μ,

$$u = \frac{dx^\alpha}{d\tau}\frac{\partial}{\partial x^\alpha} \quad \text{and} \quad s = s^\mu\frac{\partial}{\partial x^\mu} \qquad [162]$$

The derivative $D/D\tau$ along the curve corresponds to the covariant derivative ∇_u. Hence our old equation of geodesic deviation takes the new form

$$\nabla_u \nabla_u s = - R(\ , u, u, s) \qquad [163]$$

Here the empty first slot in R means that instead of mapping one 1-form and three vectors into a real number, R maps three vectors into a vector (this can be best seen from Eq. [159]; when this expression for R is allowed to act on the three vectors u, u, and s, the three 1-forms in Eq. [159] disappear, and only the vector $\partial/\partial x^\sigma$ remains).

6.8 ISOMETRIES OF SPACETIME; KILLING VECTORS

If the metric tensor of a given spacetime geometry is invariant under a coordinate transformation $x'^\mu = x'^\mu(x)$, then this coordinate transformation is said to be a symmetry, or an *isometry*, of the spacetime. For example, translations, three-dimensional rotations, and Lorentz transformations (with arbitrary direction of the velocity) are symmetries, or isometries, of flat spacetime--the metric tensor $\eta_{\mu\nu}$ remains invariant under each of these transformations.

A general coordinate transformation changes the metric tensor $g_{\mu\nu}$ into $g'_{\mu\nu}$, according to the familiar tensor transformation formula [20],

$$g'_{\alpha\beta}(x'(x)) = \frac{\partial x^\mu}{\partial x'^\alpha} \frac{\partial x^\nu}{\partial x'^\beta} g_{\mu\nu}(x) \qquad [164]$$

On the left side of this equation, the coordinate x' has been expressed as a function of x, to emphasize that the two sides of the equation are evaluated at the same point of spacetime. If we invert this formula, it gives us $g_{\mu\nu}$ in terms of $g'_{\mu\nu}$:

$$g_{\mu\nu}(x) = \frac{\partial x'^\alpha}{\partial x^\mu} \frac{\partial x'^\beta}{\partial x^\nu} g'_{\alpha\beta}(x'(x)) \qquad [165]$$

If the coordinate transformation is an isometry, the metric tensor must be invariant under this transformation, which means that the new metric tensor must be the same function of its argument as the old metric tensor, that is,*

$$g'_{\alpha\beta}(x') = g_{\alpha\beta}(x') \qquad \text{and} \qquad g'_{\alpha\beta}(x'(x)) = g_{\alpha\beta}(x'(x))$$

Thus, Eq. [165] becomes

$$g_{\mu\nu}(x) = \frac{\partial x'^\alpha}{\partial x^\mu} \frac{\partial x'^\beta}{\partial x^\nu} g_{\alpha\beta}(x'(x)) \qquad [166]$$

The isometries of a given spacetime will often be evident by inspection of the metric tensor. If not, we must try to discover them by "solving" Eq. [166] for the function $x'(x)$ that characterizes the isometry. This equation is hard or impossible to solve if $x'(x)$ is a finite coordinate transformation. But the equation becomes much simpler if

---

* Do not confuse this equation with the transformation equation for a scalar field, which is $\phi'(x'(x)) = \phi(x)$. For comparison, note that the condition for a symmetry of the scalar field is $\phi'(x') = \phi(x')$, which is equivalent to $\phi(x) = \phi(x'(x))$.

we consider the special case of an infinitesimal coordinate transformation,

$$x'^\alpha = x^\alpha + \epsilon \xi^\alpha(x) \quad \text{with } \epsilon \to 0 \qquad [167]$$

where $\xi^\alpha(x)$ is a function of position, that is, $\xi^\alpha(x)$ is a vector field. For this transformation,

$$\frac{\partial x'^\alpha}{\partial x^\mu} = \delta^\alpha_\mu + \epsilon \frac{\partial \xi^\alpha}{\partial x^\mu}$$

and Eq. [165] takes the form

$$g_{\mu\nu}(x) = \left(\delta^\alpha_\mu + \epsilon \frac{\partial \xi^\alpha}{\partial x^\mu}\right)\left(\delta^\beta_\nu + \epsilon \frac{\partial \xi^\beta}{\partial x^\nu}\right) g'_{\alpha\beta}(x'(x))$$

$$\simeq g'_{\mu\nu}(x'(x)) + \epsilon \frac{\partial \xi^\alpha}{\partial x^\mu} g_{\alpha\nu}(x'(x)) + \epsilon \frac{\partial \xi^\beta}{\partial x^\nu} g_{\mu\beta}(x'(x)) \qquad [168]$$

Here, we have neglected terms of order ϵ^2. Note that in the last two terms on the right side of Eq. [168], we have also replaced $g'_{\mu\nu}$ by $g_{\mu\nu}$. This is permissible, since the difference between these two tensors is of order ϵ; hence the difference can be neglected in these two terms, since they already contain another factor of ϵ. In order to express everything as an explicit function of x, we use the Taylor-series expansion

$$g'_{\mu\nu}(x'(x)) \simeq g'_{\mu\nu}(x) + \epsilon \xi^\alpha \frac{\partial g'_{\mu\nu}}{\partial x^\alpha} \simeq g'_{\mu\nu}(x) + \epsilon \xi^\alpha \frac{\partial g_{\mu\nu}}{\partial x^\alpha}$$

With this, Eq. [168] becomes

$$g_{\mu\nu}(x) = g'_{\mu\nu}(x) + \epsilon \xi^\alpha \frac{\partial g_{\mu\nu}}{\partial x^\alpha} + \epsilon \frac{\partial \xi^\alpha}{\partial x^\mu} g_{\alpha\nu} + \epsilon \frac{\partial \xi^\beta}{\partial x^\nu} g_{\mu\beta} \qquad [169]$$

The condition for the invariance of the metric tensor under the infinitesimal transformation is then

$$0 = \xi^\alpha \frac{\partial g_{\mu\nu}}{\partial x^\alpha} + \frac{\partial \xi^\alpha}{\partial x^\mu} g_{\alpha\nu} + \frac{\partial \xi^\beta}{\partial x^\nu} g_{\mu\beta} \qquad [170]$$

This equation is the infinitesimal version of Eq. [166]. It is convenient to write it in terms of the covariant components $\xi_\nu = g_{\alpha\nu}\xi^\alpha$ as follows:

$$0 = \xi^\alpha \frac{\partial g_{\mu\nu}}{\partial x^\alpha} + \left[\frac{\partial \xi_\nu}{\partial x^\mu} - \xi^\alpha \frac{\partial g_{\alpha\nu}}{\partial x^\mu} \right] + \left[\frac{\partial \xi_\mu}{\partial x^\nu} - \xi^\beta \frac{\partial g_{\mu\beta}}{\partial x^\nu} \right]$$

Here the three derivatives of the metric tensor combine into a Christoffel symbol, so

$$0 = \frac{\partial \xi_\nu}{\partial x^\mu} + \frac{\partial \xi_\mu}{\partial x^\nu} - 2\xi_\alpha \Gamma^\alpha{}_{\mu\nu}$$

which is the same as

$$0 = \xi_{\nu;\mu} + \xi_{\mu;\nu} \qquad\qquad [171]$$

This differential equation is called *Killing's equation.*

The condition for an infinitesimal symmetry transformation of the metric can also be expressed in terms of the *Lie derivative.* If we rewrite Eq. [169] as

$$g_{\mu\nu}(x) - g'_{\mu\nu}(x) = \epsilon\xi^\alpha \frac{\partial g_{\mu\nu}}{\partial x^\alpha} + \epsilon \frac{\partial \xi^\alpha}{\partial x^\mu} g_{\alpha\nu} + \epsilon \frac{\partial \xi^\beta}{\partial x^\nu} g_{\mu\beta} \qquad [172]$$

we see that the three terms on the right side represent the change in the tensor $g_{\mu\nu}(x)$ under the infinitesimal transformation. This change, divided by ϵ, is called the Lie derivative of $g_{\mu\nu}(x)$ with respect to ξ^μ, and it is designated by $\mathcal{L}_\xi g_{\mu\nu}(x)$:

$$\mathcal{L}_\xi g_{\mu\nu}(x) = \lim_{\epsilon \to 0} \frac{g_{\mu\nu}(x) - g'_{\mu\nu}(x)}{\epsilon} = \xi^\alpha \frac{\partial g_{\mu\nu}}{\partial x^\alpha} + \frac{\partial \xi^\alpha}{\partial x^\mu} g_{\alpha\nu} + \frac{\partial \xi^\beta}{\partial x^\nu} g_{\mu\beta}$$
$$[173]$$

The Lie derivative of an arbitrary second-rank tensor can be defined similarly, and Lie derivatives for tensors of other ranks can also be defined, by a simple calculation of the change of the tensor under an infinitesimal transformation. In terms of the Lie derivative, the condition for the invariance of the metric under an infinitesimal transformation then merely says that the Lie derivative is zero, $\mathcal{L}_\xi g_{\mu\nu} = 0$.

After this digression, let us return to the Killing equation. The solutions of this equation, called *Killing vectors,* give us the symmetry transformations of the metric, both the infinitesimal transformations and the finite transformations. It is obvious that for any finite transformation there exists an infinitesimal transformation, since the Taylor-series expansion of the finite transformation $x' = x'(x)$ yields the infinitesimal transformation $x'^\alpha = x^\alpha + \epsilon dx^\alpha/d\epsilon$, where ϵ is a parameter that characterizes the transformation (for instance, an infinitesimal angle in the case of a rotation, or an infinitesimal velocity in the case of a Lorentz transformation). We can then identify the vector field ξ^α with $dx^\alpha/d\epsilon$.

But the converse is also true: given an infinitesimal transformation we can always reconstruct the finite transformation. For this purpose, we introduce a new set of coordinates $\tilde{x}^\mu$ such that*

$$\tilde{\xi}^1 = [\text{constant}] = b \qquad [174]$$

$$\tilde{\xi}^\mu = 0 \quad \text{for } \mu \neq 1 \qquad [175]$$

This means that the $\tilde{x}^1$ coordinate line at each point has the direction of the vector ξ^α. The conditions [174] and [175] can always be satisfied. To see this, we express $\tilde{\xi}^\mu$ in terms of ξ^μ, by means of the usual transformation equation $\tilde{\xi}^\mu = \partial \tilde{x}^\mu/\partial x^\alpha \, \xi^\alpha$, and we obtain

$$\frac{\partial \tilde{x}^1}{\partial x^\alpha} \, \xi^\alpha = b \qquad [176]$$

$$\frac{\partial \tilde{x}^\mu}{\partial x^\alpha} \, \xi^\alpha = 0 \quad \text{for } \mu \neq 1 \qquad [177]$$

This is a set of linear differential equations, with coefficients $\xi^\alpha(x)$, for the functions $\tilde{x}^\mu(x)$; such equations always have a solution.

If we now write Eq. [170] in the new coordinates $\tilde{x}^\mu$ and we insert the expressions [174] and [175], we find that

$$\frac{\partial \tilde{g}_{\mu\nu}}{\partial \tilde{x}^1} = 0 \qquad [178]$$

Thus, $\tilde{g}_{\mu\nu}$ does not depend on the coordinate $\tilde{x}^1$. Consequently, any finite translation of this coordinate is a symmetry of the metric $\tilde{g}_{\mu\nu}$. This establishes that the infinitesimal symmetry transformation generates a finite symmetry transformation. Note that the argument also establishes that, with a suitable choice of coordinates, any symmetry transformation of the metric can be regarded as a translation of one of the coordinates, and the metric is then independent of this one coordinate. Checking for the independence of the metric from some coordinate is, of course, a familiar method for discovering symmetries "by inspection." For example, in this way we immediately recognize the translation symmetry of the flat spacetime metric $\eta_{\mu\nu}$ from its independence of t, x, y, and z. And we recognize the rotation symmetry of this metric about the, say, z-axis by transforming it to spherical or

---

* In what follows we arbitrarily select $\tilde{x}^1$ as the preferred coordinate. We could equally well select one of the other coordinates $\tilde{x}^0, \tilde{x}^2$, or $\tilde{x}^3$.

cylindrical coordinates, where the metric is seen to be independent of the polar angle ϕ. (However, if we want to recognize the symmetry of the metric $\eta_{\mu\nu}$ under Lorentz transformations, it is not so obvious how we ought to pick new coordinates that lead to independence from one of them; see Exercise 33.)

■ *Exercise 32.* For an infinitesimal rotation about the z-axis,

$$x'^1 = x^1 + \theta x^2 \quad \text{and} \quad x'^2 = -\theta x^1 + x^2$$

Hence $\xi^\alpha = (0, x^2, -x^1, 0)$. Integrate the equations [174] and [175] for $\tilde{x}^1$ and $\tilde{x}^2$, and show that the results are

$$\tilde{x}^1 = -b \tan^{-1} x^2 / \sqrt{(x^1)^2 + (x^2)^2} = -b\phi$$

and

$$\tilde{x}^2 = \sqrt{(x^1)^2 + (x^2)^2} = r \quad ■$$

■ *Exercise 33.* For an infinitesimal Lorentz transformation,

$$x'^0 = x^0 - vx^1 \quad \text{and} \quad x'^1 = -vx^0 + x^1$$

Hence $\xi^\alpha = (-x^1, -x^0, 0, 0)$; Fig. 6.15a displays this Killing vector field. Integrate the equations [177] and [176] for $\tilde{x}^0$ and $\tilde{x}^1$ and show that the solutions are

$$\tilde{x}^0 = \sqrt{(x^1)^2 - (x^0)^2} \tag{179}$$

$$\tilde{x}^1 = -b \tanh^{-1} x^0 / x^1 \tag{180}$$

Since the curve $\sqrt{(x^1)^2 - (x^0)^2} = [\text{constant}]$ is timelike and the curve $\tanh^{-1} x^0 / x^1 = [\text{constant}]$ is spacelike, convention requires that we redefine the coordinates $\tilde{x}^0$ and $\tilde{x}^1$:

$$\tilde{x}^0 = b \tanh^{-1} x^0 / x^1 \tag{181}$$

$$\tilde{x}^1 = \sqrt{(x^1)^2 - (x^0)^2} \tag{182}$$

Correspondingly, the Killing vector $\tilde{\xi}^\alpha$ becomes $(-b, 0, 0, 0)$. The solutions [179] - [182] apply in the wedge, or quadrant, of spacetime characterized by $x^1 > 0$ and $(x^1)^2 - (x^0)^2 \geq 0$ (see Fig. 6.15b). Find the solutions that apply in the other wedges.

Show that in the coordinates $\tilde{x}^0, \tilde{x}^1$ the spacetime interval is

$$ds^2 = \frac{(\tilde{x}^1)^2}{b^2}(d\tilde{x}^0)^2 - (d\tilde{x}^1)^2 \tag{183}$$

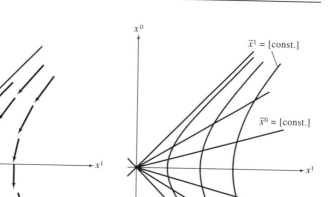

Fig. 6.15 (a) The Killing vector field $\xi^\alpha = (-x^1, -x^0, 0, 0)$. (b) Rindler coordinates in the right wedge.

As expected, the metric tensor is independent of $\tilde{x}^0$. Verify that a finite Lorentz transformation amounts to a translation of $\tilde{x}^0$.

The coordinates $\tilde{x}^0, \tilde{x}^1$ given by Eqs. [181] and [182] are called **Rindler coordinates**. The curves $\tilde{x}^1 = $ [constant] in these coordinates are the worldlines of particles moving with a constant proper acceleration ("hyperbolic motion"). The constant b is inversely proportional to the proper acceleration, $b = 1/a$. ∎

Although in principle all the symmetries of the metric can be recognized by its independence from coordinates upon making suitable choices of coordinates, this is not a convenient method for finding symmetries, since the "suitable" choices of coordinates are not always obvious. The Killing equation gives us a straightforward, systematic method for finding the symmetries. By constructing all possible solutions of this differential equation, we will find all the symmetries.

Two or more Killing vectors $\xi^\alpha_{(1)}$, $\xi^\alpha_{(2)}$,... are said to be linearly dependent if there exists a set of constant coefficients c_n such that

$$\sum_n c_n \xi^\alpha_{(n)} = 0 \qquad [184]$$

If there exist no such coefficients, the Killing vectors are said to be linearly independent. Each linearly independent Killing vector corresponds to a separate symmetry of the spacetime. The maximum number of independent solutions, and the maximum number of symmetries, is ten. This agrees with the number of symmetries of flat spacetime, which has exactly ten symmetries: four independent directions for translations, three independent directions for the axes of spatial rotations, and three independent directions for the velocities of Lorentz transformations (these symmetries constitute the ten-parameter group of general Lorentz transformations). We expect that the number of symmetries of a curved spacetime is no larger than that of flat spacetime, and that therefore curved spacetime has no more than ten symmetries (for a proof, see Weinberg, 1972). A spacetime with ten symmetries is called maximally symmetric. Apart from flat spacetime, the only maximally symmetric spacetimes are those of constant curvature, such as the isotropic and homogeneous models of the universe we will discuss in Chapter 9.

It is well known that symmetries lead to conservation laws. According to Noether's theorem, to every continuous symmetry of a physical system corresponds a conservation law. For example, in flat spacetime, the symmetry under spatial translations implies conservation of momentum, the symmetry under time translations implies conservation of energy, and the symmetry under rotations implies conservation of angular momentum. In a general, curved spacetime, we can formulate the conservation laws for the motion of a particle on the basis of Killing vectors. We can prove that if ξ_μ is a Killing vector, then, for a particle moving along a geodesic, the scalar product of this Killing vector and the momentum $P^\mu = m \, dx^\mu/d\tau$ of the particle is a constant:

$$\xi_\mu P^\mu = [\text{constant}] \qquad [185]$$

To prove this conservation theorem, we evaluate the derivative of $\xi_\mu P^\mu$ along the geodesic curve,

$$\frac{d}{d\tau} \xi_\mu P^\mu = \xi_\mu \frac{dP^\mu}{d\tau} + P^\mu \frac{d\xi_\mu}{d\tau} \qquad [186]$$

But, according to the geodesic equation and the definition of the covariant derivative,

$$\frac{dP^\mu}{d\tau} = - \Gamma^\mu_{\alpha\nu} P^\alpha \frac{dx^\nu}{d\tau}$$

and

$$\frac{d\xi_\mu}{d\tau} = \frac{\partial \xi_\mu}{\partial x^\nu} \frac{dx^\nu}{d\tau} = (\xi_{\mu;\nu} + \Gamma^\alpha_{\mu\nu} \xi_\alpha) \frac{dx^\nu}{d\tau}$$

When we substitute these expressions into Eq. [186], the two terms involving the Christoffel symbols cancel, leaving us with

$$\frac{d}{d\tau}\,\xi_\mu P^\mu = \xi_{\mu;\nu}P^\mu\frac{dx^\nu}{d\tau}$$

$$= \frac{1}{2m}(\xi_{\mu;\nu} + \xi_{\nu;\mu})P^\mu P^\nu \qquad [187]$$

This is zero in consequence of the Killing equation.

Note that if we adopt the coordinates $\tilde{x}^\mu$ that make $\tilde\xi^1 = b$ and the other components of $\tilde\xi^\mu$ zero, then $\tilde\xi^\mu \tilde{P}_\mu$ = [constant] reduces to $\tilde{P}_1$ = [constant]. This says that the component of the momentum conjugate to the coordinate $\tilde{x}^1$ is constant. In Lagrangian language, the coordinate $\tilde{x}^1$ is an ignorable coordinate, since it does not appear in the Lagrangian

$$L = m\sqrt{\tilde{g}_{\mu\nu}(d\tilde{x}^\mu/d\tau)(d\tilde{x}^\nu/d\tau)}$$

The conservation of the momentum $\tilde{P}_1 = m\tilde{g}_{1\mu}d\tilde{x}^\mu/d\tau$ conjugate to this ignorable coordinate is a corollary of the Lagrangian equations.

If the spacetime geometry is time independent, so $g_{\mu\nu}$ does not depend on x^0, then $\xi^\mu = (b, 0, 0, 0)$ is the corresponding Killing vector, and the conservation theorem [185] tells us that P_0 is constant. This is the law of conservation of energy.

This conservation law also applies to photons moving in the curved spacetime. Thus, if the spacetime geometry is time independent, the photon "energy" P_0 is constant. As we will see in Section 7.7, the redshift for photons can be extracted from this conservation law.

FURTHER READING

Introductions to Riemannian geometry and differential calculus, that is, tensor analysis in arbitrary coordinates, are given in many books on relativity. The following books rely mainly or exclusively on component notation:

Adler, R. J., Bazin, M. J., and Schiffer, M., *Introduction to General Relativity* (McGraw-Hill, New York, 1975).

Anderson, J. L., *Principles of Relativity Physics* (Academic Press, New York, 1967). (Includes a careful discussion of the spacetime structures underlying Newtonian and relativistic physics.)

Bergmann, P., *Introduction to the Theory of Relativity* (Prentice-Hall, Englewood Cliffs, 1942).

Carmeli, M., *Classical Fields: General Relativity and Gauge Theory* (Wiley, New York, 1982).

Eddington, A., *The Mathematical Theory of Relativity* (Cambridge University Press, Cambridge, 1923).

Einstein, A., *The Meaning of Relativity* (Princeton University Press, Princeton, 1955). (A very concise presentation of the mathematical formalism.)

Eisenhart, L. P., *Riemannian Geometry* (Princeton University Press, Princeton, 1925). (A classic reference on the geometry of curved spaces of an arbitrary number of dimensions.)

Fock, V., *The Theory of Space, Time, and Gravitation* (Pergamon Press, Oxford, 1964).

Landau and Lifshitz, *The Classical Theory of Fields* (Addison-Wesley, Reading, 1962).

Møller, C., *The Theory of Relativity* (Clarendon Press, Oxford, 1952).

Pauli, W., *Theory of Relativity* (Pergamon Press, London, 1958).

Sokolnikoff, I. S., *Tensor Analysis* (Wiley, New York, 1951). (A thorough introduction to general tensor analysis, with applications to physics.)

Synge, J. L., and Schild, A., *Tensor Calculus* (University of Toronto Press, Toronto, 1965). (Another excellent introduction, with applications to physics.)

Weinberg, S., *Gravitation and Cosmology* (Wiley, New York, 1972).

A neat and concise summary of tensor analysis is contained in Einstein's original paper "Die Grundlage der allgemeinen Relativitätstheorie," Ann. der Physik **49**, 769 (1916), translated in Lorentz, H. A., Einstein, A., Minkowski, H., and Weyl, H., *The Principle of Relativity* (Methuen, London, 1923).

The following books rely more heavily or exclusively on the language of differential forms:

Burke, W. L., *Spacetime, Geometry, and Cosmology* (University Science Book, Mill Valley, 1980). (An elementary introduction, with an exceptionally clear treatment of differential forms.)

Burke, W. I., *Applied Differential Geometry* (Cambridge University Press, Cambridge, 1985).

Hawking, S. W., and Ellis, G. F. R., *The Large Scale Structure of Space-Time* (Cambridge University Press, Cambridge, 1973). (A rigorous examination of the global properties of exact solutions of Einstein's equations.)

Hughston, L. P., and Tod, K. P., *An Introduction to General Relativity* (Cambridge University Press, Cambridge, 1990). (A concise mathematical treatment.)

Misner, C. W., Thorne, K. S., and Wheeler, J. A. W., *Gravitation* (Freeman, San Francisco, 1973). (Emphasizes a geometrical view of spacetime and gives an excellent introduction to differential forms, with interesting graphical representations of 1-forms and 2-forms.)

Schutz, B. F., *Geometrical Methods in Mathematical Physics* (Cambridge University Press, Cambridge, 1980).

Schutz, B. F., *A First Course in General Relativity* (Cambridge University Press, Cambridge, 1985). (A good and clear introduction to differential forms.)

Straumann, N., *General Relativity and Relativistic Astrophysics* (Springer-Verlag, Berlin, 1984).

Wald, R. M., *General Relativity* (University of Chicago Press, Chicago, 1984). (Places heavy emphasis on abstruse and rigorous mathematics.)

The article "General Relativity and Kinetic Theory" by Ehlers, J., Rendiconti S. I. F. **47**, 1 (1974), gives an excellent discussion of kinetic theory in curved spacetime and, incidentally, provides a concise introduction to differential forms and their use in the derivation of differential and integral identities.

By exploiting differential forms it is sometimes possible to simplify the calculation of the Riemann tensor. Misner, Thorne, and Wheeler, op. cit., and Hughston and Tod, op. cit., give good examples of such calculations. In the components formalism, the calculations often are more tedious (for instance, see Exercises 15 and 22). However, computer programs are available that can handle algebraic manipulations and differentiation, and that can automatically calculate the Christoffel symbols and the Riemann tensor.

For more on Lie derivatives and Killing vectors, see Carmeli, op. cit., Hawking and Ellis, op. cit., and Göckeler, M., and Schöcker, T., *Differential Geometry, Gauge Theories, and Gravity* (Cambridge University Press, Cambridge, 1987).

Ehlers, J., Pirani, F. A. E., and Schild, A., in the article "The Geometry of Free Fall and Light Propagation" in O'Raifeartaigh, L., ed., *General Relativity: Papers in Honor of J. L. Synge* (Clarendon Press, Oxford, 1972), carry out a thorough and sophisticated mathematical study of the assumptions that go into the construction of the spacetime geometry.

Ellis, G. F. R., and Williams, R. M., *Flat and Curved Spacetimes* (Clarendon Press, Oxford, 1988), present a simple geometrical and graphical introduction to spacetime; among other things, this book contains a clear discussion of Rindler coordinates.

The following articles attempt to make a case for a theory with torsion, but since they take spin as the source of torsion, they suffer from the defect described in the footnote on p. 312:

Nieh, H. T., "Possible Role of Torsion in Gravitational Theories," in Hu, N., ed., *Proceedings of the Third Marcel Grossmann Meeting on General Relativity* (Science Press and North-Holland, Amsterdam, 1983).

Nester, J. M., and Isenberg, I., Phys. Rev. D **15**, 2078 (1977).

REFERENCES

Levi-Civita, T., *The Absolute Differential Calculus* (Blackie and Son, London, 1929).
Ricci, G., and Levi-Civita, T., Math. Ann. **54**, 125 (1901).
Weinberg, S., *Gravitation and Cosmology* (Wiley, New York, 1972), p. 377.
Weyl, H., *Space-Time-Matter* (Dover, New York, 1922), p. 54.

PROBLEMS

1. Consider the two-dimensional space consisting of the surface of an ordinary sphere of radius a. If the familiar polar angles are used as coordinates in this space, then the metric tensor and the Christoffel symbols in this curved space are as given by Eqs. [67] and [23]. Find R^1_{212}, R^2_{121}, R_{11}, R_{22}, and R.

2. Consider the two-dimensional space consisting of the surface of a cylinder of

radius a. Use cylindrical coordinates z and ϕ, with $x^1 = z$ and $x^2 = \phi$. Then

$$dl^2 = g_{\mu\nu}\,dx^\mu\,dx^\nu = dz^2 + a^2 d\phi^2$$

and

$$g_{\mu\nu} = \begin{pmatrix} 1 & 0 \\ 0 & a^2 \end{pmatrix}$$

Find all components of $\Gamma^\mu_{\alpha\beta}$, $R^\mu_{\nu\alpha\beta}$, $R_{\mu\nu}$, and R. Is the geometry of the cylindrical surface the same as that of a flat two-dimensional plane? Is the topology the same?

3. Suppose that we have a *three-dimensional* spacetime with an interval

$$ds^2 = g_{\mu\nu}\,dx^\mu\,dx^\nu = dt^2 - dz^2 - [a(t)]^2 d\phi^2$$

where $a(t)$ is some increasing function of time. The spatial part of this geometry may be regarded as the surface of an expanding cylinder with a radius $a(t)$ (see Problem 2). Find the nonzero components of the Riemann tensor $R^\alpha_{\beta\mu\nu}$, where α, β, μ, $\nu = 0, 1, 2$. Show that if, and only if, $a(t) = $ [constant], the spacetime is flat.

4. Exercise 4 gives the Christoffel symbols for the surface of a sphere, regarded as a curved two-dimensional space, with coordinates $x^1 = \theta$ and $x^2 = \phi$. Suppose that a vector $(1, 0)$ is initially at the equator, on the meridian $\phi = 0$. Calculate the change of this vector if you parallel transport it along the equator to the meridian $\phi = \pi/2$, then along the meridian $\phi = \pi/2$ to the north pole, and then along the meridian $\phi = 0$ back to the equator. By what angle does the final vector differ from the initial vector?

5. The surface of a torus ("donut") is a two-dimensional curved surface. Suppose that the torus has a radius R measured from the center of symmetry to the middle line and a radius r measured from the middle line to the surface. Use the coordinates $x^1 = \phi$ and $x^2 = \theta$ to describe points on the surface; ϕ is the angle along the torus (measured around the axis of symmetry) and θ is the angle around its cross section (see Fig. 6.16).
(a) Find the metric tensor that describes the distances on the surface of the torus; find the Christoffel symbols.
(b) Suppose that a vector $a^\mu = (0, 1)$ is originally located at the point $x^\mu = (0, \pi/2)$. If this vector is parallel transported to $x^\mu = (\pi/2, \pi/2)$ along the curve $\theta = $ [constant] $= \pi/2$, what will be the components of the transported vector?
(c) Repeat the calculation for a vector $b^\mu = (1, 0)$ that is parallel transported from $x^\mu = (0, 0)$ to $x^\mu = (\pi/2, 0)$ along the curve $\theta = $ [constant] $= 0$.

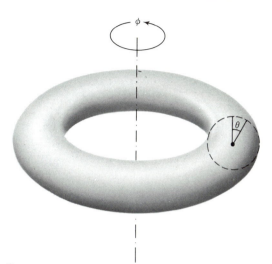

Fig. 6.16 Coordinates on a torus.

6. Suppose that the spacetime geometry is as given in Eq. [68], with the Christoffel symbols as given by Eq. [70]. Consider a contravariant vector A^μ which, in the coordinates $x^0 = t$, $x^1 = r$, $x^2 = \theta$, $x^3 = \phi$, is constant as a function of position, $A^\mu(x) = (1, 0, 0, 0)$. Evaluate the covariant derivative $A^\mu{}_{;\nu}$.

7. The surface of a cone, of half angle β at the apex, is a two-dimensional "curved" surface. A point on this surface can be described by two coordinates: $x^1 = r$ is the distance measured from the apex to the point, and $x^2 = \phi$ is the azimuthal angle measured around the z-axis (see Fig. 6.17). In terms of these coordinates, the space interval has the form $dl^2 = g_{\mu\nu} dx^\mu dx^\nu$, with $\mu, \nu = 1, 2$.
(a) What are the components of the metric tensor $g_{\mu\nu}$?
(b) Find all components of the Riemann tensor $R^\alpha{}_{\beta\mu\nu}$.
(c) Suppose that you parallel transport a vector once around the cone along the circle $r = $ [constant]. What angle will the parallel-transported vector make with the original vector?
(d) Is the geometry of this space the same as that of a flat two-dimensional space? Is the topology the same?

8. The equation $z = (x)^2 + (y)^2$ defines a paraboloid of revolution. The surface of this paraboloid is a two-dimensional curved space, whose points can be described by the coordinates $x^1 = [(x)^2 + (y^2)]^{1/2}$ and $x^2 = \phi$ (where ϕ is the usual azimuth angle measured around the z-axis). In terms of these coordinates, the space interval has the form $dl^2 = g_{\mu\nu} dx^\mu dx^\nu$, with $\mu, \nu = 1, 2$.
(a) What are the components of the metric tensor $g_{\mu\nu}$?
(b) Find all the Christoffel symbols $\Gamma^\mu_{\alpha\beta}$.
(c) Find all the components of the Riemann tensor.
(d) Show that the curve $\phi = $ [constant] is a geodesic.

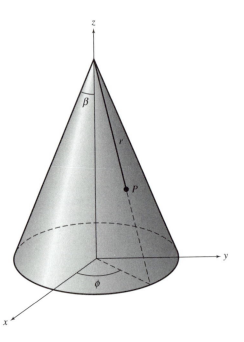

Fig. 6.17 Coordinates on a cone.

9. Suppose that the metric of a spacetime is

$$ds^2 = (x^0)^4(dx^0)^2 - 2e^{x^1}(dx^1)^2 - e^{-x^2}(dx^2)^2 - (dx^3)^2$$

Prove that all components of $R^\alpha_{\beta\mu\nu}$ are zero.

10. A curved spacetime has a spacetime interval

$$ds^2 = A(r)dt^2 - B(r)dr^2 - C(r)(r^2 d\theta^2 + r^2 \sin^2\theta \, d\phi^2)$$

where r, θ, and ϕ are spherical coordinates, and where $A(r)$, $B(r)$, and $C(r)$ are given functions of r. Show that
(a) The circumference of the circle $r = r_1$ is

$$2\pi \sqrt{C(r_1)} r_1$$

(b) The area of the sphere $r = r_1$ is

$$4\pi r_1^2 C(r_1)$$

(c) The distance between the points $r = r_1$ and $r = r_2$ on a given radial line is

$$\int_{r_1}^{r_2} \sqrt{B(r)}\ dr$$

(d) The volume of the spherical shell $r_1 < r < r_2$ is

$$4\pi \int_{r_1}^{r_2} r^2 C(r)\sqrt{B(r)}\ dr$$

11. The spacetime interval in the interior of a uniform spherical mass distribution of mass M and radius R is, in the linear approximation,

$$ds^2 = (1 - f)dt^2 - (1 + f)dr^2 - r^2 d\theta^2 - r^2 \sin^2\theta\ d\phi^2$$

where

$$f = -\frac{Gmr^2}{R^3} + \frac{3GM}{R}$$

Find a formula for the volume of the sphere $0 < r < R$. Assume that $f \ll 1$, so $\sqrt{1+f} \simeq 1 + f/2$.

12. The Schwarzschild geometry has a spacetime interval

$$ds^2 = \left(1 - \frac{2GM}{r}\right)dt^2 - \frac{dr^2}{1 - 2GM/r} - r^2 d\theta^2 - r^2 \sin^2\theta\ d\phi^2$$

Suppose that a contravariant vector initially has components $(A^0, A^1, A^2, A^3) = (1, 1, 1, 1)$ and is located at the point $r = 3GM$, $\theta = \pi/2$, $\phi = 0$ in this Schwarzschild geometry.
(a) If you parallel transport this vector instantaneously $(dt = 0)$ along a radial line to $r = 6GM$, what will be the components of the parallel-transported vector?
(b) If you parallel transport this vector at a finite speed $(dt \neq 0)$, will the result be different? Explain.

13. You parallel transport a contravariant vector A^μ around the circle $r = 6GM$, $\theta = \pi/2$ in the Schwarzschild geometry (see the preceding problem). If the vector initially has components $A^\mu = (0, 1, 0, 0)$, what will be its components after it has been transported around the circle? Assume that the transport is instantaneous $(dt = 0)$.

14. The Robertson-Walker geometry (to be discussed in Chapter 9) has a spacetime interval

$$ds^2 = dt^2 - a(t)^2[d\chi^2 + \sin^2\chi(d\theta^2 + r^2\sin^2\theta\, d\phi^2)]$$

where the coordinates are $x^0 = t, x^1 = \chi, x^2 = \theta$, and $x^3 = \phi$, and $a(t)$ is a function of time.

(a) Calculate all the Christoffel symbols $\Gamma^k{}_{lm}$ with purely spatial indices, that is, with indices from 1 to 3.

(b) A covariant vector with components $b_\mu = (0, 1, 0, 0)$ is initially located at $\chi =$ [constant] in the equatorial plane ($\theta = \pi/2$). We parallel transport this vector instantaneously ($dt = 0$) once around a closed circle in the equatorial plane ($\chi =$ [constant], $\theta = \pi/2$, $\phi = 0$ to $\phi = 2\pi$). Calculate the final vector b_μ that results from this parallel transport.

15. Eq. [19] gives the change in a covariant vector under parallel transport. Find the corresponding result for the parallel transport of a covariatn tensor of rank two.

16. Show that if the metric tensor $g_{\mu\nu}$ is diagonal, then $\Gamma^\mu{}_{\alpha\beta} = 0$ whenever μ, α, and β are all different.

17. In what follows we will use the notation g (or $|g_{\mu\nu}|$) for the determinant of the matrix $g_{\mu\nu}$.

(a) Show that the transformation law for this determinant is

$$g' = \left|\frac{\partial x^\alpha}{\partial x'^\beta}\right|^2 g$$

where $\left|\partial x^\alpha/\partial x'^\beta\right|$, called the *Jacobian*, is the determinant of the matrix $\partial x^\alpha/\partial x'^\beta$. [Hint: Take the determinant of the equation $g'_{\alpha\beta} = (\partial x^\mu/\partial x'^\alpha)(\partial x^\nu/\partial x'^\beta)g_{\mu\nu}$.]

(b) Suppose that Φ is a scalar field. Show that the integral

$$\int \Phi \sqrt{-g}\, dx^0 dx^1 dx^2 dx^3$$

evaluated over a given (four-dimensional) region of spacetime is a scalar.

18. Show that

$$\Gamma^\alpha{}_{\mu\alpha} = \frac{1}{\sqrt{-g}}\frac{\partial}{\partial x^\mu}\sqrt{-g}$$

where g is the determinant of the matrix $g_{\mu\nu}$. (Hint: Differentiation of the determinant gives $\partial g/\partial x^\mu = gg^{\alpha\beta}\partial g_{\alpha\beta}/\partial x^\mu$.)

19. Use the result of Problem 18 to show that the divergence of a vector field can be written as

$$A^\mu_{;\mu} = \frac{1}{\sqrt{-g}} \frac{\partial}{\partial x^\mu} (\sqrt{-g}\, A^\mu)$$

Hence, show that for a scalar field

$$\Phi^{,\mu}_{;\mu} = \frac{1}{\sqrt{-g}} \frac{\partial}{\partial x^\mu} (\sqrt{-g}\, g^{\mu\nu} \frac{\partial}{\partial x^\nu} \Phi)$$

20. Use the result of Problem 18 to show that if $A^{\alpha\beta}$ is an antisymmetric tensor field, then

$$A^{\alpha\beta}_{;\beta} = \frac{1}{\sqrt{-g}} \frac{\partial}{\partial x^\beta} (\sqrt{-g}\, A^{\alpha\beta})$$

21. Starting with the Christoffel symbols for flat spacetime in spherical coordinates, show that if Φ is a scalar field, then

$$\Phi^{,\mu}_{;\mu} = \frac{\partial^2 \Phi}{\partial t^2} - \frac{1}{r}\frac{\partial^2}{\partial r^2} r\Phi - \frac{1}{r^2 \sin\theta}\frac{\partial}{\partial \theta}\sin\theta\frac{\partial \Phi}{\partial \theta} - \frac{1}{r^2 \sin^2\theta}\frac{\partial^2 \Phi}{\partial \phi^2}$$

22. Contract Eq. [98] on α and σ, and extract the identity [103] from this result.

23. Calculate all the components of $R^\alpha_{\beta\mu\nu}$ for the metric corresponding to the gravitational wave given in Eq. [5.24]. Then calculate the relative acceleration between two particles instantaneously at rest separated by a displacement s^σ, according to the equation of geodesic deviation. Compare your result with the acceleration we found from the linear approximation in Chapter 5.

24. Show that for a tensor $A_{\alpha\beta}$,

$$A_{\alpha\beta;\mu;\nu} - A_{\alpha\beta;\nu;\mu} = R^\sigma_{\beta\mu\nu} A_{\alpha\sigma} + R^\sigma_{\alpha\mu\nu} A_{\sigma\beta}$$

25. Derive the following expression for the curvature scalar:

$$\sqrt{-g}\, R = \frac{\partial}{\partial x^\alpha}\left[g^{\mu\nu}\sqrt{-g}\,(-\Gamma^\alpha_{\mu\nu} + \delta^\alpha_\mu \Gamma^\beta_{\nu\beta})\right]$$

$$- g^{\mu\nu}\sqrt{-g}\,(\Gamma^\beta_{\mu\alpha}\Gamma^\alpha_{\nu\beta} - \Gamma^\alpha_{\mu\nu}\Gamma^\beta_{\alpha\beta})$$

26. Show that if $B^{\mu\nu}$ is an antisymmetric tensor field, then

$$B_{\mu\nu;\alpha} + B_{\nu\alpha;\mu} + B_{\alpha\mu;\nu} = B_{\mu\nu,\alpha} + B_{\nu\alpha,\mu} + B_{\alpha\mu,\nu}$$

and

$$B^{\mu\nu}{}_{;\mu;\nu} = 0$$

27. Show that the set of all coordinate transformations forms a group (the manifold mapping group), and show that scalars, vectors, and tensors form representations of this group.

28. Evaluate $R^k{}_{0l0}$ for the gravitational wave of the type given by Eq. [5.24]. Then use Eq. [123] to calculate the tidal force and check that your result agrees with Eqs. [5.41] and [5.42].

29. The Lie derivative of a contravariant vector χ^μ is

$$\mathscr{L}_\xi \chi^\mu = \lim_{\epsilon \to 0} \frac{\chi^\mu(x) - \chi'^\mu(x)}{\epsilon}$$

(a) Show that

$$\mathscr{L}_\xi \chi^\mu = \xi^\alpha \frac{\partial \chi^\mu}{\partial x^\alpha} - \frac{\partial \xi^\mu}{\partial x^\alpha} \chi^\alpha$$

and verify that this has the alternative form

$$\mathscr{L}_\xi \chi^\mu = \xi^\alpha \chi^\mu{}_{;\alpha} - \xi^\mu{}_{;\alpha} \chi^\alpha$$

(b) Show that $\mathscr{L}_\xi \chi^\mu = - \mathscr{L}_\chi \xi^\mu$.

30. The Lie derivative of a contravariant tensor $A^{\mu\nu}$ is

$$\mathscr{L}_\xi A^{\mu\nu} = \lim_{\epsilon \to 0} \frac{A^{\mu\nu}(x) - A'^{\mu\nu}(x)}{\epsilon}$$

(a) Show that

$$\mathscr{L}_\xi A^{\mu\nu} = \xi^\alpha \frac{\partial A^{\mu\nu}}{\partial x^\alpha} - \frac{\partial \xi^\mu}{\partial x^\alpha} A^{\alpha\nu} - \frac{\partial \xi^\nu}{\partial x^\alpha} A^{\mu\alpha}$$

and verify that this has the alternative form

$$\mathscr{L}_\xi A^{\mu\nu} = \xi^\alpha A^{\mu\nu}{}_{;\alpha} - \xi^\mu{}_{;\alpha} A^{\alpha\nu} - \xi^\nu{}_{;\alpha} A^{\mu\alpha}$$

(b) Using the results stated in the preceding problem, show that for two contravariant vectors A^μ and B^ν,

$$\mathscr{L}_\xi A^\mu B^\nu = A^\mu \mathscr{L}_\xi B^\nu + B^\nu \mathscr{L}_\xi A^\mu$$

7. Einstein's Gravitational Theory

In the preceding chapter we dealt with the kinematics of the curved spacetime geometry, that is, the *description* of the geometry and its curvature. We now come to the dynamics of the geometry, that is, the *interaction* of the geometry and matter. This interaction is the content of Einstein's equations for the gravitational field.

There are several routes that lead to Einstein's equations; they differ in their starting points. We will take as our starting point the equations of the linear approximation of Chapter 3. In addition, we will assume that the equations are of second differential order and possess general invariance. These assumptions are sufficient to determine the equations for the gravitational field completely.

That the linear equations imply the full nonlinear equations is a quite remarkable feature of Einstein's theory. Given some complicated set of nonlinear equations, it is always easy to derive the corresponding linear approximation; but, in general, if we know only the linear approximation, we cannot reconstruct the nonlinear equations. What permits us to perform this feat in gravitational theory is the requirement of general invariance. This requirement states that the equations must remain unchanged under all transformations of the coordinates.

Regarded naively, this seems to be a simple generalization of the requirement of Lorentz invariance, which states that the equations must remain unchanged under all Lorentz transformations. In his original work, Einstein sought to justify general invariance by appeal to a general principle of equivalence for reference frames in arbitrary (accelerated) motion. This was intended to be a generalization of the principle of special relativity, which asserts the equivalence of reference frames in uniform motion. Thus, Einstein's theory was intended to be a theory of "general relativity." But in fact, the theory of special relativity is already as relativistic as can be--it has the maximum amount of spacetime symmetry. The principle of general invariance is not a relativity principle; rather, the principle of general invariance is a dynamical principle that imposes restrictions on the possible interactions of geometry and matter.

Einstein also sought to justify general invariance through Mach's principle. The latter expresses the intriguing, if somewhat vague, notion that the inertia of bodies is somehow affected by the mass distribution of the universe. Although Einstein's theory does display some dependence of inertia on mass distribution, Mach's principle actually plays only a marginal role in the theory.

Since Einstein's theory of "general relativity" is no more relativistic than special relativity, it would be preferable to adopt the name *geometrodynamics* for this theory. This name, coined by Wheeler, puts the emphasis where it belongs--on the dynamical geometry that acts on and reacts to matter.

7.1 GENERAL COVARIANCE AND INVARIANCE

We begin by introducing the important concept of *covariance* of an equation*:

An equation is said to be covariant under general coordinate transformations if the form of the equation is left unchanged by the transformations.

Obviously, any equation whose right side and left side are tensors of the same type is covariant. For example, the equation

$$A^\mu = B^\mu \qquad [1]$$

where A^μ and B^μ are contravariant tensors of rank one, is covariant. A general coordinate transformation changes Eq. [1] into

---

* We have "covariant" vectors, "covariant" derivatives, and now "covariant" equations. The word "covariant" is unfortunately much overworked, and which of the three meanings is intended must be guessed from context.

$$A'^\mu = B'^\mu \qquad [2]$$

that is, the *form* of the equation is unchanged. In contrast, the equation

$$A^\mu = B_\mu \qquad [3]$$

is *not* covariant. If we try to express this equation in terms of the transformed vectors A'^μ and B'_μ, it reads

$$\frac{\partial x^\mu}{\partial x'^\alpha} A'^\alpha = \frac{\partial x'^\alpha}{\partial x^\mu} B'_\alpha \qquad [4]$$

This equation does not have the same form as Eq. [3]. Of course, we can rearrange Eq. [4] in several ways; but no matter what we do to it, we cannot get rid of the transformation coefficients $\partial x/\partial x'$ and $\partial x'/\partial x$. (However, in the exceptional case $B_\mu \equiv 0$, Eq. [3] becomes $A^\mu = 0$, which is of course covariant.)

Note that although Eq. [3] is not covariant, Eq. [4] *is* covariant. To see this, let us perform a coordinate transformation to coordinates x''^μ, and express A'^μ and B'_μ in terms of A''^μ and B''_μ:

$$\frac{\partial x^\mu}{\partial x'^\alpha} \left(\frac{\partial x'^\alpha}{\partial x''^\beta} A''^\beta \right) = \frac{\partial x'^\alpha}{\partial x^\mu} \left(\frac{\partial x''^\beta}{\partial x'^\alpha} B''_\beta \right) \qquad [5]$$

By the chain rule for derivatives, this reduces to

$$\frac{\partial x^\mu}{\partial x''^\beta} A''^\beta = \frac{\partial x''^\beta}{\partial x^\mu} B''_\beta \qquad [6]$$

which has exactly the same form as Eq. [4]. This example shows that a noncovariant equation, such as Eq. [3], can be changed into an equivalent covariant equation. Since covariance is a statement about the *form* of an equation, it is not surprising that it depends on the form in which the equation is written. In essence, we have achieved covariance for Eq. [4] by introducing the extra functions $x^\mu(x')$ into our equation; these functions play the role of extra fields in the equation. The presence of the functions $x^\mu(x')$ or $x^\mu(x'')$ in Eq. [4] or [6] reveals that the coordinates x^μ have a special significance in this equation; but this presence of special coordinates does not contradict the covariance of these equations.

We now lay down a commandment concerning the form of the equations of physics:

Principle of general covariance: All laws of physics shall be stated as equations covariant with respect to general coordinate transformations.

We will obey this commandment for the best of all reasons--it costs us nothing to do so. As we have seen, an equation that is not covariant can easily be transformed into an equivalent equation that is covariant. From a mathematical point of view, the covariance principle is therefore seen to be a triviality.

That the principle of covariance imposes no restrictions on the content of the physical laws, but only on the form in which they are written, was recognized by Kretschmann (1917) in a critical examination of Einstein's theory. The principle in no way implies that the laws must be "relativistic"; even Newton's equations can be written in generally covariant form. In reply to this criticism, Einstein wrote:

> ... Even though it is true that one must be able to bring every empirical law into generally covariant form, yet the Principle has considerable heuristic force, which has proved itself in the problem of gravitation and which relies on the following. Of two theoretical systems which agree with experience, that one is to be preferred, which from the point of view of the absolute differential calculus is the simplest and most transparent. One need only try to bring Newtonian mechanics and gravitation into the form of ... covariant equations (four-dimensional) and one will surely be convinced that the Principle rules out this theory practically if not theoretically! (Einstein, 1918)

We would like to restate Einstein's somewhat vague criterion for "simplest and most transparent" equations in a rigorous way. We can do this by appealing to the concept of *invariance* of the equations. Invariance goes beyond covariance in that it demands that not only the form, but also the content of the equations be left unchanged by the coordinate transformation. Roughly, invariance is achieved by imposing the extra condition that all "constants" in the equation remain exactly the same. For example, covariance of the Maxwell equations under Lorentz transformations requires only that the equations have the same *appearance* in all Lorentz frames; it does not require that the speed of light, which enters these equations as a constant of proportionality, be the same in all reference frames. However, invariance of the Maxwell equations does require that the speed of light be the same.

In this context, by "constants" we mean not only those things that are true numerical constants, but anything that is independent of the state of matter. A better terminology is to call such things *absolute objects*. Thus, an absolute object is a quantity, with one or more components, that is independent of the state of matter. The value of such an absolute object is completely unaffected by any changes in the condition of the matter in the universe. For example, c and $\hbar$ are absolute objects; so are $\frac{1}{4}$, $\eta_{\mu\nu}$, $\partial x^\mu/\partial x'^\alpha$ (where x is some given function of x'), etc. In contrast, *dynamical objects* are physical variables such as particle positions and momenta, field strengths, energy densities, etc., that do depend on the state of matter. Given any law of physics, we can then classify the quantities that appear in it into absolute objects and dynamical objects (Trautman, 1966; Anderson, 1967).

We can now give a somewhat more precise definition of invariance:

An equation is said to be invariant under a coordinate transformation if (i) it is covariant and (ii) all absolute objects appearing in it are left unchanged.

Note that this definition tells us that the equation [4] is not invariant. This equation is covariant, but it contains the absolute object $\partial x^\mu/\partial x'^\alpha$, and a coordinate transformation changes this object into $\partial x^\mu/\partial x''^\alpha$, in violation of item (ii).

The requirement that an equation be invariant under some transformation imposes serious restrictions on the equation. Thus, invariance under Lorentz transformations ("principle of special relativity") is a symmetry requirement which puts tight limitations on acceptable equations. We have made use of these limitations in the construction of our linear gravitational theory. Note that the Lorentz symmetry requirement determines what possible absolute objects can appear in the equations: the admissible absolute objects must remain unchanged under Lorentz transformations. The only acceptable objects are therefore scalars (for instance, c, $\hbar$, e, m) and the tensor $\eta^{\mu\nu}$.* Physically, the Lorentz symmetry requirement of special relativity means that observers cannot distinguish between different inertial frames.

The gravitational theory of Einstein is based on the following symmetry principle:

Principle of general invariance: All laws of physics must be invariant under general coordinate transformations.

This postulate demands that the laws have a higher symmetry, a symmetry that goes beyond Lorentz symmetry. One immediate consequence of the general symmetry principle is that we must regard the metric of spacetime as dependent on the state of matter; the metric tensor must be a dynamical field, satisfying some field equation of its own. The reason is obvious: the general coordinate transformations do not leave $g_{\mu\nu}$ unchanged; hence the presence of $g_{\mu\nu}$ as an absolute object contradicts the requirement that all such objects keep their values unchanged under coordinate transformations. The only alternative is to suppose that $g_{\mu\nu}$ is not an absolute object, but rather a dynamical variable.

Since we have taken some pains to make it clear that covariance is trivial (and therefore unavoidable, if the equations are suitably rewritten), but that invariance is nontrivial (and therefore avoidable),

---

* Tensors such as $\eta_{\mu\nu}\eta_{\alpha\beta}$, constructed from $\eta_{\mu\nu}$, are, of course, also acceptable.

it will be well to give some arguments in favor of invariance.

One argument is based on our belief in the simplicity of nature, a belief expressed by Newton in his First Rule for Reasoning in Philosophy, mentioned in Chapter 1. Mathematically, we may interpret a requirement for simplicity as a requirement for symmetry. Newton's First Rule then tells us that in our theoretical formulation of the laws of nature we should maximize the symmetry subject to the constraints imposed by the available experimental evidence. A symmetry should be considered "broken" only if there are experiments that bear witness against the symmetry. Since general invariance is not in conflict with any available experimental data, we feel strongly compelled to accept this symmetry.

Another argument, which does not rely quite as much on faith in the simplicity of nature, is the following: We have seen that the field equation of the linear theory is invariant under the gauge transformation (see Eq. [3.49])*

$$h^{\mu\nu}(x) \rightarrow h'^{\mu\nu}(x) = h^{\mu\nu}(x) + \partial^{\nu}\Lambda^{\mu} + \partial^{\mu}\Lambda^{\nu} \tag{7}$$

Let us compare this gauge transformation with a coordinate transformation. Consider the infinitesimal coordinate transformation

$$x'^{\mu} = x^{\mu} + \epsilon\xi^{\mu}(x) \quad \text{with } \epsilon \rightarrow 0 \tag{8}$$

where $\xi^{\mu}(x)$ is a function of position. We then obtain**

$$\frac{\partial x'^{\mu}}{\partial x^{\alpha}} = \delta^{\mu}_{\alpha} + \epsilon\xi^{\mu}_{,\alpha} \tag{9}$$

and

$$g'^{\mu\nu}(x') = \frac{\partial x'^{\mu}}{\partial x^{\alpha}}\frac{\partial x'^{\nu}}{\partial x^{\beta}} g^{\alpha\beta}(x) \tag{10}$$

$$= (\delta^{\mu}_{\alpha} + \epsilon\xi^{\mu}_{,\alpha})(\delta^{\nu}_{\beta} + \epsilon\xi^{\nu}_{,\beta})g^{\alpha\beta}(x)$$

$$\simeq g^{\mu\nu}(x) + g^{\alpha\nu}(x)\epsilon\xi^{\mu}_{,\alpha} + g^{\mu\beta}(x)\epsilon\xi^{\nu}_{,\beta} \tag{11}$$

In this equation, as in all tensor transformation equations, it is assumed that the x appearing in $g^{\mu\nu}(x)$ and in $\xi^{\mu}(x)$ on the right side is expressed as a function of x', so the right side yields a function of x' (in accord with the functional dependence indicated on the left

---

* In this equation indices are raised with the metric $\eta_{\mu\nu}$.
** This calculation is similar to that in Section 6.8. However, there we examined the change in $g_{\mu\nu}$; here, we find it more convenient to examine $g^{\mu\nu}$.

side). Alternatively, we can express the x' appearing in $g'^{\mu\nu}(x')$ on the left side as a function of x, and leave the right side as a function of x. This is advantageous if we want to see just how much the new function $g'^{\mu\nu}$ differs from the old function $g^{\mu\nu}$. For this purpose, we use the Taylor-series expansion

$$g'^{\mu\nu}(x'(x)) \simeq g'^{\mu\nu}(x) + g^{\mu\nu}{}_{,\alpha}(x)\epsilon\xi^\alpha \qquad [12]$$

which leads to

$$g'^{\mu\nu}(x) = g^{\mu\nu}(x) - g^{\mu\nu}{}_{,\alpha}(x)\epsilon\xi^\alpha + g^{\alpha\nu}\epsilon\xi^\mu{}_{,\alpha} + g^{\mu\beta}(x)\epsilon\xi^\nu{}_{,\beta} \qquad [13]$$

If $\kappa h^{\mu\nu}$ is small, the relation between the inverse metric and $h_{\mu\nu}$ is

$$g^{\mu\nu}(x) = \eta^{\mu\nu}(x) - \kappa h^{\mu\nu}(x) \qquad [14]$$

where the indices on the right side are raised with the Minkowski metric $\eta^{\mu\nu}$ (this relation can be verified by multiplying each side into the usual relation $g_{\nu\alpha} = \eta_{\nu\alpha} + \kappa h_{\nu\alpha}$; except for a negligible term of order κ^2, the result is an identity). Substituting such expressions for $g^{\mu\nu}$ and $g'^{\mu\nu}$ into Eq. [13], we obtain, except for negligible terms of order $\epsilon\kappa$,

$$- \kappa h'^{\mu\nu}(x) = - \kappa h^{\mu\nu}(x) + \epsilon\xi^{\mu,\nu} + \epsilon\xi^{\nu,\mu} \qquad [15]$$

Comparing this equation with Eq. [7], we see that the equations coincide provided we identify

$$\epsilon\xi^\mu(x) = - \kappa\Lambda^\mu(x) \qquad [16]$$

We therefore have shown that the gauge transformation [7] is nothing but an *infinitesimal coordinate transformation*, and the invariance of the linear theory under such gauge transformations is really an invariance under infinitesimal coordinate transformations. This makes it quite natural to suppose that the general nonlinear theory is invariant under general, finite coordinate transformations. Thus, we can regard finite coordinate transformations as generalized gauge transformations. From this point of view, the symmetry expressed by the principle of general invariance would be described as a gauge symmetry.

As a final argument in favor of general invariance, we note that this symmetry plays a crucial role in the general proof of the equality of inertial and gravitational mass of an arbitrary system. Here, as in Section 1.5, the gravitational mass is defined by the gravitational field that the system produces at large distances (where the gravitational field asymptotically becomes Newtonian); and the inertial mass is defined by the total energy of the system, including the energy in the gravitational field, calculated from the $_{00}$ component of the total energy-momentum tensor. We will not present the details, but it seems that without general invariance, the theoretical proof of the equality $m_I = m_G$ becomes impossible (Ohanian, 1973).

The physical interpretation of the principle of special relativity is that velocity is relative; no experiment can detect any intrinsic difference between reference frames in uniform motion with different velocities. It is tempting to give the principle of general invariance the physical interpretation that acceleration is also relative. Einstein named his theory of gravitation the theory of general relativity because he thought that (locally) the phenomena observed in a gravitational field are indistinguishable from those observed in an accelerated system of reference and that

> According to this conception one cannot speak of the *absolute acceleration* of a system of reference, just as in the ordinary theory of relativity one cannot speak of the *absolute velocity* of a system. (Einstein, 1911)

However, we saw in Section 1.9 that the tidal effects allow us to make an absolute distinction between the gravitational forces and the pseudo-forces found in accelerated reference frames. It is therefore false to speak of a general relativity of motion.

The principle of special relativity is a relativity principle, whereas the principle of general invariance is not. The lack of relativity in Einstein's theory can also be understood in geometrical terms. In flat spacetime it is meaningless to speak of absolute positions or velocities because the geometry has no identifiable features--it is exactly the same everywhere. On the other hand, in a Riemannian spacetime the features of the geometry can be used to determine positions. Astronauts can describe their location and velocity by reference to nearby bumps in the geometry. For example, astronauts in orbit near the Earth can take advantage of the lack of uniformity of the Earth's gravitational field to determine their position. Mathematically, the deviations of the geometry from uniformity are described by the Riemann tensor. Hence, we can say that the Riemann tensor makes an absolute distinction, and destroys the relativity, between different (small) reference frames in free fall. This geometrical point of view is closely related to tidal forces because, as we saw in Section 6.6, the Riemann tensor is essentially the tidal-force tensor.

Although the above arguments make it clear that Einstein's theory is not endowed with some kind of "general relativity," the theory is endowed with restricted, or special, relativity. Obviously, the principle of general invariance includes the principle of special relativity (Lorentz transformations are a special case of general coordinate transformations). Hence Einstein's theory is as "relativistic" as special relativity. In the context of the above example this means that astronauts can use the features of the geometry to measure their velocity relative to the Earth, but they cannot measure any absolute velocity for the Earth, because the Earth, or, more precisely, the solar system, is placed in a background geometry that is asymptotically flat (in this context, we ignore the gravitational field produced by the galaxy and by the average mass distribution of the universe).

The principles of special relativity and general invariance are different in kind because the symmetries involved in these principles are different in kind. In the terminology of field theory: Lorentz invariance is an *algebraic* symmetry, general invariance is a *dynamic* symmetry (Weinberg, 1970). The meaning of this is that Lorentz invariance tells us what algebraic combinations of components of (Lorentz) tensors can appear in the equations of physics; it places restrictions on *all* of the interactions that can occur in physics. General invariance tells us how the gravitational field can appear in the equations; it places restrictions only on the gravitational interactions, and it helps to determine the possible dynamical couplings between gravitation and matter.* The fundamental difference between these symmetries is related to the fact that Lorentz transformations perform the same operation at all points of spacetime (the Lorentz transformation coefficients are constants), whereas general coordinate transformations perform different operations at different points (the transformation coefficients are functions of spacetime). In the terminology of field theory, this is expressed by saying that the former are gauge transformations of the first kind, whereas the latter are gauge transformations of the second kind (Utiyama, 1956). The difference in the character of these symmetries is further illustrated by the fact that conservation laws for energy, momentum, and angular momentum may be derived from the Lorentz symmetry (Schweber, 1961), whereas no such conservation laws follow from general invariance. Instead, general invariance leads to differential identities among the field quantities, identities that hold true independently of the dynamical equations. The Bianchi identity, Eq. [6.103], is an instance of such an identity; it holds true whether or not the metric tensor satisfies Einstein's equation, and it can be derived directly from the requirement of general invariance (Soper, 1976; Anderson, 1967; Trautman, 1966).

As an example of the help offered by the principle of general invariance in the search for the couplings between gravitation and matter, consider the problem of finding the equations that describe the effect of gravitation on electromagnetism. In flat spacetime, the Maxwell equations (see Eqs. [2.106], [2.107]) are

$$F^{\mu\nu}{}_{,\mu} = 4\pi j^\nu \qquad\qquad [17]$$

---

* For the purposes of this statement we are concentrating on those aspects of general invariance that go beyond Lorentz invariance, that is, we ignore that part of general invariance that overlaps with Lorentz invariance.

$$F^{\mu\nu}{}_{,\alpha} + F^{\nu\alpha}{}_{,\mu} + F^{\alpha\mu}{}_{,\nu} = 0 \qquad [18]$$

where, as always, the commas indicate ordinary derivatives. Obviously, these equations are not invariant under general coordinate transformations. The left sides are not tensors--we know that ordinary derivatives of tensors do not give tensors. In order to make the left sides into tensors, we must replace the ordinary derivatives by covariant derivatives:

$$F^{\mu\nu}{}_{;\mu} = 4\pi j^{\nu} \qquad [19]$$

$$F^{\mu\nu;\alpha} + F^{\nu\alpha;\mu} + F^{\alpha\mu;\nu} = 0 \qquad [20]$$

These equations satisfy the principle of general invariance. The gravitational field variables appear in these equations through the Christoffel symbols that are contained in the covariant derivatives. The gravitational field therefore affects the electromagnetic field; the light deflection given in Chapter 4 can be obtained by a study of the wave solutions of Eqs. [19] and [20].

Unfortunately, although the principle of general invariance places considerable restrictions on the possible interactions of gravitation and matter, it does not determine these interactions uniquely. For example, Eq. [19] remains entirely consistent with general invariance if we add an extra term $(F^{\mu\nu}R)_{;\mu}$ to the left side.* To rule out such extra terms, it is customary to appeal to the *principle of minimal coupling*. This principle asserts that the equations of motion of matter in the presence of gravitation are to be obtained from those that hold in the absence of gravitation (equations of special relativity) by replacing $\eta^{\mu\nu}$ by $g^{\mu\nu}$ and ordinary derivatives by covariant derivatives; *no other changes are to be made*. Since in local geodesic coordinates, the covariant derivative reduces to the ordinary derivative, we can also express this principle as follows: in local geodesic coordinates the equations of motion are those familiar from special relativity.

This minimal coupling principle is not a symmetry principle, and there is not much of a justification for it, except that it keeps the equations as simple as possible. In fact, this principle is not really a principle ("general law"), because it cannot be applied to all the equations of physics. For example, if we apply the minimal coupling principle to the equation of motion of the spin of a rigid body, it would tell us that the equation of motion of the spin in local geodesic coordinates is $dS/d\tau = 0$, since this is the equation that the spin satisfies in the absence of gravitation. But this is the wrong equation of motion for the spin. We know from Section 1.9 that tidal torques are present

---

* This term satisfies the identity $(F^{\mu\nu}R)_{;\mu;\nu} = 0$, which is important for the consistency of Eq. [19].

in the true equation (Eq. [1.68]), and that the motion of the spin depends on the Riemann tensor (the connection between the tidal torque and the Riemann tensor will be further discussed in Section 7.8). Thus the minimal coupling principle is an unreliable and dangerous thing. It may be used to determine the form that laws take in the presence of gravitation only as a last resort, when we are sure that these laws cannot be determined by other means. Furthermore, it is advisable to keep in mind the precedent set by nongravitational physics. It is well known that a minimal coupling principle can be formulated for electrodynamic interactions, and that this principle is *violated* in nature (by the existence of the anomalous magnetic moments of protons and neutrons). Hence, we had better be very skeptical of the analogous principle in geometrodynamics.

7.2 EINSTEIN'S FIELD EQUATION

In our discussion of the linear theory in Chapter 3, we obtained an approximate equation for the gravitational field. Expressed in terms of the metric tensor $g_{\mu\nu} = \eta_{\mu\nu} + \kappa h_{\mu\nu}$, this equation takes the form (see Eq. [3.48]):

$$\partial^\lambda \partial_\lambda g_{\mu\nu} - \partial^\lambda \partial_\nu g_{\mu\lambda} - \partial^\lambda \partial_\mu g_{\nu\lambda} + \partial_\mu \partial_\nu g_\sigma{}^\sigma$$
$$- \eta_{\mu\nu} \partial_\lambda \partial^\lambda g_\sigma{}^\sigma + \eta_{\mu\nu} \partial^\lambda \partial^\sigma g_{\lambda\sigma} = - \kappa^2 T_{\mu\nu} \quad [21]$$

In this equation all indices are raised with the Minkowski metric $\eta^{\mu\nu}$; for example, $g_\sigma{}^\sigma \equiv \eta^{\sigma\tau} g_{\sigma\tau}$.*

As we already remarked in Chapter 3, the above equation suffers from several defects because it fails to take into account that the gravitational field gravitates. We will now derive the exact, nonlinear field equation. One possible derivation relies on a method of successive approximations to discover the exact energy–momentum tensor of the gravitational field (Gupta, 1957; Deser, 1970); the first step of this method was presented in Section 3.3. However, we will now use a different and more elegant method which relies heavily on general invariance.

To be precise, the exact, nonlinear field equation satisfied by the metric tensor $g_{\mu\nu}$ can be derived from the following assumptions:
(*i*) The equation is invariant under general coordinate transformations.
(*ii*) The equation reduces to Eq. [21] in the linear approximation.
(*iii*) The equation is of second differential order, and is linear in second derivatives.

---

* We adopt the general rule that in any equation written in the linear approximation, indices are raised (or lowered) with $\eta^{\mu\nu}$; in any other equation in this and succeeding chapters, indices are raised with $g^{\mu\nu}$.

The first two assumptions require no further comment. The third assumption expresses a prejudice of physicists in favor of equations of the general form

$$F(\psi, \partial\psi)\partial^2\psi + G(\psi, \partial\psi) = 0 \qquad [22]$$

where F and G are functions of the field ψ and of the *first* derivative $\partial\psi$, but not of the second derivative $\partial^2\psi$. This form is suggested by our experience with the dynamical equations of Lagrangian mechanics. Note that the second derivative enters linearly in Eq. [22]. Nonlinear equations of this special type are called *quasilinear*. Mathematical theorems on the uniqueness and existence of solutions, wave propagation, initial-value problem, and so on, can be proved for such quasilinear equations; very little is known about the solution of more general nonlinear equations.

To derive the gravitational field equation, we begin by asking what form the equation takes in local geodesic coordinates at some point. Since in these coordinates $g'_{\mu\nu} = 0$ and $g'_{\mu\nu,\alpha} = 0$, it is clear that the differential equation must be some expression involving *only* the second derivatives $g'_{\mu\nu,\alpha,\beta}$. By assumption (*iii*), this implies that the equation is necessarily linear. But assumption (*ii*) tells us that if the equation is linear it must have the form

$$\partial'^\lambda\partial'_\lambda g'_{\mu\nu} - \partial'^\lambda\partial'_\nu g'_{\mu\lambda} - \partial'^\lambda\partial'_\mu g'_{\nu\lambda} + \partial'_\mu\partial'_\nu g'^\sigma_\sigma$$
$$- \eta_{\mu\nu}\partial'_\lambda\partial'^\lambda g'^\sigma_\sigma + \eta_{\mu\nu}\partial'^\lambda\partial'^\sigma g'_{\lambda\sigma} = -\kappa^2 T'_{\mu\nu} \qquad [23]$$

Since [23] is the *exact* field equation in geodesic coordinates x'^μ, we can find the *exact* field equation in the original coordinates x^μ by a coordinate transformation from x'^μ to x^μ; it is of course at this stage that assumption (*i*) enters. Because the transformation $x'^\mu \to x^\mu$ from local geodesic coordinates to general coordinates is somewhat messy, it is more convenient to try to express the left side of Eq. [23] in terms of a tensor object whose transformation law is manifest. Since the Riemann tensor is the only tensor that can be constructed from linear combinations of the second derivatives of $g'_{\mu\nu}$, the left side of Eq. [23] must involve some contracted form of the Riemann tensor. To discover just what combination is required, we begin by noting that in local geodesic coordinates the Ricci tensor is

$$R'_{\mu\nu} = \tfrac{1}{2}(\partial'^\lambda\partial'_\lambda g'_{\mu\nu} - \partial'^\lambda\partial'_\nu g'_{\mu\lambda} - \partial'^\lambda\partial'_\mu g_{\nu\lambda} + \partial'_\mu\partial'_\nu g_\sigma{}^\sigma) \qquad [24]$$

■ *Exercise 1.* Show that the general equation [6.100] for $R_{\mu\nu}$ reduces to Eq. [24] if $g'_{\mu\nu} = \eta_{\mu\nu}$ and $g'_{\mu\nu,\alpha} = 0$. ■

This expression lacks the terms proportional to $\eta_{\mu\nu}$ that are present in

Eq. [23]. To obtain these terms, we add $-\frac{1}{2}\eta_{\mu\nu}R'$ to $R'_{\mu\nu}$:

$$R'_{\mu\nu} - \tfrac{1}{2}\eta_{\mu\nu}R' = \tfrac{1}{2}(\partial'^\lambda\partial'_\lambda g'_{\mu\nu} - \partial'^\lambda\partial'_\nu g'_{\mu\lambda} - \partial'^\lambda\partial'_\mu g'_{\nu\lambda} + \partial'_\mu\partial'_\nu g'_\sigma{}^\sigma$$
$$- \eta_{\mu\nu}\partial'_\lambda\partial'^\lambda g'_\sigma{}^\sigma + \eta_{\mu\nu}\partial'^\lambda\partial'^\sigma g'_{\lambda\sigma}) \qquad [25]$$

On the right sides of these equations, the indices are raised with the Minkowski metric $\eta^{\mu\nu}$. By comparing Eqs. [23] and [25], we see that our field equation in local geodesic coordinates is simply

$$R'_{\mu\nu} - \tfrac{1}{2}\eta_{\mu\nu}R' = -\tfrac{1}{2}\kappa^2 T'_{\mu\nu}$$

or

$$R'_{\mu\nu} - \tfrac{1}{2}g'_{\mu\nu}R' = -\tfrac{1}{2}\kappa^2 T'_{\mu\nu} \qquad [26]$$

Since the transformation properties of $R'_{\mu\nu}$ are those of a tensor, it is immediately obvious that the coordinate transformation from x'^μ to x^μ gives, with $\kappa^2 = 16\pi G$,

$$R_{\mu\nu} - \tfrac{1}{2}g_{\mu\nu}R = -8\pi G T_{\mu\nu} \qquad [27]$$

This is *Einstein's field equation*. Although this field equation was discovered by Einstein, it was also discovered independently by Hilbert, who therefore deserves to share the credit.* In cgs units, the field equation becomes

$$R_{\mu\nu} - \tfrac{1}{2}g_{\mu\nu}R = -\frac{8\pi G}{c^4}T_{\mu\nu} \qquad [28]$$

That the exact nonlinear equations are implied by the linear equations (the converse is of course trivial) is a remarkable feature of Einstein's theory. This tight connection between the exact equations and the linear approximation would not exist were it not for the principle of general invariance.

The second half of this chapter and the following chapters will deal with various solutions of Eq. [27]. We will obtain the exact solution for the gravitational field surrounding a static, spherically symmetric mass (Schwarzschild solution) and that surrounding a rotating mass (Kerr solution); finally, we will apply Eq. [27] to cosmology and obtain solutions that describe the evolution of the large-scale geometry of the

---

* Hilbert adopted the physical basis that Einstein had gradually laid for the theory of gravitation in work extended over several years, and he actually discovered and published the field equations a few days before Einstein (see Pais, 1982).

universe.

Unfortunately, the differential equation [27] is very complicated. The second derivatives of the metric field appear linearly, but the first derivatives enter quadratically. The dependence on $g_{\mu\nu}$ is even worse; the left side of Eq. [27] contains the inverse $g^{\mu\nu}$ (used to raise indices), and this is a horrible thing when expressed as a function of $g_{\mu\nu}$. No general procedure exists for solving the equations analytically; one guesses solutions as best one can.

The tensor $T_{\mu\nu}$ appearing on the right side of Eq. [27] is the energy-momentum tensor of matter. The energy-momentum of the gravitational field is already included (implicitly) on the left side of this equation--all the nonlinear terms on the left side represent the energy-momentum tensor of the gravitational field.

In empty space, the field equation reduces to

$$R_{\mu\nu} = 0 \tag{29}$$

This equation will give us the gravitational field in the empty space surrounding a mass distribution, for instance, the gravitational field surrounding the Sun or the gravitational field surrounding a black hole.

■ *Exercise 2.* Show this. (Hint: Contract μ, ν in Eq. [27].) ■

Note that, by raising an index, we can also change Eq. [27] into the form

$$R_\mu{}^\nu - \tfrac{1}{2}\delta_\mu^\nu R = -8\pi G T_\mu{}^\nu \tag{30}$$

Since the left side of this equation satisfies the Bianchi identity

$$(R_\mu{}^\nu - \tfrac{1}{2}\delta_\mu^\nu R)_{;\nu} = 0 \tag{31}$$

it follows that

$$T_\mu{}^\nu{}_{;\nu} = 0 \tag{32}$$

This expresses the exchange of energy momentum between matter and the gravitational field. With the definition of covariant derivative, Eq. [32] becomes

$$\partial_\nu T_\mu{}^\nu + \Gamma^\nu{}_{\alpha\nu} T_\mu{}^\alpha - \Gamma^\alpha{}_{\mu\nu} T_\alpha{}^\nu = 0 \tag{33}$$

This equation is the generalization of our approximate equation [3.68].

■ *Exercise 3.* Show that Eq. [33] implies Eq. [3.68] if we use the linear approximation for the gravitational field. [Hint: Show that in the linear approximation

$$\Gamma^\mu{}_{\alpha\beta} = \tfrac{1}{2}\eta^{\mu\nu}\kappa(h_{\nu\alpha,\beta} + h_{\beta\nu,\alpha} - h_{\alpha\beta,\nu}) \;] \;■$$ [34]

The equation of motion of a particle can be obtained from Eq. [33] by methods similar to those of Section 3.3. Of course we already know the equation of motion of a particle--it is the geodesic equation. Nevertheless, the calculation based on Eq. [33] is important in principle because we must make sure that the equations of motion are not contradictory.

To obtain the equation of motion from Eq. [33], it is simplest to go to geodesic coordinates. Then Eq. [33] reduces to

$$\partial'_\nu T'_\mu{}^\nu = 0$$ [35]

Although this equation is strictly valid only at one point, in the case of a sufficiently small (pointlike) particle we can regard it as valid throughout the volume of the particle. If we now apply the methods of Section 3.3, we find that the equation of motion is

$$\frac{dp'_\mu}{d\tau} = 0$$ [36]

Thus, as seen in the local geodesic coordinates, the energy and the momentum of the particle do not change. This agrees with the result [6.80] which we obtained from the geodesic equation. We can therefore conclude that the particle motion predicted by Eq. [33] is geodesic motion (Papapetrou, 1951).

Since in Einstein's theory all particles obey the geodesic equation of motion, it follows that the Galileo principle of equivalence is satisfied. It can also be shown that the Newton principle of equivalence is satisfied; that is, for any arbitrary system (not necessarily a particle), the gravitational mass equals the inertial mass. As already mentioned in Section 7.1, the principle of general invariance plays an important role in the proof of the equality of these masses (Ohanian, 1973).

A final general remark about the field equations [27]. Since the tensor $R_{\mu\nu}$ is symmetric, the number of equations is ten. However, not all these equations are independent; the Bianchi identity, consisting of a set of four equations, shows that we have not ten, but only six independent differential equations. Since the unknown functions are the ten components of $g_{\mu\nu}$, the Einstein equations do not determine the unknown functions entirely. We already encountered this situation in the linear case, where the solutions suffered from an ambiguity due to gauge transformations; we eliminated the ambiguity by using the Hilbert gauge condition. In the present case, the ambiguity in the solution for $g_{\mu\nu}(x)$ arises from the ambiguity in the choice of coordinates. This comes as no surprise, since the gauge transformations of the linear theory are in fact (infinitesimal) coordinate transformations. That the Einstein field equations determine the field $g_{\mu\nu}(x)$ only up to a general coordinate transformation is actually highly desirable: it would be absurd that the field equations should not only determine the geometry, but also prescribe what coordinates we must use to describe the geometry.

In order to remove the ambiguity in the solution for $g_{\mu\nu}(x)$, we can impose extra "coordinate conditions." For example, we can use the *harmonic* coordinate condition

$$g^{\mu\nu}\Gamma^{\alpha}_{\mu\nu} = 0 \qquad [37]$$

This condition replaces the Hilbert condition of the linear theory.

■ **Exercise 4.** Show that in the linear approximation, Eq. [37] reduces to the Hilbert condition. ■

Other coordinate conditions are also possible. It is often convenient to impose a purely algebraic condition on $g_{\mu\nu}$, rather than a differential equation. One such possibility is the condition for *time-orthogonal* coordinates*:

$$g_{00} = 1 , \qquad g_{0k} = 0 \qquad [38]$$

In general, the condition [38] is not very useful because the coordinate frame in which this condition holds will usually not be rigid--relative to the rectangular coordinates at infinity, the coordinate frame will have to bend and twist in a convoluted way. We will make our choice of coordinate condition whenever and however it is convenient.

7.3 ANOTHER APPROACH TO EINSTEIN'S EQUATION; THE COSMOLOGICAL TERM

We can use the principle of minimal coupling as a basis for an alternative derivation of Einstein's field equation. The argument is the following: Suppose that at some given point of spacetime, we examine the behavior of matter in local geodesic coordinates. The principle of minimal coupling tells us that, in such coordinates, the equations of motion of matter are those of special relativity, such as Maxwell's equations, the Lorentz-force equation, etc. But if these familiar equa-

---

* Also called Gaussian coordinates.

tions are valid, then they imply that the energy-momentum tensor of matter obeys the familiar conservation law

$$T_\mu{}^\nu{}_{,\nu} = 0 \qquad [39]$$

This equation is valid only at the given point, and only in geodesic coordinates. However, the principle of minimal coupling tells us that the correct conservation equation in general coordinates should differ from Eq. [39] only by the replacement of the ordinary derivative by a covariant derivative. Thus, the conservation law for the energy-momentum tensor of matter takes the form

$$T_\mu{}^\nu{}_{;\nu} = 0 \qquad [40]$$

This equation of course coincides with Eq. [32] of the preceding section. The difference is that in the preceding section we started from the linear approximation, and we obtained Einstein's field equation and Eq. [32] as a consequence, whereas now we are starting from the minimal coupling principle and Eq. [32] (or Eq. [40]), and we seek a different derivation of Einstein's field equation.

In the field equation, $T_{\mu\nu}$ should appear as source of the gravitational field. The field equation should therefore be of the form

$$G_{\mu\nu} = -8\pi G T_{\mu\nu} \qquad [41]$$

where $G_{\mu\nu}$ is some expression involving the metric and derivatives of the metric. According to Eq. [40], the object $G_{\mu\nu}$ must satisfy the identity

$$G_\mu{}^\nu{}_{;\nu} = 0 \qquad [42]$$

To proceed with this derivation of Einstein's field equation, we combine the condition [42] with the assumptions (*i*) and (*iii*) of the preceding section [but we do *not* assume (*ii*)]. General invariance demands that $G_{\mu\nu}$ be a tensor. We know from Section 6.5 that the only tensor that contains the second derivatives linearly is $R_{\mu\nu\alpha\beta}$. Hence the only tensors available for the construction of $G_{\mu\nu}$ are $R_{\mu\nu\alpha\beta}$ and $g_{\mu\nu}$. The most general second-rank tensor that can be built out of these, with $R_{\mu\nu\alpha\beta}$ entering linearly, is

$$G_{\mu\nu} = aR_{\mu\nu} + bg_{\mu\nu}R + \Lambda g_{\mu\nu} \qquad [43]$$

where a, b, and Λ are constants. Since $g_\mu{}^\nu{}_{;\nu} = 0$, the condition [42] reduces to

$$(aR_\mu{}^\nu + b\delta_\mu^\nu R)_{;\nu} = 0 \qquad [44]$$

In view of the Bianchi identity [6.103], the condition [44] will be fulfilled if $b = -\frac{1}{2}a$.

We are therefore led to a field equation

$$a(R_{\mu\nu} - \tfrac{1}{2}g_{\mu\nu}R) + \Lambda g_{\mu\nu} = -8\pi G T_{\mu\nu} \qquad [45]$$

To determine a and Λ, we must take into account that in the linear, nonrelativistic limit, Eq. [45] should agree with the equation for the Newtonian potential. It is easy to see that this requires $a = 1$ and $\Lambda = 0$, and gives us, again, the Einstein equation.

The quantity $\Lambda g_{\mu\nu}$ apearing in Eq. [45] is called the *cosmological term*, and the constant Λ is called the *cosmological constant*. Let us consider the possibility that Λ is small, but not zero. In this case, Einstein's equation becomes

$$R_{\mu\nu} - \tfrac{1}{2}g_{\mu\nu}R + \Lambda g_{\mu\nu} = -8\pi G T_{\mu\nu} \qquad [46]$$

If Λ is sufficiently small, then within the Solar System Eq. [46] will be in good agreement with the equation for the Newtonian potential. However, even a small value of Λ could have drastic effects on the evolution of the universe, as we will see in Section 9.9.

The cosmological term can be interpreted as an energy-momentum of the vacuum. To recognize this, we rewrite Eq. [46] with the cosmological term on the right side of the equation:

$$R_{\mu\nu} - \tfrac{1}{2}g_{\mu\nu}R = -8\pi G\left(T_{\mu\nu} + \frac{\Lambda}{8\pi G}g_{\mu\nu}\right) \qquad [47]$$

In this equation, $(\Lambda/8\pi G)g_{\mu\nu}$ plays the role of an extra energy-momentum tensor, that is, an energy-momentum tensor not associated with matter, but with empty (matter-free) space. Note that in local geodesic coordinates, the extra energy-momentum tensor is proportional to $\eta_{\mu\nu}$. Thus, the energy and momentum distribution is homogeneous and isotropic when viewed in these coordinates (furthermore, it is Lorentz invariant in these coordinates).

To grasp the implications of a nonzero value of Λ, let us look at the linear approximation corresponding to Eq. [47], with $a = 1$. Since Λ is certainly small, we can approximate $\Lambda g_{\mu\nu} \simeq \Lambda \eta_{\mu\nu}$, and with the usual Hilbert gauge condition (Eq. [3.50]) we obtain

$$\partial_\lambda \partial^\lambda (h_{\mu\nu} - \tfrac{1}{2}\eta_{\mu\nu} h) = -\kappa T_{\mu\nu} - \frac{2\Lambda}{\kappa}\eta_{\mu\nu} \qquad [48]$$

We can also write this equation as

$$\partial_\lambda \partial^\lambda h_{\mu\nu} = -\kappa(T_{\mu\nu} - \tfrac{1}{2}\eta_{\mu\nu} T) + \frac{2\Lambda}{\kappa}\eta_{\mu\nu} \qquad [49]$$

■ *Exercise 5.* Show this. ■

In the absence of matter, $T_{\mu\nu} = 0$, and from Eq. [49] we then obtain the following differential equation for the static Newtonian potential $\Phi = \tfrac{1}{2}\kappa h_{00}$*:

$$\nabla^2\Phi = -\Lambda \qquad [50]$$

If we compare this with the Newtonian equation $\nabla^2\Phi = 4\pi G\rho$, we recognize that the Λ term in Einstein's equation corresponds to a uniform effective mass density

$$\rho_{eff} = -\frac{\Lambda}{4\pi G} \qquad [51]$$

Thus, if Λ is positive, the vacuum has an effective mass density which is negative, and conversely. Note that these signs for the effective mass in the Newtonian equation are the opposite of what we expect if we examine the $_{00}$ component of the vacuum energy-momentum tensor $(\Lambda/8\pi G)\eta_{\mu\nu}$. The reason is that what appears as source for h_{00} in the linear equation is not T_{00}, but rather $T_{00} - \tfrac{1}{2}T = \tfrac{1}{2}(T_{00} + T_{11} + T_{22} + T_{33})$. If Λ is positive, the $_{00}$ component of the vacuum energy-momentum tensor is positive, but the 11, 22, and 33 components are negative (they play the role of a negative pressure). These negative components make the combination $\tfrac{1}{2}(\Lambda/8\pi G)(\eta_{00} + \eta_{11} + \eta_{22} + \eta_{33})$ negative, leading to a negative effective mass in the Newtonian equation.

If we arbitrarily set $\Phi = 0$ at the origin, then in spherical coordinates Eq. [50] has the solution

$$\Phi = -\frac{\Lambda}{6}r^2 \qquad [52]$$

---

* To settle the ambiguity in the units of Λ, we arbitrarily assume that, in cgs units, Eq. [50] is correct as it stands (no extra factors of c). The units of Λ are then s⁻².

■ *Exercise 6.* Show this. ■

This potential indicates that between any two particles there acts an effective harmonic-oscillator force of $\Lambda r/3$ per unit mass, attractive in the case $\Lambda < 0$, and repulsive in the case $\Lambda > 0$.

In the context of Chapter 3, we did not attempt to include a term $\Lambda \eta_{\mu\nu}$ in our field equation, because we wanted to work with weak fields in an asymptotically flat spacetime. The cosmological term violates this assumption. Obviously, at large distances the potential [52] becomes very large, and this contradicts both the weak field condition ($\Phi \ll c^2$) and the boundary condition at infinity ($\Phi \to 0$). However, a nonzero value of Λ does not entirely destroy the validity of the approximate equations of Chapter 3. If we are dealing with a central mass that is not enormously large, then there will exist a wide range of radii

$$\frac{c}{\sqrt{|\Lambda|}} \gg r \gg \frac{GM}{c^2} \qquad [53]$$

in which the spacetime is very nearly flat; in this range the linear theory is a good approximation.

For large distances, the potential [52] and the force grow very large. But for small distances they are negligible. We can set an upper limit on Λ by noting that the Newtonian inverse-square law seems to be valid for the motion of galaxies in multiple systems of galaxies. For example, Page (1952, 1960) has applied the inverse-square law to determine the mass of binary systems of galaxies, that is, pairs of galaxies in a tight orbit about each other. The masses obtained for binary systems (via the virial theorem) are in reasonable agreement with independent mass determination of the components (via the rotation curve; see Section 9.6). This implies that the deviations from the inverse-square law are not significant; that is, within regions of a size at least as large as these binary systems, the acceleration $\Lambda r/3$ is small compared with the gravitational acceleration GM/r^2. From the dimension ($r \simeq 10^{23}$ cm) and the mass ($M \simeq 10^{11} M_\odot$) of typical binary systems we can then set the limit

$$|\Lambda| < 10^{-33}/\text{s}^2 \qquad [54]$$

The effects of the Λ term will be most noticeable in the large-scale motion of the universe. Hence, cosmological data set the best limits on Λ. Our universe is expanding (see Section 9.3). The attractive Newtonian gravitational force between the masses in the universe tends to decelerate this expansion. The extra force associated with the cosmological term tends to decelerate the expansion if Λ is negative, and accelerate if Λ is positive. Consequently, a large negative value of Λ

implies a short age for the universe, and a large positive value implies a long age (this can best be seen by examining what happens if we time-reverse the motion of expansion; a negative value leads to an accelerated collapse, and a short time for the completion of this collapse). From observations on globular clusters of stars, astronomers can place a firm lower limit of 10 billion years on the age of these stars and consequently on the age of the universe. If Λ is negative, this implies the following firm observational limit on the magnitude of Λ (Carroll, Press, and Turner, 1992):

$$|\Lambda| < 2 \times 10^{-35}/s^2 \qquad [55]$$

But if Λ is positive, we cannot use the age of the universe to set a firm limit on Λ, since we do not have a firm upper limit on the age of the universe (the universe is necessarily older than any thing we find in it, and when we determine the age of any thing in the universe, we only place a lower limit on the age of the universe). If Λ is positive, the best observational limit on Λ comes from observations on gravitational lenses. In the expanding universe, the probability for the alignment of a lensing galaxy with a background quasar of a given redshift z depends strongly on the values Λ and z, especially if Λ is positive. A large positive value of Λ leads to a high density of quasars for a given redshift z and a high probability of alignment, that is, a high abundance of observable lenses. From the available statistical data on gravitational lenses, we can set an upper limit of (Carroll, Press, and Turner, 1992)

$$\Lambda < 4 \times 10^{-35}/s^2 \qquad [56]$$

Expressed as effective mass densities, the limits [55] and [56] are $|\Lambda|/4\pi G = 2 \times 10^{-29}$ g/cm^3 and $\Lambda/4\pi G = 4 \times 10^{-29}$ g/cm^3, respectively.

Although the implications of the Λ term are cosmological, the origin of this term probably is to be found in quantum theory rather than cosmology. As we saw, this term can be interpreted as an energy density of the vacuum. The quantum-theoretical vacuum is a very active place. It is crawling with electron-positron pairs, proton-antiproton pairs, photons, gravitons, and so on, which are spontaneously created and destroyed by vacuum fluctuations. It is not surprising that the vacuum should have an energy density; the surprise is rather that the energy density is as small as our limit on the cosmological term indicates: only 10^{-29} g/cm^3, or 10^{-8} erg/cm^3. Quantum-theoretical estimates suggest that the energy density should be of the order of 1 Planck mass per Planck length cubed, that is, $(\hbar c/G)^{1/2} \times (\hbar G/c^3)^{-3/2} \simeq 10^{92}$ g/cm^3, which is about 120 orders of magnitude larger than the observational limit. This indicates a defect in our theories, but it also indicates that the possibility of a nonzero value of Λ cannot be rejected peremptorily.

According to some recent speculations, the value of Λ may have been much larger (and positive) during the early stages of the evolution of the universe. Such a large value of Λ would have resulted in a fast expansion of the universe (an "inflation"). We will discuss this inflationary model of the early universe in Chapter 10.

7.4 THE SCHWARZSCHILD SOLUTION

For the discussion of the deflection, the retardation, and the redshift of light in the gravitational field of the Sun, the linear approximation of Chapter 3 was quite adequate. The same may be said for the gyroscope precession effect, which is another experimental test of gravitational theory that is now in preparation. However, the perihelion precession of planetary orbits in the field of the Sun cannot be treated by the linear approximation. We need an exact, or at least more exact, solution for the gravitational field surrounding a spherically symmetric mass distribution. This is the Schwarzschild solution. The importance of this solution goes much beyond its application in the solar system. It is believed that the gravitational collapse of any nonrotating, electrically neutral star will necessarily lead to the Schwarzschild geometry as its final result; even if the initial geometry is asymmetric, any deviations from the symmetric Schwarzschild field will ultimately be radiated away in the form of gravitational waves.

In what follows we will solve the Einstein equations in the vacuum region surrounding the mass distribution. The solution in the interior of a mass distribution can be carried out analytically only in some exceptional cases, such as the case of an ideal incompressible fluid or the case of a fluid without pressure (cloud of dust); these special cases are of little interest in the real world. The solution for the case of matter with a realistic equation of state can only be carried out numerically.

If the mass distribution is spherically symmetric, the gravitational field generated by this mass distribution must also have spherical symmetry. Furthermore, if the mass distribution is at rest and does not rotate, the field must be *static*. The latter requirement goes beyond time independence. To be precise, a field is said to be static if it is both time independent and time symmetric, that is, unchanged by time reversal. If the field is merely time independent, then it is said to be *stationary* (Møller, 1952). An example from electrodynamics will make the distinction clear: Consider the magnetic field of a loop of wire carrying a steady current. This field is stationary, that is, time independent. However, the field is not static because under time reversal the current, and hence also the magnetic field, reverses direction.

For a field with spherical symmetry it is natural to use polar coordinates t, r, θ, ϕ. The angular coordinates θ and ϕ are unambiguous;

the measurement of these coordinates depends only on our ability to divide a circumference concentric with the origin into equal parts, which we can even when we do not know the metric. The "radial" coordinate r is ambiguous because we do not yet know its precise relation to the measurement of distance. For a start, we will treat r simply as a parameter that identifies different spherical surfaces concentric with the origin. The "time" coordinate t suffers from similar ambiguities. However, we will insist that as $r \rightarrow \infty$ and the space becomes flat, increments in r and t should equal the increments in the true distance and time, respectively.

The static character and the spherical symmetry of the field impose some restrictions on the form of the spacetime interval.* The demand for a static solution forbids terms of the type $drdt$, $d\theta dt$, $d\phi dt$ in the expression for the spacetime interval, because these terms change sign when we perform the time reversal $t \rightarrow -t$. Hence the spacetime interval must be of the form

$$ds^2 = A(r)dt^2 - dl^2 \qquad [57]$$

where $A(r)$ is some function of r, and where the purely spatial interval dl^2 is some sum of products of dr, $d\theta$, $d\phi$ taken two at a time.

To discover the restrictions imposed on dl^2 by spherical symmetry, we note that r and dr are left invariant by a rotation about any axis through the center of symmetry, but that θ, ϕ, $d\theta$, and $d\phi$ are not (under a rotation, θ and ϕ change into some functions of ϕ and θ). However, the combination $d\theta^2 + \sin^2\theta \, d\phi^2$, which represents the square of the angular displacement, is left invariant under rotations, even though the individual quantities $d\theta$ and $d\phi$ are not. This is the only combination of $d\theta$ and $d\phi$ that is left invariant. To construct a spatial line element with spherical symmetry, we must use the rotationally invariant quantities r, dr, and $dr^2 + \sin^2\theta \, d\phi^2$, and nothing else. The spatial interval is therefore necessarily of the form

$$dl^2 = B(r)(dr^2 + \sin^2\theta \, d\phi^2) + C(r)dr^2$$

where $B(r)$ and $C(r)$ are some functions of r. The spacetime interval is then

$$ds^2 = A(r)dt^2 - B(r)(r^2d\theta^2 + r^2 \sin^2\theta \, d\phi^2) - C(r)dr^2 \qquad [58]$$

The unknown functions $A(r)$, $B(r)$, and $C(r)$ are to be determined by solving Einstein's differential equations.

We can simplify Eq. [58] by introducing a new radial coordinate:

---

* As we will see in Section 75, the condition of spherical symmetry by itself actually implies that the solution is static; this is Birkhoff's theorem.

$$r' = r\sqrt{B(r)} \qquad [59]$$

This has the advantage that the terms involving angular displacements reduce to $r'^2 d\theta^2 + r'^2 \sin^2\theta \, d\phi^2$, and we obtain

$$ds^2 = A'(r')dt^2 - B'(r')dr'^2 - r'^2 d\theta^2 - r'^2 \sin^2\theta \, d\phi^2 \qquad [60]$$

Here A' and B' are some new functions which, if desired, can be expressed in terms of A, B, C. Note that according to Eq. [60], the circumference of a circle of radius r' is $2\pi r'$. For instance, along a meridional circle, the element of length is $dl = \sqrt{-ds^2} = r'd\theta$, and therefore, as in flat space, $\int dl = 2\pi r'$. However, the length interval along the radial line is not dr', but $\sqrt{B'(r')}dr'$. Hence the ratio of the length of the circumference and the length of the radius differs from 2π, which gives us an immediate indication that the space is not flat.

For the solution of Eintein's equations, it is convenient to omit the primes in Eq. [60] and to write the unknown functions A' and B' as exponentials:

$$ds^2 = e^{N(r)} \, dt^2 - e^{L(r)} \, dr^2 - r^2 d\theta^2 - r^2 \sin^2\theta \, d\phi^2 \qquad [61]$$

With $x^0 \equiv t$, $x^1 \equiv r$, $x^2 \equiv \theta$, $x^3 \equiv \phi$, the metric tensor corresponding to Eq. [61] is

$$g_{\mu\nu} = \begin{pmatrix} e^N & 0 & 0 & 0 \\ 0 & -e^L & 0 & 0 \\ 0 & 0 & -r^2 & 0 \\ 0 & 0 & 0 & -r^2 \sin^2\theta \end{pmatrix} \qquad [62]$$

The unknown functions are now $N(r)$ and $L(r)$. We will use Einstein's equations to find them.

The Christoffel symbols for a metric tensor of the form [62] have already been given in Chapter 6 (see Eq. [6.70]). We repeat them here for convenience (the primes indicate derivatives; $N' = \partial N/\partial r$, $L' = \partial L/\partial r$):

$$\Gamma^0_{01} = \Gamma^0_{10} = \tfrac{1}{2}N'$$

$$\Gamma^1_{00} = \tfrac{1}{2}N'e^{N-L}$$

$$\Gamma^1_{11} = \tfrac{1}{2}L'$$

$$\Gamma^1_{22} = -re^{-L}$$

$$\Gamma^1_{33} = -r\sin^2\theta \, e^{-L}$$

$$\Gamma^2_{12} = \Gamma^2_{21} = 1/r$$

$$\Gamma^2{}_{33} = - \sin\theta \, \cos\theta$$

$$\Gamma^3{}_{13} = \Gamma^3{}_{31} = 1/r$$

$$\Gamma^3{}_{23} = \Gamma^3{}_{32} = \cot\theta \qquad [63]$$

All other Christoffel symbols are zero.

The Ricci tensor and the curvature invariant may be calculated from the Christoffel symbols (see Eq. [6.105]); only the diagonal components of $R_\mu{}^\nu$ are nonzero. The Einstein equations *in vacuo*,

$$R_\mu{}^\nu - \tfrac{1}{2}\delta_\mu^\nu R = 0 \qquad [64]$$

become

$$R_0{}^0 - \tfrac{1}{2}R = - e^{-L}\left(\frac{L'}{r} - \frac{1}{r^2}\right) - \frac{1}{r^2} = 0 \qquad [65]$$

$$R_1{}^1 - \tfrac{1}{2}R = e^{-L}\left(\frac{N'}{r} + \frac{1}{r^2}\right) - \frac{1}{r^2} = 0 \qquad [66]$$

$$R_2{}^2 - \tfrac{1}{2}R = e^{-L}\left(\frac{N''}{2} - \frac{L'N'}{4} + \frac{N'^2}{4} + \frac{N' - L'}{2r}\right) = 0 \qquad [67]$$

$$R_3{}^3 - \tfrac{1}{2}R = e^{-L}\left(\frac{N''}{2} - \frac{L'N'}{4} + \frac{N'^2}{4} + \frac{N' - L'}{2r}\right) = 0 \qquad [68]$$

These equations can be integrated quite easily. Consider Eq. [65]; it can be written

$$e^{-L}(-rL' + 1) = 1 \qquad [69]$$

This has the general solution

$$e^{L(r)} = \frac{1}{1 - C/r} \qquad [70]$$

where C is a constant.

■ *Exercise 7.* Check that Eq. [69] implies Eq. [70]. (Hint: The substitution $f(r) = e^{-L}$ simplifies the differential equation [69].) ■

By subtraction of Eq. [65] from Eq. [66] we see that

$$L' = - N' \qquad [71]$$

from which

$$L = -N + [\text{constant}] \qquad [72]$$

But at large distances $(r \to \infty)$, the metric tensor [62] must reduce to that of flat spacetime in spherical coordinates. Hence both L and N must tend to zero in this limit, and the constant must be zero,

$$L = -N \qquad [73]$$

Therefore the solution for $N(r)$ is

$$e^{N(r)} = 1 - C/r \qquad [74]$$

As a final step, we must check that the solutions [70] and [74] also satisfy the differential equations [67] and [68] which have not been used so far.

■ *Exercise 8.* Check this. ■

According to Eqs. [70] and [74], the interval for the spherically symmetric field has the form

$$ds^2 = \left(1 - \frac{C}{r}\right)dt^2 - \frac{dr^2}{1 - C/r} - r^2 d\theta^2 - r^2 \sin^2\theta \, d\phi^2 \qquad [75]$$

To find the value of C, we want to compare this expression with the result obtained from the linear theory (Eq. [4.13]),

$$ds^2 \simeq \left(1 - \frac{2GM}{r}\right)dt^2 - \left(1 + \frac{2GM}{r}\right)(dx^2 + dy^2 + dz^2) \qquad [76]$$

where M is the central mass. But we cannot compare [75] and [76] directly, because in these equations the radial variables are defined in somewhat different ways: the variable r in Eq. [75] is such that the measured circumference of a circle concentric with the origin is $2\pi r$, whereas the variable r in Eq. [76] is such that the circumference is approximately $2\pi r(1 + GM/r)$.

■ *Exercise 9.* Show this. ■

Let us change Eq. [75] by introducing a new coordinate r':

$$r' = \tfrac{1}{2}\sqrt{r^2 - Cr} + \tfrac{1}{2}r - \tfrac{1}{4}C \qquad [77]$$

or

$$r = r'\left(1 + \frac{C}{4r'}\right)^2 \tag{78}$$

This transformation gives

$$ds^2 = \left(\frac{1 - C/4r'}{1 + C/4r'}\right)^2 - \left(1 + \frac{C}{4r'}\right)^4 (dr'^2 + r'^2d\theta^2 + r'^2 \sin^2\theta \, d\phi^2) \tag{79}$$

■ *Exercise 10.* Derive Eq. [79]. ■

The coordinates used in Eq. [79] are called *isotropic*. In the weak field limit ($r' \to \infty$), Eq. [79] reduces to

$$ds^2 \simeq \left(1 - \frac{C}{r'}\right)dt^2 - \left(1 + \frac{C}{r'}\right)(dr'^2 + r'^2d\theta^2 + r'^2 \sin^2\theta \, d\phi^2) \tag{80}$$

Obviously, this equation has the same form as Eq. [76]. By comparison, we see that

$$C = 2GM$$

With this, Eq. [75] becomes

$$ds^2 = \left(1 - \frac{2GM}{r}\right)dt^2 - \frac{dr^2}{1 - 2GM/r} - r^2d\theta^2 - r^2 \sin^2\theta \, d\phi^2 \tag{81}$$

This is called the *Schwarzschild solution*.

It must be kept in mind that the mass M is the total mass of the system; the mass-energy contributed by the gravitational fields is included in M. It is clear that unless M is the total energy of the system, the principle of equivalence ($M_I = M_G$) cannot be satisfied. We will not give the proof that the gravitational mass of the system that produces the field [81] is in fact equal to the inertial mass of the system. Such a proof can be given, but requires some knowledge of the properties of the exact energy-momentum tensors of matter and of gravitation (Ohanian, 1973).

An additional remark: in the above calculations we have assumed that the cosmological constant is zero. If we want to drop this assumption, then we must replace Eq. [64] by

$$R_\mu{}^\nu - \tfrac{1}{2}\delta_\mu^\nu R = - \Lambda\delta_\mu^\nu$$

It can be shown that this equation leads to a Schwarzschild solution

$$ds^2 = \left(1 - \frac{2GM}{r} - \frac{\Lambda r^2}{3}\right)dt^2$$

$$- \frac{dr^2}{1 - 2GM/r - \Lambda r^2/3} - r^2 d\theta^2 - r^2 \sin^2\theta \, d\phi^2 \qquad [82]$$

Note that this metric does not become asymptotically flat as $r \to \infty$. However, since Λ is very small (see Section 7.3), there is a range of radii

$$\frac{c}{\sqrt{\Lambda}} \gg r \gg \frac{GM}{c^2}$$

in which the metric is nearly flat. For values of r below this range, the effect of the mass M dominates; for values of r above this range, the effect of the cosmological term dominates. (At very large values of r we must also take into account the large-scale curvature of the universe; see Chapter 9).

We know from Section 3.5 that in the Newtonian approximation the quantity $\frac{1}{2}\kappa h_{00}$ is to be identified with the Newtonian potential (see Eq. [3.102]),

$$\Phi = \tfrac{1}{2}\kappa h_{00} = \tfrac{1}{2}(g_{00} - 1)$$

According to Eq. [82], this leads to

$$\Phi = - \frac{GM}{r} - \frac{\Lambda}{6}r^2$$

The second term on the right side of this equation represents a cosmological correction to the Newtonian potential; this term agrees with Eq. [52].

For the motion of the planets, the cosmological correction is completely insignificant. We will, therefore, assume that $\Lambda = 0$ until further notice.

7.5 BIRKHOFF'S THEOREM

Our derivation of the Schwarzschild solution in the preceding section started from the assumption that the metric is spherically symmetric and static. The latter assumption is not really necessary because it is implicitly contained in the former--it can be shown that any spherically symmetric vacuum solution of Einstein's equations must be static and must agree with the Schwarzschild solution.* This is *Birkhoff's*

---

* Of course it is possible to change the *form* of any solution by a coordinate transformation. Therefore, two solutions will be said to agree if they differ by no more than a coordinate transformation; such solutions are physically identical.

theorem. As a consequence of this theorem, the field that a spherically symmetric mass distribution produces in the surrounding region is always the static Schwarzschild field, regardless of whether the mass is static, collapsing, expanding, or pulsating. (This statement relies on the implicit assumption that the mass distribution has no net electric charge; if it does, then there will exist electric fields in the surrounding space and the vacuum condition $T^{\mu\nu} = 0$ will be violated.)

To prove the theorem, we must repeat the calculation of the preceding section, but without the assumption about the static character of the solution. We saw that if the spacetime interval is spherically symmetric and static, then it must have the form given by Eq. [60],

$$ds^2 = A(r)dt^2 - B(r)dr^2 - r^2 d\theta^2 - r^2 \sin^2\theta \, d\phi^2 \qquad [83]$$

where we have omitted the primes. Since we want to drop the static assumption, we will have to let A and B depend on time. Furthermore, we will have to include an extra term proportional to $dr dt$; this term is not static, but it is spherically symmetric. In contrast, we exclude the terms $d\theta dt$ and $d\phi dt$, since these are not spherically symmetric. Thus, the most general spherically symmetric expression for the spacetime interval is

$$ds^2 = A(r, t)dt^2 - B(r, t)dr^2 - r^2 d\theta^2 - r^2 \sin^2\theta \, d\phi^2 - 2F(r, t)dr dt \qquad [84]$$

The term involving $dr dt$ can be eliminated by a change in the time coordinate. We introduce a new time coordinate t' such that

$$dt' = (Adt - Fdr)Q \qquad [85]$$

where Q is a function of r and t which is to be chosen so as to make the right side of Eq. [85] into a perfect differential.

■ *Exercise 11.* Show that this requires that Q satisfy the differential equation

$$\frac{\partial}{\partial t}(FQ) = \frac{\partial}{\partial r}(AQ) \qquad ■$$

From Eq. [85] we obtain

$$Adt^2 - 2Fdr dt = \frac{1}{Q^2 A} dt'^2 - \frac{F^2}{A} dr^2 \qquad [86]$$

and hence

$$ds^2 = \frac{1}{Q^2 A} dt'^2 - \left[B + \frac{F^2}{A} \right] dr^2 - r^2 d\theta^2 - r^2 \sin^2\theta \, d\phi^2 \qquad [87]$$

It is convenient to drop the prime on t and to write the functions as exponentials:

$$ds^2 = e^{N(r,t)} dt^2 - e^{L(r,t)} dr^2 - r^2 d\theta^2 - r^2 \sin^2\theta \, d\phi^2 \qquad [88]$$

Note that this differs from Eq. [61] only in that N and L are now functions of t as well as r.

The calculation of the curvature tensor for the metric [88] is only slightly more complicated than the corresponding calculation for the metric [61]. The resulting Einstein equations are as follows (the primes and dots indicate derivatives with respect to r and t, respectively):

$$R_0{}^0 - \tfrac{1}{2}R = - e^{-L} \left(\frac{L'}{r} - \frac{1}{r^2} \right) - \frac{1}{r^2} = 0 \qquad [89]$$

$$R_1{}^1 - \tfrac{1}{2}R = e^{-L} \left(\frac{N'}{r} + \frac{1}{r^2} \right) - \frac{1}{r^2} = 0 \qquad [90]$$

$$R_2{}^2 - \tfrac{1}{2}R = e^{-L} \left(\frac{N''}{2} - \frac{L'N'}{4} + \frac{N'^2}{4} + \frac{N' - L'}{2r} \right)$$

$$- e^{-N} \left(\frac{\ddot{L}}{2} + \frac{\dot{L}^2}{4} - \frac{\dot{L}\dot{N}}{4} \right) = 0 \qquad [91]$$

$$R_3{}^3 - \tfrac{1}{2}R = e^{-L} \left(\frac{N''}{2} - \frac{L'N'}{4} + \frac{N'^2}{4} + \frac{N' - L'}{2r} \right)$$

$$- e^{-N} \left(\frac{\ddot{L}}{2} + \frac{\dot{L}^2}{4} - \frac{\dot{L}\dot{N}}{4} \right) = 0 \qquad [92]$$

$$R_0{}^1 = e^{-L} \frac{\dot{L}}{r} = 0 \qquad [93]$$

$$R_1{}^0 = - e^{-N} \frac{\dot{L}}{r} = 0 \qquad [94]$$

Eqs. [93] and [94] imply that $\dot{L} = 0$, so L is time independent. If $\dot{L} = 0$ is substituted into Eqs. [89]–[92], these equations become identical to Eqs. [65]–[68]. Hence the solution of the equations can be carried out in exactly the same manner as before and we obtain (see Eqs. [69]-[71])

$$L' = -N' \tag{95}$$

The solution of this differential equation is

$$L = -N + h(t) \tag{96}$$

where $h(t)$ is an arbitrary function of time.
    With L as given by Eq. [70], and

$$e^N = e^{-L} e^{h(t)} = \left(1 - \frac{C}{r}\right) e^{h(t)} \tag{97}$$

we therefore find

$$ds^2 = \left(1 - \frac{C}{r}\right) e^{h(t)} dt^2 - \frac{dr^2}{1 - C/r} - r^2 \, d\theta^2 - r^2 \sin^2\theta \, d\phi^2 \tag{98}$$

This differs from the Schwarzschild solution only by the factor $e^{h(t)}$ in the first term. This time-dependent factor can be eliminated by a further transformation of the time coordinate. If we use a new coordinate t' such that

$$dt' = e^{h(t)} dt$$

then the expression [98] becomes identical to [75]. Our spherically symmetric solution [98] therefore agrees with the Schwarzschild solution.
    It is a corollary of Birkhoff's theorem that a spherically symmetric mass distribution produces no gravitational field inside an empty spherical cavity centered on the mass distribution. This result is of course well known in Newton's theory (see Exercise 1.5). In Einstein's theory, the Birkhoff theorem guarantees that the solution inside the cavity must be the Schwarzschild solution given by Eq. [75]. Since an empty cavity cannot contain any singularities, we must take $C = 0$, and hence the spacetime is flat inside the cavity.

7.6 THE MOTION OF PLANETS; PERIHELION PRECESSION

The equation of motion of a particle in a gravitational field is the geodesic equation, Eq. [6.79],

$$\frac{d^2 x^\mu}{d\tau^2} + \Gamma^\mu_{\alpha\beta} \frac{dx^\alpha}{d\tau} \frac{dx^\beta}{d\tau} = 0 \qquad [99]$$

The following quantity is a constant of the motion:

$$g_{\mu\nu} \frac{dx^\mu}{d\tau} \frac{dx^\nu}{d\tau} = [\text{constant}] = 1 \qquad [100]$$

This equation is obviously true by definition of $d\tau$. But we can also
use the equation of motion [99] to show explicitly that the left side of
Eq. [100] is a constant. Thus, Eq. [100] can be regarded as a first integral of the equations of motion.

■ **Exercise 12.** Show that $g_{\mu\nu} (dx^\mu/d\tau)(dx^\nu/d\tau)$ = [constant] by integration of
the equation of motion. [Hint: Write Eq. [99] as

$$\frac{d}{d\tau}\left[g_{\mu\nu} \frac{dx^\nu}{d\tau} \right] - \frac{1}{2} g_{\alpha\beta,\mu} \frac{dx^\alpha}{d\tau} \frac{dx^\beta}{d\tau} = 0 \qquad [101]$$

and multiply by $dx^\mu/d\tau$.] ■

In the Schwarzschild geometry, the components of the geodesic
equation are

$$\ddot{t} + N'\dot{r}\dot{t} = 0 \qquad [102]$$

$$\ddot{r} + \tfrac{1}{2}N' e^{N-L} (\dot{t})^2 + \tfrac{1}{2}L'(\dot{r})^2 - r e^{-L} (\dot{\theta})^2 - r \sin^2\theta\, e^{-L} (\dot{\phi})^2 = 0 \quad [103]$$

$$\ddot{\theta} + \frac{2}{r} \dot{r}\dot{\theta} - \cos\theta \sin\theta (\dot{\phi})^2 = 0 \qquad [104]$$

$$\ddot{\phi} + \frac{2}{r} \dot{r}\dot{\phi} + 2 \cot\theta\, \dot{\phi}\dot{\theta} = 0 \qquad [105]$$

where dots indicate derivatives with respect to τ, such as $\dot{t} \equiv dt/d\tau$, $\dot{r} \equiv$
$dr/d\tau$, etc.

■ **Exercise 13.** Derive Eqs. [102] - [105]. ■

We will assume that the orbit is in the plane $\theta = \pi/2$. Eq. [104]
shows that if $\theta = \pi/2$ and $\dot{\theta} = 0$ initially, then $\ddot{\theta} = 0$ and therefore the
orbit remains in this plane.

Eqs. [102] and [105] can now be integrated directly, giving

$$\dot{t} = \mathcal{E}e^{-N} \qquad [106]$$

and

$$\dot{\phi} = \ell/r^2 \qquad [107]$$

where $\mathcal{E}$ and ℓ are constants. These constants are proportional to the energy and the angular momentum, respectively. We can see this by calculating the energy and the angular momentum from the Lagrangian. Since

$$L = m\sqrt{g_{\mu\nu}u^\mu u^\nu} \qquad [108]$$

the energy is

$$E = P_0 = \frac{\partial L}{\partial u^0} = mg_{00}u^0 = m\,e^N\,\dot{t} \qquad [109]$$

and the angular momentum is

$$l = -P_3 = -\frac{\partial L}{\partial u^3} = -mg_{33}u^3 = mr^2\dot{\phi} \qquad [110]$$

Comparison with Eqs. [106] and [107] shows that the constants $\mathcal{E}$ and ℓ are the energy and the angular momentum per unit mass, $\mathcal{E} = E/m$ and $\ell = l/m$. That these quantities are constants of the motion is immediately obvious in the context of the Lagrangian formalism, since the Lagrangian is independent of t and of ϕ, and hence the momenta conjugate to these ignorable coordinates are constant.

Eq. [103] is somewhat harder to solve, but we can bypass it if we use the first integral of the motion given by Eq. [100]:

$$e^N (\dot{t})^2 - e^L (\dot{r})^2 - r^2(\dot{\theta})^2 - r^2 \sin^2\theta\,(\dot{\phi})^2 = 1 \qquad [111]$$

If we insert $\dot{t}$ and $\dot{\phi}$ from Eqs. [106] and [107], and $\theta = \pi/2$, $e^{-L} = e^N = 1 - 2GM/r$, we obtain

$$\frac{\mathcal{E}^2}{1 - 2GM/r} - \frac{(\dot{r})^2}{1 - 2GM/r} - \frac{\ell^2}{r^2} = 1 \qquad [112]$$

Here it is convenient to introduce the variable

$$u = \frac{1}{r} \qquad [113]$$

which is also often used in nonrelativistic celestial mechanics. Then

$$\dot{r} = \frac{dr}{d\phi}\,\dot{\phi} = -\frac{1}{u^2}\frac{du}{d\phi}\frac{\ell}{r^2} = -\frac{du}{d\phi}\,\ell \qquad [114]$$

and Eq. [112] reduces to

$$\mathcal{E}^2 - \ell^2\left(\frac{du}{d\phi}\right)^2 - \ell^2 u^2(1 - 2GMu) = (1 - 2GMu) \qquad [115]$$

By differentiating this with respect to ϕ, we obtain a second-order differential equation for the orbit:

$$\frac{d^2u}{d\phi^2} + u - \frac{GM}{\ell^2} - 3GMu^2 = 0 \qquad [116]$$

If we compare this with the corresponding orbital equation of Newtonian theory,

$$\frac{d^2u}{d\phi^2} + u - \frac{GM}{\ell^2} = 0 \qquad [117]$$

we see that the equations differ only by the term $3GMu^2$. This term represents the relativistic correction to the motion.

For the special case of a planet moving in the Schwarzschild field of the Sun, the relativistic correction to the motion is extremely small. This can be seen by comparing the second and fourth terms on the left side of Eq. [116]. These terms differ by a factor $3GMu$, or, in cgs units, $3GM/rc^2$. Even for the planet Mercury, this is a very small number; with $M = M_\odot \simeq 2.0 \times 10^{33}$ g, $r = 5.5 \times 10^{12}$ cm, it is

$$\frac{3GM}{rc^2} \simeq 10^{-7} \qquad [118]$$

Since $3GMu^2$ is small compared with the other terms, it will be sufficient to use a method of successive approximations in the solution of Eq. [116]. We first write down the solution of the Newtonian equation,

$$u = \frac{GM}{\ell^2}[1 + \epsilon\cos(\phi - \phi_0)] \qquad [119]$$

where ϵ and ϕ_0 are constants. The equation [119] is that of an ellipse with an eccentricity ϵ and a perihelion located at ϕ_0.

■ *Exercise 14.* Check that [119] solves Eq. [117]. ■

If we now replace the small term $3GMu^2$ in Eq. [116] by its Newtonian approximation as given by Eq. [119], we obtain

$$\frac{d^2u}{d\phi^2} + u - \frac{GM}{\ell^2} - \frac{3(GM)^3}{\ell^4} - \frac{6\epsilon(GM)^3}{\ell^4} \cos(\phi - \phi_0)$$
$$- 3\epsilon^2 \frac{(GM)^3}{\ell^4} \cos^2(\phi - \phi_0) = 0 \qquad [120]$$

This equation is mathematically analogous to that of a harmonic oscillator, with a restoring force $- u$, two oscillating driving forces proportional to $\cos(\phi - \phi_0)$ and $\cos^2(\phi - \phi_0)$, and two extra constant forces. The first of the oscillating driving forces is in resonance with the natural oscillations, and therefore it gradually builds up a (relatively) large secular perturbation of the orbit; but the second oscillating driving force is not in resonance, and merely produces periodic perturbations of small amplitude. We will therefore ignore the second oscillating driving force; and we will also ignore the constant term $3(GM)^3/\ell^4$, which is small by comparison with the constant term GM/ℓ^2 and produces no interesting observable effects. We are then left with

$$\frac{d^2u}{d\phi^2} + u - \frac{GM}{\ell^2} - \frac{6\epsilon(GM)^3}{\ell^4} \cos(\phi - \phi_0) = 0 \qquad [121]$$

The solution of this differential equation is

$$u = \frac{GM}{\ell^2} [1 + \epsilon \cos(\phi - \phi_0)] + \frac{3\epsilon(GM)^3}{\ell^4} \phi \sin(\phi - \phi_0) \qquad [122]$$

which we can also write in the approximate form

$$u \simeq \frac{GM}{\ell^2} \left[1 + \epsilon \cos\left(\phi - \phi_0 - \frac{3(GM)^2}{\ell^2} \phi \right) \right] \qquad [123]$$

■ *Exercise 15.* Check that Eq. [121] has the solution [122]. Check that Eq. [123] reduces to [122] if the quantity $(GM)^2\phi/\ell^2$ is small. ■

Eq. [123] represents an orbit that is a precessing ellipse. The argument of the cosine changes by 2π when ϕ changes by

$$\Delta\phi = 2\pi\left[1 - \frac{3(GM)^2}{\ell^2}\right]^{-1} \simeq 2\pi\left[1 + \frac{3(GM)^2}{\ell^2}\right] \qquad [124]$$

This shows that the angular distance between one perihelion and the next is larger than 2π by

$$6\pi\frac{(GM)^2}{\ell^2} \qquad [125]$$

This quantity gives the angular precession of the perihelion per revolution. Note that the perihelion precesses in the direction of motion, that is, the perihelion advances ($\Delta\phi > 2\pi$; see Fig. 7.1).

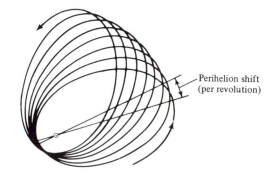

Fig. 7.1 Planetary orbit with perihelion precession. The angular displacement, reckoned around the orbit, between one perihelion and the next is $\Delta\phi > 2\pi$.

Perihelion shift (per revolution)

For an ellipse of semi-major axis a, the perihelion distance is $a(1 - \epsilon)$. By evaluating Eq. [119] at the perihelion, where $\cos(\phi - \phi_0) = 1$, we find

$$\frac{1}{a(1 - \epsilon)} = \frac{GM}{\ell^2}(1 + \epsilon) \qquad [126]$$

or

$$\frac{GM}{\ell^2} = \frac{1}{a(1 - \epsilon^2)} \qquad [127]$$

Accordingly, the angular advance of the perihelion per revolution is, in cgs units,

$$6\pi\frac{GM}{a(1 - \epsilon^2)c^2} \qquad [128]$$

For the case of Mercury, this amounts to 0.1035 arcsec per revolution, or 42.98 arcsec per century.

Table 7.1 gives the observed and predicted precessions for the inner planets, according to a recent investigation of Anderson et al. (1992).

In this table, the values for the observed perihelion precession have been corrected for the purely Newtonian perturbations of the orbits. For instance, the perturbation of Mercury's orbit by other planets contributes a precession of $\simeq 500$ arcsec per century. Furthermore, the equinoctial precession of the celestial coordinates used in planetary observations gives an apparent advance of $\simeq 5000$ arcsec per century. All these known corrections have been subtracted out in Table 7.1. No correction has been made for a possible quadrupole moment of the Sun. The agreement between observed and predicted values is as good as can be expected.

TABLE 7.1 PERIHELION PRECESSION OF PLANETS

Planet	Observed precession*	Predicted precession
Mercury	43.1 ± 0.1 arcsec/century	42.98 arcsec/century
Venus	8.65	8.62
Earth	3.85	3.84
Mars	1.36	1.35

* The values quoted for the observed precessions of the planets are not independent, since in the data analysis a single parameter is used to characterize all these precessions.

The results of Table 7.1 are based on several thousand optical observations of planetary positions obtained with Transit Circles at the U.S. Naval Observatory and also on several hundred radar-ranging observations obtained with the Arecibo, Haystack, and Goldstone radio telescopes. Either the radar pulses were reflected directly off the surfaces of the inner planets or they were reflected off Mariner or Pioneer spacecraft while in orbit around a planet. By a least-squares fit, the data were used to determine the orbital elements of all the planets, as well as the sizes of Venus, Mercury, and Mars and the electron density in the solar corona (these parameters affect the travel time of radar signals). The perturbations of the planetary orbits by each other were evaluated by numerical integration of the equations of motion over the entire twentieth century.

Whereas the relativistic perihelion precessions in the Solar System are very small, the precession can be quite large in the case of close binary star systems. For a system consisting of two white dwarfs or two neutron stars of mass 1 $M_\odot$ separated by a distance of 10^{11} cm, Eq. [128] gives a periastron advance of 3×10^{-5} radian per revolution.* In one year, such a system goes through 10^3 revolutions, and hence the periastron precession amounts to several degrees per year. The binary pulsar PSR 1913+16, discussed in Section 5.4, is an example of a close binary star system, with a large periastron precession of 4.2⁰ per year. Since the masses of the stars are not known, the measured periastron precession cannot be used as a direct test of the theoretical formula for the precession. Instead, as mentioned in Section 5.4, the measured precession is used to determine the masses of the stars, and the rate of energy loss by gravitational radiation is then calculated from these masses and the orbital parameters. Note, however, that if the theoretical formula for the periastron precession were wrong, then the masses would be wrong, and the excellent agreement between the predicted and the observed rates of change of the orbital period of PSR 1913+16 would become an accidental, and incredible, coincidence.

Although the above perturbative solution [123] of the geodesic equation is adequate for the motion of planets around the Sun, a general solution of the geodesic equation is needed for the motion of a particle around, say, a black hole. Such a solution can be given in terms of elliptic integrals. However, it is more illuminating to discuss the motion qualitatively, by means of an effective potential. For this purpose, we rewrite Eq. [112] as

$$(\dot{r})^2 + \left(1 - \frac{2GM}{r}\right)\left(1 + \frac{\ell^2}{r^2}\right) = \mathcal{E}^2 \qquad [129]$$

In this equation, the second term on the left plays the role of an effective potential. We will define this effective potential as

$$\mathcal{V}(r) = \sqrt{\left(1 - \frac{2GM}{r}\right)\left(1 + \frac{\ell^2}{r^2}\right)} \qquad [130]$$

With this definition, at large distances, our effective potential approaches $\mathcal{V}(r) \rightarrow 1 - GM/r + \ell^2/2r^2$. This coincides with the Newtonian effective potential, except for an additive constant. Eq. [129] now becomes

$$(\dot{r})^2 + \mathcal{V}^2(r) = \mathcal{E}^2 \qquad [131]$$

Note that the effective potential is zero at $r = 2GM$. According to Eq.

---

* Although Eq. [128] was derived for the motion of a small mass in the field of a much larger mass, it will give an order-of-magnitude estimate even if the two masses are comparable.

[131], a particle instantaneously at rest ($\dot{r} = 0$) at this point then has zero energy, $\mathcal{E} = 0$ and $E = 0$. Since the energy E includes the rest-mass energy of the particle, this means that for a particle instantaneously at rest at $r = 2GM$, the (negative) gravitational binding energy equals the rest-mass energy.

For a qualitative discussion of the orbital motion, we must examine the turning points. The turning points of the motion are at $\mathcal{V}(r) = \mathcal{E}$. Fig. 7.2 presents plots of $\mathcal{V}(r)$ for several values of the angular momentum ℓ. The centrifugal barrier at small values of r is of finite height, in contrast to the Newtonian case, where the centrifugal barrier tends to infinity as $r \to 0$. Hence, for some values of ℓ and $\mathcal{E}$, the interior turning point is absent--the particle is then pulled into the singularity at $r = 0$ even if it has angular momentum.

The minima in the curves in Fig. 7.2 correspond to stable circular orbits; the maxima correspond to unstable circular orbits. The smallest stable circular orbit corresponds to the inflection point in the curve for $\ell = 2\sqrt{3}$; this smallest stable circular orbit has a radius $r = 6GM$, and it has an energy $\mathcal{E} = \sqrt{8/9}$, or $E = m\sqrt{8/9}$. Note that for angular momentum $\ell < 2\sqrt{3}$ there is no interior turning point (no angular-momentum barrier); if the particle is initially moving inward, it continues to move inward with ever-increasing radial speed $|\dot{r}|$.

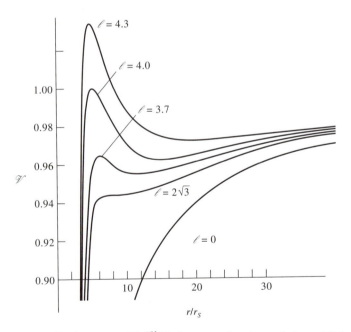

Fig. 7.2   The effective potential $\mathcal{V}(r)$ for several values of the orbital angular momentum ℓ.

7.7 THE PROPAGATION OF LIGHT; THE REDSHIFT

The trajectories of light rays that pass near the Sun are almost straight lines, and the deflection of such light rays can be adequately treated by the linear approximation we discussed in Chapter 4. However, an exact treatment of the deflection of light in the Schwarzschild geometry is of interest when the central mass is compact and the spacetime in its vicinity is strongly curved. Such extreme cases of the Schwarzschild geometry, with strong curvature, are found near neutron stars and near black holes.

The differential equation for the trajectory of a light signal in the Schwarzschild geometry can be obtained by taking a suitable limit in the equation for the trajectory of a particle, Eq. [115]. We begin by rewriting this equation as

$$\frac{\mathcal{E}^2}{\ell^2} - \left(\frac{du}{d\phi}\right)^2 - u^2(1 - 2GMu) = \frac{1 - 2GMu}{\ell^2} \qquad [132]$$

To convert this into an equation for the trajectory of a light signal, we must proceed to the limit of zero particle mass and zero proper time, $m \to 0$ and $d\tau \to 0$. This implies $\ell = r^2 d\phi/d\tau \to \infty$ and $\mathcal{E} = e^N dt/d\tau \to \infty$; however, $\mathcal{E}/\ell$ remains finite (and constant). Accordingly, Eq. [132] becomes

$$\frac{1}{b^2} - \left(\frac{du}{d\phi}\right)^2 - u^2(1 - 2GMu) = 0 \qquad [133]$$

where $b = \ell/\mathcal{E}$. By differentiating this equation with respect to ϕ, we can obtain a second-order differential equation for the trajectory (compare Eq. [116]); but we will deal with the first-order differential equation, since this is more convenient.

If the light ray is incident from a large distance, we can identify the constant b as the impact parameter. This becomes obvious in the limit $u \to 0$ (or $r \to \infty$), where Eq. [133] reduces to $1/b^2 - (du/d\phi)^2 = 0$, which has the solution $u = \phi/b + $ [constant], as expected for a distant $(u \ll 1/b)$ straight line of impact parameter b.

Eq. [133] gives us the relation between the change in azimuthal angle and the change in u, that is, $d\phi = du/\sqrt{1/b^2 - u^2(1 - 2GMu)}$. By integration, we can then find the net change $\Delta\phi$ in the azimuthal angle for a ray that arrives from some large distance, bends around the central mass, and proceeds to a large distance in a new direction. At the point of closest approach, $du/d\phi = 0$. Designating this point by $u = 1/r_0$, we see from Eq. [133] that

$$b = \frac{r_0}{\sqrt{1 - 2GM/r_0}} \qquad [134]$$

This is a cubic equation for r_0, which determines r_0 for a given value of b. Since the trajectory is symmetric about the point of closest approach, we can express the net change $\Delta\phi$ as twice the change that occurs between $u = 0$ and $u = 1/r_0$:

$$\Delta\phi = 2 \int_0^{1/r_0} \frac{du}{\sqrt{1/b^2 - u^2(1 - 2GMu)}} \qquad [135]$$

This integral can be reduced to an elliptic integral. However, it is usually easiest to obtain $\Delta\phi$ by numerical integration of Eq. [135]. Fig. 7.3 displays a plot of the bending angle $\Delta\phi - \pi$ of the light ray vs. impact parameter. This bending angle represents the amount by which the ray is deflected from a straight line (note that $\Delta\phi = \pi$ corresponds to a straight ray). As we can see from the plot, the bending angle becomes infinite as $b \rightarrow 3\sqrt{3}GM$, which corresponds to $r_0 \rightarrow 3GM$. In this limiting case, the ray revolves around the central mass many times, in a quasi-circular orbit of radius $\simeq 3\sqrt{3}GM$. The sphere of radius $r_0 = 3GM$ is called the *photosphere*. A light ray that starts at this exact radius is in an exact circular orbit. However, this circular orbit is unstable, and after a few revolutions the light ray is liable to spiral either in or out. Of course, a large bending angle and a quasi-circular orbit for light rays are possible only if the mass is sufficiently com-

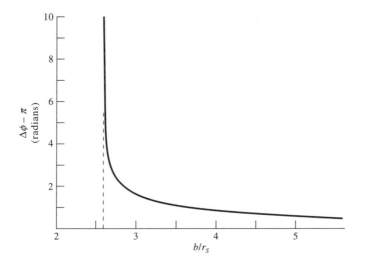

Fig. 7.3 Bending angle vs. impact parameter. (From Ohanian, 1987.)

pact, so the point of closest approach lies outside of the mass distribution. This condition is always satisfied for a black hole, in which all the mass has collapsed to a central singularity.

Fig. 7.4 shows the trajectories of several light rays that approach a compact mass from a large distance with different impact parameters. If the source of light is far to the left, and the observer far to the right, light rays can reach the observer along a path with a deflection angle of less than 2π (perhaps much less than 2π), and also along a multitude of paths with bending angles of more than 2π, which arise from the rays that circle the compact mass one, two, three, or more times. The paths can circle in either a clockwise or a counterclockwise direction. However, as seen by the observer, all such rays with large bending angles will seem to come from the immediate vicinity of the compact mass (their impact parameters of the order of $3\sqrt{3}GM$). If the observer is far from the mass, the angular separations between these images will be small, and the observer will not be able to resolve them. Hence, all these images will appear to merge into a single image seen at (approximately) the position of the compact mass.

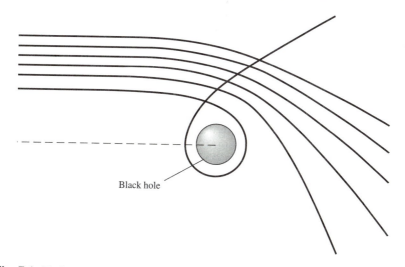

Fig. 7.4 Trajectories of light rays of different impact parameters.

The preceding discussion dealt only with the trajectories of light rays. But it is also of interest to consider the change of frequency of the light during its propagation in the Schwarzschild geometry, that is, the gravitational redshift. In Section 4.2, we already obtained a gravitational time-dilation formula. According to this formula, the ratio of the proper times that two stationary clocks, at radial positions r_1 and r_2, measure between two successive signals sent from r_1 to r_2 is (see Eq. [4.15])

$$\frac{d\tau_2}{d\tau_1} = \frac{\sqrt{g_{00}(r_2)}}{\sqrt{g_{00}(r_1)}} \qquad [136]$$

In the Schwarzschild geometry, this becomes

$$\frac{d\tau_2}{d\tau_1} = \frac{\sqrt{1 - 2GM/r_2}}{\sqrt{1 - 2GM/r_1}} \qquad [137]$$

If the two signals are two successive wave crests of a periodic light wave, then the measured frequencies are $\nu_1 = 1/d\tau_1$ and $\nu_2 = 1/d\tau_2$, and

$$\frac{\nu_1}{\nu_2} = \frac{\sqrt{1 - 2GM/r_2}}{\sqrt{1 - 2GM/r_1}} \qquad [138]$$

This is the (exact) formula for the gravitational redshift of light in the Schwarzschild geometry.

The result for the gravitational redshift of light can also be obtained by an instructive argument involving the Killing vector of the Schwarzschild geometry. Since the Schwarzschild metric is independent of time, there must be a timelike Killing vector, of the form $\xi^\mu = (1, 0, 0, 0)$ (it is easy to verify from Eq. [6.170] that whenever the metric is independent of some coordinate, a constant vector in the direction of that coordinate is a Killing vector). Note that although ξ^μ is constant, ξ_μ is not constant, that is,

$$\xi_\mu = g_{\mu\nu}\xi^\nu = (g_{00}\xi^0, 0, 0, 0) = ((1 - 2GM/r), 0, 0, 0) \qquad [139]$$

Fig. 7.5 shows the worldlines of the two clocks used for the measurement of the frequency of the light signal. Since each clock is stationary, its four-velocity is a timelike vector, or, more precisely, a timelike unit vector ($u^\mu u_\mu = 1$). This means that the four-velocity u^μ of each clock is parallel to the Killing vector ξ^μ, that is, $u^\mu \propto \xi^\mu$. To make this proportionality into an equality, we must renormalize ξ^μ so it becomes a unit vector; this requires dividing ξ^μ by $(\xi^\alpha \xi_\alpha)^{1/2}$,

$$u^\mu = \frac{\xi^\mu}{(\xi^\alpha \xi_\alpha)^{1/2}} \qquad [140]$$

The frequency of the light measured by a clock is proportional to the P_0 component of the energy-momentum vector of a photon evaluated in the local rest frame of the clock, since $h\nu = P_0$. We can also write this as $h\nu = u^\mu P_\mu$, where u^μ is the four-velocity of the clock and P_μ the energy-momentum vector of the photon. We can then write the ratio of the frequencies measured by the two clocks at the radial positions r_1 and r_2 as

$$\frac{\nu_1}{\nu_2} = \frac{u^\mu_{(1)} P_{\mu(1)}}{u^\mu_{(2)} P_{\mu(2)}}$$

$$= \frac{\xi^\mu(1) P_{\mu(1)}}{\xi^\mu(2) P_{\mu(2)}} \times \frac{[\xi^\beta(2)\xi_\beta(2)]^{1/2}}{[\xi^\alpha(1)\xi_\alpha(1)]^{1/2}} \qquad [141]$$

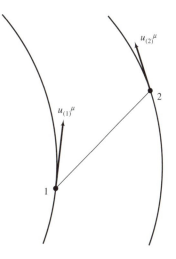

Fig. 7.5 Worldlines of two stationary clocks. The four-velocity u^μ of each clock is parallel to the timelike Killing vector ξ^μ.

But according to the conservation theorem we proved in Section 6.8 (see Eq. [6.187]), $\xi^\mu P_\mu$ is constant during the motion of the photon, and hence the first factor on the right side of Eq. [141] equals 1, and

$$\frac{\nu_1}{\nu_2} = \frac{[\xi^\beta(2)\xi_\beta(2)]^{1/2}}{[\xi^\alpha(1)\xi_\alpha(1)]^{1/2}} \qquad [142]$$

This expresses the gravitational redshift entirely in terms of the Killing vectors evaluated at the two positions. If we insert the explicit values of the Killing vectors into this equation, we find

$$\frac{\nu_1}{\nu_2} = \frac{\sqrt{1 - 2GM/r_2}}{\sqrt{1 - 2GM/r_1}} \qquad\qquad [143]$$

which is the same result as before.

7.8 GEODETIC PRECESSION

One of the characteristic features of curved space is that parallel transport of a vector alters its direction. This suggests that we should try to detect the curvature of spacetime near the Earth by examining parallel transport. But to perform such an experiment, we must first devise some physical procedure for the parallel transport of a vector. In nongravitational physics, our experience with gyroscopes tells us that the physical transport of a gyroscope suspended in frictionless gimbals results in the parallel transport of its direction of spin. However, from this we cannot immediately draw the conclusion that in gravitational physics the transport of such a gyroscope will also result in parallel transport of the direction of its spin.

To find under what conditions transport of a gyroscope results in parallel transport, we begin with the equation of motion for the spin of a rigid body according to Newton's theory. From Section 1.9 we know that a rigid body in a gravitational field is subject to a tidal torque (see Eq. [1.66]), which leads to a rate of change of the spin,

$$\frac{dS^n}{dt} = \epsilon^{kln} R'^k_{0s0}(-I^{ls} + \frac{1}{3}\delta_l{}^s I^{rr}) \qquad\qquad [144]$$

Here R'^k_{0s0} is the Riemann tensor evaluated in the rest frame of the gyroscope. The presence of the Riemann tensor on the right side of Eq. [144] tells us that the equation of motion for the spin does not obey the principle of minimal coupling, and that the transport of the spin of a gyroscope does not imitate parallel transport. However, if we use a spherical gyroscope, with $I^{ls} \propto \delta_l{}^s$, then the tidal torque in Eq. [144] vanishes, and the Newtonian equation of motion becomes $dS^n/dt = 0$. This Newtonian equation remains valid in curved spacetime, in a reference frame in free fall along a geodesic (Fermi coordinates). The Newtonian time coordinate t must now be interpreted as the proper time measured along the geodesic,

$$\frac{dS^n}{d\tau} = 0 \qquad\qquad [145]$$

Accordingly, in the freely falling reference frame, the spin of our spherical gyroscope remains constant in magnitude and direction. This means the spin moves by parallel transport.

A warning: the agreement between parallel transport and gyroscope transport holds only along a geodesic, that is, the gyroscope has to be in free fall. If an extra nongravitational force pushes on the gyroscope and makes it move along some other worldline, not a geodesic, then we cannot introduce local geodesic coordinates at every point of this worldline, and then the equation of motion for the spin does not reduce to the simple form [145]. For example, it is well known that if a gyroscope in *flat* spacetime is transported along a nongeodesic curve, corresponding to accelerated motion, then it will change its direction relative to the fixed stars, and it will therefore deviate from the direction given by parallel transport (in flat spacetime, parallel transport leaves the components of a vector constant in an inertial rectangular coordinate system; thus, it leaves the direction of the vector constant relative to the fixed stars). This precession of an accelerated gyroscope in flat spacetime is called the *Thomas precession*.

In general coordinates, the equation for the parallel transport of the spin vector S^μ involves the Christoffel symbols:

$$\frac{dS^\mu}{d\tau} = - \Gamma^\mu_{\alpha\beta} S^\alpha \frac{dx^\beta}{d\tau} \qquad [146]$$

From this equation we can calculate how the spin of our gyroscope changes direction while the gyroscope moves along some free-fall trajectory. To be specific, let us assume that the gyroscope is in a circular orbit of radius r_0 around the Earth. Experimentally, we measure the change of the direction of spin relative to the fixed stars; this is equivalent to finding the change with respect to a fixed set of coordinates at infinity. Rectangular coordinates are more convenient for the calculation of the change of direction than polar coordinates, since a change of the rectangular components of the spin vector can be immediately attributed to the curvature of spacetime, whereas a change of the polar components contains contributions from both the curvature of the coordinates and the curvature of spacetime. We will therefore use the "isotropic" rectangular coordinates x, y, z introduced in Section 7.4. Since it will suffice to calculate the change of direction to lowest order in GM, we will not need the exact Schwarzschild solution [79], but only the approximate solution [76]:

$$ds^2 \simeq \left(1 - \frac{2GM}{r}\right)dt^2 - \left(1 + \frac{2GM}{r}\right)(dx^2 + dy^2 + dz^2) \qquad [147]$$

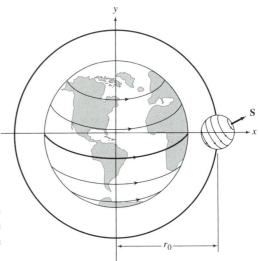

Fig. 7.6 Gyroscope orbiting the Earth in the x-y plane. At one instant, the gyroscope is at the point $x = r_0, y = 0, z = 0$.

We assume that the orbit is in the x-y plane (see Fig. 7.6). For convenience we will evaluate [146] at one point of the orbit, say, the point $x = r_0$, $y = 0$, $z = 0$. All points on a circular orbit are equivalent, and from the rate of change of the spin at one point we can deduce the rate of change at any other point. At the chosen point, the Christoffel symbols are, to lowest order in GM,

$$\Gamma^0{}_{01} = \Gamma^0{}_{10} = GM/r_0{}^2$$
$$\Gamma^1{}_{00} = GM/r_0{}^2$$
$$\Gamma^1{}_{11} = -GM/r_0{}^2$$
$$\Gamma^1{}_{22} = GM/r_0{}^2$$
$$\Gamma^1{}_{33} = GM/r_0{}^2$$
$$\Gamma^2{}_{12} = \Gamma^2{}_{21} = -GM/r_0{}^2$$
$$\Gamma^3{}_{13} = \Gamma^3{}_{31} = -GM/r_0{}^2 \qquad [148]$$

All the other Christoffel symbols are zero.

■ *Exercise 16.*   Obtain these values of the Christoffel symbols from the spacetime interval [147]. Neglect all terms of order $(GM)^2$ or higher. ■

For the spatial components of the spin, Eq. [146] reads

$$\frac{dS^k}{d\tau} = -\Gamma^k{}_{\alpha\beta}S^\alpha\frac{dx^\beta}{d\tau} \qquad [149]$$

To evaluate the right side, we need the four-velocity, which is $dx^\beta/d\tau \simeq (1, 0, v, 0)$. We also need the S^0 component of the spin. To find this component, we note that in the rest frame of the gyroscope, $S'^0 = 0$ and $dx'^\beta/d\tau = (1, 0, 0, 0)$, and therefore $g'_{\mu\nu} S'^\mu dx'^\nu/d\tau = 0$. Consequently, it must also be true that in our isotropic coordinates $g_{\mu\nu} S^\mu dx^\nu/d\tau = 0$, from which

$$S^0 = -\frac{1}{g_{00}}\left[S_x g_{11}\frac{dx}{d\tau} + S_y g_{22}\frac{dy}{d\tau} + S_z g_{33}\frac{dz}{dt}\right]$$

$$\simeq S_y \qquad \qquad [150]$$

We can now evaluate the right side of Eq. [149], and we obtain

$$\frac{dS_x}{d\tau} = -2\frac{GM}{r_0^2}S_y v \qquad\qquad [151]$$

$$\frac{dS_y}{d\tau} = \frac{GM}{r_0^2}S_x v \qquad\qquad [152]$$

$$\frac{dS_z}{d\tau} = 0 \qquad\qquad [153]$$

■ *Exercise 17.* Obtain these equations from Eq. [149]. ■

These equation are valid only at the choosen point $x = r_0$, $y = 0$, $z = 0$. But we can rewrite these equations in a form valid at any point around the orbit if we recognize that the three equations [151]-[153] can be combined into a single (three-dimensional) vector equation:

$$\frac{d\mathbf{S}}{d\tau} = -2\mathbf{v}\cdot\mathbf{S}\,\nabla\Phi + \mathbf{v}\,\mathbf{S}\cdot\nabla\Phi \qquad\qquad [154]$$

where $\Phi = -GM/r_0$ is the Newtonian potential.

Since both the velocity $\mathbf{v}$ and the gradient $\nabla\Phi$ vary with position around the orbit, the behavior of $\mathbf{S}$ given by Eq. [154] is rather complicated, with small periodic oscillations in both the magnitude and the direction of $\mathbf{S}$. However, we are not interested in such periodic wobbles, but only in the long-term, secular change in $\mathbf{S}$. We can calculate this secular change by taking the average of Eq. [154] over an orbit. For this purpose, we express $\mathbf{v}$ and $\nabla\Phi$ as functions of time:

$$\mathbf{v} = -\hat{\mathbf{x}}v\sin\omega t + \hat{\mathbf{y}}v\cos\omega t \qquad\qquad [155]$$

$$\nabla\Phi = \hat{\mathbf{x}}\frac{GM}{r_0^2}\cos\omega t + \hat{\mathbf{y}}\frac{GM}{r_0^2}\sin\omega t \qquad\qquad [156]$$

If we insert these expressions into the right side of Eq. [154], and we average over one period of the orbit, we find

$$\langle \frac{d\mathbf{S}}{d\tau} \rangle = \frac{3}{2} \frac{GM}{r_0^2} v(- \hat{\mathbf{x}} S_y + \hat{\mathbf{y}} S_x) \tag{157}$$

We can rewrite this result more neatly as follows:

$$\langle \frac{d\mathbf{S}}{d\tau} \rangle = \frac{3}{2} \frac{GM}{r_0^3} (\mathbf{r} \times \mathbf{v}) \times \mathbf{S} \tag{158}$$

■ *Exercise 18.*   Perform the time average in Eq. [154], and verify that Eq. [158] is equivalent to Eq. [157]. ■

From Eq. [158] we see that, on the average, **S** precesses about the axis of the orbit with an angular velocity

$$\mathbf{\Omega}_G = \frac{3}{2} \frac{GM}{r_0^3} \mathbf{r} \times \mathbf{v} \tag{159}$$

This is the *geodetic precession*, or the *de Sitter precession*.

The geodetic precession has sometimes been described as analogous to the Thomas precession of special relativity. But this analogy is somewhat misleading: The Thomas precession in the flat spacetime of special relativity results from nongeodesic motion, that is, it results from accelerated motion brought about by the push of some force (without torque) acting on a gyroscope or on a spinning particle, such as an electron. To find a true analog of the Thomas precession in curved spacetime, we would have to examine the behavior of a gyroscope when some extra, nongravitational force makes it deviate from geodesic motion. A Thomas precession would then arise as an extra contribution to the precession, in excess of the contribution from parallel transport.

Although we derived Eq. [159] for a circular orbit, it can be shown that the result is valid in general. Note that the time-average rate of change of the spin can be written in the concise form

$$\langle \frac{d\mathbf{S}}{d\tau} \rangle = \frac{3}{2} \frac{GM}{mr_0^3} \mathbf{L} \times \mathbf{S} \tag{160}$$

where $\mathbf{L} = m\mathbf{r} \times \mathbf{v}$ is the orbital angular momentum of the gyroscope. In Newtonian mechanics, we would interpret the right side of Eq. [160] as a torque due to spin–orbit coupling with a potential (in cgs units)

$$U = \frac{3}{2} \frac{GM}{mc^2 r_0{}^3} \mathbf{L \cdot S} \qquad\qquad [161]$$

■ *Exercise 19.* Show that the "torque" on the right side of Eq. [160] is correctly given by $- \partial U/\partial \theta$ (where θ is the angle between **L** and **S**). ■

A spinning particle with the potential [161] will suffer not only a spin precession, but also a translational acceleration caused by a spin-orbit force. This suggests that a spinning body in a gravitational field will deviate from geodesic motion. This conjecture is confirmed by a careful examination of the equation for the translational equation of motion for spinning bodies in curved spacetime (Papapetrou, 1951). Although the deviation from geodesic motion is quite small and practically undetectable, it is of some theoretical importance in that it shows that Galileo's principle of equivalence fails for spinning particles. For example, electrons, protons, neutrons, etc., do not move exactly on geodesics in a gravitational field.

■ *Exercise 20.* What is the ratio of $\partial U/\partial r$ to $GMm/r_0{}^2$ for an electron at the surface of the Earth? ■

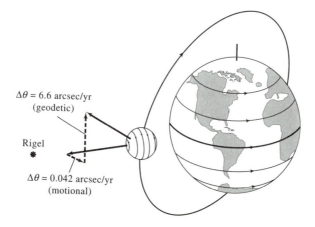

Fig. 7.7 A gyroscope in a 650-km polar orbit. The spin of the gyroscope is (initially) in the plane of the orbit. The geodetic precession causes the spin to swing through 6.6 arcsec per year, in the direction of the orbital motion. In addition, the Lense-Thirring precession causes the spin to swing through 0.042 arcsec per year, in the direction perpendicular to the orbit.

For a gyroscope in orbit 650 km above the Earth, with the spin in the plane of the orbit (see Fig. 7.7), the geodetic precession [159] amounts to

$$\frac{3}{2}\frac{GM}{c^2r_0{}^2}\sqrt{\frac{GM}{r_0}} \simeq 6.6 \text{ arcsec per year} \qquad [162]$$

Incidentally: the calculation of the rate of change of spin in Eqs. [148]-[158] did not use the exact Schwarzschild solution and could have been carried out in Chapter 4 in the context of the linear theory. But at that stage we did not have the necessary mathematical tools to deal with parallel transport.

For a gyroscope in orbit around the Earth, there is an additional precession effect caused by the rotation of the Earth. The gravitational field of the Earth is not exactly a Schwarzschild field; rather, it is the field of a rotating mass (see Eq. [4.94]). If we take into account the extra "gravimagnetic" terms in the metric associated with the rotation of the Earth, we find an extra precession, called the *Lense-Thirring precession* (see Eq. [4.100]),

$$\mathbf{\Omega}_{LT} = \frac{G}{r_0{}^3}\left[\frac{3}{r_0{}^2}\langle \mathbf{r}\ \mathbf{r}\cdot\mathbf{S}_E\rangle - \mathbf{S}_E\right] \qquad [163]$$

where $\mathbf{S}_E$ is the spin angular momentum of the Earth.

In Newtonian terms, the precession [163] may be thought of as due to a spin-spin coupling between the spins of the gyroscope and the Earth. For a gyroscope in a polar orbit at 650 km, the average precession rate that results from Eq. [163] is $\simeq 0.042$ arcsec per year, in the same direction as the Earth's rotation.

■ *Exercise 21.* Perform the time average in Eq. [163] for a circular polar orbit and show that the result is

$$\mathbf{\Omega}_{LT} = \frac{G}{2r_0{}^3}SE \qquad ■$$

In Section 4.7 we obtained this precession by analogy with the precession of a magnetic moment in a magnetic field. But the result can also be obtained by evaluating the change of the spin vector during parallel transport in the field of a rotating mass.

■ *Exercise 22.* Repeat the calculations [148]-[158], taking into account the extra off-diagonal terms in the metric [4.94]. ■

The development of an experiment to measure both the geodetic and the Lense-Thirring precession with gyroscopes was initiated by

Fairbank, Everitt et al. at Stanford University in the 1970s, and this experiment is expected to be ready for operation in 1995 (Turneaure et al., 1986). This Stanford Relativity Gyroscope Experiment, also called Gravity Probe B, will use gyroscopes installed on a satellite to be launched into a polar orbit from the space shuttle. The rotors of the gyroscopes are quartz spheres, 3.8 cm in diameter, suspended in their housings by electrostatic forces supplied by capacitor plates (see Fig. 7.8). In order to avoid any extra precession produced by the tidal torque [144], the quartz spheres must be accurately round and homogeneous; deviations from spherical shape larger than 10^{-6} cm would be unacceptable. The quartz spheres are coated with a thin layer of superconducting niobium. When the spheres are spinning, the rotating superconducting niobium generates a magnetic moment (London moment) parallel to the spin. Any change in the direction of the magnetic moment is detected by the current it induces in a stationary pickup loop surrounding the sphere; this induced current serves as an indicator of the change in the direction of spin. The gyroscopes are surrounded by a large dewar filled with superfluid liquid helium,which keeps the temperature of the instrument package at 1.8 K (see Fig. 7.9). The longitudinal axis of the satellite is held in a fixed orientation by an attitude control system consisting of thrusters governed by a telescope locked on a guide star (Rigel) located in the plane of the orbit. The precession of the gyroscopes relative to the body of the satellite is therefore equivalent to their precession relative to an inertial reference frame at infinity (however, a correction for the aberration of starlight and for the relativistic deflection of light by the gravitational field of the Earth must be applied). The satellite is also equipped with a drag-free control system consisting of thrusters that compensate for residual atmospheric friction. The thrusters are governed by a proof mass floating freely in a chamber within the satellite, where it is shielded from atmospheric friction. Sensors detect any incipient deceleration of the body of the satellite relative to this free-floating mass and send suitable signals to the thrusters so as to keep the body of the satellite at rest relative to the free-floating mass.

When de Sitter (1916) first considered the geodetic precession effect, there were no prospects of a test with gyroscopes. Instead, de Sitter proposed that the effect could be detected by a precession that the gravitational field of the Sun produces in the orbit of the Moon around the Earth. We can regard the Earth-Moon system as a gyroscope in free fall in the gravitational field of the Sun. The axis of spin of this gyroscope, that is, the axis of the geocentric orbit of the Moon, then suffers a geodetic precession according to Eq. [159]. The predicted precession rate amounts to 0.0192 arcsec per year. Although this precession is too small to be detected by conventional astronomical observations, it can be detected by lunar laser-ranging, which permits us to determine any change in the direction of the Moon's perigee with extreme precision. Analysis of the accumulated lunar laser-ranging data confirms the predicted value of the geodetic precession to

Fig. 7.8   Quartz sphere used as rotor for the gyroscope and its housing. (Courtesy C. W. F. Everitt, Stanford University.)

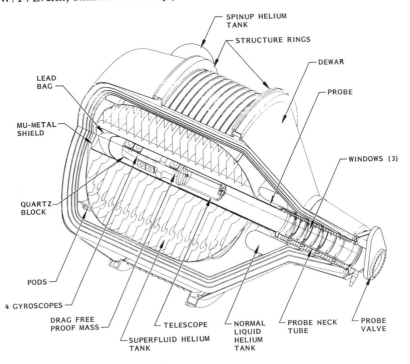

Fig. 7.9   Instrument package and dewar for the Stanford Relativity Gyroscope Experiment. (Courtesy C. W. F. Everitt, Stanford University.)

within the observational error, about 2% (Bertotti et al., 1987; Dickey et al., 1989).

The geodetic precession effect has also been detected in the Hulse-Taylor binary pulsar. This spinning neutron star is effectively a gyroscope, and its axis of spin is expected to precess as it orbits in the gravitational field of its companion star (Damour and Ruffini, 1974). In consequence of the precession, the pulsar radio beam gradually dips in relation to the Earth. Monitoring of the pulse strength received at the Earth over several years has revealed a gradual change in the intensity profile of the pulses, consistent with the expected precession rate of 1.21⁰ per year (Weisberg, Romani, and Taylor, 1989). However, since we do not know the detailed angular shape of the beam, we cannot make any quantitative comparison between theory and observation.

7.9 MACH'S PRINCIPLE

The conclusion that we drew from the principle of general invariance --that the metric $g_{\mu\nu}$ must be a dynamical field responsive to the state of matter--agrees with certain ideas put forward by Mach in his critical discussion of Newton's concept of absolute space.

In the view of Newton, acceleration is absolute. The inertial pseudo-forces give an absolute measure of acceleration in space. In his *Principia* Newton stated:

> The effects which distinguish absolute from relative motion are the forces of receding from the axis of circular motion . . . If a vessel, hung by a long cord, is so often turned about that the cord is strongly twisted, then filled with water, and held at rest together with the water; thereupon, by the sudden action of another force, it is whirled about the contrary way, and while the cord is untwisting itself, the vessel continues for some time in this motion; the surface of the water will at first be plane, as before the vessel began to move; but after that, the vessel, by gradually communicating its motion to the water, will make it begin sensibly to revolve, and recede by little and little from the middle, and ascend to the sides of the vessel, forming itself into a concave figure (as I have experienced), and the swifter the motion becomes, the higher will the water rise, till at last, performing its revolutions in the same times with the vessel, it becomes relatively at rest in it. This ascent of the water shows its endeavor to recede from the axis of its motion; and the true and absolute circular motion of the water, which is here directly contrary to the relative, becomes known, and may be measured by this endeavor. At first, when the relative motion of the water in the vessel was greatest, it produced no endeavor to recede from the axis; the water showed no tendency to the circumference, nor any ascent towards the sides of the vessel, but remained of a plain surface, and therefore its true circular motion had not yet begun. But afterwards, when the relative motion

of the water had decreased, the ascent thereof towards the sides of the vessel proved its endeavor to recede from the axis; and this endeavor showed the real circular motion of the water continually increasing, till it had acquired its greatest quantity, when the water rested relatively in the vessel. (Newton, 1686)

In rebuttal, Mach argued that the inertial forces should not be regarded as indicating motion in absolute space, but rather as indicating motion relative to the masses in the entire universe:

> The behavior of . . . bodies relative to [a given system of reference] may be traced to their behavior relative to the distant celestial bodies. If we were to claim that we know more about the motion of bodies than the hypothetical behavior relative to the celestial bodies suggested by experience, we would render ourselves culpable of dishonesty. When accordingly, we say that a body maintains its direction and speed in space, then this is nothing but an abbreviated reference to a consideration of the entire universe . . .
>
> . . . The experiment of Newton with the rotating vessel only teaches us that the relative rotation of the water against the walls of the vessel does not generate any appreciable centrifugal forces, but that the relative rotation against the mass of the earth and of the celestial bodies does generate such a force. Nobody can predict how the experiment would turn out quantitatively and qualitatively, if the walls of the vessel were to get thicker and more massive, ultimately a few miles thick . . . (Mach, 1933)

In essence, Mach's view is that the inertial forces associated with the motion of a given mass are generated by all the other masses in the universe. Consequently, Wheeler proposed to formulate Mach's principle as follows (Wheeler, 1964):

The inertial properties of an object are determined by the energy-momentum throughout all space.

On the basis of Mach's ideas we would expect that if a large mass is accelerated, it will tend to "drag" the inertial reference frame of nearby regions along and induce inertial forces in nearby bodies. If the large mass has a linear acceleration, nearby bodies should experience a force in the same direction as this acceleration; if the large mass rotates, nearby bodies should experience centrifugal and Coriolis forces (Einstein, 1955). Of course these effects are not expected to be of large magnitude--the mass of a planet, a star, or even a galaxy is very small compared with the total mass in the universe. But if we perform our experiments very near a large, massive body, the effects may be detectable.

Einstein's theory of gravitation agrees with Mach's principle in that it predicts such drag effects. For example, the rotation of the Earth is predicted to drag the local inertial reference frames in its vicinity along. The Lense-Thirring precession of the spin of a gyroscope can be regarded as a manifestation of such a drag effect. As we saw in the preceding section, near the surface of the Earth, a free gyroscope with its spin perpendicular to the axis of rotation of the Earth is predicted to precess in the direction of rotation (see Fig. 7.7). An apparently paradoxical feature of the Lense-Thirring effect is that for a gyroscope located over the equator and with its spin parallel to the axis of rotation of the Earth, the drag induces a precession *opposite* to S_E (with $r \cdot S_E = 0$, Eq. [163] gives $\Omega_{LT} \propto - S_{E_0}$). This was explained by Schiff (1960) in terms of the decrease of the drag effect with distance: for a gyroscope over the equator, the side of the rotor nearest the Earth is dragged with the Earth more than the side farthest from the Earth; hence the induced rotation is opposite to the rotation of the Earth. Crudely, we can think of the Earth and the gyroscope as two adjacent wheels, with their rims coupled (weakly) by "friction"; the wheels will then rotate in opposite directions. For a gyroscope placed over the pole, the precession is of course in the direction of S_E.

Naively, we might also expect that Mach's principle should lead to another effect: when bodies are placed near a large mass, their inertial mass should increase, with all the masses increasing in proportion (Einstein, 1955). However, from an operational point of view such a phenomenon makes no sense. Masses are defined relative to a standard of mass, and therefore a proportional change in all masses is meaningless.

Similar arguments (Dicke, 1961) can be given to rule out the conjecture that a large mass might produce an anisotropy in the inertia of bodies moving in its vicinity. For example, one might conjecture that the large mass at the center of our galaxy increases the inertial mass for, say, motion towards or away from the center, and reduces it for motion in the transverse direction. But mass is a Lorentz *scalar*, and a dependence of mass on direction makes no sense at all. Hence the Hughes-Drever experiment, and other such experiments (Hughes, 1964) which sought to test whether the motion of electrons in lithium atoms and other atoms displays any anisotropy with respect to the Galactic center, should be regarded as a test of the isotropy of electromagnetic forces rather than inertial forces. The negative results of these experiment tell us that, on the freely falling Earth, the dielectric properties of space are isotropic.

We saw that Einstein's theory displays certain specific "Machian" effects. More generally, we can say that in Einstein's theory the inertial properties of space are determined (locally) by the metric tensor through the geodesic equation. This led Einstein to formulate Mach's principle as a connection between metric and energy-momentum (Einstein, 1955, p. 107): *The spacetime geometry is entirely determined by the energy-momentum tensor of matter.*

However, this formulation runs into a difficulty. The equations that relate the geometry to the energy-momentum are the Einstein equations. These are differential equations for $g_{\mu\nu}$, and in order to determine their solution uniquely it is necessary to specify not only the source terms, that is, the energy-momentum tensor, but also the boundary conditions on $g_{\mu\nu}$. The need for boundary conditions can be avoided only in the case of a (spatially) closed universe, that is, a universe that has no boundaries. Einstein regarded this as a strong argument in favor of a closed universe (Einstein, 1955, p. 107).

Furthermore, Wheeler has pointed out that even in a closed universe it is necessary to specify more than just the energy-momentum tensor. In the general time-dependent case, the Einstein equations determine a unique solution only if the *initial values* for $g_{\mu\nu}$ and $\partial g_{\mu\nu}/\partial t$ are given. Thus, in the context of Einstein's theory, Wheeler proposed a more precise formulation of Mach's principle, as follows (Wheeler, 1964, p. 333):

The spacetime geometry is uniquely determined by the energy-momentum and by the initial conditions on the geometry.

In a broad sense, Mach's principle merely says that matter acts on the geometry. Since we already know, from Einstein's equations, how the metric tensor is related to the distribution of energy-momentum, Mach's principle does not really tell us anything new, and it plays only a secondary, supportive role in our development of gravitational theory.

FURTHER READING

The clear distinctions between covariance and invariance and between absolute and dynamical objects are due to J. L. Anderson and are rigorously spelt out by him in his book, *Principles of Relativity Physics* (Academic Press, New York, 1967), and also in his article "Relativity Principles and the Role of Coordinates in Physics," in Chiu, H., and Hoffmann, W. F., eds., *Gravitation and Relativity* (Benjamin, New York, 1964). Note that whereas the article uses the misleading term *relativity group*, the book uses the much better terms *symmetry group*, or *invariance group*.

Wigner, E. P., *Symmetries and Reflections* (Indiana University Press, Bloomington, 1967) gives a nice qualitative description of the differences between algebraic (or geometric) symmetries and dynamical symmetries.

An interesting discussion of general invariance as a gauge symmetry, analogous to the gauge symmetry of electrodynamics, is contained in the article "Invariant Theoretical Interpretation of Interaction" by Utiyama (1956).

Pais, A., *Subtle is the Lord* (Clarendon Press, Oxford, 1982) provides a fascinating historical account of Einstein's gradual discovery of his equations.

The term "geometrodynamics" was originally introduced by Wheeler in a broad sense to describe a theory that attempts to interpret matter as an excited state of the geometry. According to this view "there is nothing in the world except empty curved space. Matter, charge, electromagnetism, and other fields are manifestations of the bending of space. *Physics is geometry*" (Wheeler, 1962). The program of this aspiring project is outlined in the books Wheeler, J. A., *Einsteins Vision* (Springer-

Verlag, Berlin, 1968) and *Geometrodynamics* (Academic Press, New York, 1962), and also in the article "Geometrodynamics" by Fletcher, J. G., in Witten, L, ed., *Gravitation: An Introduction to Current Research* (Wiley, New York, 1962).

Fock, V., *The Theory of Space, Time, and Gravitation* (Pergamon, Oxford, 1964), strongly condemns Einstein for describing the theory of gravitation as "general relativity." Although Fock's criticisms are excessive, they do help to clear the air.

A complete list of the various routes that lead to Einstein's equations is given in Misner, C. W., Thorne, K. S., and Wheeler, J. A., *Gravitation* (Freeman, San Francisco, 1973). This book also offers a direct geometrical interpretation of the Einstein equations, based on differential forms. This interpretation is explained in greater detail in Wheeler, J. A., *A Journey into Gravitation and Spacetime* (Freeman, New York, 1990), which is addressed at a "general," mathematically disadvantaged, audience. By an ingenious geometrical argument, Wheeler succeeds in extracting the exact Schwarzschild solution from his form of the Einstein equations, without having to solve any differential equations.

Somewhat different treatments of Einstein's equations, the Schwarzschild solution, and other solutions will be found in the books recommended in the reading list of Chapter 6.

The "interior" Schwarzschild solution (that is, the solution for the metric inside a spherically symmetric ball of incompressible fluid) is given in several of these books, for instance, Møller, C., *The Theory of Relativity* (Clarendon Press, Oxford, 1952) and Tolman, R. C., *Relativity, Thermodynamics and Cosmology* (Clarendon Press, Oxford, 1934). This solution is of little astrophysical relevance.

A rich collection of articles dealing with a variety of topics in Einstein's theory is contained in Held, A., *General Relativity and Gravitation* (Plenum Press, New York, 1980).

Comprehensive discussions of the initial-value problem for Einstein's equations and the implications of causality for the topology of spacetime are offered by Hawking, S. W., and Ellis, G. F. R., *The Large Scale Structure of Space-Time* (Cambridge University Press, Cambridge, 1973). This book presupposes some knowledge of differential geometry and topology. Somewhat briefer discussions are given by Wald, R. M., *General Relativity* (University of Chicago Press, Chicago, 1984).

The derivation of the general equations of motion of test particles, with and without spin, from the Einstein field equations is presented in the articles "Spinning Test Particles in General Relativity I" by Papapetrou, A., Proc. Roy. Soc. London **A209**, 248 (1951) and "Spinning Test Particles in General Relativity II" by Corinaldesi, E., and Papapetrou, A., Proc. Roy. Soc. London **A209**, 259 (1951). A different and older method is presented in the article "The Motion of Test Particles in General Relativity" by Infeld, L., and Schild, A., Rev. Mod. Phys. **21**, 408 (1949).

A careful mathematical treatment of Fermi coordinates is given in Synge, J. L., *Relativity: The General Theory* (North-Holland, Amsterdam, 1971). The physical interpretation of the curvature tensor and a method for measuring it are also described in this book. Other methods for the measurement of the curvature tensor are described by Bertotti, B., "The Theory of Measurement in General Relativity," in Møller, ed., *Evidence for Gravitational Theories* (Academic Press, New York, 1962) and by Wigner, op. cit.

The general properties of the motion of test particles in a Schwarzschild geometry are described in Rees, M. R., Ruffini, R., and Wheeler, J. A., *Black Holes, Gravitational Waves, and Cosmology* (Gordon and Breach, New York, 1974) and in Misner, Thorne, and Wheeler, op. cit.

Landau, L. D., and Lifshitz, E. M., *The Classical Theory of Fields* (Addison-Wesley, Reading, 1962), obtain the perihelion precession by an elegant method based on the Hamilton-Jacobi equation.

A nice explanation of how the curvature of the spatial part of the Schwarzschild geometry contributes to the perihelion precession and the light deflection is given by Rindler, W., *Essential Relativity* (Springer-Verlag, New York, 1977).

The theory of the geodetic and Lense-Thirring precessions of a gyroscope is well explained in the article "Motion of a Gyroscope According to Einstein's Theory of Gravitation" by Schiff (1960). Details on the proposed experimental test will be found in the article by Turneaure, J. P., Everitt, C. W. F., Parkinson, B. W., et al., "The Gravity-Probe-B Relativity Gyroscope Experiment," in Ruffini, R., ed., *Proceedings of the Fourth Marcel Grossmann Meeting on General Relativity* (Elsevier, Amsterdam, 1986).

Mach, E., *Die Mechanik* (Brockhaus, Wiesbaden, 1933; translated as *The Science of Mechanics*, see the reading list in Chapter 1) contains a profound criticism of Newton's concepts of space, time, and mass. This book exerted a strong influence on Einstein.

Mach's principle is discussed from somewhat different points of view in the following articles:

Wheeler, J. A., "Mach's Principle as Boundary Condition for Einstein's Equations," in Chiu, H.-Y., and Hoffman, W. F., eds., *Gravitation and Relativity* (Benjamin, New York, 1964).

Dicke, R. H., "The Many Faces of Mach," in Chiu, H.-Y., and Hoffman, W. F., eds., *Gravitation and Relativity* (Benjamin, New York, 1964).

Dicke, R. H., "Mach's Principle," in Klauder, J. R., ed., *Magic Without Magic: John Archibald Wheeler* (Freeman, San Francisco, 1972).

Elementary discussions are given in Sciama, D. W., *The Unity of the Universe* (Doubleday, Garden City, 1961) and Sciama, D, W., *The Physical Foundations of General Relativity* (Doubleday, Garden City, 1969); the first is philosophical and the second speculates on a very literal interpretation of Mach's principle that lies outside of geometrodynamics.

An impressive collection of nearly 500 challenging problems, and their solutions, covering many puzzling questions in special relativity, general relativity, gravitation, astrophysics, and cosmology is contained in Lightman, A. P., Press, W. H., Price, R. H., and Teukolsky, S. A., *Problem Book in Relativity and Gravitation* (Princeton University Press, Princeton, 1975).

REFERENCES

Anderson, J. D., Campbell, J. K., Slade, M. A., Jurgens, R. F., Lau, E. L., Newhall, X X, and Standish, E. M., "Recent Developments in Solar-System Tests of General

Relativity," in Sato, H., and Nakamura, T., eds., *Sixth Marcel Grossmann Meeting on General Relativity* (World Scientific, Singapore, 1992); see also Anderson, J. D., Keesey, M. S. W., Lau, E. L., Standish, E. M., and Newhall, X X, Acta Astronautica **5**, 43 (1978).

Anderson, J. L., *Principles of Relativity Physics* (Academic Press, New York, 1967).

Bertotti, B., Ciufolini, I., and Bender, P. L., Phys. Rev. Lett. **58**, 1062 (1987).

Carroll, S. M., Press, W. H., and Turner, E. L., Ann. Rev. Astron. Astrophys. **30**, 499 (1992).

Damour, T., and Ruffini, R., Comptes Rendus **279A**, 971 (1974).

Deser, S., Gen. Rel. and Grav. **1**, 9 (1970).

de Sitter, W., Mon. Not. Roy. Astron. Soc. **77**, 155 and 481 (1916).

Dicke, R. H., Phys. Rev. Lett. **7**, 359 (1961).

Dickey, J. O., Newhall, X. X., and Williams, J. G., Adv. in Space Res. **9**, 75 (1989).

Einstein, A., Ann. Physik **35**, 898 (1911); a translation is given in Lorentz, H. A., Einstein, A., Minkowski, H., and Weyl, H., *The Principle of Relativity* (Methuen, London, 1923).

Einstein, A., Ann. Physik **55**, 241 (1918). Quoted with the permission of the Estate of Albert Einstein.

Einstein, A., *The Meaning of Relativity* (5th ed., Princeton University Press, Princeton, 1955), p. 100.

Gupta, S. N., Rev. Mod. Phys. **29**, 334 (1957).

Hughes, V. W., in Chiu, H.-Y., and Hoffman, W. F., eds., *Gravitation and Relativity* (Benjamin, New York, 1964).

Kretschmann, E., Ann. Physik **53**, 575 (1917).

Lense, J., and Thirring, H., Phys. Zeits. **19**, 156 (1918).

Lorentz, H. A., Einstein, A., Minkowski, H., and Weyl, H., *The Principle of Relativity* (Methuen, London, 1923). Quoted with the permission of the Estate of Albert Einstein.

Mach, E., *Die Mechanik* (9th ed., Brockhaus, Wiesbaden, 1933), pp. 226-27.

Møller, C., *The Theory of Relativity* (Clarendon Press, Oxford, 1952), p. 250.

Newton, I., *Principia* (University of California Press, Berkeley, 1962), vol. 1, pp. 10-11. Originally published by the University of California Press; reprinted by permission of The Regents of the University of California.

Ohanian, H. C., J. Math. Phys. **14**, 1892 (1973).

Ohanian, H. C., Am J. Phys., **55**, 428 (1987).

Page, T. L., Ap. J. **116**, 63 (1952); ibid., **132**, 910 (1960).

Papapetrou, A., Proc. Roy. Soc. London **A209**, 248 (1951).

Shapiro, I. I., Counselman, C. C., and King, R. W., Phys. Rev. Lett. **36**, 555 (1976).

Schiff, L. I., Proc. Natl. Acad. Sci. (U.S.) **46**, 871 (1960).

Schweber, S. S., *An Introduction to Relativistic Quantum Field Theory* (Harper and Row, New York, 1961), p. 207.

Soper, D. E., *Classical Field Theory* (Wiley, New York, 1976).

Standish, E. M., "Radar and Spacecraft Ranging to Mercury between 1966 and 1988," in International Astronomical Union, Fifth Asian-Pacific Regional Astronomy Meeting, p.21.

Turneaure, J. P., Everitt, C. W. F., and Parkinson, B. W., et al., "The Gravity-Probe-B Relativity Gyroscope Experiment," in Ruffini, R., ed., *Proceedings of the Fourth Marcel Grossmann Meeting on Relativity* (Elsevier, Amsterdam, 1986).

Trautman, A., Usp. Fiz. Nauk. **89**, 3 (1966); translated in Sov. Phys. Uspekhi **9**, 319 (1966).

Utiyama, R., Phys. Rev. **101**, 1597 (1956).

Weisberg, J. M., Romani, R. W., and Taylor, J. H., Astrophys. J. **347**, 1030 (1989).

Wheeler, J. A., *Geometrodynamics* (Academic Press, New York, 1962), p. 225.

Wheeler, J. A., "Mach's Principle as Boundary Condition for Einstein's Equations," in Chiu, H.-Y., and Hoffman, W. F., eds., *Gravitation and Relativity* (Benjamin, New York, 1964).

Weinberg, S., in Deser, S., Grisaru, M., and Pendleton, H., eds., *Lectures on Elementary Particles and Quantum Field Theory* (MIT Press, Cambridge, 1970), vol. 1.

PROBLEMS

1. Show that none of the following equations is covariant (A^μ and B^μ are vectors, $A_{\mu\nu}$ is a tensor):

$$A_\mu = \frac{1}{B^\mu}$$
$$g_{\mu\nu}A^\alpha = \Gamma^\alpha_{\mu\nu}$$
$$A^{\mu\nu}_{,\sigma} = 0$$
$$A^0 = B^1$$

2. Show that Newton's equation of motion, $(\partial/\partial x^0)p^k = F^k$, can be put into the generally covariant form

$$\frac{\partial x'^\alpha}{\partial x^0}\frac{\partial}{\partial x'^\alpha}\left(\frac{\partial x^k}{\partial x'^\mu}p'^\mu\right) = \left(\frac{\partial x^k}{\partial x'^\mu}\right)F'^\mu$$

[Hint: Define $p'^\mu \equiv (\partial x'^\mu/\partial x^l)p^l$ and $F'^\mu \equiv (\partial x'^\mu/\partial x^l)F^l$.]

3. Show that Einstein's equation can also be written as

$$R_{\mu\nu} = -\frac{8\pi G}{c^4}\left(T_{\mu\nu} - \tfrac{1}{2}g_{\mu\nu}T\right)$$

4. Show that Eq. [82] is a solution of the Einstein equation with a nonzero cosmological constant.

5. The total mass and radius of some multiple systems of galaxies are listed in Table 7.2. For a galaxy at the outer edge of each of these systems, calculate the ratio of the Newtonian gravitational force on one member of such a system to the

"harmonic-oscillator force" associated with the cosmological term (see Eq. [52]). Assume that $\Lambda = 10^{-35}/\text{s}^2$. Is the cosmological term important?

TABLE 7.2 SYSTEMS OF GALAXIES

Multiple system	Number of galaxies	M	r
Typical binary	2	$10^{11} \, M_\odot$	10^{23} cm
M 81 (inner part)	5	$4 \times 10^{11} \, M_\odot$	10^{24} cm
Coma cluster	1000	$3 \times 10^{15} \, M_\odot$	10^{25} cm

6. Show that in a region in which the energy-momentum tensor differs from zero, the divergence of the tidal force (in geodesic coordinates) is

$$\frac{\partial f'^k}{\partial \xi'^k} = -8\pi G(T'_{00} - \tfrac{1}{2}T')m$$

Check that this agrees with Exercise 1.13.

7. Consider a spacetime with a spacetime interval

$$d\tau^2 = e^{-2ax} \, dt^2 - dx^2 - dy^2 - dz^2$$

where a is a constant.

(a) Find all the nonzero Christoffel symbols and find the geodesic equation for $d^2x/d\tau^2$.

(b) Show that for a particle of instantaneously zero velocity ($dx/dt = dy/dt = dz/dt = 0$), the acceleration $d^2x/d\tau^2$ is constant (independent of position),

$$\frac{d^2x}{d\tau^2} = a$$

Because of this, the spacetime may be said to describe a uniform gravitational field in the x direction.

(c) Find the components $R^1{}_{001}$, $R^0{}_{110}$, $R^1{}_{010}$, and $R^0{}_{101}$ of the Riemann tensor and show that the other components of this tensor are zero.

(d) From the Einstein equations, find the components T_{00}, T_{11}, T_{22}, and T_{33} of the energy-momentum tensor that generates this gravitational field. Is this a physically acceptable energy-momentum tensor?

8. Find a formula for the actual measured volume in the spherical shell $r_1 < r < r_2$ in a space with the Schwarzschild metric [81]. If the shell is taken around the Sun with $r_1 = R_\odot$ and $r_2 = 2R_\odot$, by how many cubic centimeters does the volume differ from the result expected in a flat space? What fractional deviation does this represent?

9. Consider two concentric coplanar circles, r = constant, in the Schwarzschild geometry. Suppose that the measured lengths of their circumferences are L_1 and L_2.
(a) What is the radial coordinate difference Δr between these circles? What is the measured radial distance between them?
(b) Take two circles around the Sun with $L_1 = 2\pi R_\odot$ and $L_2 = 4\pi R_\odot$. By how many centimeters does the measured radial distance between them differ from the result expected in a flat space?

10. An observer at $r = r_1$ in a Schwarzschild field sends a light signal in the radial direction toward $r = r_2$ (where $r_2 > r_1$).
(a) What is the instantaneous coordinate velocity dr/dt of the signal?
(b) Suppose the signal is reflected at $r = r_2$ and returns to r_1. How long, as measured in t time, does the signal take to return?
(c) How long does the signal take to return according to the clock of the observer at $r = r_1$?

11. Compare the exact time-dilation formula with the approximate formula given by the linear approximation, Eq. [4.16],

$$\frac{d\tau_2}{d\tau_1} \simeq 1 - \frac{GM}{r_2} + \frac{GM}{r_1}$$

What is the fractional difference between the time dilations predicted by the exact and approximate formulas if the first clock is at the surface of the Sun and the second at infinity? What if the first clock is at the surface of a neutron star, with $M = M_\odot$ and R = 10 km?

12. Show that for a particle in an orbit in a Schwarzschild field

$$\frac{d\phi}{dt} = \frac{\ell}{r^2 \mathcal{E}}\left(1 - \frac{2GM}{r}\right)$$

and show that for a *circular* orbit

$$\mathcal{E}^2 = \left(1 - \frac{2GM}{r}\right)\left(1 + \frac{\ell^2}{r^2}\right)$$

Use these results to prove that a particle in a stable circular orbit of radius r_0 (see Problem 16 for the value of r_0) satisfies Kepler's law

$$\left(\frac{d\phi}{dt}\right)^2 r_0^3 = GM = \text{[constant]}$$

13. A particle is to be launched in the outward radial direction from the point $r =$

$4GM$ in the Schwarzschild spacetime.

(a) With what speed $dr/d\tau$ must the particle be launched from its starting point if it is to reach $r = 8GM$ with zero speed?

(b) How much proper time does the particle take for this trip?

14. Suppose that a given gravitational field has a spacetime interval (in isotropic coordinates)

$$ds^2 = \left[1 - \frac{2GM}{r'} + \frac{aG^2M^2}{r'^2}\right]dt^2$$

$$- \left[1 + \frac{2GM}{r'} + \frac{bG^2M^2}{r'^2}\right](dr'^2 + r'^2d\theta^2 + r'^2\sin^2\theta\, d\phi^2)$$

where a and b are numerical constants. Making suitable approximations, find the perihelion precession for a planet that moves in this gravitational field. Show that if $a = 2$, then your result agrees with Eq. [125]. Show that this is true regardless of the value of b. This proves that in order to obtain the correct perihelion precession, it is sufficient to know the terms of order $(GM)^2$ in g_{00} and the terms of order GM in the other components of the metric. Transform the approximate (second-order) solution of Problem 4.7 into isotropic coordinates and show that it gives the correct perihelion precession.

15. Estimate the periastron precession, in degrees per year, for each of the first three binary systems listed in Table 5.2.

16. A stable circular orbit must correspond to a minimum in the effective potential $\mathcal{V}(r)$.

(a) Show that the minimum in $\mathcal{V}(r)$ occurs at

$$r_0 = \frac{6GM}{1 - \sqrt{1 - 12(GM/\ell)^2}}$$

From this, conclude that there exists no stable circular orbit with a radius less than $6GM$.

(b) For an orbit that is almost circular, the energy $\mathcal{E}$ is slightly above the minimum in the effective potential. The radius therefore executes small oscillations between an interior and an exterior turning point. Find the frequency for such small radial oscillations about a circular orbit. By comparing this frequency with the angular frequency $d\phi/dt$, deduce the perihelion precession rate.

17. An artificial satellite is in a circular orbit of radius $r = 6GM$ around a Schwarzschild black hole.

(a) What is the orbital period of this satellite as measured by a clock located at large distance? As measured by a clock located at a fixed position at $r = 6GM$? As measured by a clock aboard the satellite? (Hint: Use Kepler's law; see Problem 12.)

(b) What is the orbital speed of this satellite as measured with a clock located at large distance? As measured with a clock located at a fixed position at $r = 6GM$?

18. (a) Show that for a particle instantaneously at rest in the Schwarzschild field, the radial acceleration in Schwarzschild coordinates is

$$\ddot{r} = - \frac{GM}{r^2}$$

This equation looks exactly like the corresponding Newtonian equation.
(b) Show that in local geodesic coordinates at rest in the Schwarzschild field, the acceleration is

$$\ddot{l} = - \frac{GM}{r^2} \frac{1}{\sqrt{1 - r_S/r}}$$

(Hint: If the acceleration in Schwarzschild coordinates is a^μ, then the acceleration in local geodesic coordinates is $\sqrt{a_\mu a^\mu}$. Why?)

19. The eccentricity of the orbit of the pulsar PSR 1113+16 is $\epsilon = 0.617$ and the orbital period is 2.79×10^4 s. The periastron advance is 4.2^0 per year.
(a) Carefully draw the orbit with this eccentricity at $t = 0$ and at $t = 1$ year, super-imposing the second drawing on the first. How many orbital revolutions intervene between the two orbits that you have drawn?
(b) Assuming that the mass of the pulsar is $1.4\ M_\odot$, that of the companion is also $1.4\ M_\odot$, and that the semi-major axis of the orbital ellipse is 9×10^{10} cm, what is the ratio of the gravitational time-dilation factors for the pulsar pulses at periastron and at apastron?

20. Suppose that a light signal orbits around a black hole in a circular orbit of radius $r_0 = 3GM$ (where r is the usual radial Schwarzschild coordinate). For a light signal in an orbit of this radius, use the Schwarzschild metric to deduce the orbital period measured by a distant observer. Also deduce the orbital period measured by a (stationary) observer at $r = r_0 = 3GM$.

21. Show that the circular orbit of radius $r_0 = 3GM$ we found in Section 7.7 for a light signal in the Schwarzschild geometry is unstable; that is, show that if u is slightly larger (smaller) than $(3GM)^{-1}$, then the "acceleration" $d^2u/d\phi^2$ is positive (negative).

22. Show that the solution of the cubic equation [134] is

$$r_0 = \frac{2b}{\sqrt{3}} \cos \left[\frac{1}{3} \cos^{-1} \left(-\frac{3\sqrt{3}GM}{b} \right) \right]$$

and verify that $r_0 = 3GM$ corresponds to $b = 3\sqrt{3}GM$.

23. Suppose that the rotor of a gyroscope has the shape of an ellipsoid of revolution with semiaxes a and $b = a(1 + \delta)$, where $\delta \ll 1$. If the gyroscope is in orbit around the Earth at a height of 650 km, *estimate* the magnitude of the tidal torques that act on it. For what value of δ will the tidal torque be comparable to the "geodetic torque" given by Eq. [158]?

24. For a gyroscope in a satellite in a circular orbit around the Earth at a height of 650 km, evaluate the average of Ω_{LT} (see Eq. [163]) over one revolution. Do this both for a polar orbit and for an equatorial orbit. Express your answer in seconds of arc per year. (The moment of inertia of the Earth is $I = 0.331 M_\oplus R_\oplus^2$ where $M_\oplus$ is the mass and $R_\oplus$ the equatorial radius of the Earth.)

25. Find a potential with spin-spin coupling such that the torque produced by this potential gives the precession Ω_{LT} of Eq. [163]. Compare the translational force that your potential exerts on an electron at the surface of the Earth with the ordinary gravitational force. How do these forces compare at the surface of a neutron star with mass $\simeq M_\odot$, radius $\simeq$ 10 km, and a rotation period of 0.033 s?

26. Some years ago, Japanese scientists measured the weight of a spinning gyroscope with a balance, and they claimed that this gyroscope displayed a weight loss of 0.5% when its spin was antiparallel to the spin of the Earth (but this claim was refuted by later, more careful experiments). Estimate the rate of rotation of a gyroscope of diameter 10 cm if such an effect is to arise from the spin-spin coupling that produces the Lense-Thirring precession.

27. According to Mach's ideas, a light ray passing near a rotating body should suffer an extra deflection because it tends to be carried along by the rotation. That part of the wavefront which is nearer the star tends to be carried along more than the part which is farther away; this results in an extra deflection of the wavefront. Are the results of Problem 4.33 in agreement with Mach's ideas?

28. (a) Suppose that initially we have a spherical mass m_1 of radius R_1 surrounded by a Schwarzschild field. Next, a uniform, thin, spherical shell of radius R_2 is placed concentrically around the mass m_1. Given that the total mass of the configuration (spherical mass plus shell) is m_2, find the solution of the Einstein equations in the space $r > R_2$ outside the shell and in the space $R_1 < r < R_2$ between the shell and the spherical mass; use Schwarzschild coordinates so the solution has the form of Eq. [75]. Explain why g_{00} and g_{11} have discontinuities at $r = R_2$. [Hint: What does Eq. [65] say about L' if the right side of this equation is not zero, but instead has a sharp peak (a delta function) representing the energy density $T_0{}^0$ of the thin shell?]

(b) Show that the discontinuity in g_{00} can be eliminated by changing the time coordinate in the region $R_1 < r < R_2$ by a constant factor.

8. Black Holes and Gravitational Collapse

ABANDON ALL HOPE YE WHO ENTER HERE.
Dante Alighieri, *The Inferno*

The dimensionless quantity GM/rc^2 may be regarded as a measure of the strength of the gravitational field. This quantity enters into the formulas for light deflection, light retardation, gravitational redshift, perihelion precession, etc. The small magnitude of the relativistic gravitational effects in the Solar System is related to the small magnitude of this quantity; even at the solar surface, GM/rc^2 is only 2×10^{-6}. Large relativistic effects are found in the gravitational field in the neighborhood of an extremely compact mass, where GM/rc^2 can attain values of the order of magnitude of 1. For example, near such a compact mass, at a radius $r = 3GM/c^2$, the deflection of light in the Schwarzschild field becomes so large that a light signal will move in a closed circular orbit around the central mass.*

The relativistic effects become spectacular when $r = 2GM/c^2$. The gravitational fields at this radius are so strong that nothing can escape from their grip. Light signals, particles, and even spaceships with the most powerful engines are inexorably pulled inward.

The existence of a "no-escape" radius for light is suggested by the following naive calculation based on Newton's gravitational theory: At a radius r, the escape velocity for a particle projected upward is $v =$

---

* However, this orbit is unstable (see Problem 7.21).

$\sqrt{2GM/r}$. Since this expression is independent of the mass of the particle, we may hope that it also applies to the case of light. For $v = c$, we then obtain a "no-escape" radius of $r = 2GM/c^2$. This result was obtained long ago by Michell and by Laplace, who speculated that a sufficiently massive and compact star might appear dark and black (Michell, 1784; Laplace, 1795). By dumb luck, this result agrees exactly with the result of a careful relativistic calculation (see below). Although the naive calculation gives the correct value for the "no-escape" radius, it gives the wrong picture of what happens. Light signals emitted at or inside the "no-escape" radius do not first rise, then stop, and then fall back. Rather, they fall immediately and never begin to move in the outward direction. The same is true of particles, and even of particles with propulsive devices (spaceships). If light cannot escape, no particle can, because the speed of any particle must remain below the speed of light, and the worldline must remain within the light cone.

A (finite) region of space into which signals can enter, but from which no signal can ever emerge, is called a black hole.

8.1 SINGULARITIES AND PSEUDOSINGULARITIES

For a spherically symmetric and stationary gravitational field, we found that the spacetime geometry is the Schwarzschild geometry with the spacetime interval

$$ds^2 = \left[1 - \frac{2GM}{r}\right]dt^2 - \frac{dr^2}{1 - 2GM/r} - r^2 d\theta^2 - r^2 \sin^2\theta \, d\phi^2 \qquad [1]$$

By inspection, we see that this spacetime interval develops a "singularity" as $r \to 2GM$. At this radius,

$$g_{00} = 1 - \frac{2GM}{r} \to 0 \qquad [2]$$

and

$$g_{11} = - \frac{1}{1 - 2GM/r} \to \infty \qquad [3]$$

In cgs units, the critical radius is

$$r_S = \frac{2GM}{c^2} \qquad [4]$$

This is called the *Schwarzschild radius*, or the *gravitational radius* of the mass M. For a mass $M = M_\odot = 2 \times 10^{33}$ g, this radius is

$$r_S = 3.0 \text{ km} \tag{5}$$

If a body has an actual radius larger than r_S, then the singularities [2] and [3] need not concern us, since the solution [1] is applicable only in the region exterior to the body. If a body with $M \simeq M_\odot$ is to have a radius smaller than r_S, it must have a density of $\simeq 10^{16}$ g/cm^3, which is even larger than the density of a neutron star. In fact, as we will see, a body that has collapsed to a radius smaller than r_S is unable to come to equilibrium and will continue to collapse--the gravitational forces are so strong that nothing can resist them.

Note that the critical density at which a body has a size smaller than its Schwarzschild radius r_S decreases with mass. For example, suppose that the nucleus of a galaxy consists of 10^{11} stars, each of a mass $\simeq M_\odot$, uniformly distributed over a spherical volume. In this case the critical density is only 10^{-6} g/cm^3; at this density, the typical distance between adjacent stars in the galactic nucleus is about the same as the Earth-Sun distance.

For the discussion of the general features of the Schwarzschild solution, we will assume that the body has collapsed completely, so the mass density is zero at all points (except perhaps at $r = 0$). This means we treat the Schwarzschild solution as an exact vacuum solution of Einstein's equations.

The surface $r = r_S$ is a surface of *infinite redshift*. A clock placed at rest near $r = r_S$ shows a proper time

$$d\tau = \sqrt{1 - 2GM/r}\, dt \tag{6}$$

which approaches zero as $r \to r_S$; that is, the clock runs infinitely slow compared with a clock at a larger distance. Thus, if an astronaut at rest near $r = r_S$ sends out light pulses, or signal rockets, with a time interval of 1 s (as shown by his clock) between one pulse and the next, then an observer at large distance will receive these pulses with a time interval much larger than 1 s (as shown by her clock) between successive pulses.

Eq. [6] gives an infinite time dilation for a clock at rest at $r = r_S$. Actually, a clock cannot remain at rest on this surface. The vanishing of $d\tau$ is characteristic of the worldline of a light signal, and hence only a light signal, aimed in the outward direction, can remain at rest at r_S. The infinite-redshift surface is sometimes called the *static limit*, because material particles cannot remain at rest on it.

If r is in the range $r_S > r > 0$, then the Schwarzschild solution is free of singularities. However, in this region

$$g_{00} = 1 - \frac{r_S}{r} < 0 \qquad [7]$$

and

$$g_{11} = -\frac{1}{1 - r_S/r} > 0 \qquad [8]$$

Thus, the signs of g_{00} and g_{11} are now the opposite of what is normal. In the region $r < r_S$, t is a *spacelike* coordinate, and r is a *timelike* coordinate. Since the metric is a function of r, and since r now measures the progress of time, it follows that when $r < r_S$, the metric is actually time dependent. We should be a bit more careful: the static character of the metric in the region $r > r_S$ is a consequence of our particular choice of coordinates; if we were to use new coordinates x'^μ which deform in time with respect to the t, r, θ, and ϕ coordinates, then the metric as seen in these new coordinates would of course be time dependent. However, the assertion that the metric in the interior region $r < r_S$ is not static holds true in all possible coordinates, the reason being that regardless of how the new coordinates x'^μ are defined in terms of the old t, r, θ, and ϕ coordinates, an advance of time ($ds^2 > 0$) is impossible unless $dr^2 > 0$; hence advance of time necessarily entails a change in r, and hence a change in the metric.

It is important to recognize that the Schwarzschild "singularity" at $r = r_S$ is not a physical singularity. The "singularity" in Eqs. [2] and [3] is a spurious singularity, or a pseudosingularity; it arises from an inappropriate choice of coordinates, and can be eliminated by a change of coordinates. An astronaut falling toward and crossing the surface $r = r_S$ will not feel anything unusual. In his immediate vicinity physics will go on in the way it always does in a freely falling reference frame. Of course, there will be tidal forces, and these tidal forces will grow stronger and stronger as the astronaut falls deeper and deeper toward the center. However, the tidal forces remain finite at $r = r_S$; only at $r = 0$ do the tidal forces become infinite and indicate the presence of a real physical singularity.

Mathematically, the absence of any genuine singularity at $r = r_S$ can be seen from an examination of the Riemann curvature tensor; it turns out that, even though in Schwarzschild coordinates g_{00} and g_{11} misbehave, all components of $R^\alpha{}_{\beta\mu\nu}$ are finite at $r = r_S$ when calculated in local geodesic coordinates. For example, consider the component $R^0{}_{101}$. This component is given by (see Eq. [6.104]):

$$R^0{}_{101} = -\tfrac{1}{2}N'' + \tfrac{1}{4}L'N' - \tfrac{1}{4}(N')^2$$

$$= \frac{r_S}{r^3} \frac{1}{1 - r_S/r} \qquad [9]$$

At $r = r_S$, this function is singular. However, the singularity is spurious, just as that in Eq. [3]. To recognize this, let $x^\mu = (t_0, r_0, \pi/2, \phi_0)$ be a given point in the equatorial plane, and introduce geodesic coordinates x'^μ at this point by means of the transformation

$$x'^0 = (t - t_0)\sqrt{1 - r_S/r_0} + \dots$$

$$x'^1 = \frac{r - r_0}{\sqrt{1 - r_S/r_0}} + \dots$$

$$x'^2 = r_0(\theta - \pi/2) + \dots$$

$$x'^3 = r_0(\phi - \phi_0) + \dots \qquad [10]$$

The dots in these equations stand for quadratic terms, such as those appearing in Eq. [6.35]. These quadratic terms are, of course, crucial to achieve the geodesic coordinate condition $\Gamma'^\alpha_{\mu\nu} = 0$; but we are now interested only in the transformation law for the Riemann tensor, and this law does not depend on the quadratic terms. The transformation coefficients at $x'^\mu = 0$ are

$$\frac{\partial x^0}{\partial x'^0} = \frac{1}{\sqrt{1 - r_S/r_0}} \qquad \frac{\partial x^1}{\partial x'^1} = \sqrt{1 - r_S/r_0}$$

$$\frac{\partial x^2}{\partial x'^2} = \frac{1}{r_0} \qquad \frac{\partial x^3}{\partial x'^3} = \frac{1}{r_0} \qquad [11]$$

All other coefficients are zero.

■ *Exercise 1.* Show that this coordinate transformation changes the Schwarzschild metric (at the point $t_0, r_0, \pi/2, \phi_0$) into $g'_{\mu\nu} = \eta_{\mu\nu}$, and therefore eliminates the singularity in the metric. ■

Note that the coordinate transformations given by Eq. [10] are singular if r_0 coincides with r_S. The elimination of the "singularity" of the Schwarzschild solution hinges on the use of a singular coordinate transformation. We will of course adopt the view that the coordinates that go bad at $r = r_S$ are the Schwarzschild coordinates x^μ, specifically, the coordinates t and r. The geodesic coordinates x'^μ can be given the physical interpretation of belonging to a reference frame in free fall (see Section 6.3), and they must therefore necessarily be good coordinates.

For the transformed Riemann tensor we find

$$R'^0{}_{101} = \frac{\partial x'^0}{\partial x^\alpha} \frac{\partial x^\beta}{\partial x'^1} \frac{\partial x^\mu}{\partial x'^0} \frac{\partial x^\nu}{\partial x'^1} R^\alpha{}_{\beta\mu\nu}$$

$$= \frac{\partial x'^0}{\partial x^0} \frac{\partial x^1}{\partial x'^1} \frac{\partial x^0}{\partial x'^0} \frac{\partial x^1}{\partial x'^1} R^0{}_{101}$$

$$= \left(1 - \frac{r_S}{r_0}\right) \frac{r_S}{r_0{}^3} \frac{1}{1 - r_S/r} = \frac{r_S}{r_0{}^3} \qquad [12]$$

If we now let r_0 approach r_S, the function given by Eq. [12] remains finite. This shows that in the x'^μ coordinates, the $R'^0{}_{101}$ component of the Riemann tensor is free of singularities. It is easy to check that all other components of the Riemann tensor are also free of singularities. The tidal force therefore remains finite at $r = r_S$.

Note that at $r_0 = 0$, the expression [12] diverges. Hence the tidal forces diverge at the "center," and we have a true singularity. Since near this point, r is a timelike coordinate and t a spacelike coordinate, this singularity happens at a given instant of time ($r = 0$) in all of space (at all values of t). Thus the time-dependent, dynamic geometry in the interior region evolves into a singularity and comes to an end. But the geometry on the outside remains static forever.

In some respects the Schwarzschild pseudosingularity at $r = r_S$ is similar to the pseudosingularity found when the metric of flat space-time is expressed in Rindler coordinates (see Fig. 8.1). As we saw in Section 6.8, the Rindler coordinates $\tilde{x}^0$ and $\tilde{x}^1$ are related to the iner-tial rectangular coordinates x^0 and x^1 by

$$\tilde{x}^0 = b \tanh^{-1} x^0/x^1 \qquad [13]$$

$$\tilde{x}^1 = \sqrt{(x^1)^2 - (x^0)^2} \qquad [14]$$

In terms of these coordinates $\tilde{x}^0$ and $\tilde{x}^1$, the spacetime interval of flat spacetime, $ds^2 = (dx^0)^2 - (dx^1)^2$, takes the form (see Eq. [6.183])

$$ds^2 = \frac{(\tilde{x}^1)^2}{b^2}(d\tilde{x}^0)^2 - (d\tilde{x}^1)^2 \qquad [15]$$

For $\tilde{x}^1 \to 0$, the $\tilde{g}_{00}$ component of the Rindler metric develops a sin-gularity similar to the singularity in Eq. [2],

$$\tilde{g}_{00} = \frac{(\tilde{x}^1)^2}{b^2} \to 0 \qquad [16]$$

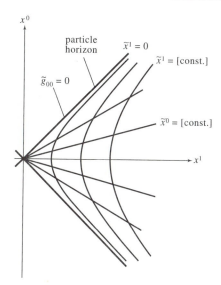

Fig. 8.1 Grid of Rindler coordinates $\tilde{x}^0$ and $\tilde{x}^1$ in flat spacetime. A particle at rest in Rindler coordinates at $\tilde{x}^1$ = [constant] has accelerated ("hyperbolic") motion relative to the inertial coordinates x^0, x^1. In the limiting case $\tilde{x}^1$ = 0, the particle has the worldline of a light signal.

This means that a clock located at $\tilde{x}^1 = 0$ has an infinite redshift relative to a clock located elsewhere. We can readily understand this infinite redshift by noting that the worldline $\tilde{x}^1 = 0$ is actually the worldline of a light signal; the infinite redshift is simply the familiar special-relativistic time dilation for a clock moving at a speed approaching the speed of light. As in the case of the Schwarzschild spacetime, where no particle can remain at rest at $r = r_S$, no particle can remain at rest at $\tilde{x}^1 = 0$, since this would require that it move at the speed of light relative to the inertial coordinates x^0, x^1.

The singularity [16] in Rindler coordinates at $\tilde{x}^1 = 0$ arises from an inappropriate choice of coordinates. It is a pseudosingularity--flat spacetime has no physical singularity at $\tilde{x}^1 = 0$ or anywhere else. The pseudosingularity can be eliminated by a coordinate transformation back to the inertial coordinates. The example of Rindler coordinates shows very clearly that the presence of surfaces of infinite redshift depends on the choice of coordinates.

8.2 THE BLACK HOLE AND ITS HORIZON

Although the region $r < r_S$ has no unusual properties of a local kind (except at $r = 0$, where there is a singularity), it does have some very unusual properties of a global kind. As we will see from a careful analysis of the spacetime geometry, the region $r < r_S$ is a *black hole*. By this is meant that no signal of any kind can emerge from the region $r < r_S$ and reach the region $r > r_S$. The surface $r = r_S$ (regarded as a surface in spacetime; see Figs. 8.2a,b) is the boundary between those spacetime points that will become observable at some time, and those that will never be observable by outside observers. This boundary of the black hole is called the *event horizon*. It is the place beyond which we cannot see, which is reminiscent of the horizon on a large lake or on the ocean. The surface $r = r_S$ acts as a "one-way membrane," through which signals can be sent in, but not out. This is a global (or nonlocal) property because in order to test it, we must examine the propagation of light signals and other signals and check what happens to them in the long run.

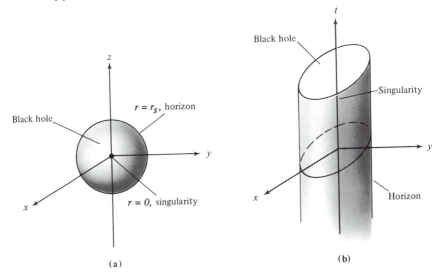

Fig. 8.2 (a) The Schwarzschild black hole and its event horizon displayed in x, y, z coordinates. The event horizon is a sphere of radius r_S. Note that r_S is merely the value of a radial coordinate, not a measured distance. (b) The Schwarzschild black hole and its event horizon displayed in x, y, t coordinates. The event horizon is a cylinder of radius r_S.

Note that in flat spacetime in Rindler coordinates, there is also some kind of event horizon. From the discussion in the preceeding section, we know that observers at rest in Rindler coordinates (and permanently accelerated relative to inertial coordinates) perceive a surface of infinite redshift, $\widetilde{x}^1 = 0$ in Fig. 8.1. For these observers, this surface of infinite redshift is also a horizon, since no signal from the region above $\widetilde{x}^1 = 0$ can ever reach the region below $\widetilde{x}^1 = 0$ (the worldlines of light signals starting anywhere in the region above $\widetilde{x}^1 = 0$ are parallel to the boundary line $\widetilde{x}^1 = 0$, and they never intersect any of the worldlines $\widetilde{x}^1 = [\text{constant}]$). But this Rindler horizon is perceived only by a special class of observers, at rest in Rindler coordinates. The worldline of any observer at rest in inertial coordinates crosses the Rindler horizon, and such an observer can see the region of spacetime beyond. Thus, the Rindler horizon is an artifact resulting from the accelerated motion of the observers, whereas the Schwarzschild horizon must be thought of as an attribute of the Schwarzschild geometry, since it is perceived by all outside observers. This does not mean that at the event horizon the Schwarzschild geometry possesses any kind of distinctive local property, such as a local singularity. But it does mean that a broad class of observers agree on the existence and location of the surfaces across which two-way communication is impossible. Event horizons refer to global rather than local properties of spacetime, but that does not make event horizons any less real.

Although the surface of infinite redshift and the horizon coincide in the case of the Schwarzschild solution, there exist other solutions of Einstein's equations for which these surfaces are distinct (for example, the Kerr solution of Section 8.4). If the redshift, for static observers, is infinite on some given surface, this does not by itself necessarily prevent communication across that surface. It could happen that for a sender located inside the infinite-redshift surface, the redshift, relative to an observer at large distance, is again reduced to a finite value. Also, we must keep in mind that the total redshift of a signal depends not only on the location of the sender in the gravitational field, but also on the velocity of the sender. For a sender in motion, in some suitable direction, the Doppler shift may compensate for part, or all, of the gravitational redshift. Thus, an "infinite-redshift surface" is defined only relative to a special class of senders and receptors. By contrast, an event horizon is an intrinsic feature of spacetime, which does not directly depend on the state of motion of the observers. All external observers will agree that the surface $r = r_S$ in the Schwarzschild spacetime is an event horizon (although some observers, such as observers with a permanent acceleration, may perceive additional Rindler horizons).

To understand how signals are cut off at the Schwarzschild horizon $r = r_S$, consider a light signal propagating in the radial direction. The velocity of this light signal, with respect to t, r coordinates, follows from Eq. [1] by taking $ds^2 = 0$:

$$\frac{dr}{dt} = \pm\left(1 - \frac{r}{r_S}\right) \qquad [17]$$

When the signal approaches $r = r_S$, this coordinate velocity tends to zero. Fig. 8.3 shows the forward light cones obtained from Eq. [17] at different values of r. In the exterior of the black hole ($r > r_S$), the axis of the light cones is parallel to the t-xis. In the interior of the black hole ($r < r_S$), the axis of the light cones is parallel to the r-axis. The strange orientation of the light cones in the interior region is a simple consequence of reversal of the character of the coordinates in this region: r is a timelike coordinate and t is a spacelike coordinate. Note that in this region, the quantity dt/dr gives what we would normally call the "velocity," that is, the ratio of spacelike increment to timelike increment.

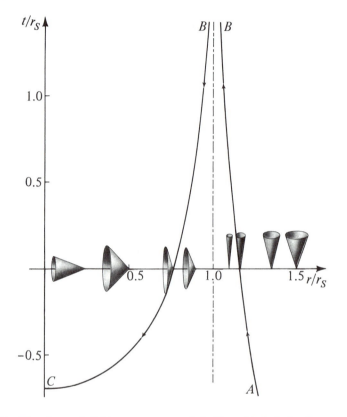

Fig. 8.3 The forward light cones near and inside a black hole. As $r \to \infty$, the light cone assumes its usual shape and direction, that is, $dr/dt = \pm 1$. The curve $ABBC$ is the worldline of an ingoing light signal.

The existence of an event horizon at $r = r_S$ is obvious from an inspection of the light cones in Fig. 8.3. Any kind of signal must necessarily travel in a spacetime direction that lies within a light cone. Since the light cones in the black-hole region are oriented toward $r = 0$, any signal in this region is unavoidably pulled toward decreasing values of r and can never leave the black hole.

Note that the light cones are tangent to the surface $r = r_S$ (indicated by the dashed line in Fig. 8.3); this means that, viewed in spacetime, the horizon is a null surface. This is a general property of event horizons, since a light signal that starts exactly on a horizon and is aimed in the outward direction is sandwiched between those signals just outside of the horizon (which escape outward) and those just inside the horizon (which fall inward); the light signal therefore propagates neither out nor in--it hovers in place forever, and thereby indicates that the place at which it is hovering is a null surface. Such a hovering light signal on the horizon of a black hole should not be confused with a signal in a circular orbit around the black hole; the hovering signal has no circular motion, and it is completely stationary.

Although signals cannot emerge from the black hole, they can enter it freely. The curve in Fig. 8.3 is the worldline of a light signal that travels inward. This curve is obtained by integrating Eq. [17]. The signal follows the worldline AB to $t = \infty$, and then it follows the worldline BC to $r = 0$. As measured in t, r coordinates, the signal velocity tends to zero as $r \rightarrow r_S$, and the signal takes an infinite t time to reach $r = r_S$. Hence from the point of view of an observer at infinity, whose clocks give t time, the signal never reaches the horizon.

To make this concrete, suppose that the black hole is surrounded by some dust which scatters a small part of the light signal so the position of the signal becomes visible to an observer at infinity. Then this observer will see the lighted spot asymptotically approach the Schwarzschild radius without ever reaching it. This is, in part, a manifestation of the gravitational time-dilation effect. From the point of view of an astronaut in free fall in the vicinity of $r = r_S$, no such slowing down occurs; the signal always has the speed of light relative to him; the signal and the astronaut (!) both cross into the black hole in a finite proper time.

If the observer at infinity keeps in touch with the astronaut by radio and television, she finds that the motion and metabolic rate of this astronaut are slowed down in much the same way as the motion and vibration rate of the light signal. The astronaut appears to go into "slow motion" and his movements freeze asymptotically. The last syllable he sends through his radio before crossing into the black hole is drawn out to infinite length when received by the outside observer; the words he speaks after that remain inside the black hole.

Not only will all signals of an astronaut who has entered the black hole remain trapped, but the astronaut himself is trapped. His worldline intersects the singularity at $r = 0$, and when he comes near that place, the large tidal forces rip him apart. Since all worldlines within the astronaut's future light cone terminate on $r = 0$, the collision with the singularity cannot be avoided. Not even a spaceship with the most powerful rocket engine can resist the pull of gravity in a black hole. (In fact, the use of rocket engines inside a black hole is not recommended; it can never delay the collision with the singularity, only speed it up! See Problem 5.)

Incidentally: We can also conclude that if a spherical star is compressed, by some astrophysical process, to a radius smaller than its Schwarzschild radius, then gravitational collapse necessarily ensues. We need only consider a particle on the surface of the star and remember that, according to the Birkhoff theorem, the exterior geometry must be that of Schwarzschild. Hence the particle on the surface has the same equation of motion as a spaceship with a rocket motor in an empty Schwarzschild spacetime--we may treat the pressure with which the star pushes outward on the particle as mathematically analogous to the thrust of the rocket motor. The surface of the star must, therefore, necessarily fall toward the singularity $r = 0$, just as the spaceship does. Gravitational collapse is inevitable.

In the rest frame of a falling astronaut the amount of proper time needed to enter a black hole and crash into the singularity at $r = 0$ is not only finite, but also quite short. For a typical worldline, the proper time between r_S and $r = 0$ is of the order of r_S/c. This means that once an astronaut crosses the horizon of a black hole with $M \simeq M_\odot$, he has only $\simeq 10^{-5}$ s to live.

Although the Schwarzschild coordinates suffer from the inconvenience of a singularity in the t time at $r = r_S$ (see Fig. 8.3), this does not affect the calculation of the proper time needed to fall into a black hole and into the singularity. Eq. [7.112] gives us $dr/d\tau$ for an astronaut in free fall. For purely radial motion, $\ell = 0$ (see Eq. [7.107]) and hence Eq. [7.112] reduces to

$$\left(\frac{dr}{d\tau}\right)^2 = \frac{r_S}{r} - (1 - \mathcal{E}^2)$$

This can be integrated easily:

$$\tau = [\text{constant}] + \frac{r(r_S/r - 1 + \mathcal{E}^2)^{1/2}}{1 - \mathcal{E}^2}$$

$$+ \frac{r_S}{(1 - \mathcal{E}^2)^{3/2}} \tan^{-1}\left[\frac{r_S/r - 1 + \mathcal{E}^2}{1 - \mathcal{E}^2}\right]^{1/2} \qquad [18]$$

■ *Exercise 2.* Perform the integration and obtain Eq. [18]. ■

For a finite change in r, the corresponding change in proper time is finite. For example, suppose the astronaut is initially at rest at the radius $r_0 = r_S/(1 - \mathcal{E}^2)$; this radius is larger than r_S and hence outside of the black hole. It then only takes him a proper time

$$\Delta\tau = r_S \frac{\pi/2}{(1 - \mathcal{E}^2)^{3/2}} = r_S \frac{\pi}{2}\left(\frac{r_0}{r_S}\right)^{3/2} \qquad [19]$$

to fall freely all the way to the radius $r = 0$. If the initial value r_0 is near r_S, then $r_0/r_S \simeq 1$ and $\Delta\tau \simeq \pi r_S/2$, which is of the order of r_S, as claimed above.

Although event horizons represent physical and coordinate-independent properties of a spacetime, it is often easier to recognize the presence, or absence, of a horizon by particular, cleverly chosen coordinates. Thus, the absence of an event horizon in flat spacetime is obvious in inertial coordinates, but not so obvious in the Rindler coordinates used in the metric [15]. For the case of the Schwarzschild geometry, the coordinates that make things easy are the Kruskal coordinates, which we will describe in the next section.

8.3 THE MAXIMAL SCHWARZSCHILD GEOMETRY

The Schwarzschild coordinates t, r are singular at $r = r_S$. There exist several coordinate systems that are well behaved at $r = r_S$. But the price we must pay for such well-behaved coordinates is that the metric will not appear static, not even in the exterior region.

A very convenient set of coordinates are the *Kruskal coordinates* v, u, also called Kruskal-Szekeres coordinates (Kruskal, 1960). In the interior of the black hole, these coordinates are defined as follows:

$$v = \sqrt{1 - r/r_S}\; e^{r/2r_S} \cosh t/2r_S \qquad [20]$$
$$\text{for } r < r_S$$
$$u = \sqrt{1 - r/r_S}\; e^{r/2r_S} \sinh t/2r_S \qquad [21]$$

For the exterior region, the definitions are

$$v = \sqrt{r/r_S - 1}\; e^{r/2r_S} \sinh t/2r_S \qquad [22]$$
$$\text{for } r > r_S$$
$$u = \sqrt{r/r_S - 1}\; e^{r/2r_S} \cosh t/2r_S \qquad [23]$$

The inverse transformation is given (implicitly) by

$$(r/r_S - 1)\, e^{r/r_S} = u^2 - v^2 \qquad [24]$$
$$\text{for } r < r_S$$
$$t = 2r_S \tanh^{-1} u/v \qquad [25]$$

and

$$(r/r_S - 1)e^{r/r_S} = u^2 - v^2 \qquad\qquad [26]$$

$$\text{for } r > r_S$$

$$t = 2r_S \tanh^{-1} v/u \qquad\qquad [27]$$

■ *Exercise 3.* Show that Eqs. [24]-[27] agree with [20]-[23]. ■

The coordinate v is timelike and u is spacelike. Note that the functions that appear on the right sides of Eqs. [24]-[27] are similar to the functions that appear in the transformation to Rindler coordinates, in Eqs. [13] and [14]. This similarity is no accident. Since we know that the Schwarzschild pseudosingularity at $r = r_S$ is similar to the Rindler pseudosingularity, we expect that a transformation of roughly the same form as the *inverse* of the Rindler transformation (from Rindler coordinates to inertial coordinates) will help to remove the pseudosingularity.

In Kruskal coordinates, the spacetime interval takes the new form

$$ds^2 = \frac{4r_S{}^3}{r} e^{-r/r_S} (dv^2 - du^2) - r^2 d\theta^2 - r^2 \sin^2\theta \; d\phi^2 \qquad [28]$$

where r is to be regarded as a function of v and u, as given by Eq. [24] or [26].

The new metric has no singularity at $r = r_S$. Of course, there remains a singularity at $r = 0$; this is unavoidable, since the curvature becomes infinite at that point.

■ *Exercise 4.* Derive Eq. [28]. ■

Let us now disregard displacements in the θ and ϕ directions, and concentrate on the radial and time dependence of the geometry. Fig. 8.4 shows the u, v coordinate grid and the corresponding values of the r, t coordinates. In this diagram, r and t are measured in units of r_S; thus, $r = 1.5$ means $r = 1.5r_S$, etc. As in the case of Rindler coordinates, the curves $r = $ [constant] are hyperbolas, and the curves $t = $ [constant] are straight lines through the origin. It is obvious from the diagram that the coordinate transformation from u, v to r, t is singular at $r = r_S$. Thus, the *line* $u = v$ is mapped into the *point* $r = r_S$, $t = \infty$. The elimination of the Schwarzschild pseudosingularity is accomplished precisely by the singular character of the coordinate transformation.

One of the nice features of the Kruskal coordinates is that the worldlines of radial light signals (that is, light signals with $d\theta = d\phi = 0$) are straight lines at 45° with respect to the u-, v-axes, just like the worldlines of light signals in flat spacetime.

■ *Exercise 5.* Show that the worldline of such a light signal must have $du = \pm dv$. ■

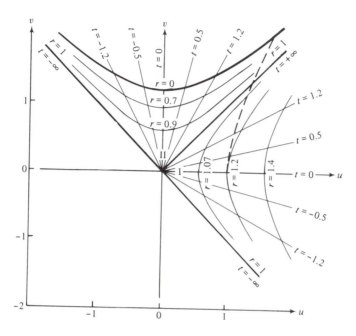

Fig. 8.4 Schwarzschild spacetime in Kruskal coordinates. The u, v coordinates are dimensionless. The r, t coordinates are given in units of r_S. The dashed line is the worldline of a particle falling toward $r = 0$.

The worldlines of other freely falling particles are not straight lines. For example, the dashed line in Fig. 8.4 indicates a typical worldline for a particle (or an astronaut) falling into the black hole.

The surface $r = r_S$ divides spacetime into two regions: the black hole (region II) and the asymptotically flat space surrounding it (region I). The existence of an event horizon at $r = r_S$ is immediately obvious from inspection of the spacetime diagram, Fig. 8.4: a light signal sent out by the astronaut after he crosses $r = r_S$ will proceed at 45^0 in this spacetime diagram, and ultimately intersect $r = 0$. The signal always moves toward decreasing values of r, and it can never reach $r = r_S$. Signals other than light must necessarily lie within the forward light cone, and they will be even less able to escape. Signals from outside the black hole can of course cross $r = r_S$ without trouble, that is, the surface of the black hole acts as a "one-way membrane."

Note that the singularity $r = 0$ displayed in Fig. 8.4 is not a point-like singularity. Rather, $r = 0$ is a spacelike surface. As seen in u, v coordinates, the singularity happens at different points of space at different times. The singularity happens at the point u when the v time is

$$v = \sqrt{1 + u^2} \qquad [29]$$

Thus, at large values of u, the singularity is delayed. We recall that, as seen in Schwarzschild coordinates, the singularity happens at the same time ($r = 0$) at all points of space ($-\infty < t < \infty$). This difference in behavior simply reflects a difference in the definition of simultaneity.

In Fig. 8.4, the spacetime diagram has been left blank below the line $u = -v$. It is obvious that the only region that can be explored by an astronaut who starts his expedition in the exterior (region I) of the black hole is the region $u > -v$. The region $u < -v$ is necessarily outside of his light cone. However, the latter region could still have a physical significance if signals sent by somebody or something in this region cross the line $u = -v$ and reach an astronaut in the region $u > -v$. Examination of the Kruskal metric shows that in fact the line $u = -v$ is in no way singular, and geodesics can cross it without any trouble. This indicates that the diagram in Fig. 8.4 is incomplete.

A more complete diagram is given in Fig. 8.5. This diagram is as complete as can be. Technically, a manifold is said to be *maximal* if every geodesic either is of infinite extent in both directions (has no beginning and no end), or else ends or begins on a singularity. A manifold is said to be geodesically *complete* if all geodesics are of infinite length. In physical terms, this means that in a complete spacetime manifold, every particle (if stable, and if it does not collide and

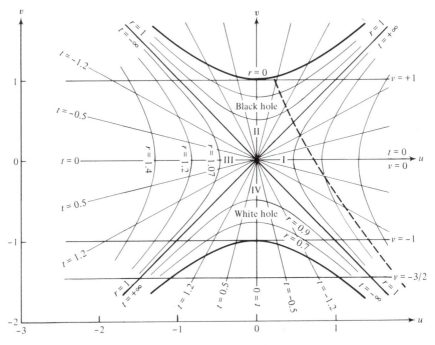

Fig. 8.5 The maximal Schwarzschild spacetime in Kruskal coordinates.

react with another particle) has been in the spacetime forever and remains in it forever. In a maximal spacetime, particles can appear or disappear, but only at singularities. We will always demand that spacetime be maximal, because it makes no sense whatsoever that the worldline of a particle should suddenly come to an end (that is, leave or enter our universe) in the middle of nowhere. The spacetime shown in Fig. 8.5 is maximal, but not complete; that shown in Fig. 8.4 is neither maximal nor complete.

Note that several sets of r, t coordinates are needed to cover all parts of spacetime in Fig. 8.5, but one set of u, v coordinates is sufficient. For a given value of u, the spacetime has two singularities $r = 0$; one is at $v = \sqrt{1 + u^2}$, the other at $v = -\sqrt{1 + u^2}$. The spacetime also has two asymptotically flat regions, one at $u \to \infty$, the other at $u \to -\infty$; these two regions are labeled I and III in Fig. 8.5.

■ **Exercise 6.** Write down the coordinate transformations from r, t to u, v coordinates for the regions III and IV. ■

By examination of Fig. 8.5, we can gain some insight into the time evolution of the Schwarzschild spacetime, using v as our time variable. Several lines $v = [\text{constant}]$ have been drawn through Fig. 8.5. For instance, at time $v = -3/2$, we have a space that is asymptotically flat (as $u \to \infty$) with a singularity at $u = \sqrt{5}/2$. At the same time we have a second asymptotically flat space ($u \to -\infty$) with a singularity at $u = -\sqrt{5}/2$. This second space is in a different universe; alternatively, it may be regarded as a region in our own universe, if this second region of our universe is very far away from the first. At the time $v = -1$, both of these singularities coalesce, and we obtain two asymptotically flat spaces joined together by a region in which the geometry deviates greatly from flatness. Such a region is called a *wormhole* (also called an Einstein-Rosen bridge, or a Schwarzschild throat). Whether we regard this wormhole as connecting two universes, or as connecting two remote parts of the same universe, the topology in either case is non-Euclidean. Between the times $v = -1$ and $v = +1$ there is no singularity, only a wormhole in which the curvature, although perhaps large, is everywhere finite. At time $v = +1$ the connection between the two asymptotically flat spaces disappears, and we are left with two separate spaces, each with a singularity. These spaces continue to evolve parallel in time. Note that the time evolution described here is perfectly symmetric under time reversal (reversal of v to $-v$).

The second asymptotically flat space (region III) is outside of the forward light cone of the first asymptotically flat space (region I), and conversely. This implies that communication between the two universes is impossible. However, if astronauts from both universes jump into the black hole (region II), then they can meet, embrace, and die together.

What prevents signals from passing through the wormhole is the time evolution of the interior geometry. The wormhole "pinches off" so fast that not even a light signal can pass through before the singularity happens. If we want to interpret the two asymptotically flat regions as remote parts of the same universe, then the impossibility of signal transmission through the wormhole is crucial for the preservation of causality (Fuller and Wheeler, 1962). It is obvious that if remote parts of the universe could communicate via a short wormhole, the effective signal velocity would exceed the speed of light.

Region IV is called a *white hole*. According to the above discussion such a white hole may be regarded as the time reverse of a black hole. Signals from region IV can go out to the exterior regions I and III, but no signal from the latter regions can ever be sent into region IV. This must not be interpreted as implying that a white hole is intrinsically luminous. We are dealing with vacuum solutions of Einstein's equations; light signals are to be regarded as test signals. Light will come out of a white hole only if somebody places a light source in the white hole. There is no reason to suspect that anybody has done this or could do this. The white-hole portion of the complete Kruskal solution is probably as irrelevant to our universe as are the advanced solutions of Maxwell's equations. In electromagnetism, the mathematical existence of these solutions is required by the time-reversal symmetry of the Maxwell equations, but their physical existence would require some very peculiar boundary conditions at the beginning of the universe (incoming waves that converge on particles and are *exactly* absorbed by them).

So far we have examined the spacetime geometry in the u-v (or r-t) plane. Let us now concentrate on the purely spatial part of the geometry, that is, the geometry at a given fixed time. We are then dealing with a curved three-dimensional space with coordinates r (or u), θ, and ϕ. Since it is impossible to make a drawing of three-dimensional curved space, we prefer to look at the geometry in the two-dimensional curved "plane" $\theta = \pi/2$ (because of spherical symmetry, the geometry is the same in all "planes" passing through the center of symmetry). Fig. 8.6a shows this curved surface at the time $v = 0$.* The curved two-dimensional space r, ϕ is shown embedded in a three-dimensional space r, ϕ, z; here, the z coordinate is an artificial coordinate which has nothing to do with the z coordinate of real space. The surface is a paraboloid of revolution; the shape of the surface has been chosen in such a way that the distances measured along the curved surface with, say, a string agree with the distances measured in the r-ϕ plane of the Schwarzschild geometry.

---

* In general, the times v = [constant] and t = [constant] do not coincide (see Fig. 8.4). The case $v = 0$, $t = 0$ is exceptional; these times coincide.

■ **Exercise 7.** For $v = t = 0$ and $\theta = \pi/2$, the Schwarzschild geometry in the r-ϕ plane is described by the distance interval

$$dl^2 = \frac{dr^2}{1 - r_S/r} + r^2 d\phi^2$$

In order to embed this curved two-dimensional surface in flat three-dimensional space, we introduce an extra coordinate z and write the distance interval as (r, ϕ, z are regarded as cylindrical coordinates)

$$dl^2 = dz^2 + dr^2 + r^2 d\phi^2$$

$$= dr^2 \left[1 + \left(\frac{dz}{dr} \right)^2 \right] + r^2 d\phi^2$$

By comparing these two expressions, we obtain the following differential equation for the surface:

$$\left(\frac{dz}{dr} \right)^2 = \frac{1}{1 - r_S/r} - 1$$

$$= \frac{1}{r/r_S - 1}$$

Solve this equation and show that the surface is a paraboloid of revolution, $r = r_S + z^2/4r_S$. ■

The upper half of the figure ($u > 0$) may be thought of as showing the deformation of the flat r-ϕ plane with which we start out in the absence of gravitation. To be precise, imagine that the r-ϕ plane of flat space (Fig. 8.6b) has the hole $r < 1$ cut out of it (the region $r < 1$ is nonexistent at "time" $v = 0$; as we can see from Fig. 8.6, the surface $v = 0$ does not contain values of r smaller than $r = 1$). Then gravitation deforms this plane into the funnel-like structure that is the upper part of the curved surface shown in Fig. 8.6a. The lower half arises from a similar deformation of the r-ϕ plane of the second universe (Fuller and Wheeler, 1962).

The two universes of Fig. 8.6a are joined along the circle $0 \leq \phi \leq 2\pi$. If we want to take into account the θ coordinate, then we must imagine similar circles for different values of θ. This means that the two universes are actually joined along a spherical surface $0 \leq \phi \leq 2\pi$, $0 \leq \theta \leq \pi$ (a "two-sphere").

At later times ($v > 0$) or earlier times ($v < 0$) the wormhole is narrower (in Fig. 8.5, the surface $v > 0$ contains values of r smaller than $r = 1$), and at time $v = \pm 1$ the wormhole pinches off at the center and we are left with two separate surfaces, each with a cusp. Fig. 8.7 gives

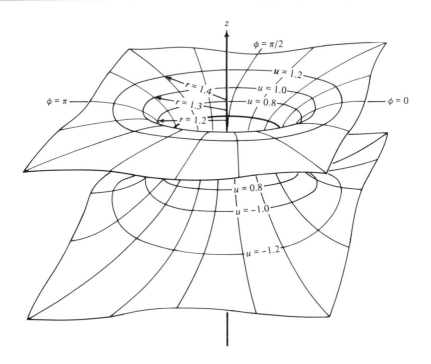

Fig. 8.6(a) The geometry of the surface $v = 0$, $\theta = \pi/2$ (or, equivalently, $t = 0$, $\theta = \pi/2$). Only u and ϕ (or, equivalently, r and ϕ) vary along this surface. The surface is embedded in ordinary three-dimensional space, and it has been constructed in such a fashion that the distances measured along the surface (with, say, a string) directly give the true distances that would be measured in the Schwarzschild geometry. (After Fuller and Wheeler, 1962.)

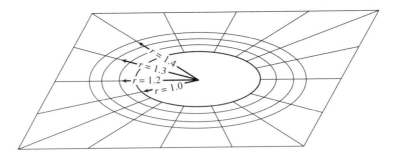

Fig. 8.6(b) The geometry of the same surface when the mass is zero (flat surface).

schematic diagrams that take us through a sequence of values from $v <$ −1 to $v > 1$. The middle diagram is a cross section through Fig. 8.6a; the other diagrams indicate how the structure shown in Fig. 8.6a evolves.

These diagrams give us a clear picture of how the wormhole pinches off. Any signal sent into the wormhole from one side or the other fails to pass through because it is trapped in the cusp. The suggestion has been made that an advanced civilization might keep the wormhole open, and avoid the formation of a horizon and a singularity, by artificial means (Morris, Thorne, and Yurtsever, 1988). But this can be achieved only if the wormhole is constructed out of a distribution of matter with negative energy density. Ordinary matter does not have a negative energy density; quantum fields might perhaps have the required negative energy density, but even if they do, they are not likely to extend over a macroscopically large volume. If wormholes could be kept open and travel through them were possible, then they could be exploited for time travel. A particle that takes a shortcut through a wormhole to travel from one point in space to a distant point effectively attains a speed in excess of the speed of light. As is well known, such a superluminal, or tachyonic, speed results in travel backward in time when observed from a new reference frame (Lorentz frame) relative to which the wormhole is in translational motion.*

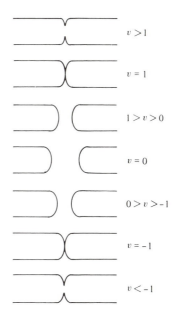

$v > 1$

$v = 1$

$1 > v > 0$

$v = 0$

$0 > v > -1$

$v = -1$

$v < -1$

Fig. 8.7 Evolution of the geometry in v time (schematic).

Finally, it must be kept in mind that although the Kruskal diagram describes a complete (or at least as complete as possible) solution of Einstein's equations, only a part of this solution will usually be rele-

---

* Morris, Thorne, and Yurtsever present a more complicated scheme for achieving time travel with a wormhole, a scheme with a host of extra complications related to an acceleration imposed on one end of the wormhole.

vant in the real world. For example, in the formation of a black hole by gravitational collapse of a star, the Schwarzschild vacuum solution applies only to those regions of spacetime outside of the star. If the dashed line in Fig. 8.4 is the worldline of the surface of the collapsing star, then the Schwarzschild solution is valid everywhere to the right of the dashed line. The region to the left of the dashed line is occupied by the matter of the star and is described by some new solution of Einstein's equations with $T^{\mu\nu} \neq 0$; the Kruskal diagram does not apply in this region. Obviously regions III and IV are never of any relevance to the problem of gravitational collapse. Since the complete Kruskal diagram is irrelevant to black holes formed by gravitational collapse, it can apply only to black holes created jointly with the universe, that is, black holes that have existed ever since the beginning of time.

An annoying feature of the Kruskal coordinates is that the metric depends on time (it depends on v), both inside and outside of the black hole, whereas the Schwarzschild coordinates avoid a dependence on time outside of the black hole. For some purposes it is convenient to introduce a set of coordinates that achieves a compromise between Kruskal coordinates and Schwarzschild coordinates--the metric does not depend on time, and is free of some, but not all, of the pseudosingularities at $r = r_S$. Such a set of coordinates are the Eddington-Finkelstein coordinates, often used for the solution of the wave equation in the Schwarzschild geometry. The timelike Eddington-Finkelstein coordinate $\tilde{v}$ is defined as follows:

$$\tilde{v} = t + r + r_S \ln|r/r_S - 1| \tag{30}$$

The other coordinates r, θ, ϕ remain the same as in the Schwarzschild case. The spacetime interval in these coordinates $\tilde{v}$, r, θ, ϕ is

$$ds^2 = \left(1 - \frac{r_S}{r}\right)d\tilde{v}^2 - 2d\tilde{v}\,dr - r^2d\theta^2 - r^2\sin^2\theta\,d\phi^2 \tag{31}$$

Note that the spatial part of the metric has no singularity, but the time part of the metric retains the pseudosingularity at $r = r_S$.

■ *Exercise 8.* Derive Eq. [31]. ■

The worldline of a radially ingoing light signal has the equation $\tilde{v} =$ [constant], and the worldline of an outgoing light signal has the equation $d\tilde{v}/dr = 2/(1 - r_S/r)$. Thus, the surfaces $\tilde{v} =$ [constant] are simply the worldlines of ingoing light signals.

■ *Exercise 9.* Derive these results for the motion of light signals. ■

Fig. 8.8 displays the Schwarzschild spacetime in $\tilde{v}$, r coordinates. The lines $\tilde{v}$ = [constant] have been drawn at 45°, so that the light cones at large r have the usual orientation. The existence of the event horizon at $r = r_S$ is, again, obvious by inspection of the light cones.

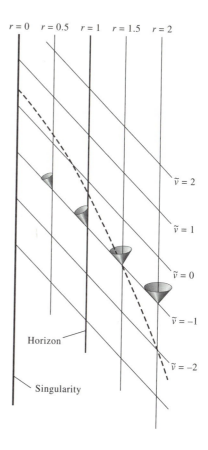

Fig. 8.8 The Schwarzschild spacetime in Eddington-Finkelstein coordinates $\tilde{v}$, r. The coordinates are given in units of r_S. The dashed line is the worldline of an infalling particle.

8.4 THE KERR SOLUTION AND THE REISSNER-NORDSTRÖM SOLUTION

Next to the Schwarzschild solution, the two most important solutions of Einstein's equations are the Kerr solution and the Reissner-Nordström solution. These two solutions represent, respectively, the curved spacetime geometries surrounding a rotating mass and an electrically charged mass.

In the linear approximation, the stationary solution for the gravitational field surrounding a rotating mass was given in Section 4.7. The corresponding exact solution of the nonlinear Einstein equations is given by the following expression for the spacetime interval (Kerr, 1963; Boyer and Lindquist, 1967):

$$ds^2 = dt^2 - \frac{\rho^2}{\Delta} dr^2 - \rho^2 d\theta^2 - (r^2 + a^2) \sin^2\theta \, d\phi^2$$

$$- \frac{2GMr}{\rho^2} (dt - a \sin^2\theta \, d\phi)^2 \qquad [32]$$

where ρ^2 and Δ are functions of r and θ,

$$\rho^2 \equiv r^2 + a^2 \cos^2\theta \qquad [33]$$

$$\Delta \equiv r^2 - 2GMr + a^2 \qquad [34]$$

and a and M are constants. This is the *Kerr solution*. It is straightforward, and very tedious, to verify that Einstein's equations *in vacuo* are satisfied by this metric.

The Kerr solution is stationary, but not static; that is, the metric is not a function of t, but fails to be invariant under time reversal because of the presence of the term $\propto d\phi dt$. The solution is rotationally symmetric about the z-axis; that is, it is independent of the angle ϕ.

For large r, Eq. [32] reduces to

$$ds^2 \simeq \left(1 - \frac{2GM}{r}\right)dt^2 - \left(1 + \frac{2GM}{r}\right)dr^2 - r^2 d\theta^2 - r^2 \sin^2\theta \, d\phi^2$$

$$- \frac{2GMa^2}{r} \sin^4\theta \, d\phi^2 + \frac{4GMa}{r} \sin^2\theta \, d\phi dt \qquad [35]$$

In this expression, all terms $\propto 1/r$ have been kept, and all terms $\propto 1/r^2$ have been omitted. The next to last term on the right side of Eq. [35] can be neglected, because it is smaller by a factor $2GMa^2 \sin^2\theta/r^3$ than the term that precedes it. To recognize the last term, we express it in rectangular coordinates. With $x = r \sin\theta \cos\phi$, $y = r \sin\theta \sin\phi$, we obtain

$$r^2 \sin^2\theta \, d\phi = xdy - ydx \qquad [36]$$

and therefore Eq. [35] is the same as

$$ds^2 \simeq \left(1 - \frac{2GM}{r}\right)dt^2 - \left(1 + \frac{2GM}{r}\right)dr^2 - r^2 d\theta^2 - r^2 \sin^2\theta \, d\phi^2$$

$$+ \frac{4GMa}{r^3}(xdy - ydx)dt \qquad [37]$$

The first four terms in Eq. [37] are exactly the same as in the Schwarzschild case and permit us to identify M as the total mass of the system. The last term corresponds to off-diagonal components in the metric. Expressed in rectangular coordinates, these off-diagonal terms are

$$g_{01} = g_{10} = -2GMa \, \frac{y}{r^3}$$

$$g_{02} = g_{20} = +2GMa \, \frac{x}{r^3} \qquad [38]$$

Comparison with the weak field solution for a rotating mass (Eq. [4.94]) shows that the quantity Ma is to be identified as the spin angular momentum of the system,

$$S_z = Ma \qquad [39]$$

Hence the parameter a is the angular momentum per unit mass. If $a > 0$, then the direction of rotation is counterclockwise around the $+z$-axis. Note that in Eq. [32], the time reversal $t \rightarrow -t$ has exactly the same effect as the change $a \rightarrow -a$; this simply corresponds to reversal of the direction of rotation.

The Kerr solution is not the only solution for the field surrounding a rotating mass. In general, such solutions will depend on the exact shape of the mass. Different mass distributions will generate different solutions, and these differences will show up at large r as different multipole moments in the gravitational fields. However, the Kerr solution is the unique stationary solution that represents a rotating black hole, that is, a rotating system that has collapsed inside its horizon. As we will see in Section 8.8, the gravitational collapse of a rotating, electrically neutral star will ultimately lead to the Kerr geometry.

The coordinates t, r, θ, ϕ used in Eq. [32] are Schwarzschild-like. In fact, in the limit $a \rightarrow 0$, the expression [32] for the Kerr solution agrees exactly with the expression [1] for the Schwarzschild solution.

■ *Exercise 10.* Check this. ■

However, if $a \neq 0$, then the coordinates t, r, θ, ϕ are not spherical coordinates, but quasi-spheroidal coordinates. These coordinates are defined in terms of rectangular coordinates by (Boyer and Lindquist, 1967)

$$x = (r^2 + a^2)^{1/2} \sin\theta \, \cos[\phi - f(r)] \qquad [40]$$

$$y = (r^2 + a^2)^{1/2} \sin\theta \, \sin[\phi - f(r)] \qquad [41]$$

$$z = r \cos\theta \qquad [42]$$

where

$$f(r) = a \int_{\infty}^{r} dr(r^2 - 2GMr + a^2) \tan^{-1}\frac{a}{r} \qquad [43]$$

If $M = 0$, then $f(r) = 0$, and the quasi-spheroidal coordinates defined by Eqs. [40]-[42] become ordinary spheroidal coordinates. In the limit $r \rightarrow \infty$, the coordinates r, θ, ϕ approach spherical coordinates.

Plotted in the rectangular axes x, y, z, the surfaces $r =$ [constant] are confocal ellipsoids, and the surfaces $\theta =$ [constant] are hyperboloids of one sheet. The surfaces $\phi =$ [constant] are rather complicated "bent planes" which gradually become flat at large values of r (in ordinary spheroidal coordinates, these surfaces would be ordinary planes). Fig. 8.9 shows the lines $r =$ [constant] and $\theta =$ [constant] in the x-z plane. At large values of r, the ellipses tend to circles and the hyperbolas straighten out to give us ordinary polar coordinates. Note that Fig. 8.9 adopts the convention that if the upper part of a given hyperbola is characterized by the parameter θ, then the lower part is characterized by $\pi - \theta$. This makes the correspondence between spheroidal and spherical coordinates more obvious; the discontinuity in θ at $r = 0$ is of course a purely mathematical device with no physical implications whatsoever. Finally, it must be emphasized that Fig. 8.9 is intended to display only the relationships among x, y, z and the r, θ, ϕ coordinates; this figure does not in any way display the geometry of the curved space.

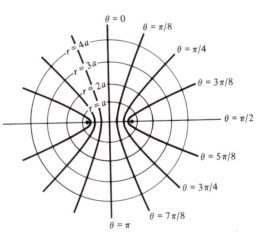

Fig. 8.9 Spheroidal coordinates in the x-z plane. The curves $r =$ [constant] are confocal ellipses with foci at $x = \pm a$; the curves $\theta =$ [constant] are confocal hyperbolas. The spheroidal coordinates in three dimensions are obtained by rotating this pattern about the z-axis.

In contrast to the Schwarzschild and the Kerr solutions, which are vacuum solutions of Einstein's equations, the Reissner-Nordström solution is not a vacuum solution. An electrically charged mass will be surrounded by an electric field, which gives rise to a nonzero energy-momentum tensor throughout space. We must then solve the coupled

Einstein and electromagnetic equations (Eqs. [7.19], [7.20]):

$$R_\mu{}^\nu - \tfrac{1}{2}\delta_\mu^\nu R = -8\pi G T_\mu{}^\nu \qquad [44]$$

$$F^{\mu\nu}{}_{;\nu} = 0 \qquad [45]$$

$$F^{\mu\nu;\alpha} + F^{\nu\alpha;\mu} + F^{\alpha\mu;\nu} = 0 \qquad [46]$$

where $T_\mu{}^\nu$ is the energy-momentum tensor [2.127] of the electromagnetic field. For the stationary spherically symmetric solution of these coupled equations, the spacetime interval will, again, have the general form of Eq. [7.61], but the functions $N(r)$ and $L(r)$ will differ from those we found in the Schwarzschild case. The electric field will be in the radial direction, and the electromagnetic field tensor will therefore have the form (in spherical coordinates)*

$$F^{\mu\nu} = \begin{bmatrix} 0 & -E(r) & 0 & 0 \\ E(r) & 0 & 0 & 0 \\ 0 & 0 & 0 & 0 \\ 0 & 0 & 0 & 0 \end{bmatrix} \qquad [47]$$

The solution of Eqs. [44]-[46] can be performed in a straightforward manner, as in the Schwarzschild case. The results for the spacetime interval and the electric field are

$$ds^2 = \left(1 - \frac{2GM}{r} + \frac{GQ^2}{r^2}\right) dt^2$$
$$- \frac{dr^2}{1 - 2GM/r + GQ^2/r^2} - r^2 d\theta^2 - r^2 \sin^2\theta \, d\phi^2 \qquad [48]$$

and

$$E(r) = \frac{Q}{r^2} \qquad [49]$$

This solution contains two constants of integration, M and Q. These constants must be interpreted as the mass and the electric charge (in Eqs. [48] and [49], these constants have been written in a form that anticipates their interpretation). This interpretation becomes obvious if

---

* In flat spacetime, the electric field can be expressed in the alternative forms $E = F^{01}$ or $E = F_{10}$. In curved spacetime, the components F^{01} and F_{10} are, in general, not equal; hence it is not immediately clear how the electric field is to be related to the field tensor. However, in the case of the Reissner-Nordström solution it turns out that F^{01} and F_{10} are equal, and we can identify either one as the electric field.

we compare Eq. [48] with the Schwarzschild solution, and if we compare Eq. [49] with the usual expression for the electric field of a spherical charge in flat spacetime. Surprisingly, the expression [49] for the electric field in the curved spacetime has exactly the same inverse-square dependence on r as in flat spacetime, even in the strongly curved region near $r = 0$ (however, in curved spacetime, r does not represent the measured radial distance). This inverse-square dependence is actually a direct consequence of Gauss' law. The electric flux through all spherical surfaces $r = $ [constant] must be the same. Since the area of such a surface is $4\pi r^2$, as in flat spacetime, the electric field strength must be inversely proportional to r^2, as in flat spacetime.

■ *Exercise 11.* Show that, with the Reissner-Nordström metric given by Eq. [48], $F^{01} = F_{10}$. ■

■ *Exercise 12.* Show that if $Q^2 > GM^2$, then the Reissner-Nordström metric is completely free of singularities except at r = 0. ■

8.5 HORIZONS AND SINGULARITIES IN THE ROTATING BLACK HOLE

The Kerr geometry is much more complicated than the Schwarzschild geometry, and depends drastically on whether $GM > |a|$, $GM = |a|$, or $GM < |a|$. We will first discuss the case $GM > |a|$. One of the least serious complications of the Kerr geometry is the presence of an infinite-redshift surface, where $g_{00} = 0$. According to Eq. [32] this corresponds to

$$g_{00} = 1 - \frac{2GMr}{\rho^2} = 0 \qquad [50]$$

that is,

$$r^2 + a^2 \cos^2\theta - 2GMr = 0$$

or

$$r = GM \pm \sqrt{G^2M^2 - a^2 \cos^2\theta} \qquad [51]$$

There are two distinct infinite-redshift surfaces; they have been drawn (in quasi-spheroidal coordinates) in Fig. 8.10. As in the Schwarzschild case, these surfaces do not correspond to any physical singularity. The vanishing of g_{00} merely tells us that a particle cannot be at rest (with $dr = d\theta = d\phi = 0$) at these surfaces; only a light signal emitted in the radial direction can be at rest.

In the region between these two surfaces, the value of g_{00} is negative. Hence t is not a timelike coordinate, and the t independence of the metric does not necessarily imply that the geometry is truly time independent. At a radius $r < GM - \sqrt{G^2M^2 - a^2 \cos^2\theta}$, the coordinate t again reverts to the character of a time coordinate.

In the Kerr geometry, the infinite-redshift surfaces do not coincide with event horizons. We can calculate the (coordinate) velocity of a light signal by setting $ds^2 = 0$ in Eq. [32]. For a signal in the equatorial plane ($\theta = \pi/2$, $d\theta = 0$), we obtain

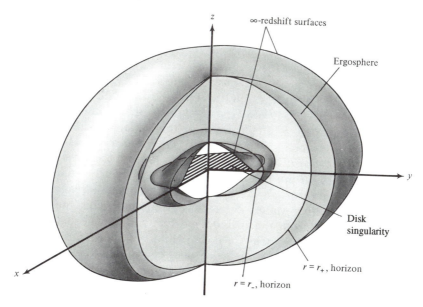

(a)

Fig. 8.10(a) The horizons and infinite-redshift surfaces of a rotating black hole with $GM = 1.2a$, represented in x, y, z coordinates. The horizons $r = r_+$ and $r = r_-$ are ellipsoids.

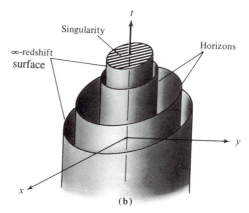

Fig. 8.10(b) The horizons and infinite-redshift surfaces of a rotating black hole represented in x, y, t coordinates.

$$\left(\frac{dr}{dt}\right)^2 = \frac{\Delta}{\rho^2}\left\{1 - (r^2 + a^2)\left[1 + \frac{2GMra^2}{\rho^2(r^2 + a^2)}\right]\left(\frac{d\phi}{dt}\right)^2 - \frac{2GMr}{\rho^2} + \frac{4GMr}{\rho^2}a\frac{d\phi}{dt}\right\}$$

[52]

If we evaluate this on the infinite-redshift surfaces, we find

$$\left(\frac{dr}{dt}\right)^2 = \frac{\Delta}{2GMr}\left[2a\frac{d\phi}{dt} - (r^2 + a^2)\left(1 + \frac{a^2}{r^2 + a^2}\right)\left(\frac{d\phi}{dt}\right)^2\right]$$

[53]

where it is understood that r has one of the values given by Eq. [51]. For a signal in the radial direction $d\phi/dt = 0$, and hence $dr/dt = 0$; such a signal cannot escape.

For a light signal in the x-y plane, with $d\phi/dt \neq 0$, Eq. [53] does not determine dr/dt as a function of r. The velocity of the signal must then be calculated by integrating the geodesic equation so as to obtain dr/dt and $d\phi/dt$. Eq. [53] is a first integral of the equation of motion and may be regarded as a constraint on the initial velocities. It is the *only* constraint. Hence, we can arbitrarily assign dr/dt a positive initial value and adjust $d\phi/dt$ so as to satisfy Eq. [53] (note that excessively large values of dr/dt are not acceptable because we are dealing with a quadratic equation for $d\phi/dt$, and we must obtain a real solution). Obviously, a positive value of $ad\phi/dt$ is required. This implies that signals can escape from the infinite-redshift surfaces if sent outward with a component of velocity in the direction of rotation of the Kerr field.

Where are the event horizons? Their location is given by the condition that $dr/dt = 0$ for arbitrary values of $d\theta$ and $d\phi$. If we set $ds^2 = 0$ in Eq. [32] and solve for $(dr/dt)^2$, we obtain an expression similar to Eq. [52], but with some extra terms involving θ and $d\theta/dt$. In any case, Δ appears as an overall multiplicative factor on the right side, just as it does in Eq. [52]. Hence the condition $dr/dt = 0$ becomes

$$\Delta = 0 \qquad\qquad\qquad [54]$$

that is,

$$r^2 - 2GMr + a^2 = 0$$

or

$$r = GM \pm \sqrt{G^2M^2 - a^2} \qquad\qquad\qquad [55]$$

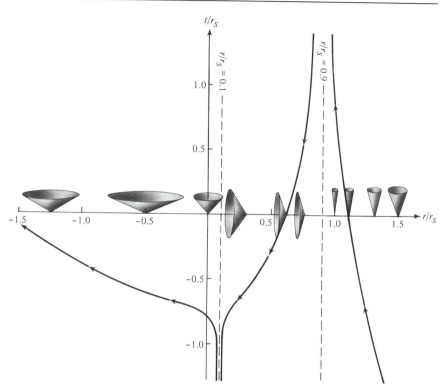

Fig. 8.11 The forward light cones on the axis of rotation near and inside a rotating black hole with $GM = 1.666a$ (this value of GM has been chosen for convenience in plotting; it has no special significance). The curve is the worldline of an ingoing light signal.

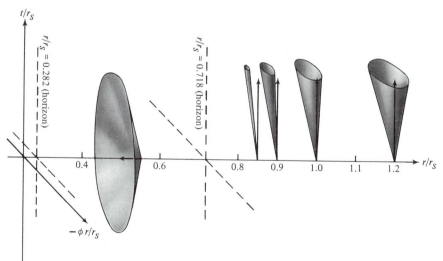

Fig. 8.12 The forward light cones in the equatorial plane near and inside a rotating black hole with $GM = 2.222a$ (again, this value of GM has been chosen for convenience in plotting; it has no special significance). The rotation is counterclockwise.

It is customary to define

$$r_+ \equiv GM + \sqrt{G^2M^2 - a^2} \qquad [56]$$

and

$$r_- \equiv GM - \sqrt{G^2M^2 - a^2} \qquad [57]$$

At $r = r_+$ and at $r = r_-$ a light signal necessarily has zero velocity in the radial direction; thus, the light cones lie along the surfaces $r =$ [constant], and light cannot escape from these surfaces. These horizons are shown in Fig. 8.10; at the poles ($\theta = 0$, $\theta = \pi$), the infinite-redshift surfaces meet the horizons. The one-way membrane $r = r_+$ defines the outer boundary of the rotating black hole. The one-way membrane $r = r_-$ is a second, inner boundary.

In the region between the two event horizons, $g_{11} > 0$ and hence r is a timelike coordinate. As in the Schwarzschild case, the r dependence of the metric therefore implies that the metric in this region is necessarily dynamic.

Figs. 8.11 and 8.12 show the light cones at different values of the radius; these figures should be compared with Fig. 8.3 for the Schwarzschild case.

Neither at the infinite-redshift surfaces nor at the horizons does the Kerr geometry develop a true singularity; in local geodesic coordinates, the curvature remains finite. The only true singularity occurs at $r = 0$. At this location the curvature tensor diverges. To understand how this singularity arises, let us examine g_{00} in the vicinity of $r = 0$. According to Eq. [32], we have

$$g_{00} = 1 - \frac{2GMr}{r^2 + a^2 \cos^2\theta} \qquad [58]$$

For $\theta = 0$, this expression can be regarded as a function of z,

$$g_{00} = 1 - \frac{2GMz}{z^2 + a^2} \qquad [59]$$

In Fig. 8.13, this function is plotted vs. z. Note that at $z = 0$, $\partial g_{00}/\partial z$ is not zero. This implies that the value of $\partial g_{00}/\partial z$ depends on how we approach the point $z = 0$. If we approach from above ($z > 0$), then $\partial g_{00}/\partial z < 0$; if we approach from below ($z < 0$), then $\partial g_{00}/\partial z > 0$. Thus, $\partial g_{00}/\partial z$ has a discontinuity at $z = 0$. It follows that the second derivative $\partial^2 g_{00}/\partial z^2$ does not exist at $z = 0$; more precisely, the second derivative diverges at $z = 0$.* The behavior of the second derivatives

---

* The second derivative at $z = 0$ is a delta function.

of g_{11}, g_{33}, and g_{03} is similar. Since the curvature tensor involves these second derivatives, we can conclude that the curvature tensor is singular at $r = 0$. This singularity is a physical singularity, which cannot be blamed on our choice of coordinates; the spheroidal coordinates r, θ, ϕ and the metric [32] are perfectly regular when $r = 0$ (except when $\theta = \pi/2$).

Note that in our spheroidal coordinates, $r = 0$ is a disk centered on the origin. This can be seen from Eqs. [40]-[42]; in rectangular coordinates, $r = 0$ corresponds to $x^2 + y^2 = a^2 \sin^2\theta$, $z = 0$. Since θ varies between 0 and $\pi/2$, this is equivalent to $x^2 + y^2 \leq a^2$, $z = 0$. The edge of the disk is at $x^2 + y^2 = a^2$; that is, it is at the foci of the ellipses in Fig. 8.9. If we take the limit $a \to 0$, then the disk singularity shrinks and ultimately reduces to the Schwarzschild singularity $r = 0$.

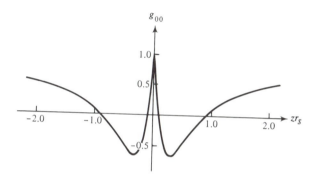

Fig. 8.13 The function g_{00} evaluated along the z-axis.

What is the physical meaning of the disk singularity? Einstein's equations tell us that an infinite value of the curvature and of $R_{\mu\nu} - \frac{1}{2}g_{\mu\nu}R$ corresponds to an infinite value of the energy-momentum density. Hence, the disk $r = 0$ contains an infinite density of matter. The Kerr solution can then be interpreted as giving the spacetime geometry in the region surrounding a thin rotating disk of matter. The Kerr solution also has an alternative interpretation without such a disk of matter, but this interpretation requires that our asymptotically flat space be joined, at the disk $r = 0$, to another asymptotically flat space. We will deal with this extension of the Kerr geometry in Section 8.6.

The region between the outer surface of infinite redshift and the outer horizon (see Fig. 8.10) is called the *ergosphere* (Rees, Ruffini, and Wheeler, 1974). A procedure discovered by Penrose (1969) makes it possible to extract energy from the rotating black hole by means of particles that are moved in and out of the ergosphere. The key to this procedure is that some orbits in the ergosphere have a negative total energy, that is, the gravitational binding energy exceeds the sum of rest-mass and kinetic energies. We recall that the energy is related to the Lagrangian by

$$P_0 = \frac{\partial L}{\partial u^0}$$

With $L = m\sqrt{g_{\mu\nu}u^\mu u^\nu}$, we obtain

$$P_0 = \frac{mg_{0\alpha}u^\alpha}{\sqrt{g_{\mu\nu}u^\mu u^\nu}} = mg_{0\alpha}u^\alpha \qquad [60]$$

For a freely falling particle, this energy is conserved, since the gravitational field is independent of x^0 (see the discussion following Eq. [3.94]). With the values of $g_{0\alpha}$ from Eq. [32], we find

$$P_0 = m(g_{00}u^0 + g_{03}u^3) \qquad [61]$$

In the ergosphere, $g_{00} < 0$ and $g_{03} > 0$. Furthermore, $u^0 > 0$ and $u^3 > 0$ (see Fig. 8.12 and note that u^μ must be in the forward light cone). Hence the two terms in Eq. [61] are of opposite signs, and the energy will be negative if the first term dominates.

■ *Exercise 13.* Find a set of values for the components of u^μ such that $u^\mu u_\mu = 1$ and $P_0 < 0$. Assume that $\theta = \pi/2$ and $2GM > r > r_+$. ■

To extract energy from a black hole, we proceed as follows. We let a spaceship fall from infinity into the ergosphere along an orbit of positive energy, and there, by means of a spring-loaded device, eject a brick from the spaceship in such a way that the brick enters an orbit of negative energy while the spaceship, by recoil, enters an orbit of increased positive energy. The brick then falls into the black hole, and the spaceship returns to infinity. Since the total energy of the system consisting of black hole, spaceship, and brick is conserved, and the brick carries some amount of negative energy into the black hole, the spacehip must carry a corresponding amount of positive energy away to infinity, that is, the spacehip emerges with more energy than it had initially--the spaceship extracts energy. (Note that the energy budget for the spaceship includes the rest-mass and the initial kinetic energy of the brick and the energy initially stored in the spring-loaded device; the Penrose process also converts these energies into kinetic energy of the spaceship, but this conversion does not alter the net energy of the spaceship and does not represent an energy extraction.) The extra positive energy with which the spaceship returns to infinity arises from the rotational kinetic energy of the black hole. When the black hole captures the negative-energy brick, the rate of rotation of the black hole decreases slightly, and the mass of the black hole decreases slightly, by an amount equal to the extracted energy.

Christodolou and Ruffini (1970, 1971) have shown that there is an upper limit on the amount of energy that can be extracted from a black hole. The limit is

$$\Delta M = M - M_{ir} \qquad [62]$$

where M_{ir} is the *irreducible mass*:

$$M_{ir} = \frac{1}{2G}[(GM + \sqrt{G^2M^2 - a^2})^2 + a^2]^{1/2} \qquad [63]$$

The upper limit given by Eq. [62] corresponds to a complete extraction of the rotational energy; we are then left with a Schwarzschild black hole of mass M_{ir}.

These results follow directly from Hawking's theorem: *In any process involving one or more black holes, the total area of the horizon surfaces can never decrease.* (Hawking and Ellis, 1973; Hawking, 1973). The proof of this theorem uses some clever global topological techniques that are beyond the scope of our discussion. We will not give the proof, but we will use the theorem. In the case of the Kerr black hole, the area of the horizon $r = r_+$ is

$$A = 8\pi GM(GM + \sqrt{G^2M^2 - a^2}) \qquad [64]$$

■ *Exercise 14.* Show that Eq. [64] gives the area of the surface $r = r_+$. (Hint: The area is given by the integral

$$A = \iint \sqrt{g_{22}} \sqrt{g_{33}} \, d\theta d\phi$$

where g_{22} and g_{33} are evaluated at $r = r_+$.) ■

With this expression for A, it is easy to show that for arbitrary infinitesimal changes δM and δa, the corresponding changes in A and M_{ir} are related as follows:

$$\delta(M_{ir})^2 = \frac{1}{16\pi G^2} \delta A \qquad [65]$$

Hawking's theorem tells us that $\delta A \geq 0$, and consequently $\delta M_{ir} \geq 0$. The black-hole mass M can be written in the form

$$M^2 = M_{ir}^2 + \frac{S^2}{(4GM_{ir})^2} \qquad [66]$$

where $S = aM$ is the spin of the black hole. Since M_{ir} must either remain constant or increase, Eq. [66] tells us that any decrease in M must come from a decrease of S; furthermore, Eq. [66] tells us that the minimum attainable value of M is M_{ir}.

■ *Exercise 15.* Derive Eq. [65]. ■

The Penrose process is rather contrived, and we do not expect that it will occur spontaneously in nature. However, there is a closely related process which is somewhat less contrived and might occur spontaneously. This process involves the scattering of a wave--electromagnetic, gravitational, or whatever--by a black hole. When such a wave is incident on a rotating black hole, the ergosphere selectively scatters the partial waves of different angular momenta (different multipoles) contained in the incident wave. Calculations of the scattering indicate that the partial waves that have angular momentum in the same direction as the black hole are reinforced by the scattering, and they emerge with more energy than they had initially (Zeldovich, 1972). The extra energy arises from a slight reduction of the rate of rotation of the black hole. This process is called *superradiance*. The energy gain depends on the type of wave; for example, a gravitational wave can emerge with up to 39% more energy.

This increase of the intensity of a wave by scattering is analogous to the Klein paradox of relativistic quantum mechanics, according to which an electron wave incident on a potential step of height in excess of $2m_e c^2$ can be reflected with an increased amplitude (Bjorken and Drell, 1964). The explanation is that the incident electron wave triggers the formation of electron-antielectron pairs when the height of the step is large enough to supply the energy for this pair formation. Likewise, the explanation for superradiance is that the incident wave triggers pair formation in the ergosphere. One member of the pair can enter an orbit of negative energy and the other an orbit of positive energy in such a way that the total energy is conserved. The former then falls into the black hole while the latter escapes to infinity. The net result is that particles steadily stream out of the ergosphere while the energy of the black hole steadily decreases. Note that the energy loss of the black hole is not due to particles emerging from the horizon; rather, it is due to the capture of negative-energy particles that originate in the ergosphere (Eardley and Press, 1975; Unruh, 1974).

8.6 THE MAXIMAL KERR GEOMETRY

The Kerr geometry has some weird features which we can discover by tracing worldlines of particles through it. Consider a worldline that starts outside the rotating black hole and is everywhere timelike ($ds^2 > 0$). This worldline represents the motion of a particle with propulsion or perhaps the motion of a spaceship with rocket motors. Such a particle can travel on any worldline we choose, subject only to the requirement that the worldline be timelike and in the forward light cone. Assuming that the motion is along the axis of rotation of the black hole ($\theta = 0$), a possible worldline is shown in Fig. 8.14. The worldline begins outside of the black hole, crosses the horizons $r = r_+$ and $r = r_-$, and approaches near $r = 0$. After a short rest at this position, the particle moves in the outward direction and approaches the surface $r = r_-$ from the inside. It reaches this surface in a finite proper time. But the particle cannot continue across $r = r_-$ because this is a one-way membrane. It is a peculiar feature of the Kerr geometry that the one-way membrane can be approached from inside, yet it cannot be crossed (as $r \to r_-$ from the inside, the light cones become narrow and thin, similar to the light cones as $r \to r_+$ from the outside; see Fig. 8.11). Physically, we cannot accept that the worldline of the particle simply comes to an end--there is no singularity at the horizon. To avoid this geodesic incompleteness, we have only one alternative: we must suppose that upon crossing $r = r_-$ in the outward direction, the particle finds itself in another universe in which the surface $r = r_-$ belongs to a white hole rather than a black hole. Thus, in this second universe the surface is still a one-way membrane, but it now is one way out rather than one way in. Of course, the white hole must have the same mass and spin as the black hole. The particle then continues toward $r = r_+$, which is to be regarded as another one-way-out membrane, and finally emerges from the white hole and reaches the asymptotically flat regions of the second universe.

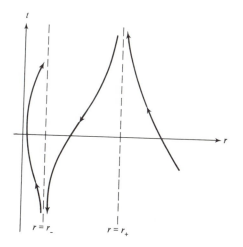

Fig. 8.14 A possible worldline for a particle moving along the axis of a black hole.

But the maximal Kerr geometry does not end with two universes. In a universe with a rotating black hole, we can introduce Kruskal-like coordinates. The explicit coordinate transformations are messy, and we will not try to write them down. Once these coordinates have been introduced, we can argue, just as in the Schwarzschild case, that geodesic completeness requires that a universe that contains a black hole must also contain a corresponding white hole, and, vice versa, a universe that contains a white hole must also contain a black hole (compare Fig. 8.5). Continuing with our description of the travels of the particle, we conclude that the universe cum white hole in which the particle finds itself must also contain a black hole. Assume the particle heads for this black hole and drops in. It will then emerge in a third universe, and so on.

These arguments suggest that the maximal Kerr geometry has a very complicated structure. A careful analysis of the geometry shows that the geodesically most complete extension of the Kerr solution consists of an infinite sequence of universes, each containing a rotating black hole and white hole. The universes are joined by one-way "tunnels" whose entrances and exits are these black and white holes, respectively.

Fig. 8.15 shows part of this infinite sequence (Carter, 1966). To fit all the universes on the page, the asymptotically flat regions have been compressed so as to appear finite. This compression has been accomplished by a conformal coordinate transformation that maps the infin-

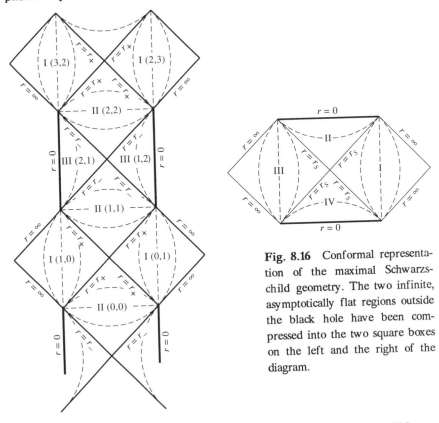

Fig. 8.16 Conformal representation of the maximal Schwarzschild geometry. The two infinite, asymptotically flat regions outside the black hole have been compressed into the two square boxes on the left and the right of the diagram.

Fig. 8.15 Conformal representation of the maximal Kerr geometry, with $GM > a$. The multiplicity of infinite, asymptotically flat regions outside the black hole have been compressed into the square boxes on the left and the right of the diagram. (After Carter, 1966.)

ite spacetime into a finite region, but leaves the worldlines as straight lines at 45° with respect to the axes. For comparison, Fig. 8.16 shows the Kruskal diagram subjected to the same kind of conformal transformation. Such diagrams, showing the compressed spacetime, are called *Penrose diagrams*. The squares labeled I(m, n) in Fig. 8.15 are asymptotically flat universes in which the coordinate r lies in the region $r_+ < r < \infty$ (in this region, r is spacelike). The squares labeled II(k, k) correspond to the region $r_- < r < r_+$ that lies between the inner and outer horizons (in this region, r is timelike). Note that II(1, 1) is a black hole as seen from the universes I(1, 0) and I(0, 1), whereas II(2, 2) is a white hole as seen from I(3, 2) and I(2, 3). The triangles III(m, n) correspond to the region $0 < r < r_-$.

Some geodesics of freely falling particles are shown in Fig. 8.17. The shape of the geodesic depends on the energy. Note that only geodesics of sufficiently high energy can reach $r = 0$; the others have a turning point at $r > 0$.

Any geodesic that strikes the disk singularity $r = 0$ will of course terminate. Is there any way we can avoid having this singularity in our geometry? It turns out to be possible to eliminate the singularity by continuation of the metric to *negative* values of r. If we insist that r is always positive, then the derivative of g_{00} has a discontinuity at $r = 0$ (see Fig. 8.11). But if we extrapolate Eq. [50] to negative values of r,

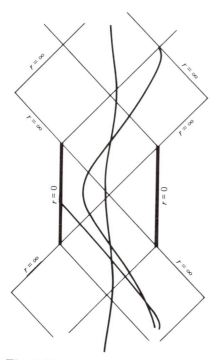

Fig. 8.17 Typical worldlines in the maximal Kerr geometry. (After Carter, 1966.)

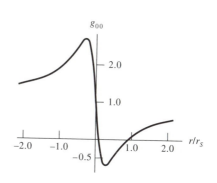

Fig. 8.18 The function g_{00} extrapolated to negative values of r along the z-axis.

then there is no discontinuity (see Fig. 8.18). Thus, we assume that when a worldline crosses $r = 0$, it enters a space with negative values of r; conversely, a worldline that approaches $r = 0$ from the negative side enters into the region of positive r. Of course, the space with negative values of r must form a separate universe, distinct from ours; the universes are joined at $r = 0$.

Fig. 8.19 shows the structure of the complete Kerr geometry, including the spaces of negative values of r. In these spaces r has the range $0 > r > -\infty$. It is obvious from Eq. [30] that as $r \to -\infty$, the spacetime becomes flat. Since the values of r given by Eqs. [43] and [47] are positive, there are no infinite-redshift surfaces or horizons in these new universes.

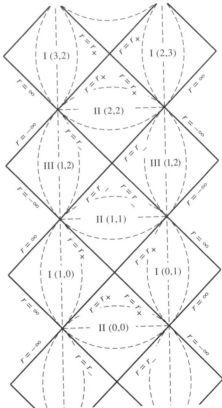

Fig. 8.19 Conformal representation of the maximal Kerr geometry, extended to negative values of r.

Note that these diagrams show only the two-dimensional r-t spacetime; they can be used only to analyze the motion of a particle that stays on the axis of symmetry. Particles that are not on the axis can also go from one universe to the other provided they cross the disk $r = 0$; that is, their orbit must thread through the ring singularity $r = 0$, $\theta = \pi/2$ (see Fig. 8.10a). An orbit that avoids this disk remains within the same universe.

To arrive at a physical interpretation of the negative values of r, we note that in Eq. [32], a change in the sign of r is equivalent to a change in the sign of M. An observer in the new universe will of course insist on interpreting his observations in terms of a positive value of r, and he will therefore measure M to be negative. A particle of positive mass near such a Kerr black hole will experience a repulsive gravitational force that pushes it away from $r = 0$.

It must be emphasized that the continuation to negative values of r is nothing but a farfetched speculation. This speculation should probably be discarded on physical grounds. Negative values of the mass (or total energy) are very objectionable. If particles of negative mass could be created, the vacuum would be extremely unstable, because quantum-mechanical vacuum fluctuations could continuously create an endless number of pairs of positive- and negative-mass particles. In the absence of negative mass, such an instability is prevented by energy conservation.

There is also another objection to a continuation to negative values of r: in some regions of negative r, the worldline of a particle can be made to close upon itself, that is, the particle can return into its own past (Carter, 1968). This can be seen from Eq. [32] by examining the coefficient of $d\phi^2$:

$$g_{33} = -(r^2 + a^2) \sin^2\theta \left[1 + \frac{2GMa^2r}{\rho^2(r^2 + a^2)} \sin^2\theta \right] \qquad [67]$$

The term in brackets becomes negative if

$$r < -\frac{(r^2 + a^2 \cos^2\theta)(r^2 + a^2)}{2GMa^2 \sin^2\theta} \qquad [68]$$

When this condition is satisfied, ϕ becomes a timelike coordinate. Hence a circle around the symmetry axis, with $r = $ [constant], $\theta = $ [constant], $t = $ [constant], is a possible particle worldline. If a particle moves along this circle, it returns to its starting point when ϕ increases by 2π. Note that the motion proceeds without any increase in t time, that is, not only is the "orbit" (in three dimensions) closed, but also the worldline (in four dimensions).

Although it is necessary that any closed timelike worldline enter the region [68], it is not necessary that such a worldline lie entirely within this region. For example, consider a worldline that emerges from the region [68], proceeds to the asymptotically flat region $r \rightarrow -\infty$, plunges back into the region [68], and there is made to close upon itself. Since, in the asymptotically flat region, t necessarily increases along a timelike worldline, we must arrange the worldline in the region [68] in such a way that t decreases along it. This is possible because there exist timelike worldlines with $ad\phi/dt > 0$. For such a worldline we can attain a decrease in t if the worldline goes around the axis of rotation in a direction opposite to that of the rotation ($dt < 0$ and $ad\phi < 0$).

■ *Exercise 16.* By examination of the terms in Eq. [32] involving dt^2, $d\phi^2$, and $d\phi dt$, show that ds^2 can be positive for negative dt and negative $ad\phi$. ■

The worldline described in the preceding paragraph has the paradoxical property that the particle emerges from the Kerr hole before it enters it. The time interval between emergence and entrance can be made arbitrarily large by completing many orbits around the symmetry axis before leaving the region [68]. Thus, the Kerr hole can act as a "time machine"; by passing through the machine, a particle travels into the past.

Such a time machine violates causality and results in a logical contradiction. To put the paradox in a very glaring form, suppose we regard the particle as a signal (for instance, a signal rocket) that is emitted, at time $t = 0$, by an apparatus located in the asymptotically flat universe and is received by this apparatus at some earlier time, say at $t = -2$. Suppose the apparatus is programmed with the following instructions: emit the signal if the signal is not received before $t = 0$, do not emit the signal if the signal is received before $t = 0$. These instructions are in logical contradiction with emission at $t = 0$ and reception at $t = -2$. This causality paradox is a strong argument against the existence of regions of negative r. If we suppose that only the regions with positive r in Fig. 8.15 exist, then the paradox disappears.

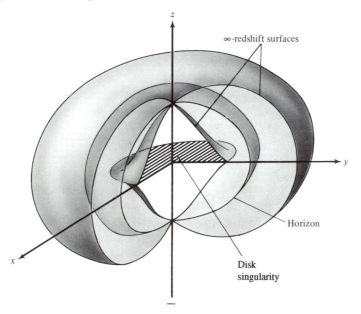

Fig. 8.20 The horizons and infinite-redshift surfaces of a rotating black hole with $GM = a$.

Incidentally: There is another causality violation which is sometimes discussed in connection with Fig. 8.15. Since the regions I(1,0) and I(3,2), for instance, have exactly identical geometries, it is tempting to identify them. A particle that drops into the rotating black hole would then simply emerge from the rotating white hole in the *same* universe. Such identification would be helpful in reducing the number of universes. But we run into the causality problem: some timelike worldlines that enter the black hole would emerge from the white hole at an earlier time. This type of causality violation has been called trivial, because it is the consequence of unfortunate and unnecessary topological identifications. In contrast, the causality violation at negative r is nontrivial (Carter, 1968).

So far, our discussion of the Kerr geometry has dealt only with the case $|a| < GM$. In the case $|a| = GM$, the two horizons coincide (see Fig. 8.20), and in the case $|a| > GM$, there is no horizon. Of course, there still is a ring singularity at $r = 0$, $\theta = \pi/2$. A singularity *not* surrounded by a horizon is called a *naked singularity*. A rotating "hole"* of large spin, $|a| > GM$, has such a naked singularity. Figs. 8.21 and 8.22 show the geometry for these two cases.

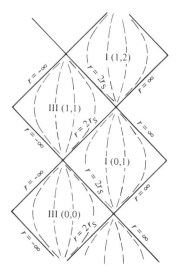

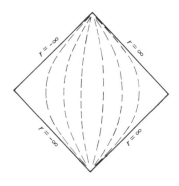

Fig. 8.22 Conformal representation of the maximal Kerr geometry, with $GM < a$. (After Carter, 1966.)

Fig. 8.21 Conformal representation of the maximal Kerr geometry, with $GM = a$. (After Carter, 1966.)

Although naked singularities do exist mathematically, as solutions of the Einstein equations, we do not know whether they are ever found in the real world. Penrose has conjectured that the complete

---

* If there is no horizon, we will not call it a *black* hole.

gravitational collapse of a (nonsingular) mass never results in a naked singularity, but rather in a singularity hidden within a horizon. This conjecture is known as the *cosmic censorship conjecture*. This conjecture does not forbid the existence of naked singularities in the mathematical solutions of Einstein's equations; it forbids only the formation of such a naked singularity in gravitational collapse.

The cosmic censorship conjecture remains unproved. Some attempts have been made at constructing counterexamples to the conjecture, by numerical integration of the Einstein equations for the gravitational collapse of various mass distributions. For instance, the gravitational collapse of an elongated (cigar-shaped) mass distribution appears to generate a naked singularity (Shapiro and Teukolsky, 1991). However, the numerical integration becomes inaccurate and breaks down when the curvature tensor becomes large, and this makes it impossible to establish beyond all reasonable doubt that the curvature does indeed become infinite and that no horizon develops at some future time.

The Kerr solution for $|a| \geq GM$ has a naked singularity, but the formation of this singularity in collapse is probably prevented by an instability: the solution is probably unstable for $|a| \geq GM$, and a collapsing mass with a large spin probably breaks up into several pieces during collapse (whether the Kerr solution is in fact stable or unstable is not yet known). The Reissner-Nordström solution with a large electric charge ($Q^2 > GM^2$) also has a naked singularity. But a spherical mass with $Q^2 > GM^2$ is indeed unstable, since the large electric repulsive force overcomes the attractive gravitational force, and this means that the mass will not collapse into the Reissner-Nordström configuration.

From a practical point of view, the complete Kerr geometry is as irrelevant to physics as the complete Schwarzschild geometry. If a rotating black hole forms by gravitational collapse, then the interior region will be occupied by matter, and the pure vacuum Kerr solution does not apply. Hence the plurality of universes in the Kerr geometry has nothing to do with reality unless a rotating black hole was, somehow, created when the universe began. And even if such an object was created, it is unlikely that it has survived: there is evidence that the "tunnels" and white holes of the maximal Kerr geometry are violently unstable.

Finally, note that in the case of collapse of a spherical, nonrotating star, the Schwarzschild solution fails to hold in the interior, but the Birkhoff theorem guarantees that it will at least hold in the exterior. In the case of collapse of a rotating star, there is no such theorem, and the exterior solution need not be that of Kerr. Only a long time after the collapse, when everything becomes stationary, will the exterior solution tend toward those of the Kerr solution.

8.7 BLACK-HOLE THERMODYNAMICS; THE HAWKING PROCESS

The horizon, or one-way membrane, of a black hole acts as a perfect absorber--it permits anything to enter, but does not permit anything to leave. We therefore expect that it could serve as an ideal heat sink for the operation of a thermodynamic engine. We expect that when we dump heat on the horizon, in the form of thermal radiation at some given temperature, it will be completely absorbed. With such an ideal heat sink, we can operate a thermodynamic engine with 100% efficiency, that is, we can accomplish the complete conversion of heat into work. The following *Gedanken*-experiment describes a simple machine that attempts to accomplish this goal.

At large distance from a black hole, fill a box with thermal radiation at some temperature T (see Fig. 8.23). Close the box and slowly lower it toward the horizon, by means of a rope attached to a winch. As the box descends, the gravitational potential energy of the radiation is converted into useful mechanical work (the gravitational potential energies of the box and the rope are irrelevant, since we will ultimately have to lift the box and rope back to their initial positions). At the horizon of a Schwarzschild black hole, the gravitational potential energy of a mass m of radiation is $-mc^2$ (see the discussion following Eq. [7.131]); that is, all of the energy of the radiation will have been converted into useful work. We now open the bottom of the box, and dump the radiation into the black hole. Then we raise the empty box to its initial position, and thereby complete one cycle of the operation of our thermodynamic engine. The net result of this

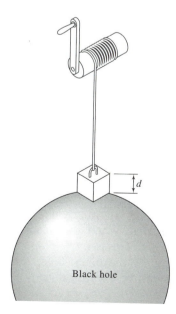

Fig. 8.23 A box attached to a winch is filled with thermal radiation and then lowered until its bottom reaches the horizon of the black hole.

cycle would seem to be the complete conversion of heat into mechanical work. Such a complete conversion would lead us to the conclusion that the thermodynamic temperature of the black hole is exactly zero.

However, this argument neglects to take into account the size of the box. Since thermal radiation of temperature T typically has a wavelength of about $\hbar c/kT$ (by Wien's law), a box that contains such thermal radiation must have a size d at least as large as the typical wavelength,

$$d \simeq \hbar c/kT \qquad [69]$$

When the bottom of the box reaches the horizon, the center of gravity of the box will be at a distance of $d/2$ above the horizon, and the potential energy of the radiation will be $-mc^2 + mgd/2$, where g is the acceleration of gravity at the surface of the black hole. From Newtonian theory, we can estimate this acceleration of gravity as $g = GM/r_S{}^2 = c^4/4GM$ (by dumb luck, the exact relativistic value of g agrees with this Newtonian value; see Problem 7.18). The mechanical work we gain by dumping the radiation is therefore not mc^2, but

$$W = mc^2 - m\frac{c^4}{4GM}\frac{d}{2} \simeq mc^2\left(1 - \frac{\hbar c^3}{8GMkT}\right) \qquad [70]$$

The efficiency of our thermodynamic engine is then

$$e = \frac{W}{mc^2} \simeq 1 - \frac{\hbar c^3}{8GMkT} \qquad [71]$$

According to the definition of the thermodynamic temperature, this efficiency can be expressed as $1 - T_{BH}/T$, where T_{BH} is the temperature of the black hole. The formula [71] therefore leads to the conclusion that the temperature of the black hole is

$$T_{BH} \simeq \frac{\hbar c^3}{8GMk} \qquad [72]$$

Note that this temperature is inversely proportional to the surface gravity of the black hole.

The result [72] for the temperature of a black hole was first obtained by Bekenstein (1973, 1974), who proposed to associate not only a temperature with the black hole, but also an entropy, so that the laws of thermodynamics can be applied to processes involving black holes. When our black hole absorbs an amount of heat Q, its entropy increases by Q/T_{BH}. The heat Q absorbed equals the increase of mass of the black hole, that is, $Q = \delta Mc^2$. Hence the entropy increase can be expressed as

$$\delta S = \frac{\delta Mc^2}{T_{BH}} \simeq \frac{8kGM\delta M}{\hbar c} \qquad [73]$$

Alternatively, this can be expressed in terms of the change of the surface area of the black hole, $\delta A = 8\pi r_S \delta r_S = 32\pi G^2 M \delta M / c^4$:

$$\delta S \simeq \frac{kc^3}{4\pi\hbar G}\delta A \qquad [74]$$

This simple proportionality between δS and δA indicates that the entropy of a black hole must be proportional to its surface area,

$$S \simeq \frac{kc^3}{4\pi\hbar G}A \qquad [75]$$

The formation of a black hole therefore entails a large increase of entropy. Such an increase can be made plausible by the information-theoretic interpretation of entropy. When the black hole forms, or whenever we increase the size of the black hole by dumping any kind of matter into it, we lose information about the trapped matter. This loss of information corresponds to an increase of entropy (Bekenstein, 1973, 1974, 1980; Hawking, 1976).

We can now formulate the first law of thermodynamics for a black hole in the usual way: the increase δMc^2 in the internal energy equals the sum of the heat absorbed by the black hole and the mechanical work performed on the black hole by external forces. Since the black hole behaves like a rigid body, the only way to increase its internal mechanical energy is by spinning it up by an external torque. In this case $\delta W = \Omega \delta J$, where Ω is the angular velocity of the black hole and J its spin angular momentum.* Hence the first law of thermodynamics becomes

$$\delta Mc^2 = T_{BH}\, \delta S + \Omega \delta J \qquad [76]$$

The second law of thermodynamics for black holes is a direct consequence of Hawking's theorem for the increase of the surface area in any process involving one or more black holes. Since the entropy is proportional to the surface area, Hawking's theorem ensures the increase of entropy.

In the above discussion, we treated the black-hole temperature T_{BH} as a thermodynamic temperature. But it must also be a radiation temperature, that is, the black hole must emit thermal radiation of this characteristic temperature. If it did not, we could immerse the black hole in a bath of radiation of lower temperature, $T < T_{BH}$, and the

---

* In contrast to the notation of Section 8.4, we now use the symbol J for the spin.

black hole would then absorb this radiation without emitting any. Thus, while the radiation bath loses an amount of entropy $\delta Q/T$, the black hole would gain only a smaller amount of entropy $\delta Q/T_{BH}$, with a violation of the second law of thermodynamics.

The explicit proof that black holes emit thermal radiation was given by Hawking, who discovered that in the curved spacetime of a black hole, radiation is generated by a quantum process. Hawking found an elegant way to calculate the spectrum of the radiation from the behavior of quantum fields in curved spacetime (Hawking, 1974, 1975). Earlier attempts at such calculations foundered on the ambiguities of what boundary conditions to apply at the horizon and at the central singularity. Hawking cleverly bypassed these ambiguities by tracing the evolution of the quantum fields in time, from an initial, well-defined vacuum state before the beginning of gravitational collapse and before the formation of the black hole. He demonstrated that when the black hole forms and settles into its final, stationary state, the quantum fields settle into a state that involves a steady outward emission of radiation from the horizon toward infinity. The energy spectrum of this radiation is thermal, with a temperature

$$T_{BH} = \frac{\hbar c^3}{8\pi GMk} \tag{77}$$

This exact result for the temperature of the black hole is consistent with the approximate result [72] obtained from thermodynamic arguments.

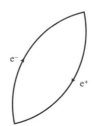

Fig. 8.24   Feynman diagram illustrating the spontaneous creation of a particle-antiparticle pair and its subsequent annihilation.

The Hawking emission process seems to contradict the fundamental property of the horizon--nothing should emerge from the horizon. Actually the thermal radiation does not come from inside the black hole, but it is created by quantum fluctuations at or near its surface. We know from quantum field theory that the vacuum is a restless and violent place, where particles are continuously created and destroyed. Fig. 8.24 shows such a creation and destruction event: a particle-antiparticle pair, such as an electron-antielectron pair, is spontaneously created at a spacetime point, and this pair is destroyed at a later point. If this event happens in the normal vacuum, far from a black hole, it merely produces an unobservable, small-scale fluctuation in the electric current density. But if this event happens just outside the horizon of a black hole, the antiparticle (or the particle) might enter the horizon and fall into the singularity. If the antiparticle is in a state of negative energy $-E$, the particle will be left behind in a state of positive energy $+E$, and it will then be free to move outward, away from the black hole, and reach the detection instruments of an observer. The net effect is that a more or less steady stream of particles of positive energy flows outward from the region of the horizon, while the black hole absorbs a stream of antiparticles of negative energy, and therefore decreases its mass. An explicit calculation by Damour and Ruffini (1976) shows that the thermal spectrum of the emerging, liberated, particles arises from a "barrier-penetration" factor. The region in the immediate vicinity of the horizon strongly attenuates the ingoing antiparticle wave (in the calculation, the ingoing antiparticle wave of negative energy is treated as an outgoing particle wave of positive energy proceeding backward in time; see Fig. 8.25). In the limiting case of large energy ($E \gg \hbar c^3/GM$), it turns out that the attenuation suffered by the antiparticle wave in crossing the horizon is $e^{-4\pi GME/\hbar c^3}$; thus, the probability for the antiparticle to penetrate into the black hole, and leave the particle liberated, is $e^{-8\pi GME/\hbar c^3}$. From quantum field theory, it is known that all quantum states contribute equally to the particle-antiparticle vacuum fluctuations, so the flux of antiparticles incident on the horizon is proportional to the number of

Fig. 8.25 In this diagram, the spacetime inside and outside a black hole is described by Eddington-Finkelstein coordinates. A particle-antiparticle pair is created spontaneously at the horizon. The antiparticle falls into the black hole, and the particle travels away. An antiparticle of negative energy can be treated as a particle of positive energy proceeding backward in time, from the singularity to the horizon. The net wavefunction, indicated schematically, is a wave of positive energy that proceeds from the singularity to the horizon, and from there to infinity. Near the horizon, this wave goes through an infinite number of oscillations and its amplitude is reduced by a penetration factor. (After Damour and Ruffini, 1976.)

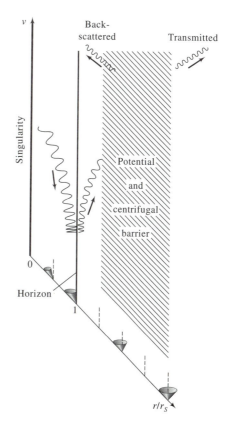

quantum states, that is, $4\pi p^2 dp/h^3$. If we multiply this incident flux by the probability for penetration of the horizon, we obtain the outgoing flux of liberated particles:

$$[\text{outgoing flux of particles}] \propto \frac{4\pi p^2 dp}{h^3} \, e^{-8\pi GME/\hbar c^3} \qquad [78]$$

As expected, this has the form of a thermal spectrum, in the limiting case of large energy ($E \gg kT$). The penetration factor $e^{-8\pi GME/\hbar c^3}$ plays the role of the Boltzmann factor $e^{-E/kT}$, with a temperature $T = T_{BH} = \hbar c^3/8\pi GMk$.

If the energy is not large compared with kT_{BH}, then the penetration factor is somewhat more complicated, and depends on whether the particles are fermions or bosons; but in any case, the resulting spectrum is thermal.

Actually, not all the particles liberated at the horizon manage to escape to infinity. Some of these particles are backscattered by the

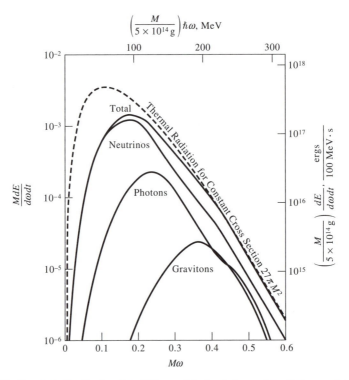

Fig. 8.26 Spectral distribution for different kinds of thermal radiation emitted by a black hole. The scales on the left and the bottom edges of the plot are expressed in units with $G = \hbar = c = 1$. (After Page, 1976a.)

gravitational potential surrounding the black hole, and they disappear into the black hole. This merely means that the black hole and its surrounding potential form a thermal radiator with a less than perfect emissivity, a thermal "gray" body rather than a thermal blackbody. The backscattering modifies the spectrum of the radiation that finally escapes to infinity, reducing it considerably below the spectrum of a perfect blackbody (see Fig. 8.26).

Numerically, the black-hole temperature can be expressed as

$$T_{BH} = (1.2 \times 10^{26} \text{ K}) \frac{1}{M/(1 \text{ g})} \qquad [79]$$

Thus, the temperature of a black hole of a mass equal to that of the Sun ($M_\odot = 2 \times 10^{33}$ g) is only 10^{-7} K. But a very small black hole, or minihole, of a mass of 10^{14} g, would have a temperature of 10^{12} K, and the typical thermal energy of the radiated particles would be about $kT_{BH} \simeq 100$ MeV. In general, the typical thermal energy determines what kinds of particles can be radiated. Particles of rest mass m cannot be radiated in significant numbers unless the typical thermal energy is of the order of mc^2. Thus, a black hole of mass equal to that of the Sun can radiate only particles of mass zero, that is, photons, neutrinos, and gravitons; but a minihole of mass 10^{14} g can also radiate electrons, mu mesons, and pions. Black holes of masses significantly smaller than a solar mass cannot be formed by the gravitational

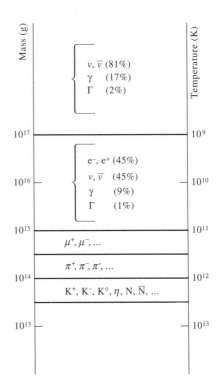

Fig. 8.27 Different kinds of particles radiated by a black hole as a function of its mass.

collapse of a star; such miniholes can form only in the early stages of the universe, from fluctuations in the very dense primordial matter. Fig. 8.27 lists the different kinds of particles radiated by a black hole as a function of its mass.

By Stefan's law the total power radiated by a black hole is of the order of

$$P \simeq [\text{area}] \times \sigma T_{BH}{}^4 \simeq 4\pi r_S{}^2 \times \sigma T_{BH}{}^4 \simeq \frac{10^{47} \text{ erg/s}}{M^2/(1 \text{ g})^2} \quad [80]$$

where σ is the Stefan-Boltzmann constant. The exact numerical factor in this equation depends on the various species of particles being radiated and on their backscattering (Page, 1976a,b).

From Eq. [80] we can calculate the rate of decrease of the mass of the black hole:

$$\frac{dM}{dt} = -\frac{P}{c^2} \simeq -\frac{10^{26} \text{ g/s}}{(M/1 \text{ g})^2} \quad [81]$$

As the mass decreases, the rate of radiation increases. When the mass becomes small, the rate of radiation becomes explosive. However, it is not known what happens when the mass reaches a value smaller than $\simeq \sqrt{\hbar c/G} \simeq 10^{-5}$ g (the Planck mass), where the time scale for the change of mass will be smaller than the Schwarzschild time, that is, $|dM/dt| \geq M/(r_S/c)$. Under these conditions, the spacetime geometry cannot be treated as a stationary or quasi-stationary background, and the dynamical changes in the geometry will begin to play a crucial role. Although it is widely believed that the minihole will release all its mass in a final explosion and disappear, the details are not understood.

From Eq. [81] we can estimate the lifetime of a black hole. If the initial mass is M, the time until it disappears or almost disappears is approximately

$$t \simeq \frac{M}{|dM/dt|} \simeq 10^{-26} \text{ s} \times \frac{M^3}{(1 \text{ g})^3} \quad [82]$$

According to this estimate, a black hole of initial mass 10^{14} g would have a lifetime of about 10^{10} years, equal to the age of the universe. Thus, such a black hole, formed during the early stage of the universe, would be reaching its final explosive phase today.

The observational search for radiation from black holes has concentrated on gamma rays. A black hole of 10^{14} g spends most of its lifetime at a temperature of about 10^{12} K, and at this temperature it produces gamma rays of about 100 MeV. If there is a more or less uniform distribution of such black holes all over the universe, we should see a diffuse background of 100-MeV gamma rays all over the sky. The observational limit on gamma rays sets an upper limit on the density of black holes of about 10^4 per (pc)^3. A tighter limit has been set by examining the sky for localized bursts of high-energy gamma rays, 1000 MeV or more, produced during the final explosive phase of a black hole. The observations set an upper limit of 0.04 explosion per (pc)^3. These upper limits on the abundance of miniholes imply that their contribution to the overall mass density in the universe is small-- no more than 1 part in 10^8 of the total mass is in the form of mini-holes.

Although the negative observational evidence for radiation from miniholes is disappointing, we can draw some interesting conclusions from the absence of such miniholes. The preponderance of normal matter over miniholes tells us that conditions in the early universe were not favorable for the formation of miniholes. Since the miniholes formed from fluctuations, this means that either the fluctuations were not very violent or else the primordial matter offered strong resistance to compression (had a "stiff" equation of state).

8.8 GRAVITATIONAL COLLAPSE AND THE FORMATION OF BLACK HOLES

In normal stars, such as the Sun, the inward gravitational pull is held in equilibrium by the thermal pressure of the gas. This thermal pressure will be sufficient to resist the gravitational pull only if the star is hot enough. The star can therefore remain in equilibrium as long as the thermonuclear reactions in its core supply enough heat, that is, as long as the energy released in these reactions compensates for the energy lost by radiation at the surface. In a star that has exhausted its supply of nuclear fuel, the thermal pressure will ultimately disappear, and the star will collapse under its own weight. The collapse may be sudden (implosion) or gradual (contraction), but in any case it can be halted only if an alternative mechanism for generating sufficient pressure becomes available at high density.

In white-dwarf stars and in neutron stars such an alternative mechanism is available: these stars are so dense that the quantum mechanical zero-point pressure becomes dominant. Essentially, a degenerate Fermi gas of electrons supplies the equilibrium pressure in a white dwarf and a Fermi gas of neutrons that in a neutron star.

At the white-dwarf densities ($\rho \geq 10^5$ g/cm^3) the electrons are detached from their nuclei and move quite freely throughout the volume of the star. The star consists of interpenetrating gases of electrons and nuclei. The zero-point pressure of the electron gas gives the main contribution to the pressure, and the nuclei give the main contribution to the mass density. The equation of state (pressure as function of density) based on this model permits equilibrium configurations, provided that the total mass is below a critical upper limit. If the mass exceeds this critical limit, then the electron pressure cannot support the star. The trouble is that the Fermi energy increases with density, and ultimately the electron gas becomes relativistic; such a relativistic Fermi gas, in contrast to the nonrelativistic gas, cannot supply enough pressure for equilibrium.

The following rough calculation helps to make this clear. Let us pretend that the gas has uniform density throughout the star. The Fermi momentum for a gas of N electrons in a volume V is

$$p_F = (3\pi^2\hbar^3 N/V)^{1/3} \tag{83}$$

In the nonrelativistic case, the Fermi energy equals $p_F{}^2/2m_e$ (where m_e is the electron mass); in the extreme relativistic case, the Fermi energy equals cp_F. The average energy per electron is of the same order of magnitude as the Fermi energy, and hence we may regard the quantity

$$N(3\pi^2\hbar^3 N/V)^{2/3}/2m_e \qquad \text{(nonrelativistic)}$$
$$\tag{84}$$
$$N(3\pi^2c^3\hbar^3 N/V)^{1/3} \qquad \text{(extreme relativistic)}$$

as a rough estimate of the total zero-point energy. The Newtonian gravitational binding energy is of the order of magnitude of

$$- \frac{GM^2}{R} \tag{85}$$

where M is the mass of the star and R the radius. Hence, the total energy, ignoring rest masses, is roughly

$$\frac{N^{5/3}(9\pi\hbar^3/4)^{2/3}}{2m_e R^2} - \frac{GM^2}{R} \qquad \text{(nonrelativistic)}$$
$$\tag{86}$$
$$\frac{N^{4/3}(9\pi c^3\hbar^3/4)^{1/3}}{R} - \frac{GM^2}{R} \qquad \text{(extreme relativistic)}$$

where we have taken $V = 4\pi R^3/3$. We must now try to adjust the density, or the radius R, so that a star of given N and M is placed in an equilibrium configuration. For stable equilibrium, the energy [86] should have a minimum as a function of R. In the nonrelativistic case, a minimum exists. But in the extreme relativistic case, there is no minimum, since both the zero-point energy and the gravitational energy are proportional to the same power of R. Essentially, the trouble is that although in the extreme relativistic case the Fermi energy increases with decreasing R, it does not increase strongly enough--the relativistic gas is too soft to resist the gravitational compression.

■ *Exercise 17.* Check that the energy [86] has a minimum at

$$R = \frac{N^{5/3}(9\pi\hbar^3/4)^{2/3}}{GM^2 m_e} \tag{87}$$

in the nonrelativistic case, and that it has no minimum in the extreme relativistic case. ■

The critical mass at which the equilibrium disappears may be estimated by asking when the electron gas turns relativistic. This happens when $p_F \simeq m_e c$. According to Eq. [83], this corresponds to

$$R = \frac{(9\pi\hbar^3 N/4)^{1/3}}{m_e c} \tag{88}$$

By comparing Eqs. [87] and [88], we find that the critical mass is given by

$$M_{crit}^2 \simeq \frac{cN^{4/3}(9\pi\hbar^3/4)^{1/3}}{G} \tag{89}$$

If the nuclei in the white dwarf are helium, then $N \simeq M/2m_n$, or $N \simeq M_{crit}/2m_n$, where m_n is the neutron mass. From Eq. [89], we then obtain a critical mass

$$M_{crit} \simeq \sqrt{\frac{9\pi}{4^3}} \left[\frac{\hbar c}{Gm_n^2} \right]^{3/2} m_n \tag{90}$$

Numerically, $(9\pi/4^3)^{1/2}(\hbar c/Gm_n)^{3/2}m_n = 2.4 \times 10^{33}$ g $\simeq 1\ M_\odot$. Thus, our rough calculation indicates that the critical mass should be of the order of one solar mass.

For a realistic model of white dwarfs, with nuclear matter in the form of helium nuclei and with a pressure distribution appropriate for hydrostatic equilibrium, a more precise calculation of the critical mass gives a value of $1.44 M_\odot$. This result was first obtained by Chandrasekhar (1931), and the critical mass, beyond which the zero-point pressure of electrons becomes insufficient for equilibrium, is known as the *Chandrasekhar limit*.

The density of a neutron star is much higher than that of a white dwarf. The density of a neutron star is comparable with nuclear densities ($\simeq 2 \times 10^{14}$ g/cm³), and hence the star may be described as a single giant nucleus. The bulk of the star consists of a Fermi gas of neutrons, with some few electrons and protons. At the center is a small core with heavy particles (hyperons). Like the white dwarf, the neutron star has a limiting value of the mass, beyond which the zero-point pressure of the neutrons becomes insufficient for equilibrium. Our rough calculation for the critical mass of a white dwarf also applies to a neutron star. Note that the mass of the electron, or the mass of whatever particle generates the zero-point pressure, cancels in Eq. [89]; thus, according to Eq. [90], the values of the critical masses of white dwarfs and neutron stars should be about the same, about 1 $M_{\odot}$. But when we take into account the pressure distribution required for hydrostatic equilibrium, these critical masses will differ somewhat. According to a first calculation by Oppenheimer and Volkoff (1939), in which the nuclear forces between the neutrons were neglected, the critical mass of a neutron star is about $0.7M_{\odot}$, called the Oppenheimer-Volkoff limit. Later calculations by Wheeler and colleagues (Harrison et al., 1965) sought to take into account the nuclear forces, and found a slightly larger limiting mass. Recent calculations lead to a limiting mass of $1.6M_{\odot}$ to $2M_{\odot}$.

The higher values of the limiting mass result from assuming a hard repulsive core in the nuclear interaction, which makes the nuclear material very stiff at high densities. But no matter how much the equation of state at high densities is modified by interactions, it can be shown that there always is an upper limit to the stiffness and to the mass that can be supported. The reason for this is that the stiffness of a material is directly related to the speed of sound waves in the material--the stiffer the material, the higher the speed of sound. The requirement that the speed of sound be no more than the speed of light then sets a limit on the stiffness of the material. According to Rhoades and Ruffini (1974), this leads to the conclusion that the mass of a neutron star can never exceed $3.2M_{\odot}$, independent of any assumption about the details of the equation of state at high densities.

Fig. 8.28 summarizes equilibrium configurations of matter at high densities as calculated by numerical integration of the relativistic hydrostatic equations (Rees, Ruffini, and Wheeler, 1974). The curve gives the mass of the star as a function of the density at the center. For comparison, the dashed curve gives the mass according to a calculation based on Newtonian physics, without relativity. In the calculation of the curves it has been assumed that the matter has reached the endpoint of thermonuclear evolution--the star is completely burnt out and cold.

For the ideal white dwarfs in Fig. 8.28, in which the nuclear matter is assumed to have reached the state of least nuclear energy,* the Chandrasekhar limit is somewhat lower than $1.44M_{\odot}$.

The configurations between $\rho_0 \simeq 10^8$ and 10^{13} g/cm^3 in Fig. 8.28 correspond to an increasing fraction of neutrons in the atomic nuclei.

---

* The nucleus with the least nuclear energy is ^{56}Fe.

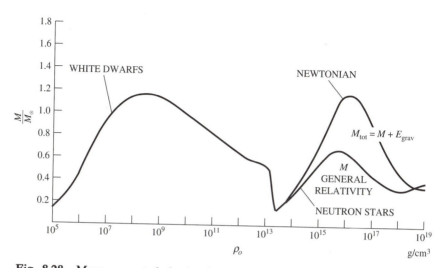

Fig. 8.28 Mass vs. central density for the possible equilibrium configurations of matter. (After Rees, Ruffini, and Wheeler, 1974.)

(Incidentally: In these configurations the equilibrium is unstable and therefore of little practical interest.) At a density above $\simeq 10^{13}$ g/cm^3, we reach the realm of the neutron stars, in which neutrons are by far the most abundant kind of particle. As indicated by Fig. 8.28, the masses of stable neutron stars lie in the range from $\simeq 0.2 M_\odot$ to $\simeq 0.8 M_\odot$.

The absence of equilibrium configurations with masses exceeding these limits has drastic consequences for stellar evolution. A star with a mass below the Chandrasekhar limit evolves along a sequence of equilibrium configurations. When the nuclear fuel is exhausted it settles down in a white-dwarf state. The luminosity of such a white dwarf is entirely due to the residual thermal energy; as the energy is lost by radiation, the star grows dimmer and dimmer, ultimately becoming a "black star" (the cooling time is typically of the order of 10^9 years).

On the other hand, when the nuclear fuel is exhausted in a star with a mass above the Chandrasekhar limit, the pressure will become insufficient to prevent gravitational collapse. At late stages of its evolution, such a massive star will have a dense core, and the collapse begins in this core--the core implodes. If the mass of the core is sufficiently small, then the implosion can be halted when neutron-star densities are reached. The sudden halt of the implosion creates a violent outward shock wave, which carries away the kinetic energy of the infalling material. This shock wave is analogous to the shock wave observed in a water pipe when the flow of water is suddenly halted by quickly shutting off a valve; the kinetic energy of the moving water is then transferred to a shock wave which travels backward along the pipe, halting the flow of water ("water hammer"). As the shock wave of the imploding star travels outward, it blows away the outer layers of the star. The outward explosion of these layers shows up on the sky as a supernova (a type II supernova). For example, the famous Crab pulsar is a neutron star that was formed in such a supernova explosion; the Crab nebula, which surrounds it, consists of hot gases which once constituted the outer layers of the original massive star (of a mass of $\simeq 4M_\odot$). If the mass of the collapsing core exceeds the critical mass for a neutron star, then the implosion cannot be halted. The complete gravitational collapse of the core produces a black hole.

Supernova explosions are one possible mechanism for the formation of black holes. Mass exchange in a compact binary star system is another mechanism. In such a system, a neutron star is in an orbit near the surface of a more or less normal companion star. If the neutron star is near enough to the companion, it will pull material off its surface and gradually increase in mass. Once the mass increases beyond the critical limit, the neutron star will collapse to a black hole.

The first calculations of complete gravitational collapse, with formation of a Schwarzschild horizon, were performed by Oppenheimer and Snyder (1939), who assumed that the collapsing material consists of a sphere of gas without pressure (a cloud of dust), so each particle is in free fall along a radial geodesic. Although this is an unrealistic assumption, their calculation illustrates the general behavior of the spacetime geometry during the approach to the horizon. In a sense the collapse never ends. An outside observer would say that the surface of the collapsing star asymptotically approaches the Schwarzschild radius, but never reaches it in a finite time. In this sense, the object may be appropriately called a "frozen star" rather than a black hole. However, the "frozen star" very soon becomes practically indistinguishable from a black hole. Calculations of the intensity of the light emitted by the surface of a collapsing star show that the brightness seen by an outside observer decreases sharply while the redshift increases (the brightness decreases exponentially with time, with a characteristic "half-life" of the order of GM; Thorne, 1972). Furthermore, the gravitational and electromagnetic fields surrounding the collapsing mass asymptotically approach those corresponding to a black hole. The evolution of the initial gravitational and electromagnetic fields into the final black-hole fields will in general be accompanied by radiation, because the collapsing mass has to shed any gravitational or electromagnetic multipole fields in excess of those characteristic of a black hole. For example, detailed calculations by Price (1972) show that if the initial fields deviate slightly from the spherically symmetric Schwarzschild field, then these multipole fields are radiated away as the collapse proceeds. Some of the fields are radiated downward, into the black hole; some are radiated outward, forming a pulse of gravitational and electromagnetic radiation. Within a finite time, the electromagnetic and gravitational fields surrounding the collapsing mass approach so close to those characteristic of a black hole that the quantum uncertainties in the measurement process prevent an outside observer from detecting any difference at all (Ohanian and Ruffini, 1974).

When the external gravitational and electromagnetic fields of a collapsing mass ultimately reach a stationary state, the fields will be uniquely characterized by three parameters: mass (M), spin angular momentum (S), and electric charge (Q). This general theorem was established by a sequence of several complementary theorems on the uniqueness of static and stationary black-hole solutions of Einstein's equations (Robinson, 1975; Hawking, 1972). For instance, the Kerr solution is the unique black-hole solution with $S \neq 0$, $Q = 0$; and the Reissner-Nordström solution is the unique black-hole solution with $S = 0$, $Q \neq 0$. The unique solution with $S \neq 0$ and $Q \neq 0$ is known as the Kerr-Newman solution (the Kerr solution and the Reissner-Nordström solution are special cases of the Kerr-Newman solution). As far as an outside observer is concerned, any two black holes with the same mass, angular momentum, and charge are therefore absolutely identical, regardless of how the two black holes were created. When a black hole swallows up some matter, only its mass, spin angular momentum, and charge change. All other properties of matter, such as intrinsic multipole moments, baryon number, lepton number, etc., are "forgotten" by the black hole. If we drop anything into a black hole, there is no way of telling from the fields surrounding the black hole at a later time what was dropped into the black hole. The absence of distinctive fields--such as independent gravitational or electromagnetic multipole fields, or nuclear force fields--in the region surrounding a black hole led Wheeler to remark that "a black hole has no hair" (Rees, Ruffini, and Wheeler, 1974).

Although the collapse of a mass never quite ends for an outside observer, it does end very soon for an observer who rides along with the collapsing mass. Within a finite proper time, this observer reaches the central singularity. In the special case of spherical symmetry or axial symmetry, the existence of the Schwarzschild singularity ($r = 0$) or of the Kerr singularity ($r = 0$, $\theta = \pi/2$) follows directly from the solution of Einstein's equations. In the general case, the existence of singularities is indicated by the Hawking-Penrose theorem, which roughly says that a singularity will develop whenever the spacetime contains a closed two-dimensional *trapped surface* (Hawking and Ellis, 1973). By such a trapped surface is meant a spacelike surface such that light rays emitted perpendicularly from the surface converge, regardless of whether the rays are emitted in the outward or the inward direction. In flat spacetime, there can be no such trapped surface; for example, light rays emitted from a spherical surface converge if they are emitted in the inward direction, but they diverge if emitted outward. However, in a strongly curved spacetime, even the light rays emitted in the outward direction may converge. For instance, the spherical surface r = [constant] $< r_S$, $0 \le \theta \le \pi$, $0 \le \phi \le 2\pi$ in the Schwarzschild geometry is a trapped surface. Light rays emitted perpendicular to this surface, in the positive or the negative t direction,* necessarily move toward smaller values of r (see Fig. 8.3), and therefore their transverse separation decreases, since it equals $r \times$ [angular separation].

It is known that trapped surfaces will form in all cases of gravitational collapse that are reasonably close to spherical collapse, and it is believed that trapped surfaces will form even if the collapse is very different from spherical collapse. The Hawking-Penrose theorem then tells us that singularities will also form.

What is the nature of these singularities? In the context of the above theorem, a singularity is simply a region in which geodesics terminate. What is not known is how they terminate. The most obvious possibility is that the curvature becomes infinite; this is, of course, exactly what happens at $r = 0$ in the Schwarzschild geometry. In any case, the presence of a singularity signals the breakdown of classical physics, and quantum effects will then presumably play a crucial role. [It is tempting to conjecture that the Hawking-Penrose theorem of classical geometrodynamics is as irrelevant to the real world as the Earnshaw theorem of classical electrodynamics (Jeans, 1925) according to which atoms are unstable and must collapse to a singularity. It may well be that when quantum effects are taken into account, the collapsed matter attains some nonsingular final state.]

The formation of a singularity is expected to be accompanied by the formation of a horizon around the singularity, that is, the formation of a black hole. This expectation is based on Penrose's cosmic censorship conjecture (see Section 8.6), which forbids the formation of naked singularities, that is, singularities not surrounded by horizons. For an outside observer, the horizon that forms around the singularity then hides the singularity from view.

8.9 IN SEARCH OF BLACK HOLES

Since black holes are an end product of stellar evolution, they could be quite abundant. It is even possible that in some galaxies most of the mass is found in the form of black holes. For example, in elliptical galaxies, the observed ratio of total mass to luminosity is $\simeq 70$ times that of the Sun. Since the Sun is a fairly typical star, most of the matter in these galaxies must be "dark matter" of very low luminosity. Several forms of dark matter have been hypothesized (see Table 10.2), among them black holes. The observed light distribution and the observed velocity dispersion in bright elliptical galaxies may be used to set limits on the amount of mass that can be in the form of black holes. If there is a large central black hole, then its mass cannot exceed

---

* For $r < r_S$, t is a spacelike coordinate, and the t direction is perpendicular to the θ and the ϕ directions.

$10^{10}M_\odot$. However, a much larger amount of mass can exist in the form of a large number of somewhat smaller black holes, of $\simeq 10^3 M_\odot$ each, distributed over the galaxy.

An isolated black hole does not emit light and would not be directly observable. In principle, if a black hole is located on, or very near, the line of sight from the Earth to a star, then the gravitational-lensing effect brings about a large change in the apparent position of the star (see Section 4.5). However, the probability for such an alignment is very small.

Favorable conditions for the observation of a black hole occur if there is a nearby source of gas or plasma, such as a companion star or an interstellar cloud. The black hole can then capture gas and plasma, and while this material falls down toward the horizon, it heats by friction and emits thermal radiation (the radiation is emitted by the infalling material before it reaches the horizon; any radiation emitted after the material has crossed the horizon remains trapped in the black hole). This accretion process results in an exceptionally efficient conversion of gravitational potential energy into radiation.

For maximum efficiency of radiation, the gas ought to approach the black hole with some angular momentum. This will prevent the gas from falling straight in on a radial line; instead, the gas will first be captured into a circular orbit, and then gradually spiral downward into smaller and smaller circular orbits while it sheds its angular momentum. Gas in such circular orbits around a black hole (or a compact star, such as a neutron star) forms an *accretion disk* (see Fig. 8.29). Since the orbital angular velocities for adjacent circular orbits are different, parcels of gas in such adjacent orbits experience a viscous drag. This friction generates heat, and also transfers orbital angular momentum from the inner regions of the disk to the outer (ultimately, this orbital angular momentum must be transferred from the outer edge of the accretion disk to the surrounding material). The mechanism that generates the viscous drag is poorly understood; it probably involves a combination of turbulence that tends to mix adjacent parcels of gas and magnetic fields whose field lines link adjacent parcels.

Accretion by a Schwarzschild black hole can convert up to 5.7% of the rest-mass energy of the infalling material into radiation (see below); accretion by a Kerr black hole can convert up to 42%. This is to be compared with the maximum energy released in nuclear reactions in stars, which is only about 0.7% of the rest-mass energy.

To see how much energy is converted into heat in the accretion disk of a Schwarzschild black hole, consider the total change in orbital energy of a parcel of gas as it spirals inward from the outer rim of the disk to the inner. This change is $E(r_2) - E(r_1)$, where $E(r) = m\mathcal{V}(r)$ is the energy for a circular orbit, which includes the rest-mass energy m of the parcel (see Eq. [7.131]). At the outer radius, we can approximate $E(r_1) \simeq m$, since the outer radius of the disk is much larger than the inner radius, and the binding energy at the outer radius is insignificant. At the inner radius, $r_2 = 6GM$, for which Eq. [7.131] gives $E_2 = (1- \sqrt{8/9})m$. Hence the amount of energy that is converted into heat is $-E(r_2) + E(r_1) = \sqrt{8/9}\,m = 0.057m$, or 5.7% of the rest-mass energy of the parcel.

The temperature attained by the accretion disk depends on several properties of the disk, such as its thickness and opacity. Detailed calculations indicate that, for a black hole of $M \simeq M_\odot$, temperatures of 10^8 or 10^9 K are attained. At these temperatures the disk will radiate X rays of a characteristic energy of 10 to 100 keV.

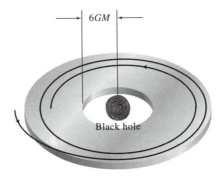

Fig. 8.29 Accretion disk around a black hole. New material enters the accretion disk at its outer rim and gradually spirals inward, toward the black hole. The accretion disk ends at $r = 6GM/c^2$, which is the radius of the smallest stable circular orbit around a black hole. Beyond this radius, the material plunges in quickly, and therefore the density of material is drastically reduced.

The high efficiency for energy release by the accretion of a black hole suggests that this mechanism could provide a suitable power source for quasars (Zeldovich and Novikov, 1964; Salpeter, 1964; Lynden-Bell, 1969). As is well known, a quasar releases energy at a prodigious rate, typically more than 10^{48} erg/s, or $10M_\odot$ per year. To account for such a large radiated power, we need a source of exceptional efficiency. A black hole of about 10^7 or $10^8 M_\odot$ at the center of the quasar could satisfy the power requirements if it accretes gas, dust, and maybe entire stars at the rate of about a few hundred $M_\odot$ per year. Alternative power sources, such as dense star clusters or multiple supernova explosions, have been proposed, but their lower efficiency for the conversion of rest mass into radiation would demand improbably large masses at the cores of quasars (Blandford, 1987).

Some galaxies with exceptionally active nuclei, such as radio galaxies and Seyfert galaxies, which emit almost as much radiation as quasars, might also be powered by accreting black holes. It has even been proposed that ordinary galaxies might contain black holes in their nuclei. Fig. 8.30 (see color plate) shows a large disk of cool dust and gas orbiting the nucleus of the galaxy NGC 4261; the bright spot at the center of this disk is believed to indicate a hot accretion disk belonging to a large black hole in the galactic nucleus.

Radio observations of the center of our own Galaxy reveal a radio source which has been interpreted as an accreting black hole. The Doppler shifts of infrared spectral lines emitted by gas orbiting this black hole indicate a mass of about $3 \times 10^6 \ M_\odot$. However, there is also an unusual infrared source nearby, which probably consists of a cluster of 100 or so stars. The presence of such a cluster is inconsistent with the presence of a massive black hole, since the tidal forces exerted by the black hole would quickly destroy the cluster. Further observations are required to clear up this inconsistency.

Because of the lack of data on the inner structure of quasars, we cannot use the existence of quasars as conclusive evidence for the existence of black holes. To obtain such conclusive evidence, we must rely on compact binary systems consisting of a compact body--a neutron star or a black hole--and a more or less normal star in orbit around each other. The compact body will capture gas from its normal companion and form an accretion disk which emits a copious flux of X rays (see Fig. 8.31). Such X-ray emission is a distinctive feature of compact binary systems--observationally they are seen as X-ray binaries.

The observation of X-ray sources from the Earth is complicated by atmospheric absorption. The X-ray telescopes must be lifted above the atmosphere in high-flying rockets or in satellites. The first X-ray sources were discovered by Giacconi et al. (1962) with detectors flown on rockets. Later, X-ray telescopes were installed on satellites, such as the *UHURU* satellite and the *Einstein* satellite. By now, several hun-

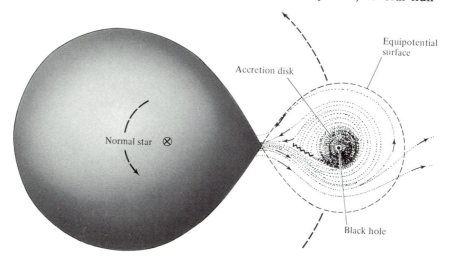

Fig. 8.31 Binary system consisting of a black hole with a normal star as a companion. The black hole draws gas from the normal star and forms an accretion disk. (After Eardley and Press, 1975.)

dred X-ray sources have been discovered, most lying within our galaxy. In some cases, the accurate measurement of the position of the source has made it possible to identify the optical companion of the compact star (see Fig. 8.32). Measurements of the Doppler shifts of the spectral lines of the optical companion then give us the orbital period and the orbital velocity (projected along the line of sight). The orbital data, in conjunction with other data--such as the intensity modulation of the light (the light curve), the spectral classification, the luminosity, and the distance--permit us to determine the masses of the compact star and the companion star.

In most cases, the mass determination yields a mass of about 1 $M_\odot$ or less, which is consistent with the mass expected for a neutron star. The neutron-star interpretation for the compact star in these binary systems is supported by the observational evidence in two ways: the sources have a large X-ray luminosity (typically 10^{37} or 10^{38} erg/s for X rays in the range of 2 to 10 keV), which requires a large mass of small size, with a deep gravitational potential well; and the X-ray luminosity of some of the sources displays pulsations, with a pulsation period of the order of seconds, in agreement with the typical periods of rotation of neutron stars (pulsars).

TABLE 8.1 BINARY SYSTEMS WITH BLACK-HOLE CANDIDATES*

X-ray source	Orbital period	Optical companion	M_{opt}	M_X
Cyg X-1	5.6 days	HDE 226868	$> 20 M_\odot$	$> 7 M_\odot$
LMC X-3	1.7	B3V spectral type	> 4	> 6
0620-00	0.32	K5V spectral type	> 0.4	> 4
LMC X-1	4.2	07III spectral type	> 2 ?	> 4
CAL 87	0.44	F/G spectral type	> 0.4 ?	> 4 ?

* Based on Blandford (1987) and Cowley (1992).

However, in several exceptional cases, the mass determinations give results in excess of the maximum permitted mass for a neutron star. These exceptional cases are interpreted as black holes. The most reliable results, and the most conclusive evidence for black holes, have been obtained from observations conducted on the five binary X-ray sources Cyg X-1, LMC X-3, 0620-00, LMC X-1, and CAL 87 (see Table 8.1). Among these, the strongest case for a black hole is provided by Cyg X-1. The data available from spectroscopic observations of the optical companion (see Fig. 8.32) include the orbital period, amplitude of periodic Doppler shifts (arising from orbital motion), reddening of light by interstellar dust (an indication of distance), effective temperature, brightness, and periodic variation in brightness (arising from the tidal distortion of the shape of the star). These data allow several more or less independent calculations of a lower limit for the mass M_X of the compact body. All these calculations agree that its mass is several times the solar mass. The most comprehensive analysis of the data actually sets a lower limit of $7M_\odot$, and assigns a most probable mass of about $16M_\odot$ (Gies and Bolton, 1986). This means that the mass is considerably in excess of the Oppenheimer-Volkoff limit for neutron stars--about $2M_\odot$ for a reasonable equation of state, and $3.2M_\odot$ for the most extreme equation of state (see Section 8.8). This means that the compact body in this system is not made of neutrons or of any other kind of normal matter. The only remaining alternative is a black hole.

Calculations of the X-ray flux and spectrum emitted by gas in the accretion disk surrounding a black hole give results consistent with the observed flux and spectrum of Cyg X-1. The accretion model can also

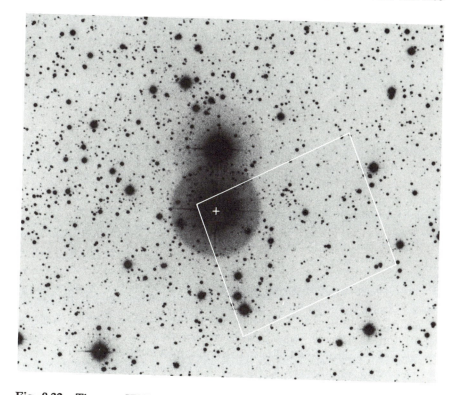

Fig. 8.32 The star HDE 226868 is at the center of this photograph. The X-ray source Cygnus X-1 is believed to orbit this star; the box indicates the uncertainty in the position of the X-ray source. The system also emits radio waves; the cross indicates the uncertainty in the position of the radio source. The distance from the Earth is about 10^4 light-years, and hence the size of the image is not related to the size of the star; rather, it is due to the effects of overexposure. (Hale Observatories photograph by J. Kristian.)

account, in a qualitative way, for the observed irregular changes (flicker) in the X-ray flux on a time scale of about a millisecond. This time scale is the orbital time for a parcel of gas at the inner edge of the accretion disk.

■ *Exercise 18.* The smallest stable circular orbit around a black hole has a radius of $6GM$ (see Section 7.6). Use Kepler's law $(T/2\pi)^3 = r^3/GM$ to show that the period T for a circular orbit of radius $6GM$ is of the order of 10^{-3} s for $M \simeq 10M_\odot$. This period does not have to be corrected for redshift because it is already expressed in t time (see Problem 7.12). ■

For the other X-ray sources listed in Table 8.1, the lower limits on the mass are not quite as firm as for Cyg X-1, but they are very plausible (these lower limits can be circumvented only by highly contrived scenarios). Besides these five strong black-hole candidates, there also are a dozen of weaker candidates, such as GX339-4, GS1124-68, SS433, etc. However, for these other candidates, no mass determinations are yet available, and their candidacy rests merely on similarities between their X-ray or optical characteristics and those of Cyg X-1 (see Cowley, 1992).

FURTHER READING

An excellent introduction to the physics of Schwarzschild and Kerr black holes can be found in Misner, C. W., Thorne, K. S., and Wheeler, J. A., *Gravitation* (Freeman, San Francisco, 1973); the general black hole with both spin and electric charge is also treated there. Rees, M. J., Ruffini, R., and Wheeler, J. A., *Black Holes, Gravitational Waves, and Cosmology* (Gordon and Breach, New York, 1974) also contains much information on these topics.

Further details are given in DeWitt, C., and DeWitt, B., eds., *Black Holes* (Gordon and Breach, New York, 1973). This contains numerous lectures and articles. It includes:

Hawking, S. W., "The Event Horizon." (General definition of horizons and of black holes.)

Carter, B., "Black Hole Equilibrium States." (Analytic and geometric properties of the Kerr solution with both angular momentum and electric charge; integration of the geodesic equation; general theorems regarding black holes.)

Bardeen, J. M., "Rapidly Rotating Stars, Disks, and Black Holes." (The Kerr solution treated as the limiting case of the field surrounding a rotating disk of matter.)

Novikov, I. D., and Thorne, K. S., "Astrophysics of Black Holes." (Physics of the accretion disk formed by matter as it falls into the hole.)

Ruffini, R., "On the Energetics of Black Holes." (Critical mass for the gravitational collapse of a neutron star, emission of gravitational and electromagnetic radiation by a particle orbiting or falling into a black hole.)

More systematic and comprehensive expositions of the physics of compact objects are offered by Shapiro, S. L., and Teukolsky, S. A., *Black Holes, White Dwarfs, and Neutron Stars* (Wiley, New York, 1983) and by Zel-ovich, Y. B., and Novikov, I. D., *Relativistic Astrophysics, Vol. 1: Stars and Relativity* (University of Chicago Press, Chicago, 1971). A shorter, less ambitious, but well-balanced exposition is given by Demianski, M., *Relativistic Astrophysics* (Pergamon Press, Oxford, 1985).

The behavior of geodesics and of electromagnetic and gravitational perturbations in the space surrounding a Schwarzschild, Reissner-Nordström, or Kerr black hole is calculated in exhausting detail in Chandrasekhar, S., *The Mathematical Theory of Black Holes* (Clarendon Press, Oxford, 1983). The scattering of a wide variety of waves (electromagnetic, gravitational, neutrino, scalar) when incident on a black hole is crisply treated in Futterman, J. A. H., Handler, F. A., and Matzner, R. A., *Scattering from Black Holes* (Cambridge University Press, Cambridge, 1988).

When a black hole interacts with external electric or magnetic fields, its horizon effectively behaves like a membrane, endowed with conductivity, viscosity, and surface pressure. These effective characteristics of the horizon are explored in Thorne, K. S., Price, R. H., and MacDonald, A., eds., *Black Holes: The Membrane Paradigm* (Yale University Press, New Haven, 1976).

Hawking, S. W., and Ellis, G. F. R., *The Large Scale Structure of Space-Time* (Cambridge University Press, Cambridge, 1973), provide a rigorous mathematical discussion of the symmetry, topology, causal structure, and singularities of the Schwarzschild, Reissner-Nordström, Kerr, and other geometries. They also develop general theorems (Hawking-Penrose theorems) on the formation of singularities in spacetime.

The equilibrium (or lack of equilibrium) of large masses under the influence of their gravitational self-attraction is described at length in Harrison, B. K., Thorne, K. S., Wakano, M., and Wheeler, J. A., *Gravitation Theory and Gravitational Collapse* (University of Chicago Press, Chicago, 1965). A short, very simple, and very clear presentation of the essential physics behind the equilibrium problem is given in Wheeler, J. A., "The Superdense Star and the Critical Nucleon Number," in Chiu, H., and Hoffmann, W. F., eds., *Gravitation and Relativity* (Benjamin, New York, 1964).

Our knowledge of what happens in gravitational collapse is summarized in Thorne, K. S., "Nonspherical Gravitational Collapse: A Short Review," in Klauder, J. R., ed., *Magic Without Magic: John Archibald Wheeler* (Freeman, San Francisco, 1972).

Eardley, D. M., and Press, W. H., "Astrophysical Processes near Black Holes," Ann. Rev. Astron. Astrophys. **13**, 381 (1975), give a comprehensive review of the formation of black holes and of the observable effects of the black hole on its environment.

Ruffini, R., and Gursky, H., *Neutron Stars, Black Holes, and Binary X-ray Sources* (Reidel, Amsterdam, 1975), discuss the generation by binary systems and their observation.

There are several excellent articles addressed to a wider audience that discuss gravitational collapse and black holes in a more or less qualitative way:

Chandrasekhar, S., "The Black Hole in Astrophysics: The Origin of the Concept and Its Role," Contemp. Physics **14**, 1 (1974).

Penrose, R., "Gravitational Collapse: The Role of General Relativity," Nuovo Cimento, vol. I, Numero speciale, 252 (1969).

Ruffini, R. "Neutron Stars and Black Holes in Our Galaxy," N.Y. Academy of Sciences **35**, 196 (1973).

For an interesting historical account of the development of our ideas about black holes, their thermodynamics, and their emission of radiation, see Israel, W., "Dark stars: the evolution of an idea," in Hawking, S. W., and Israel, W., *Three Hundred Years of Gravitation* (Cambridge University Press, Cambridge, 1987); also see Sciama, D. W., "Black Holes and their Thermodynamics," Vistas in Astronomy **19**, 385 (1976).

A detailed quantum-theoretic treatment of the Hawking process is supplied in Birrell, N. D., and Davies, P. C. W., *Quantum fields in curved space* (Cambridge University Press, Cambridge, 1982).

Rees, M. J., Ann. Rev. Astron. and Astrophys. **22**, 471 (1984), discusses the possible role of black holes as energy sources in active galactic nuclei, including quasars.

A review of neutron-star mass limits is given by Baym, G., and Pethick, C., Ann. Rev. Astron. Astrophys. **17**, 415 (1979).

Recent summaries of the observational evidence for black holes in binary systems are given by Blandford, R. D., "Astrophysical black holes," in Hawking and Israel, op. cit., and by Cowley, A. P., Ann. Rev. Astron. Astrophys. **30**, 287 (1992). The latter reference contains an extensive list of black-hole candidates.

REFERENCES

Bekenstein, J. D., Phys. Rev. D, **7**, 2333 (1973); ibid., **9**, 3292 (1974).

Bekenstein, J. D., Physics Today, January, 1980.

Bjorken, J. D., and Drell, S. D., *Relativistic Quantum Mechanics* (McGraw-Hill, New York, 1964).

Blandford, R. D., "Astrophysical Black Holes," in Hawking, S. W., and Israel, W., eds., *Three hundred years of gravitation* (Cambridge University Press, Cambridge, 1987).

Blumenthal, G. R., and Tucker, W. H., Ann. Rev. of Astron. and Astrophys. **12**, 66 (1974).

Boyer, R. H., and Lindquist, R. W., J. Math. Phys. **8**, 265 (1967).

Carter, B., Phys. Rev. **141**, 1242 (1966).

Carter, B., Phys. Rev. **174**, 1559 (1968).

Chandrasekhar, S., Astrophys. J. **74**, 592 (1931).

Christodolou, D., Phys. Rev. Lett. **25**, 1596 (1970).

Christodolou, D., and Ruffini, R., Phys. Rev. **D4**, 3552 (1971).

Cowley, A. P., Ann. Rev. Astron. Astrophys. **30**, 287 (1992).

Damour, T., and Ruffini, R., Phys. Rev. **14**, 332 (1976); see also Damour, T., "Klein Paradox and Vacuum Polarization," in Ruffini, R., ed., *Proceedings of the First Marcel Grossmann Meeting on General Relativity* (North-Holland, Amsterdam, 1977).

Eardley, D. M., and Press, W. H., Ann. Rev. Astron. Astrophys. **13**, 381 (1975).

Fuller, R. W., and Wheeler, J. A., Phys. Rev. **128**, 919 (1962).

Giacconi, R., Gursky, H., Paolini, F. R., and Rossi, B. B., Phys. Rev. Lett. **9**, 439 (1962).

Gies, D. R., and Bolton, C. T., Astrophys. J. **304**, 371 (1986).

Harrison, B. K., Thorne, K. S., Wakano, M., and Wheeler, J. A., *Gravitational Theory and Gravitational Collapse* (Chicago University Press, Chicago, 1965).

Hawking, S. W., Commun. Math. Phys. **25**, 152 (1972).

Hawking, S. W., Proc. N.Y. Acad. Sci. **224**, 268 (1973).

Hawking, S. W., Nature **248**, 30 (1974).

Hawking, S. W., Commun. Math. Phys. **43**, 199 (1975).

Hawking, S. W., Phys. Rev. D **13**, 191 (1976).

Hawking, S. W., and Ellis, G. F. R., *The Large Scale Structure of Space-Time* (Cambridge University Press, 1973), p. 266.

Jeans, J., *The Mathematical Theory of Electricity and Magnetism* (5th ed., Cambridge University Press, 1925), pp. 168, 631, 632.

Kerr, R. P., Phys. Rev. Lett. **11**, 237 (1963).

Kruskal, M. D., Phys. Rev. **119**, 1743 (1960).

Laplace, P. S., *Exposition du system du monde* (1795).

Lynden-Bell, D., Nature **223**, 690 (1969).

Michell, J., Trans. Royal Soc. London **74**, 35 (1784).

Misner, C. W., Thorne, K. S., and Wheeler, J. A., *Gravitation* (Freeman, San Francisco, 1973).

Morris, M. S., Thorne, K. S., and Yurtsever, U., Phys. Rev. Lett. **61**, 1446 (1988); see also Morris, M. S., and Thorne, K. S., Am. J. Phys. **56**, 5 (1988).

Ohanian, H. C., and Ruffini, R., Phys. Rev. D**10**, 3903 (1974).

Oppenheimer, J. R., and Snyder, H., Phys. Rev. **56**, 455 (1939).

Oppenheimer, J. R., and Volkoff, G. M., Phys. Rev. **55**, 374 (1939).

Page, D. N., Phys. Rev. D **13**, 198 (1976a); ibid. **14**, 3260 (1976b).

Penrose, R., Nuov. Cim. Ser. I, **1**, 252 (1969).

Price, R. H., Phys. Rev. D **5**, 2419 (1972); ibid. **5**, 2439 (1972).

Rees, M., Ruffini, R., and Wheeler, J. A., *Black Holes, Gravitational Waves, and Cosmology* (Gordon and Breach, New York, 1974).

Rhoades, C. E., and Ruffini, R., Phys. Rev. Lett. **32**, 324 (1974).

Robinson, D. C., Phys. Rev. Lett. **34**, 905 (1975).

Ruffini, R., "On the Energetics of Black Holes," in DeWitt, C., and DeWitt, B. S., eds., *Black Holes* (Gordon and Breach, New York, 1973).

Salpeter, E. E., Astrophys. J. **140**, 796 (1964).

Sciama, D. W., Vistas in Astronomy, **19**, 385 (1976).

Shapiro, S. L., and Teukolsky, S. A., Phys. Rev. Lett. **66**, 994 (1991).

Thorne, K. S., "Nonspherical Gravitational Collapse–A Short Review," in Klauder, J. R., ed., *Magic Without Magic: John Archibald Wheeler* (Freeman, San Francisco, 1972).

Unruh, W. G., Phys. Rev. D**10**, 3194 (1974).

Unruh, W. G., and Wald, R. M., Phys Rev. D **25**, 942 (1982).

Zeldovich, Ya. B., Zh. Eksp. Teor. Fiz. **62**, 2076 (1972); translation in Soviet Phys. JETP **35**, 1085 (1972).

Zeldovich, Ya. B., and Novikov, I. D., Sov. Phys. Dok. **158**, 811 (1964).

PROBLEMS

1. Estimate the tidal force (stretching force) that acts on an astronaut of normal size and mass as he falls, feet first, through the horizon of a black hole of mass $M = 2 \times 10^{33}$ g.

2. In flat spacetime we can introduce a rotating coordinate system with coordinates $x'^\mu = (t, \rho, \phi, z)$ defined by

$$
\begin{aligned}
t &= t \\
x &= \rho \cos(\phi + \omega t) \\
y &= \rho \sin(\phi + \omega t) \\
z &= z
\end{aligned}
$$

The coordinates ρ, ϕ, z are cylindrical coordinates rotating with angular velocity ω. Express the spacetime interval in these coordinates. Show that $\rho = 1/\omega$ is an infinite-redshift surface and a static limit. Show that the coordinates are not viable for $\rho > 1/\omega$.

3. A flash of light is emitted at $r = 2r_S$, $\theta = \pi/2$ in a Schwarzschild geometry. Make a polar plot of the coordinate velocity $[(dr/dt)^2 + r^2(d\phi/dt)^2]^{1/2}$ at $r = 2r_S$ as a function of the "emission angle." The "emission angle" α is defined by

$$
\tan \alpha = r d\phi/dr
$$

4. Calculate the scalar invariant $R^\alpha{}_{\beta\mu\nu} R_\alpha{}^{\beta\mu\nu}$ for the Schwarzschild metric and verify it remains finite at $r = r_S$ and diverges at $r = 0$.

5. (a) Show that in the interior of a black hole ($r < 2GM$), the "radial" worldlines with $dt = d\phi = d\theta = 0$ are geodesics, that is, solutions of the geodesic equation. Hence the straight lines $t = $ [constant] in the interior of the black hole in the Kruskal diagram are geodesics.
(b) Among these geodesics is the line $t = 0$, or $u = 0$ (see Fig. 8.4). Suppose that a particle falls along this geodesic from $r = 2GM$ to $r = 0$. How much proper time elapses during this motion?

6. Suppose that a spaceship enters a Schwarzschild black hole on a radial line with a radial velocity

$$
\frac{dr}{dt} = - \frac{1 - r_S/r}{\sqrt{(r/r_S)^2 - r/r_S + 1}}
$$

(a) Show that the worldline of the spaceship is timelike for the given dr/dt

(therefore this dr/dt is permissible).

(b) How much proper time elapses as the spaceship moves from $r = 4r_S$ to $r = 0$?

(c) In order to move with the given dr/dt, must the spaceship use its rocket motor? (Hint: Solve this problem in r-t coordinates.)

7. An astronaut in command of a spaceship equipped with a powerful rocket motor enters the horizon $r = r_S$ of a Schwarzschild black hole. Prove that in a proper time no larger than $(\pi/2)r_S$, the astronaut reaches the singularity $r = 0$. Prove that in order to avoid the singularity as long as possible, the astronaut ought to move in a purely radial direction. [Hint: For purely radial motion, with $dr < 0$ and $dt = d\phi = d\theta = 0$, the increment in proper time is

$$d\tau = - \frac{dr}{\sqrt{r_S/r - 1}} \ , \text{ for } r \le r_S$$

Integrate this between $r = r_S$ and $r = 0$ to obtain

$$\Delta\tau = \frac{\pi}{2} r_S$$

Finally, check that if dt, $d\theta$, $d\phi$ are different from zero, then the increment $d\tau$, for a given value of $-dr$, is necessarily smaller than the value given above.]

Show that in order to achieve this longest proper time the astronaut must use his rocket motor in the following way: outside the horizon, he must brake his fall so as to arrive at $r = r_S$ with nearly zero radial velocity; inside the horizon he must shut off his motor and fall freely. (Hint: Show that $\Delta t = (\pi/2)r_S$ corresponds to free fall from $r = r_S$; see Eq. [19].)

8. The coordinate $\tilde{v}$ introduced by Eq. [30] is called the "ingoing" Eddington-Finkelstein coordinate. Alternatively, we can introduce an "outgoing" Eddington-Finkelstein coordinate by

$$\tilde{u} = t - r - r_S \ln|r/r_S - 1|$$

Express the spacetime interval for the Schwarzschild geometry in the coordinates $\tilde{u}$, r, θ, ϕ. Show that an outgoing light signal has the worldline $\tilde{u} = $ [constant]. Draw a diagram similar to Fig. 8.5 describing the Schwarzschild geometry in the coordinates $\tilde{u}$, r. (Hint: The spacetime geometry is that of a white hole, not a black hole.)

9. Given a surface $f(x^0, x^1, x^2, x^3) = 0$ in a four-dimensional spacetime, we can define the *normal* to the surface by

$$n_\mu = \frac{\partial f}{\partial x^\mu}$$

The surface is said to be a *null surface* if $n_\mu n^\mu = 0$. Show that the surface $r - r_S = 0$ is a null surface in the Schwarzschild geometry. Show that the surfaces $r - r_\pm = 0$ are null surfaces in the Kerr geometry.

10. Show that in *classical mechanics* the angular velocity of rotation of a rigid body can be expressed as

$$\Omega_x = \frac{\partial E}{\partial S_x}, \quad \Omega_y = \frac{\partial E}{\partial S_y}, \quad \Omega_z = \frac{\partial E}{\partial S_z}$$

where E is the energy (kinetic energy of rotation) of the body.

This suggests that we define the angular velocity of a Kerr black hole as follows*:

$$\Omega_z = \frac{\partial M}{\partial S_z}$$

Differentiate Eq. [66] with respect to S (holding M_{ir} constant), and show that the angular velocity of rotation of the black hole is

$$\Omega_z = \frac{a}{r_+^2 + a^2}$$

11. Calculate the Christoffel symbols for the Reissner-Nordström solution [48] and verify that the electromagnetic field tensor $F^{\mu\nu}$ given by Eq. [47] satisfies the field equations [45] and [46].

12. (a) Show that the Reissner-Nordström solution has no horizon if $Q^2 > GM^2$.
(b) Find the radius of the horizon of the Reissner-Nordström solution if $Q^2 < GM^2$.

13. (a) According to Eq. [87], what is the radius of a white dwarf of mass $M = M_\odot$? Assume that the nuclear matter is in the form of helium nuclei.
(b) By carrying out a calculation similar to that given in Eqs. [83] - [87], show that the radius of a neutron star is of the order of

$$R = \frac{N^{5/3}(9\pi\hbar^3/4)^{2/3}}{GM^2 m_p}$$

According to this equation, what is the radius of a neutron star of mass $M = M_\odot$?

---

* See Misner, Thorne, and Wheeler (1973) for a rigorous justification of this definition.

14. Suppose that a black hole is surrounded by gas which, initially at rest at a large distance, falls radially into the black hole. For the purpose of this problem, pretend that gravitation can be described with sufficient accuracy by the Newtonian $1/r^2$ field and that the supply of gas at large distance is inexhaustible. Show that the density of the gas increases as $r^{-3/2}$ as it falls. [Hint: At any given radius the density is constant in time because gas leaving is replaced by gas arriving, that is, $\partial\rho/\partial t = 0$. Show that this implies $(1/r^2)(\partial/\partial r)r^2\rho v = 0$, where v is the velocity at radius r.]

9. Cosmology

WE ARE ALL IN THE GUTTER, BUT SOME OF US ARE
LOOKING AT THE STARS.
>   Oscar Wilde, *Lady Windermere's Fan*

As we begin the study of the universe, a fundamental question immediately comes to mind: is it legitimate to apply the laws of physics that are known to hold on and near the Earth to faraway regions of the universe? In a broad sense the answer must necessarily be in the affirmative. For suppose that the laws of physics in two regions of the universe were different; say, in one region gravitation is a tensor field and in another a scalar field. Then at the boundary between these regions physics would make no sense--there is an unacceptable discontinuity.

Such arguments based on continuity cannot rule out the possibility that the laws contain some parameters with a smooth and gradual variation in space or time. In general terms, functions of space and time are fields, and the question therefore is: do the laws of physics contain a dependence on some cosmological, or "background," fields? If there are such extra fields then we will not say that the laws of physics are different in different regions of the universe; rather we will include the description of these fields in our laws, and simply say that physics is more complicated than we were wont to believe.

One possible manifestation of a cosmological field might be changes in "constants." For instance, the gravitational constant or the fine-structure constant might vary with position or with time. However, observation indicates that the general features of the universe are pretty much the same irrespective of the direction in which we look. This isotropy constitutes strong evidence against a dependence of "constants" on position. For example, regions of the universe with different values of G would expand at different rates, and we would observe an anisotropy in the expansion rate and a consequent anisotropy in the density of galaxies. Likewise, a dependence of atomic constants on position would affect the characteristics of the emitted light, and galaxies at different positions of the sky would display different, and mysterious, spectral lines, in contradiction with observation.

Another fundamental question that comes to mind is related to an exclusive property of the universe: there is one and only one universe. As long as we are dealing with physical systems smaller than the universe, we can regard differential equations as a suitable expression of the laws of nature. Differential equations, such as Maxwell's equations, are summaries of our knowledge of the dynamics of physical systems. We can test these differential equations against experiment by observing systems that have been started with different initial conditions. But when we are dealing with the universe, the initial conditions are beyond our control, and we cannot carry out such a test. Under these circumstances, would it not be better to give up any attempt at formulating a differential equation of motion for the universe and, instead, invent some new law of motion completely unrelated to Einstein's differential equations? What compels us to accept Einstein's equations for the universe as a whole is that physics on a small scale determines physics on a large scale. Given that Einstein's equations hold in all regions of, say, the size of the Solar System, it then follows that they also hold across the boundaries of these regions and throughout the universe. Einstein's equations therefore determine not only the local gravitational effects produced by the Sun and planets, but also the gravitational effects of the universe as a whole.

There is, however, one possible modification that we have to take into account when we write Einstein's equations for the universe. A small cosmological term (see Section 7.3) would have no appreciable effects inside the Solar System, but could have significant effects on the evolution of the universe. We will therefore include such a term in Einstein's equations for the universe.

9.1 THE LARGE-SCALE STRUCTURE OF THE UNIVERSE

The stars around us that are visible to the naked eye are all part of a large cloud with spiral arms called the Galaxy of the Milky Way. This Galaxy contains about 10^{11} stars and measures about 3×10^4 pc

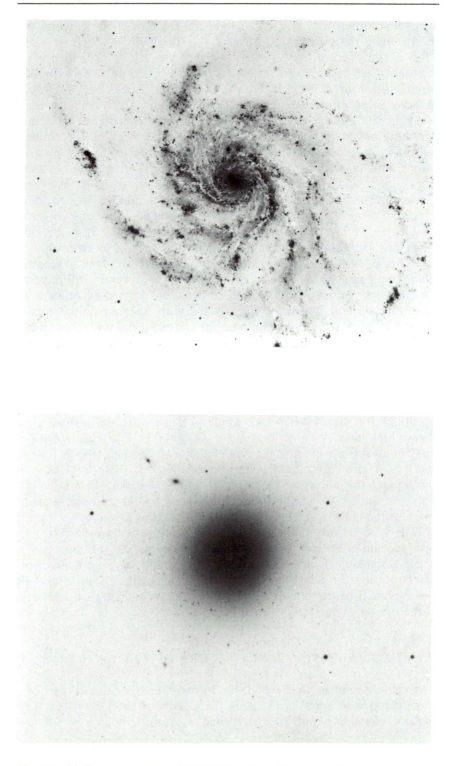

Fig. 9.1 (a) The spiral galaxy NGC 5457 in Ursa Major. (b) The elliptical galaxy NGC 4486 in Virgo (Hale Observatories; negative photographs).

across.* There are many other galaxies--about 10⁹ are within the reach of our telescopes. Some are spirals similar to the Milky Way (see Fig. 9.1a); some are ellipticals (see Fig. 9.1b); and some are irregular. Several million galaxies have been cataloged and plotted by astronomers. Fig. 9.2 shows the distribution of galaxies over a portion of the sky.

Fig. 9.2 Distribution of galaxies over a portion of the sky according to the APM catalog. This plot does not show individual galaxies, but only the density of galaxies (bright regions have high density, dark regions low density). The plot covers about 1/10 of the sky, and includes about 2 million galaxies at distances of up to about 200 Mpc. (From Maddox et al, 1990.)

By inspection of Fig. 9.2, we can see that the galaxies are not randomly distributed over the sky. On a scale of a few degrees, we can see lumps and filaments where galaxies appear to be concentrated. However, some of these apparent concentrations are the result of accidental overlaps of galaxies at different distances along the line of sight. To discover the genuine structural details in the distribution of galaxies, we need to take into account the radial distances of the galaxies and we need to examine the three-dimensional distribution. We can then discern that the distribution of galaxies is indeed lumpy--the galaxies tend to associate in clusters consisting of a few tens of galaxies, or a few hundred, or a few thousand. Our Milky Way Galaxy is a member of a small cluster, the Local Group, consisting of about 30 galaxies; this group measures about 1 Mpc across. By comparison

---

* The parsec (pc) is a unit of distance, 1 pc = 3.26 light-years = 3.08 × 10¹⁸ cm. For cosmological distances, the megaparsec (Mpc) is preferred.

with some of the large, rich clusters, the Local Group is puny. For example, the Virgo cluster contains over 2000 galaxies and is about 5 Mpc across.

Clusters of galaxies are often associated in superclusters. Our Local Group is part of the Local Supercluster, which also contains the Virgo cluster. This supercluster is about 50 Mpc across.

The clustering of galaxies can be described quantitatively by a galaxy correlation function, or two-point function (Peebles, 1980, 1993). If the distribution of galaxies were random, with a mean density of n galaxies per unit volume, then the relative (not normalized) probability for finding a neighbor galaxy in a volume dV at a distance r from an average galaxy would be $dP = ndV$, which is independent of r. If the distribution is not random, with a tendency for galaxies to cluster, then the probability for finding such a neighbor galaxy is larger than ndV and it depends on r. We write this probability as

$$dP = ndV[1 + \xi(r)] \qquad [1]$$

where the correlation function $\xi(r)$ characterizes the excess probability. For large values of r, we expect that $\xi(r)$ decreases toward zero, since the correlations among the positions of the galaxies disappear at large distances. The observational data on the galaxy distribution fit a simple power law for $\xi(r)$ (Peebles, 1993),

$$\xi(r) = \left(\frac{5 \text{ Mpc}}{r}\right)^{1.8} \quad \text{for } r < 20 \text{ Mpc} \qquad [2]$$

Note that $\xi(r) = 1$ for $r = 5$ Mpc. This means that at this distance the probability per unit volume for finding a neighbor galaxy of an average galaxy is twice as large as for a random distribution.

The clustering of clusters of galaxies can also be described by a correlation function. Surprisingly, the observational data on clusters again fit a power law, with the same power -1.8 as for galaxies (Bahcall, 1988):

$$\xi(r) = \left(\frac{26 \text{ Mpc}}{r}\right)^{1.8} \quad \text{for } r < 100 \text{ Mpc} \qquad [3]$$

There is also some indication that superclusters tend to cluster. The correlation function for this clustering of superclusters is not known, because the observational data are indequate; but possibly the correlation function also fits a power law, with the same power -1.8.

Systematic surveys reveal that on a scale of 50 to 100 Mpc, the universe has a sponge-like or foam-like structure. The universe is permeated by large voids, and the superclusters and clusters are arranged in relatively thin sheets around the periphery of these voids.

Fig. 9.3 shows a thin slice through the universe, extending over a sector 135° wide. The rich knot of galaxies near the center of this slice is part of the Coma cluster. A large void can be seen to the left of this knot; this void has a diameter of about 50 Mpc. Note that this "void" is not quite empty; however, the density of galaxies in the "void" is drastically lower than the average density. A dense sheet of galaxies, called the Great Wall, lies along the farthest edge of this void and extends across the entire slice.

On a scale of a few hundred Mpc, the universe appears fairly uniform. We can discern this large-scale uniformity in the plot of galaxies

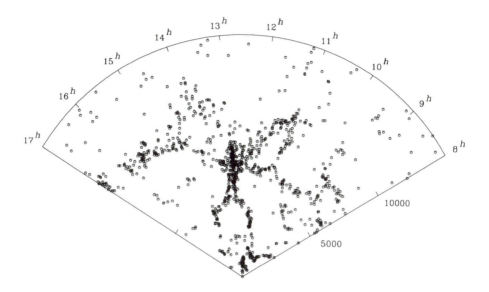

Fig. 9.3 A slice through the distribution of galaxies. This slice includes a region 12° thick, extending over 135° (or 9 hours) of the sky, out to a radial distance of 150 Mpc. (From Geller and Huchra, 1988.)

in Fig. 9.2, where all regions more than a few degrees across seem to have about the same density, except for fine detail. This plot includes galaxies out to distances of 200 Mpc, and therefore gives evidence that on this scale of distance the universe is isotropic. From the large-scale isotropy of the galaxy distribution we can infer large-scale homogeneity. The reason is that isotropy about the position of the Earth requires that the density of galaxies be some function $n(r)$, independent of angular variables. Unless $n(r)$ is a constant, this would imply that the universe has a unique center of symmetry ($r = 0$) and that the Earth is at this center. In accord with the Copernican tradition, we will make the reasonable assumption that the Earth does *not* occupy such a privileged position. Then the function $n(r)$ must be a constant, which means that, on a large scale, the distribution of galaxies is homogeneous.

Further evidence for large-scale isotropy (and, therefore, homogeneity) comes from observations of distant radio sources and of the diffuse X-ray background. The distant radio sources--radio galaxies and quasars--are distributed randomly over the sky. Some of these radio sources are about 3000 Mpc away, and their random distribution therefore confirms isotropy on a very large scale. Likewise, the diffuse X-ray flux originating from distant sources is distributed uniformly over the sky, with fluctuations of only a few percent on a scale of a few degrees.

But the strongest evidence for isotropy is supplied by the observations on the cosmic microwave background radiation, which is a relic from the early, hot stage of the universe, or Big Bang. As we will see in Section 9.5, this radiation is distributed uniformly over the sky to better than a few parts in 10^5, and this indicates that the universe was extremely uniform in its early stages.

The large-scale isotropy and homogeneity of the universe imply that all observers, wherever they are located in the universe, see the same large-scale density of galaxies. Thus, on a large scale, the most remarkable feature of the universe is its lack of distinguishing features. The assertion that all positions in the universe are essentially equivalent, except for local irregularities, is known as the *Cosmological Principle*.

For a direct test of homogeneity of the galaxy distribution plotted in Fig. 9.2, we would need distance measurements of these galaxies. But such distance measurements (either by direct methods or via the redshift; see Section 9.3) have been performed on only a small fraction of the galaxies. Alternatively, we can test the large-scale homogeneity of the universe by simply counting the number of galaxies of a given brightness. If the number of galaxies per unit volume is constant, then the total number within a distance l increases as l^3. Consider those galaxies that have a given intrinsic luminosity $\mathcal{L}$ (erg/s); the apparent brightness (erg/cm²·s) of such a galaxy, or the incident flux measured at the Earth, is then*

$$S = \frac{\mathcal{L}}{4\pi l^2} \qquad [4]$$

According to this, $l^3 \propto S^{-3/2}$, and therefore the number of such galaxies with an apparent brightness in excess of S is proportional to $S^{-3/2}$,

$$N(>S) \propto S^{-3/2} \qquad [5]$$

---

* This formula is based on the assumption that space is flat; the effects of the curvature of the universe become noticeable only at distances larger than 1000 Mpc.

This proportionality does not depend on the value of $\mathscr{L}$ and hence applies to the totality of all galaxies. Thus, a count of numbers of galaxies as a function of brightness can be used as a direct observational test of homogeneity.

For the brighter, nearer galaxies, surveys by Hubble (1926, 1936) established that the counts were in good agreement with Eq. [5], which provided the first direct observational evidence for the large-scale homogeneity of the universe. Later surveys extended the counts to fainter, faraway galaxies. Over a wide range of brightnesses, the counts remain in good agreement with Eq. [5] (Peebles, 1993). However, noticeable deviations occur for the faintest observable galaxies. This is interpreted as a combined effect of curvature of spacetime and evolution of the luminosity of the galaxies--we see these distant galaxies as they were a long time ago, when their luminosity was different from what it is now, and such a change of luminosity with time invalidates Eq. [5].

For radio sources, such as radio galaxies and quasistellar objects (quasars), the counts are *not* in agreement with Eq. [5]. There is an excess of radio sources of low brightness. This is believed to be an evolutionary effect. A change of luminosity with time invalidates Eq. [5], and so does a change of density of sources with time. For instance, if there was a higher density of radio sources in the past, then the counts will exceed $S^{-3/2}$ at small S.

The evidence for evolutionary effects in radio sources can be used as an argument against the steady-state theory of the universe that was proposed by Bondi and Gold (1948) and by Hoyle (1948). In this theory, the universe has no beginning and no end, and the average properties of the universe are constant not only in space, but also in time (the Perfect Cosmological Principle). In such a universe, new radio sources must be continually created to replace those that fade away or move away, and the average luminosity and density remain constant.

The conclusion that radio-source counts contradict the steady-state theory was first reached by Ryle and Scheuer in 1955. Later, more precise counts by Ryle and others have confirmed this conclusion (Ryle, 1968; Wall, Pearson, and Longair, 1980, 1981; Peacock and Gull, 1981). For quasars, the analysis of counts has also been carried out using their optical, rather than radio, brightness. According to Schmidt (1970, 1972), the counts indicate that the density of quasars must have been much higher in the past than it is now. An analysis of counts with respect to redshift (which is a measure of distance; see Section 9.3) also confirms this conclusion (Schmidt, 1968; Osmer, 1982). This again contradicts the steady-state theory.

9.2 COSMIC DISTANCES

The methods used by astronomers to measure distances to objects in the sky vary with the distance involved. For the nearest stars, at distances of up to ≃30 pc, triangulation with the diameter of the Earth's orbit as base line is adequate. Some other geometric methods give the slightly larger distances to nearby clusters of stars, for example, the Hyades cluster at ≃41 pc. All such trigonometric and other geometric methods are called *parallax* methods, since they hinge on measuring a displacement (apparent or real) of the stars across the sky.

Beyond a few hundred pc, distances are usually determined by *photometric* methods, by the apparent brightness of a standard light source, or "standard candle." If a light source, such as a star, has an

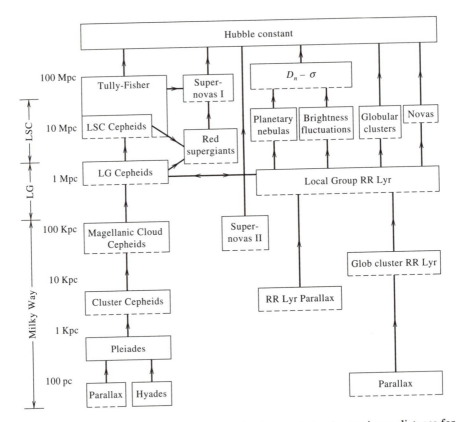

Fig. 9.4   The cosmological distance ladder. Each rung shows the maximum distance for which a particular distance indicator can be used. The distance indicators for each rung are calibrated by means of the indicators in the next lower rung. (Based on Jacoby et al., 1992.) The ranges of distances for the Milky Way, the Local Group, and the Local Supercluster are indicated in the left margin.

intrinsic luminosity $\mathcal{L}$ (erg/s), then the apparent brightness S (erg/cm²·s) measured at the Earth is inversely proportional to the square of the distance,

$$S = \frac{\mathcal{L}}{4\pi l^2}$$ [6]

A measurement of the apparent brightness S therefore gives the distance l provided the intrinsic luminosity $\mathcal{L}$ is known.

For distances within our Galaxy and within our Local Group, stars serve as standard light sources. But for distances beyond 10 Mpc, individual stars are too faint to be distinguished, and larger and more luminous standard light sources are needed, such as clusters of stars or entire galaxies. The standard light sources used for large distances must be calibrated by comparing them with the standard light sources for shorter distances. This step-by-step scheme for the calibration of standard light sources is called the *cosmological distance ladder* (see Fig. 9.4). The lower rungs of the ladder, reaching to the edge of the Local Group, rely on Cepheid stars and RR Lyrae stars, which are calibrated by comparison with distances obtained by parallax methods. The higher rungs, reaching to the edge of the Local Supercluster, rely on red supergiant stars, H II regions, globular clusters, novas, planetary nebulas, spiral galaxies (characterized by their Tully-Fisher relation; see below), and surface-brightness fluctuations in galaxies, all of which are calibrated by comparison with Cepheids and RR Lyrae stars. And the highest rungs, reaching beyond the Local Supercluster, rely on brightest galaxies (in clusters), supernovas, and elliptical galaxies (characterized by their D_n - σ relation; see below), which are calibrated by comparison with the standard sources of the third rung.

The following is a brief description of these different kinds of standard light sources used for distance determinations (for more details and references, see Huchra, 1992; Jacoby et al., 1992; Rowan-Robinson, 1985).

Cepheids and RR Lyrae Stars. Cepheids are variable supergiant stars, whose luminosity pulsates with a period of between 2 and 150 days. The time-average luminosity is correlated with the period: the long-period Cepheids are intrinsically the most luminous, and the short-period Cepheids are the least luminous. The empirical period-luminosity relation is plotted in Fig. 9.5. According to this plot, a simple measurement of the period of a Cepheid gives us its luminosity, and therefore permits us to adopt it as a standard light source. Cepheids are exceptionally bright, 10^3 to 10^4 $\mathcal{L}_\odot$,* and they are therefore useful as standard light sources for distances up to 10 Mpc.

---

* This luminosity is only a rough number. Actual distance determinations are based on the photographic magnitudes M_V, M_B, and M_U. These magnitudes are defined in terms of the luminosities in narrow ranges of wavelengths selected by filters, rather than the total luminosity.

RR Lyrae stars also are variable stars, but with a pulsation period of between 0.4 and 1 day. They have a time-average luminosity of about 40 $\mathscr{L}_\odot$. Since RR Lyrae stars are not as bright as Cepheids, they cannot be detected as far away as Cepheids, and they are somewhat less useful as distance indicators.

Brightest Red Supergiants. In the galaxies of the Local Group and other nearby groups, the brightest red supergiant star in a galaxy usually has a luminosity of about 9×10^4 $\mathscr{L}_\odot$. This means that we can regard the brightest red star in a galaxy as a standard light source, which we can use for a distance determination. However, this method is subject to a statistical bias: when we look at more distant galaxies, and thereby enlarge our sample of galaxies, we are likely to stumble across more and more freak instances of even brighter red supergiants. If we mistakenly assume that these freaks have the standard luminosity, we will underestimate their distance. Because of this statistical bias, the use of brightest stars (and similar methods based on selecting an extreme property) has fallen out of favor.

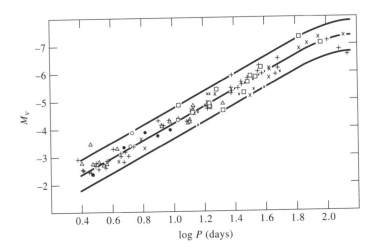

Fig. 9.5 Period-luminosity relation for Cepheid stars. The luminosity is expressed in terms of the photographic magnitude, $M_V \propto \log \mathscr{L}$. (After Rowan-Robinson, 1985.)

H II *Regions.* These are large clouds of interstellar hydrogen, ionized and made luminous by the ultraviolet radiation emitted by a hot star located within the cloud. In some galaxies, these ionized clouds have sizes of several hundreds of pc. If we consider only the brightest H II regions in the galaxy, we find that the cloud size, the luminosity, and the velocity dispersion (that is, the spread of velocities of parcels of gas in the cloud) are correlated with the luminosity of the galaxy. These correlations have been exploited in three different methods of distance determination: the distance is calculated from the inferred diameter of the cloud and its observed angular diameter; or the cloud is used as a standard light source for the distance calculation; or the velocity dispersion is used to infer the luminosity of the galaxy, and the galaxy is then used as a standard light source for the distance calculation.

Globular Clusters. Globular clusters are roughly spherical swarms of 10^4 to 10^5 stars, usually found in the outlying regions of galaxies (see Fig. 9.6). The brightest globular cluster found in a galaxy can be used as a standard light source. In a recent refinement of this method, the luminosity distribution function for the ensemble of all the globular clusters in the galaxy is plotted and parametrized by a smooth curve (a Gaussian distribution function). The standard light source is then taken to have a luminosity equal to the maximum, or the "turnover," of this smooth curve. This refined data analysis leads to somewhat more consistent results, less influenced by the presence of one or a few exceptional clusters with abnormal luminosities.

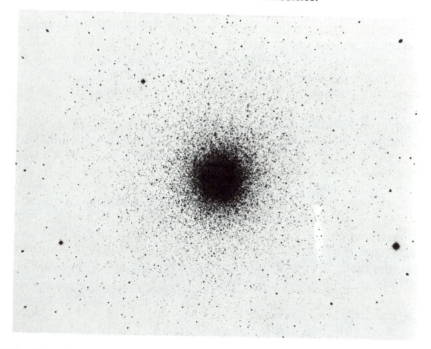

Fig. 9.6 Globular cluster NGC 5272 in Canes Venatici. (Hale Observatories; negative photograph.)

Novas. Novas are variable stars that display a brief, explosive increase of luminosity that lasts only a few days and then decays gradually. The outburst of a nova is brought about by violent thermonuclear burning ignited on the surface of the star when a critical mass of gas accumulates on the surface by accretion from a companion star. The maximum luminosity attained by a nova is correlated with the decay time of its luminosity. Thus, a measurement of the decay time permits us to calibrate the nova as a standard light source. Novas attain luminosities of up to $7 \times 10^5 \, \mathcal{L}_\odot$, which make them suitable as distance indicators far beyond the Local Group.

Spiral and Elliptical Galaxies and Their Spectral Line Widths. Tully and Fisher (1977) discovered that the width of the 21-cm spectral line emitted by clouds of hydrogen in spiral galaxies increases with the luminosity of the galaxy, and can be used to infer the luminosity. Such a broadening of the spectral lines was later also found in the optical part of the spectrum. The broadening of the spectral lines arises from Doppler shifts caused by the orbital motion of the gas clouds around the galactic center, and the broadening is consequently often called a *velocity dispersion*. The most luminous galaxies have the largest masses, and hence the largest orbital speeds and the largest broadening.

For elliptical galaxies, Faber and Jackson (1976) discovered a similar correlation between the width of stellar absorption lines and the luminosity, so the same method can be applied to these galaxies. Recent versions of this method incorporate both the luminosity and the surface brightness of the galaxy into a parameter D_n, which correlates with the velocity dispersion (the D_n - σ relation).

Planetary Nebulas. These are expanding spherical shells of gas ejected by some hot stars. The gas fluoresces while it is illuminated by the light from the hot star. Planetary nebulas are fairly abundant (typically, a few hundred per galaxy), and their light, in the form of sharp spectral lines, is easily observed and measured. However, individual planetary nebulas do not serve as good distance indicators, since they have a wide range of luminosities. Instead, the luminosity distribution function for the ensemble of all the planetary nebulas in a galaxy can be plotted and calibrated against other distance indicators; thus, the luminosity distribution effectively plays the role of standard light source.

Surface-Brightness Fluctuations. When we observe a distant elliptical galaxy, which is too far away to permit us to resolve individual stars, we merely see a dense luminous cloud, with some surface brightness. However, the surface brightness is not quite uniform. The local concentration of light from discrete stars within the galaxy, especially giant stars, gives the surface a mottled appearance. This mottling is called surface-brightness fluctuation, sometimes also called incipient resolution. The amount of fluctuation depends on the distance: the closest galaxies display the strongest fluctuation; the most distant galaxies display the least fluctuation. If we superimpose an array of squares (pixels) on the surface, we can characterize the surface-brightness fluctuation quantitatively by the rms deviation of the brightnesses of the squares from the mean brightness. Other things being equal, this rms deviation can be shown to decrease in inverse proportion to the distance *l*. Hence, we can determine the distance by a measurement of the rms fluctuation. Note that for the application of the method, we need a calibration of the strength of the fluctuation in the elliptical galaxies, not a calibration of their luminosity. The method can also be applied to the central bulges of spiral galaxies, which look like dense luminous clouds.

Supernovas. Supernovas display an explosive increase in luminosity, like novas, but they attain a much higher luminosity and they take longer to decay. Supernovas are of two types, I and II. As mentioned in Section 8.8, supernovas of type II result from the catastrophic gravitational collapse of a star of large mass that has exhausted its nuclear fuel. Supernovas of type I result from the explosion of a white dwarf whose carbon core is ignited when the star is overheated by gas accumulated on its surface by accretion from a companion star. Both types of supernovas can be used as distance indicators. Supernovas of type I are especially convenient to use as standard light sources, because the maximum luminosity they attain during their explosion has a more or less standard value of $5 \times 10^9 \mathscr{L}_\odot$.

Brightest Galaxies. The brightest galaxies in rich clusters are usually found to be giant ellipticals of type E, with a more or less fixed luminosity. For instance, the luminosity of the brightest galaxies in the Virgo cluster, the Coma cluster, and the Centaurus cluster is about $9 \times 10^{10} \mathscr{L}_\odot$, and this can be used as a standard luminosity for the brightest galaxy in any remote cluster.

When we use the standard light sources in the construction of the cosmological distance ladder (Fig. 9.4), each step of the ladder relies on the preceding step. Any error in one of the methods of distance determination will affect the calibration of the methods used for larger distances. For instance, it was shown by Baade in 1952 that the period-luminosity relation for Cepheids that had been in use until then was in error. A recalibration indicated that the distances previously determined by this method had to be increased by a factor of 2. This implied that all extragalactic distance determinations also had to be increased by this factor.

As emphasized by Rowan-Robinson (1985) and Huchra (1987), the accumulation of systematic errors implies that at the upper end of the distance ladder, the measured distances have an accumulated uncertainty of about 25%. Astronomers hope that observations with the new Hubble Space Telescope will reduce the errors in the distances to 10% or less.

9.3 THE COSMOLOGICAL REDSHIFT; HUBBLE'S LAW

Observed from the Earth, the spectral lines of almost all galaxies display a shift toward lower frequencies, that is, a redshift. This redshift is conventionally described by a *redshift parameter z*, defined as

$$z = \frac{\lambda - \lambda_0}{\lambda_0} \qquad [7]$$

where λ is the wavelength received at the Earth and λ_0 the wavelength emitted by the atoms in the remote galaxy.

The physical interpretation of the redshift is that it is a Doppler shift produced by the motion of the galaxies away from the Earth. If the speed of the galaxy is small compared with the speed of light, the fractional change in wavelength of the light is given by the simple Doppler formula

$$\frac{\lambda - \lambda_0}{\lambda_0} = \frac{v}{c} \qquad [8]$$

Thus, the speed of recession of the galaxy, in units of c, equals the redshift parameter z. For example, the galaxies in the Virgo cluster have a redshift $z = 0.004$, and therefore a speed of recession of $0.004c$, or 1200 km/s. Many galaxies have much larger redshifts. The galaxy with the largest redshift discovered so far has $z = 3.8$. Quasars attain even larger redshifts; the quasar with the largest redshift has $z = 4.9$. For such large values of z, the simple Doppler formula [8] fails--it becomes necessary to take into account relativistic effects and the large-scale geometry of the universe.

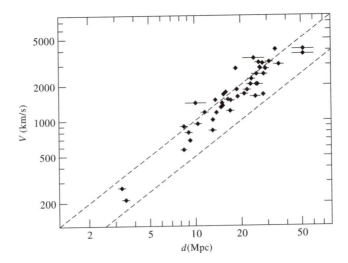

Fig. 9.7 Measured redshift vs. distance for some galaxies. The redshift increases with distance. (From Jacoby et al., 1992.)

The redshift of those galaxies that are nearby is small, and the red-shift of those galaxies that are far away is large. Fig. 9.7 is a plot of the measured redshift of some galaxies versus their distance. From this plot we see that the magnitude of the redshift is directly proportional to the distance, but the large scatter in the observational data makes it difficult to determine the constant of proportionality. For the galaxies plotted in Fig. 9.7, $z \ll 1$, and hence the simple Doppler formula [8] is valid. A redshift directly proportional to the distance then indicates a recession velocity directly proportional to the distance. This propor-tionality was discovered by Hubble in 1929, and it is known as *Hubble's law*,

$$v = H_0 l \qquad [9]$$

The constant of proportionality H_0 is called *Hubble's constant*. It is usually expressed in units of kilometers per second per megaparsec (km/s·Mpc).

The search for an accurate value of H_0 has a long and checkered history. To determine the value of H_0 we must examine the recession velocities and the distances of remote galaxies. The measurement of the recession velocities, via the redshift, has next to no error. How-ever, galaxies have peculiar velocities, that is, irregular deviations from the Hubble motion of recession, caused by the gravitational attraction of other galaxies or clusters in their vicinity. For instance, our own Local Group is thought to have acquired a peculiar "infall" velocity toward the large Virgo cluster, caused by the gravitational attraction of the large mass in Virgo, and also a peculiar velocity toward the even larger Hydra-Centaurus cluster ("the Great Attractor"), farther away than Virgo. Because of this peculiar velocity, the recession velocity of Virgo relative to us is somewhat smaller than the Hubble recession. For an accurate determination of H_0, we must correct all the observed recession velocities according to our peculiar velocity. Alternatively, we can try to minimize the influence of the peculiar velocity by dealing exclusively with galaxies much farther away than Virgo. The Hubble recession velocities of such galaxies are much larger than any peculiar velocity, and this permits us to neglect the latter.

The measurements of the distances are afflicted with large errors because of the uncertainties in the calibrations of the standard light sources, as described in the preceding section. The large errors in the distance measurements lead to correspondingly large errors in H_0.

Hubble originally found a value of about 500 km/s·Mpc for H_0. But his distance measurements contained several errors, and recent results for H_0 are five or ten times smaller. Thorough and extensive series of distance determinations were completed by Sandage and Tammann and by de Vaucouleurs et al. in the 1980s. Their results for H_0 were (for references, see Rowan-Robinson, 1985; Huchra, 1992;

de Vaucouleurs, 1993)

$$H_0 = 40 \text{ to } 60 \text{ km/s·Mpc} \quad \text{(Sandage and Tammann)} \quad [10]$$

$$H_0 = 80 \text{ to } 100 \text{ km/s·Mpc} \quad \text{(de Vaucouleurs)} \quad [11]$$

The difference between these results arose from different choices of methods used for distance determinations, different ways of calibrating the standard light sources, and differences in how the observed brightnesses of these sources were corrected for absorption by interstellar dust.

More recent distance determinations have tended to give results intermediate between those of Sandage and de Vaucouleurs. According to recent reviews of the best available data (Jacoby et al., 1992; Huchra, 1992), a wide variety of methods for distance determination favor a value in the range from 70 to 90 km/s·Mpc.

A novel method for the determination of H_0 exploits observations of gravitational lenses which form multiple "images" of distant quasars. The light from the quasar to the Earth then travels along different paths through the lens, with a different path length and travel time for each image. The path differences depend on the distance scale, and therefore a measurement of the time delay between light signals reaching us along different paths can be used to determine the distance scale, if the geometry of the paths is known. As mentioned in Section 4.5, recognizable time-delayed signals, resulting from luminosity changes, have been observed in the double images of the "twin" quasar Q0957+561, and the analysis of the expected time delays in this system leads to the result (Hewitt, 1993)

$$35 \leq H_0 \leq 90 \text{ km/s·Mpc} \quad [12]$$

The large spread of values in this result is due to our lack of knowledge of the precise mass distribution in the gravitational lens.

In view of the uncertainties in the value of H_0, cosmologists have adopted the expedient of writing H_0 in the form

$$H_0 = 100h \text{ km/s·Mpc} \quad [13]$$

where h is a fudge factor whose value lies somewhere between the extremes of 0.4 and 1 indicated by Eqs. [10] and [11].

With a known or an assumed value of H_0, Hubble's law provides a convenient and quick method for finding the distance of a galaxy from a simple redshift measurement. Astronomers often use Hubble's law for this purpose, when they do not want to trouble with any of the direct methods for distance measurement. For instance, the radial distances in Fig. 9.3 were found from redshifts.

Although according to Eq. [9] galaxies are moving radially away from us, this does not mean that the Earth occupies a preferred position in the universe. Let us write Eq. [9] in vector notation:

$$\mathbf{v} = H_0\mathbf{r} \tag{14}$$

Consider now an observer on some galaxy at a position $\mathbf{r}'$. Relative to the Earth this observer has a velocity

$$\mathbf{v}' = H_0\mathbf{r}' \tag{15}$$

Hence, relative to this observer, another galaxy at a position $\mathbf{r}'$ has a velocity

$$\mathbf{v} - \mathbf{v}' = H_0(\mathbf{r} - \mathbf{r}') \tag{16}$$

Since $\mathbf{r} - \mathbf{r}'$ is simply the position vector of the other galaxy relative to this observer, Eq. [16] shows that there is a general motion of galaxies radially away from this observer, just as there is such a motion away from an observer on Earth.

Hence Eq. [14] indicates a general expansion of the universe, with distances increasing in proportion. However, note that this increase of distances applies only to distances between galaxies, and only when they are not gravitationally bound to each other. The distances within a galaxy or the distances within a (bound) cluster of galaxies do not increase.

Incidentally: A nonlinear relation between $\mathbf{v}$ and $\mathbf{r}$ would lead us to the conclusion that the Earth *does* occupy a preferred position in the universe, since then the right sides of Eqs. [14] and [15] would not combine by simple vector addition, and expansion of the universe would look different from the Earth and from some other galaxy. This is a strong argument in favor of the linear Hubble law.

An important question is how the recession velocities change in time. Since the attractive gravitational forces oppose the expansion, we expect a decrease in the expansion rate. If the recession velocities decrease strongly enough, then the universe will reach a state of maximum size at some time in the future and then begin to contract.

Let us calculate the expected deceleration in the expansion rate. For the calculation of the average motion of the galaxies, we can approximate the mass distribution of the universe by an average uniform mass density ρ. Consider a spherical volume of radius l, small compared with the size of the observable universe, but large enough so irregularities in the mass density can be neglected. The mass distribution that surrounds this volume and extends throughout the rest of the universe can be regarded as a spherical shell of very large thickness; such a shell produces no gravitational force in its interior (see Exercise 1.5). Hence, the gravitational force acting on a particle on the surface of our spherical volume depends only on the mass m in this spherical volume, and acceleration of this particle is

$$\frac{dv}{dt} = - \frac{Gm}{l^2}$$ [17]

With $m = 4\pi l^3 \rho / 3$, we then obtain

$$\frac{dv}{dt} = - \frac{4\pi G\rho}{3} l$$ [18]

This equation shows that the change in the relative velocity of two galaxies is proportional to the distance between them. Since Hubble's law tells us that the present velocities are proportional to the distances, it follows that the velocities remain proportional to the distance at all times. We can therefore write

$$v(t) = H(t)l(t)$$ [19]

where $H(t)$ is the Hubble "constant" at time t [in contrast, $H(t_0)$, or H_0, is the Hubble constant at the present time].

The rate of change of $H(t)$ is directly related to the mass density. To obtain this relation, we need only differentiate Eq. [19]:

$$\frac{dv}{dt} = Hv + \frac{dH}{dt} l$$

$$= \left(H^2 + \frac{dH}{dt} \right) l$$ [20]

It is customary to define a *deceleration parameter*,

$$q \equiv - \left(1 + \frac{1}{H^2} \frac{dH}{dt} \right)$$ [21]

in terms of which

$$\frac{dv}{dt} = - qH^2 l$$ [22]

Note that if the expansion is slowing down, then $q > 0$. If we compare Eqs. [18] and [22], we find

$$-qH^2 = -\frac{4\pi G}{3}\rho$$

that is,

$$q_0 = \frac{4\pi G}{3H_0{}^2}\rho_0 \tag{23}$$

where the subscript $_0$ indicates that the quantities are to be evaluated at the present time.

It is customary to define a *density parameter* (or *closure parameter*) Ω_0,

$$\Omega_0 = \frac{8\pi G}{3H_0{}^2}\rho_0 \tag{24}$$

in terms of which Eq. [23] becomes

$$q_0 = \tfrac{1}{2}\Omega_0 \tag{25}$$

As we will see in later sections, in principle, the value of the density parameter discriminates between models of the universe that continue to expand forever and models that come to a stop and then recollapse. The quantity $3H_0{}^2/8\pi G$ has the dimensions of a density, and is called the *critical density*. Its numerical value is

$$\rho_{crit} = \frac{3H_0{}^2}{8\pi G} = 1.88 \times 10^{-29}\,h^2\ \text{g/cm}^3 \tag{26}$$

Note that Eq. [23] is equally valid in Newton's theory and in Einstein's theory. The derivation of this equation depends on the absence of gravitational fields in a spherical cavity surrounded by a spherical mass distribution. According to Birkhoff's theorem (see Section 7.5), this is just as true in geometrodynamics as it is in Newtonian theory. Furthermore, if we assume that the cavity is of small dimension, then the velocity and acceleration will be small, and therefore the nonrelativistic equation of motion will be a valid approximation.

There is, however, one modification of Eq. [23] which we may want to consider. If the Einstein equations contain a cosmological constant, then we must take this into account. We know from Section 7.3 that the cosmological term has the effect of a uniform mass density $-\Lambda/4\pi G$ (see Eq. [7.51]). Thus, the mass density ρ in Eq. [18] must be replaced by $\rho - \Lambda/4\pi G$:

$$\frac{dv}{dt} = - \frac{4\pi G}{3} \left(\rho - \frac{\Lambda}{4\pi G} \right) l \qquad [27]$$

and our final result becomes

$$q_0 = \frac{\Omega_0}{2} - \frac{\Lambda}{3H_0{}^2} \qquad [28]$$

This equation will play an important role in the comparison of the theoretical models of the universe with observational data (see Section 9.11).

The direct observational determination of the deceleration parameter is very difficult. In principle, it is possible to determine q_0 by looking for systematic deviations from Hubble's law at large distances. Since we see remote galaxies as they were at the time when they emitted the light that reaches us now, such galaxies indicate the state of the universe a long time ago. If q_0 is, say, positive and the expansion rate has been decreasing, we expect remote galaxies to display redshifts in excess of those given by Hubble's linear law. (Actually, q_0 is not the only source for deviations from Hubble's linear law. Since the distance in this law is defined by the brightness of a standard light source, we must take into account the modification of the inverse-square law [6] relating brightness to distance in a curved spacetime, and also the decrease of the energy of the light brought about by the reduction of the frequency of the light wave. In general, for a given cosmological model, the distance can be expressed as a nonlinear function of z, in which q_0 enters as a parameter.)

In practice, because of the large uncertainties in the distance scale, it is not feasible to detect the systematic deviations from Hubble's linear law in a simple plot of redshift vs. distance. Several alternative methods for the determination of q_0 have been tried. One method relies on a plot of brightness of standard light sources, such as brightest galaxies, vs. redshift z. The brightness is inversely proportional to l^2; and, if there were no deviations from Hubble's law, l would be directly proportional to z; hence the brightness would be inversely proportional to z^2. The deviation from this inverse proportionality to z^2 depends on the cosmological model and on q_0. Astronomers have made several attempts at finding a value for q_0 by seeking such deviations in plots of brightness vs. z. Unfortunately, the deviation, if any, occurs only at large values of z, where we are looking at galaxies that emitted their light a long time ago. An evolution in the luminosity of galaxies with time then also contributes deviations. There is no general agreement about what, if any, corrections need to be made for such galactic evolution, and therefore all of the results that have been obtained for q_0 by this method are questionable. The only reliable conclusion that can be drawn from these attempts is that $|q_0|$ is no larger than 1 or 2.

If we assume that $\Lambda = 0$, then we do not need to determine q_0 directly from observational data. Instead, we can calculate it from the mass density, by means of Eq. [24]. Alternatively, we can calculate q_0 from the age of the universe and the value of H_0. In the Friedmann cosmological model with a given value of H_0, the value of q_0 is uniquely related to the age of the universe (the Friedmann model is uniquely characterized by *two* parameters).

9.4 THE AGE OF THE UNIVERSE

If we extrapolate the observed expansion of the universe backward in time, we find that the early universe must have been extremely dense and extremely hot. This suggests an explosive origin of the universe. It seems that initially all the matter was concentrated in a very dense fireball (the fireball had no preferred position in the universe; rather, it filled the entire universe). This fireball blew apart in a violent, primordial explosion, or *Big Bang*. As the fireball expanded and cooled, slight fluctuations in density grew into local condensations which evolved into gravitationally bound systems, such as galaxies and bound clusters of galaxies. These bound systems continued the average motion of expansion, away from each other. Thus, the general motion of recession of the galaxies is directly related to the initial velocities that these galaxies had when they were first formed.

We can use Hubble's law to estimate how long ago the universe began. Under the assumption that the recession velocity v in Eq. [9] remains constant for any given galaxy, the backward extrapolation of the motion of such a galaxy tells us that the time needed for complete collapse is $v/l = H_0^{-1}$. This time is called the *Hubble age*. The value of H_0 given in Eq. [13] implies a Hubble age

$$ H_0^{-1} = \frac{1}{100h} \text{ s·Mpc/km} = h^{-1} \times 9.78 \times 10^9 \text{ years} \qquad [29] $$

The true age of the universe is presumably somewhat less than the Hubble age, because the recession velocities in the past were somewhat larger than they are now. For instance, in a cosmological model of zero curvature (see Section 9.11), the actual age can be shown to be $\frac{2}{3}$ of the Hubble age. It is therefore of considerable interest to compare the Hubble age with estimates of the actual age as directly determined from observational data. There are several methods that permit us to estimate the actual age of the universe.

By radioactive dating methods, the age of solid bodies in the Solar System (rocks on Earth and Moon, meteorites) is known to be (4.6 ±

0.1) $\times$ 10^9 years (Wasserberg et al., 1977). But the age of solid bodies in the Solar System sets only a lower limit on the age of the universe. A better limit is set by the age of the chemical elements. Almost all elements, with the exception of hydrogen, some of the helium, and a few other light elements, were formed by nuclear reactions in the interior of stars. For instance, the uranium found on the Earth was produced by massive stars that lived, and died, before the Solar System was born. Astrophysical theory allows us to calculate the relative abundance with which the isotopes ^{235}U and ^{238}U must have been produced in the nuclear reactions (Cowan, Thielemann, and Truran, 1991):

$$[^{235}U/^{238}U]_i = 1.16 \qquad\qquad [30]$$

At present, the measured abundance ratio of these isotopes is

$$[^{235}U/^{238}U]_0 = 7.25 \times 10^{-3} \qquad\qquad [31]$$

The decrease of the relative abundance is due to the higher decay rate of the isotope ^{235}U as compared with the isotope ^{238}U; the decay rate of the former isotope is $1/(1.015 \times 10^9 \text{ yr})$; that of the latter is $1/(6.446 \times 10^9 \text{ yr})$.* If we assume that all of the uranium and thorium were produced suddenly and changed only by radioactive decay ever since, it is a simple matter to calculate that the time needed to change [30] into [31] is

$$t_0 = \frac{\ln[^{235}U/^{238}U]_i - \ln[^{235}U/^{238}U]_0}{1/1.015 - 1/6.446} \times 10^9 \text{ yr} = 6.1 \times 10^9 \text{ yr}$$

However, for a more accurate calculation, we must take into account that the production of the isotopes occurred gradually, and, for a while, the abundance was determined by the balance between generation and decay. This complicates the analysis considerably, and increases the age implied by Eqs. [30] and [31] by a factor of 2 or 3.

The ages of some other elements can be calculated by the same method, with similar results. These calculations place the age of elements in our Galaxy somewhere between 10×10^9 years and 20×10^9 years (Cowan, Thielemann, and Truran, 1991; Tonry, 1992).

The ages of globular star clusters in our Galaxy provide us with another estimate of the age of the universe. The stars in such a cluster were all born at once, presumably soon after the Galaxy formed. The stars burn hydrogen into helium by thermonuclear reactions; and, as

---

*   The numbers in the denominators are the mean lifetimes of these isotopes. The relation between half-life and the mean lifetime is [half-life] = 0.693 $\times$ [mean lifetime].

long as the supply of nuclear fuel lasts, they lead a placid, stable existence. For such hydrogen-burning stars, the luminosity and the surface temperature (or, equivalently, the spectral class) are directly related. The relationship is displayed by the Hertzsprung-Russell diagram of Fig. 9.8. The stable, hydrogen-burning stars fit on a smooth curve in this diagram, called the *main sequence*.

When the stars reach old age and run low on nuclear fuel, they depart from the main sequence in the Hertzsprung-Russell diagram. The first stars to depart are the most luminous, most massive stars at the upper end of the main sequence. Thus, as a globular cluster ages, its main sequence is progressively vacated from the top down (the main sequence "turns off" or "burns down"). From astrophysical calculations of stellar evolution we know how long a star of given mass remains on the main sequence, and we can therefore predict how far the main sequence will burn down as a function of time. By inspection of the Hertzsprung-Russell diagram of the globular cluster, we can identify the turn-off point of its main sequence, and we can deduce the age. This method has been applied to a number of globular clusters; some are old, some are young. For the oldest globular clusters, the estimates of the age lie somewhere between 15×10^9 and 18×10^9 years (Rood, 1990; Vandenberg, 1990; Sandage and Cacciari, 1990).

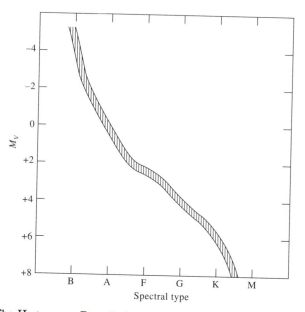

Fig. 9.8 The Hertzsprung-Russell diagram. (After Rowan-Robinson, 1985.)

The ages of white-dwarf stars provide us with another estimate of the age of the universe. These are burnt-out stars that have exhausted their nuclear fuel. The light radiated by these stars must therefore come from the heat stored in their bodies, and they must cool and become fainter as they age. The temperature and the luminosity decrease in a predictable way as a function of time. The oldest white dwarfs are the coolest and faintest. Temperature determinations of these coolest, faintest white dwarfs indicate an age of 8 to 12×10^9 years (Winget et al., 1987).

In principle, the ages of the stars and of the chemical elements give us only a lower limit for the age of the universe. The universe could be much older than the stars and the chemical elements. However, according to the standard model of the Big Bang, star formation and nucleosynthesis began soon ($\simeq 10^8$ years) after the Big Bang. Hence the ages of elements, globular clusters, and white dwarfs may be regarded as equal to the age of the universe, to within the stated uncertainties. The age of the universe arrived at by these observational data is consistent with the Hubble age, just as the Big Bang theory leads us to expect.

9.5 THE COSMIC THERMAL RADIATION

At a very early epoch, the dense and hot universe must have contained thermal blackbody radiation at a very high temperature. As the universe expanded, the density of this thermal radiation must have decreased, but the radiation should still be present, filling the interstellar and intergalactic space around us even now.

The expansion of the universe, and the expansion of the radiation, leads to a reduction of the temperature of the radiation. To see what happens to the thermal radiation in some expanding region of space, imagine that this region, of size much smaller than the size of the universe, is enclosed in a box with perfectly reflecting walls. Suppose that this box expands at the same rate as the universe. Under these assumptions, the radiation in the region inside the box will behave in the same way as the radiation in some region outside the box. The reason is that while in the case of a region surrounded by walls, photons are reflected and cannot escape, in the case of a region without walls, any photons that do escape are, on the average, simply replaced by similar photons that enter the region from the surrounding space.

It is now easy to calculate what happens to the radiation temperature in the expanding box. At temperature T, the number of photons in some normal mode of oscillation of the box is given by statistical mechanics as

$$n(\nu) = \frac{1}{e^{h\nu/kT} - 1} \tag{32}$$

where ν is the frequency of the mode. If the linear dimension of the box increases by a factor α, then the wavelength of the mode must increase by the same factor, and hence the frequency decreases to the value $\nu' = \nu/\alpha$. If the expansion proceeds sufficiently slowly (adiabatically), the number of photons remains constant as their frequency changes. Hence after expansion by the factor α, Eq. [32] will give the number of photons at the frequency ν':

$$n(\nu') = \frac{1}{e^{h\nu/kT} - 1} \tag{33}$$

$$= \frac{1}{e^{\alpha h\nu'/kT} - 1} \tag{34}$$

Obviously Eq. [34] still corresponds to a thermal distribution of photons, but with the new temperature

$$T' = T/\alpha \tag{35}$$

This shows that expansion by, say, a factor of 2 changes thermal radiation of a given temperature into thermal radiation of one-half the initial temperature.

In an attempt at a detailed calculation of nucleosynthesis in the early universe, Gamow and his collaborators (Alpher, Bethe, and Gamow, 1948; Gamow, 1948) estimated the temperature of the primordial fireball as $\simeq 10^9$ K at a time of $\simeq 200$ s after the Big Bang. From this Alpher and Herman (1948) estimated the present temperature of the remaining thermal radiation at $\simeq 5$ K (for details, see Section 10.4). Such a low temperature implies that the maximum in the spectral distribution of the radiation (Planck's law) is in the millimeter wavelength range, which is the wavelength range of microwaves.

Some early evidence for a nonzero temperature of interstellar space had been uncovered by astronomers (Adams, 1941; McKellar, 1941), by spectroscopic examination of interstellar clouds of cyanogen gas. But, at the time, this was not recognized as indicating the presence of microwave radiation. The first direct observational evidence for a uniform distribution of microwave radiation over the sky was found in 1965 by Penzias and Wilson with a very sensitive horn antenna originally constructed for a satellite communication system. This microwave radiation was identified as the cosmic fireball radiation by Dicke et al. (1965). These first measurements were carried out at a wavelength of several centimeters, and they gave a radiation temperature of about 3 K.

Since then many measurements of the intensity of the radiation have been performed at many different wavelengths. For the range of wavelengths between $\simeq 100$ and 0.3 cm, ground-based radiometers may be used, but for wavelengths below 0.3 cm the radiometers must be carried above the atmosphere by a rocket, balloon, or satellite because the atmosphere absorbs radiation at these wavelengths.

Ground-based measurements and rocket and balloon measurements between 1965 and 1989 indicated a Planck spectrum, but with some possible deviations at short wavelengths. All of these measurements were superseded by the new measurements performed with the COBE satellite (*Cosmic Background Explorer*) launched in 1989, which showed that the radiation has a perfect Planck spectrum, with a temperature of 2.736 K (see Fig. 9.9).

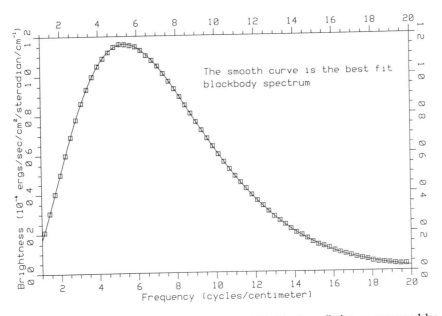

Fig. 9.9 The spectrum of the cosmic thermal blackbody radiation as measured by the COBE satellite. The smooth curve is the Planck spectrum for $T = 2.736$ K. (From Mather et al., 1990.)

The distribution of the radiation over the sky is nearly isotropic, but displays two large-scale systematic deviations. The first of these deviations is of the form $\cos\theta$ (a "dipole" term), which can be interpreted as resulting from the motion of the Earth relative to the blackbody radiation. Because of Doppler shifts, such a motion of the Earth increases the temperature of the radiation incident on the Earth from the front, and reduces the temperature of the radiation incident from behind. This results in a small additive perturbation in the temperature, of the form $\cos\theta$, where the angle θ is measured from the line of motion of the Earth. The observed temperature variation is about 5×10^{-3} K, and indicates a velocity of about 630 km/s. Note that this determination of the velocity of the Earth in no way contradicts the principle of relativity, since the measurement is not made relative to empty space, but relative to the photon gas in the blackbody radiation.

The other large-scale deviation from isotropy is concentrated along the Milky Way, and is attributed to radiation emitted by gas and by electrons (from ionized hydrogen) in our Galaxy (Fig. 9.10a; see color plate).

Many experimenters have sought to detect small-scale deviations from isotropy in the blackbody radiation. Such deviations, if any, reveal density fluctuations in the universe at an early time, when the photons in the radiation last interacted with the matter in the universe and began their unimpeded travel to the Earth. No deviations from isotropy have been detected on angular scales from 0.5 arcminute to 1^0; within this range of angular scales, the upper limit on deviations is about 10^{-4} K (Wilkinson, 1986). However, with a differential radiometer on the COBE satellite, Smoot et al. (1992) detected some deviations from isotropy on an angular scale of about 10^0. The deviations are of the order of 10^{-5} K (Fig. 9.10b; see color plate). The presence of these deviations was confirmed by measurements with balloon-borne radiometers (Ganga et al., 1993).

This means that the universe was not perfectly isotropic at the time the radiation was generated; but the deviations from isotropy were extremely small--they were smaller than 1 part in 10^4. Since the last interaction between a typical photon in the radiation and the normal matter (electrons and protons) occurred when the universe was about 200,000 years old, the distribution of such ordinary matter must have been very nearly isotropic at that time. This is the strongest argument for the overall isotropy (and homogeneity) of the universe.

9.6 THE MASS DENSITY; DARK MATTER

To apply Einstein's equations to the dynamics of the universe, we need to know the energy-momentum tensor of the universe and, in particular, the mass density. In view of the large-scale homogeneity of the universe, it will be sufficient to deal with the average mass density, that is, the mass density averaged over volumes of (100 Mpc)^3 or more.

Astronomers find it convenient to express the average mass density as a product of two factors, the average luminosity density and the average mass-to-light ratio for galaxies,

$$\rho_0 = [\text{luminosity density}] \times M/\mathscr{L} \qquad [36]$$

The luminosity density is obtained from a simple count of galaxies per unit volume and their luminosity. According to several surveys, the value of this factor is 1.0×10^8 $h\mathcal{L}_\odot/\text{Mpc}^3$ (Efstathiou, Ellis, and Peterson, 1988; the factor h enters here because it determines the distance scale).

To find the average mass-to-light ratio $M/\mathcal{L}$, we need to determine the masses of various kinds of galaxies. The masses of spiral galaxies, in which stars and gas move in nearly circular orbits about the galactic center, can be determined by measuring the orbital velocities of stars and gas as a function of radius ("rotation curve"). If the mass distribution of the galaxy can be approximated as spherical or ellipsoidal, the orbital velocity at a radius r is given by Newton's equation of motion

$$v^2r = GM(r) \qquad [37]$$

where $M(r)$ is the mass included within the sphere or the ellipsoid of radius r.* Thus, the mass of a galaxy can be determined by a measurement of the orbital velocity of a star or a cloud of gas at the outer edge of the galaxy. This orbital velocity can be readily measured by means of the Doppler shifts of spectral lines of the star or the cloud of gas. From such measurements, astronomers find that the visible, luminous portions of galaxies have mass-to-light ratios of the order of (Faber and Gallagher, 1979; Lauer, 1985)

$$M/\mathcal{L} = (10h \text{ to } 20h) \times M_\odot/\mathcal{L}_\odot \qquad [38]$$

If all of the mass of a galaxy were in its visible, luminous portions, then we would expect that orbital velocities beyond the visible outer edge of the galaxy should decrease according to $v^2 \propto 1/r$, since $M(r)$ would remain constant beyond the outer edge. However, measurements on isolated stars and clouds of gas in this region show that the orbital rotation velocity at and beyond the visible outer edge of the galaxy tends to a constant value, with no indication of a decrease with radius, not even at the largest radii for which data are available (see Fig. 9.11). Such a flat rotation curve indicates that there is extra, invisible, dark mass beyond the visible outer edge of the galaxy. The dark mass is believed to form a spherical halo around the galaxy. To give a flat, constant rotation curve, the mass of this halo must increase with radius, $M(r) \propto r$, and the density of the halo must therefore be proportional to $1/r^2$.

We do not know what the dark mass, or the "missing mass," in halos consists of. It has been conjectured that it could consist of black holes, cold white dwarfs, or brown dwarfs (bodies like Jupiter, too small to support thermonuclear reactions). Alternatively, it could consist of clouds of elementary particles, such as neutrinos (if the

---

*   The disk of a spiral galaxy can be approximated as a very flattened prolate ellipsoid, for which Newton's theorem is valid.

neutrino mass is nonzero), or some kind of exotic particles, such as axions, photinos, gravitinos, higgsinos, etc. We will consider some of these speculations regarding the nature of the dark matter in Section 10.5 (see Table 10.2).

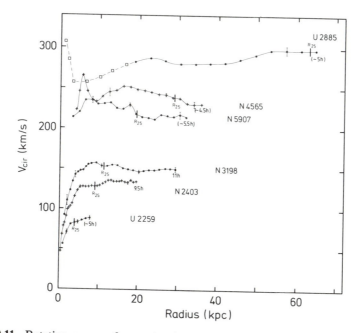

Fig. 9.11 Rotation curves of several spiral galaxies. These curves show the orbital velocities of clouds of hydrogen gas; the velocities were measured by the Doppler shifts of the 21-cm line. The vertical markers indicate the outer edges of the visible portions of these galaxies. (From Sancisi and van Albada, 1987.)

For our determination of the average mass density of the universe, we need the total masses of galaxies, including the masses of their halos. Rotation curves cannot be measured far enough outward beyond the edges of galaxies to detect the extent of the halos. However, we can determine the total masses of galaxies by measuring the velocities of galaxies in binary systems, in small groups, and in clusters. In such systems, the rms velocity of the galaxies and the mean inverse distance between the galaxies are related to the total mass by the *virial theorem* of statistical mechanics. If, for the sake of simplicity, we assume that all the galaxies in the system have equal masses, the virial theorem can be put in the form

$$\langle v^2 \rangle = \tfrac{1}{2}GM \langle r^{-1} \rangle \qquad [39]$$

This equation is a generalization of the well-known result that for a particle moving in a circular orbit in a $1/r$ potential, the kinetic energy equals one-half of the absolute value of the potential energy. In the case of binary systems or small groups containing only a few galaxies, this equation is valid only in a statistical sense, when averaged over an ensemble consisting of a large number of such systems.* In the case of clusters containing many galaxies, the equation is valid when averaged over an ensemble of clusters, but it is also valid for each individual cluster, provided the galaxies in the cluster have orbited around each other long enough for their velocities to have attained statistical equilibrium.

■ *Exercise 1.* Show that for a binary system consisting of two galaxies of equal masses $\frac{1}{2}M$ in circular orbits about each other, the virial theorem [39] is a direct consequence of Newton's equation of motion. ■

With the measured velocities and positions of galaxies, the virial theorem leads to mass-to-light ratios of about $100h\ M_{\odot}/\mathscr{L}_{\odot}$ in binary systems and about $300h\ M_{\odot}/\mathscr{L}_{\odot}$ in groups and in large, rich clusters, such as the Coma cluster (for recent reviews of mass determinations, see Davis, 1987; Mushotzky, 1991; and Peebles, 1993). Since most of the galaxies are found in groups, we can adopt $300h \times M_{\odot}/\mathscr{L}_{\odot}$ as a representative value of the mass-to-light ratio for the entire population of galaxies.

If we multiply this average mass-to-light ratio by the average luminosity density, we obtain the average mass density contributed by galaxies and the halos of dark matter associated with them:

$$\rho_0 = [\text{luminosity density}] \times M/\mathscr{L} \qquad [40]$$

$$= 1.0 \times 10^8\ h\ \mathscr{L}_{\odot}/\text{Mpc}^3 \times 300h\ M_{\odot}/\mathscr{L}_{\odot}$$

$$= 3 \times 10^{10}\ h^2 \times M_{\odot}/\text{Mpc}^3$$

$$= 2 \times 10^{-30}\ h^2\ \text{g/cm}^3 \qquad [41]$$

The corresponding density parameter is

$$\Omega_0 = \frac{\rho_0}{3H_0{}^2/8\pi G} = \frac{2 \times 10^{-30}\ h^2\ \text{g/cm}^3}{1.88 \times 10^{-29}\ h^2\ \text{g/cm}^3}$$

$$= 0.1 \qquad [42]$$

Note that the fudge factor h cancels in Eq. [42]; thus, the uncertainty in H_0 does not affect the density parameter (although it affects the

---

* Since the angle between the orbital plane and our line of sight is not known, it is also necessary to average over this angle.

density ρ_0; see Eq. [41]). However, the uncertainties in the mass determinations do affect Ω_0, and a rough estimate of the probable error is a factor of 2, which means that Ω_0 lies somewhere between 0.05 and 0.2.

Instead of applying the virial theorem to specific, identifiable groups or clusters, we can also apply it to the generic clumps in the galaxy distribution indicated by the correlation function $\xi(r)$. On a scale r, the excess density indicated by the correlation function is $\delta\rho_0 \simeq \xi(r)\rho_0$, and hence a clump of size r has a mass

$$M \simeq \frac{4\pi}{3} \xi(r)\rho_0 r^3 \qquad [43]$$

Assuming that the clump is bound, we can apply the virial theorem:

$$\langle v^2 \rangle \simeq \frac{GM}{r} \simeq \frac{4\pi}{3} \xi(r)\rho_0 r^2 \qquad [44]$$

This equation is called the *cosmic virial theorem*. A more rigorous analysis (Peebles, 1993) gives a slightly different numerical factor for the right side of Eq. [44], but confirms that this equation is qualitatively correct. According to this equation, the mass density ρ_0 can be calculated from the known correlation function and the measured rms value of the peculiar velocities in the clump. This cosmic virial theorem leads to a value $\Omega_0 \simeq 0.1$ (Efstathiou and Silk, 1983), which is consistent with the value we obtained above by applying the virial theorem to groups and clusters. This agreement is expected, since the correlation function is merely a mathematical representation of groups and clusters.

An entirely different determination of ρ_0 is based on the measured velocity of our Galaxy relative to the cosmic blackbody radiation. As mentioned in Section 9.5, our Galaxy, and our Local Group, has a velocity of 630 km/s relative to the cosmic blackbody radiation. This large peculiar velocity can be interpreted as an "infall" velocity, brought about by the gravitational attraction exerted by one or more mass concentrations in the sky, such as the Virgo cluster and the Hydra-Centaurus cluster. Note that the mass that causes such an "infall" velocity is the excess mass, that is, the mass in excess of the mean mass density in the universe. From galaxy counts, the fractional excess density in the Virgo cluster has been estimated as $\delta\rho_0/\rho_0 \simeq 1.4$. The component of our peculiar velocity in the direction of Virgo is 170 km/s, whereas our recession velocity relative to Virgo is 1200 km/s. Thus, since the beginning of the universe, the gravitational attraction of Virgo has produced a fractional velocity perturbation of $\delta v/v = 170/1200 \simeq 0.14$. From perturbation theory (Peebles, 1993), it can be shown that this velocity perturbation is directly proportional to the fractional density excess in Virgo, with a constant of proportionality that depends on Ω_0:

$$\frac{\delta v}{v} = \frac{1}{3}(\Omega_0)^{0.6} \frac{\delta \rho_0}{\rho_0} \qquad [45]$$

With the stated values of $\delta v/v$ and $\delta \rho_0/\rho_0$, this equation yields $\Omega_0 \simeq 0.14$.

The above determinations of the average mass density refer only to the luminous mass in galaxies and the dark mass in the halos of galaxies. Besides this, there could be extra dark mass uniformly distributed throughout the universe. The only observational limit on such a uniform dark mass distribution comes from the deceleration parameter q_0. According to Section 9.3, we may take this observational limit to be roughly $q_0 \leq 1$. With $\Lambda = 0$, Eq. [25] tells us that $\Omega_0 = 2q_0$, and hence the observational limit on Ω_0 is roughly $\Omega_0 \leq 2$.

A different upper limit on Ω_0 can be extracted from a detailed analysis of the process of nucleosynthesis in the early universe. As we will see in Section 10.4, the early universe cannot produce the right amount of deuterium unless $\Omega_0 h^2 \leq 0.010$, which implies $\Omega_0 \leq 0.06$ if h has its smallest possible value ($h = 0.4$). However, this upper limit applies only to the density of baryonic matter (protons and neutrons); it does not apply to the density of nonbaryonic matter, such as black holes, gravitational radiation, neutrinos, or exotic elementary particles.

Table 9.1 lists the known and conjectural contributions to the average mass density of the universe.

Theoretical considerations concerning the early universe strongly suggest that the most plausible value of Ω_0 is exactly 1 (see Section 10.6). If our universe actually has this value of Ω_0, then the upper limit on baryonic matter set by deuterium production implies that by far most of the matter in the universe must be something very different from the familiar form of matter we find in our immediate environment.

Besides the forms of matter listed in Table 9.1, the universe contains some amount of electromagnetic energy. For example, the cosmic blackbody radiation at 2.74 K contributes 4×10^{-34} g/cm^3, and radio waves, starlight (optical), and X rays contribute smaller amounts. Besides, there are cosmic rays with a density no larger than $\simeq 10^{-35}$ g/cm^3.

TABLE 9.1 THE AVERAGE MASS DENSITY IN THE
UNIVERSE

Form of matter	Ω_0
Luminous mass in galaxies	0.003 to 0.006
Luminous mass and dark mass in galaxies and halos	0.05 to 0.2
Baryons, upper limit from deuterium production	$0.010/h^2$
All forms, upper limit from $q_0 \leq 1$	2

9.7 COMOVING COORDINATES; THE ROBERTSON-WALKER GEOMETRY

In order to formulate the Einstein equations for the dynamics of the universe, we will ignore the small-scale deviations from uniformity. We want to concentrate on the large-scale features of the geometry, and we will assume that on the large scale the universe is isotropic and homogeneous. The mass in the universe can then be treated as a fluid, with a constant density throughout. The galaxies, or clusters of galaxies, may be regarded as the particles out of which this fluid is made. This means that we discount local irregularities in the density and the motion of galaxies and take into account only the motion of uniform expansion.

By appealing to the symmetry (homogeneity and isotropy) of the universe, we can simplify the spacetime interval and the metric tensor considerably. We begin by specifying the coordinates we will use: they will be *comoving* coordinates. By this is meant that the spatial coordinates partake of the uniform motion of expansion of the matter in the universe. If we ignore the small irregularities in the motion of galaxies (local deviations from uniform expansion), we may say that each galaxy carries its spatial coordinates with it; the coordinate points move with the galaxies as the latter fall freely in the background gravitational field of the universe. The coordinate interval between any two galaxies then remains forever constant, and the expansion of the universe results not from a change in the coordinate position of the galaxies, but rather from a change in the metric of spacetime.

For the time coordinate x^0 we will use the proper time measured by clocks carried by the galaxies. We will assume that these clocks not only run at the same rate, but also that they are synchronized. In general, synchronization is a rather tricky problem, but in a uniform universe a simple operational procedure for synchronization is available. An observer on a galaxy A sends out a flash of light when his clock reads time t_0; an observer on galaxy B also sends out a flash of light when her clock reads t_0. The flash from A is received at B when the clock at the latter point reads t_B; the flash from B is received at A when the clock there reads t_A. The clocks are then synchronized if $t_A = t_B$. This method depends on the uniformity of the universe; homogeneity ensures that both light signals will have the same travel time. An alternate synchronization method is also available: all observers set their clocks so they read zero at the instant of the Big Bang.

We can show that with the above choice of coordinates,

$$g_{00} = 1 \qquad\qquad [46]$$

$$g_{0k} = 0 \qquad\qquad [47]$$

This means that our comoving coordinates are time orthogonal (see Eq. [7.38]).

Eq. [46] follows from the assumption that the clocks used to measure x^0 are at rest in our comoving coordinates. For such a clock $dx^1 = dx^2 = dx^3 = 0$, and hence

$$d\tau^2 = g_{00}(dx^0)^2$$

This establishes Eq. [46] because, by definition of x^0, we have $dx^0 = d\tau$. To establish Eq. [47], we recall from Section 6.4 that a nonzero value of g_{0k} indicates a desynchronization of the clocks located at different positions. Since we have taken pains to synchronize the clocks at all positions in our universe, it follows that g_{0k} must be zero.

■ **Exercise 2.** Since we have chosen our coordinates so galaxies are forever at rest, we must check that this is consistent with the geodesic equation of motion. To see that a galaxy at rest at some initial time will remain at rest, consider the geodesic equation of motion for a particle at rest, with $dx^\mu/d\tau = (1, 0, 0, 0)$,

$$\frac{d^2x^k}{dt^2} = -\Gamma^k{}_{00} \qquad\qquad [48]$$

and show that

$$\Gamma^k{}_{00} = \tfrac{1}{2}g^{kl}(g_{l0,0} + g_{0l,0} - g_{00,l}) = 0 \quad ■ \qquad\qquad [49]$$

In view of Eqs. [46] and [47], the metric of spacetime reduces to

$$ds^2 = (dx^0)^2 - g_{kn}dx^k dx^n \qquad\qquad [50]$$

It is customary to write this as

$$ds^2 = dt^2 - dl^2 \qquad\qquad [51]$$

where $dt = dx^0$ is simply the time interval and

$$dl^2 \equiv - g_{kn} dx^k dx^n \tag{52}$$

is the distance interval. In order to emphasize the three-dimensional nature of the distance interval, we will write Eq. [52] as

$$dl^2 = {}^{(3)}g_{kn} dx^k dx^n \tag{53}$$

where

$$ {}^{(3)}g_{kn} \equiv - g_{kn} \tag{54}$$

The 3×3 tensor ${}^{(3)}g_{kn}$ describes the geometry of three-dimensional space at a given instant of time.

As a first step toward the construction of the geometry of space-time, we construct the geometry of three-dimensional space. This space is supposed to be homogeneous and isotropic: the geometry should not distinguish between different points or between different directions about a point. In order to discover the consequences of these symmetry requirements, let us concentrate on the curvature tensor ${}^{(3)}R_{mnsk}$ of the three-geometry.

It follows from the isotropy and homogeneity of the three-geometry that ${}^{(3)}R_{mnsk}$ must have the form

$$ {}^{(3)}R_{mnsk} = K({}^{(3)}g_{ms} {}^{(3)}g_{nk} - {}^{(3)}g_{mk} {}^{(3)}g_{ns}) \tag{55}$$

where K is some constant. We can justify Eq. [55] as follows: At a given point, introduce local geodesic coordinates so the metric becomes ${}^{(3)}g'_{mk} = \delta_m^k$. The isotropy of space demands that it must not be possible to distinguish between different directions by their curvature; that is, the curvature tensor ${}^{(3)}R'_{mnsk}$ must be unchanged by rotations of the geodesic coordinates. Since the unit tensor δ_k^s is the only tensor unchanged by rotations, the curvature tensor must be some combination of unit tensors,

$$ {}^{(3)}R_{mnsk} = K\delta_m^s \delta_n^k + K_1 \delta_m^k \delta_n^s + K_2 \delta_m^n \delta_k^s \tag{56}$$

The antisymmetry relation ${}^{(3)}R_{mnsk} = - {}^{(3)}R_{mnks}$ requires that $K_1 = -K$ and $K_2 = 0$; hence,

$$ {}^{(3)}R_{mnsk} = K(\delta_m^s \delta_n^k - \delta_m^k \delta_n^s) \tag{57}$$

Transformation of this equation from geodesic coordinates back to the original coordinates gives the form [55]. The quantity K must be a constant (independent of position) in order to satisfy the requirement of homogeneity. This conclusion also follows from isotropy: if there were some point at which the gradient of K is different from zero, then the direction of this gradient would define a preferred direction in space, in contradiction with isotropy. Thus, isotropy at all points implies that K is a constant, which means it implies homogeneity.

■ *Exercise 3.* Show that

$$^{(3)}R_{ns} = -2K\,^{(3)}g_{ns} \tag{58}$$

$$^{(3)}R = -6K \quad ■ \tag{59}$$

Since the curvature tensor $^{(3)}R_{mnsk}$ can be expressed in terms of the metric $^{(3)}g_{kn}$ and the first and second derivatives of this metric, Eq. [55] may be regarded as a differential equation for $^{(3)}g_{kn}$. We want to find a solution of this equation; for this purpose we consider the cases of positive, negative, and zero curvature separately.

(i) Positive Curvature. Rather than try to solve Eq. [55] by brute force, we will exploit geometrical arguments. First consider the analogous two-dimensional problem. What two-dimensional space has uniform (that is, homogeneous and isotropic) curvature? Obviously, the surface of an ordinary sphere has this property; in mathematics, this surface is usually called a two-sphere.*

Correspondingly, to obtain a three-dimensional space of uniform curvature, we take the surface of a four-dimensional hypersphere; this surface is called a three-sphere. The equation of the surface of a four-dimensional hypersphere is, in rectangular coordinates,

$$(x^1)^2 + (x^2)^2 + (x^3)^2 + (x^4)^2 = a^2 \tag{60}$$

where a is the "radius" of the sphere. The distance between any two nearby points on the surface is

$$dl^2 = (dx^1)^2 + (dx^2)^2 + (dx^3)^2 + (dx^4)^2 \tag{61}$$

Note that the coordinate x^4 has nothing to do with time; it is simply an extra, unphysical, coordinate that must be introduced if we want to pretend that the curved three-dimensional space is a subspace of a flat Euclidean space.**

Eq. [60] can be used to eliminate the unphysical coordinate x^4. This yields a distance interval

---

* Flat two-dimensional space is a special case of a two-sphere of infinite radius.

** The three-dimensional space is *embedded* in the four-dimensional space (compare Exercise 8.7).

$$dl^2 = (dx^1)^2 + (dx^2)^2 + (dx^3)^2 + \frac{(x^1dx^1 + x^2dx^2 + x^3dx^3)^2}{a^2 - (x^1)^2 - (x^2)^2 - (x^3)^2} \quad [62]$$

In this expression there appear only the physical coordinates x^1, x^2, x^3. This expression gives us the desired metric for a space of uniform curvature.

We can easily check that the curvature tensor for the distance interval [62] does satisfy Eq. [55] with

$$K = \frac{1}{a^2} \quad [63]$$

■ *Exercise 4.* Check this. [Hint: It is sufficient to check Eq. [63] near the origin; since the geometry described by Eq. [62] is uniformly curved (by construction), what holds at one point must hold at all. In the vicinity of the origin it is sufficient to approximate

$$^{(3)}g_{kn} \simeq \delta_k^n + x^k x^n / a^2 \quad] \quad ■ \quad [64]$$

A geometry with a positive value of K is said to have *positive curvature*. To study this geometry further, we begin by introducing "polar" coordinates in the usual way:

$$x^1 = r \sin\theta \cos\phi$$

$$x^2 = r \sin\theta \sin\phi$$

$$x^3 = r \cos\theta \quad [65]$$

In terms of these coordinates, the distance interval [62] takes the form

$$dl^2 = \frac{dr^2}{1 - r^2/a^2} + r^2 d\theta^2 + r^2 \sin^2\theta \, d\phi^2 \quad [66]$$

Note that for each set of values of the "rectangular" coordinates x^k and for each value of the "radial" coordinate r there are actually two distinct points on the three-sphere (the coordinates do double duty). If we start at the "top" of the three-sphere ($x^4 = a$), move in the x^1 direction, and continue straight ahead, we reach the equator ($x^4 = 0$), then the bottom of the sphere ($x^4 = -a$). If we continue straight ahead, we again pass through the equator and finally return to the top. The values of x^1 for the top, the first equatorial crossing, the bottom, and the second equatorial crossing are, respectively, $x^1 = 0$, $x^1 = a$, $x^1 = 0$, and $x^1 = -a$; and the values of r are, respectively, $r = 0$, $r = a$, $r = 0$, and $r = a$ (see Fig. 9.12). That there are two distinct points with the same values of x^1 and r need not bother us too much; we can resolve the ambiguity by, say, writing coordinates for points in the upper hemisphere in red ink, and points in the lower in blue.

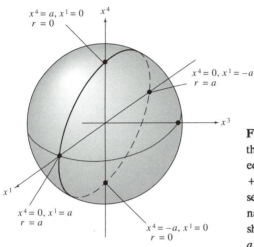

Fig. 9.12   Section $x^3 = 0$ through the three-sphere. The equation for this section is $(x^1)^2 + (x^2)^2 + (x^4)^2 = a^2$; thus, this section is the surface of an ordinary sphere. The heavy line shows a path that goes from $x^4 = a$ to $x^4 = -a$ to $x^4 = a$.

To gain some feeling for our space of positive curvature, let us look at the radius and the circumference of a circle placed in this space. For convenience, take the circle defined by $r =$ [constant] around the origin. This circle has as radius the distance between 0 and r,

$$l = [\text{radius}] = \int_0^r \frac{dr'}{\sqrt{1 - r'^2/a^2}} = a \sin^{-1}\frac{r}{a} \qquad [67]$$

The circumference of the circle is obtained in the usual way. For example, if the circle is in the plane $\theta = \pi/2$,

$$[\text{circumference}] = \int_0^{2\pi} r \sin\theta \, d\phi = 2\pi r \qquad [68]$$

The ratio of radius to circumference is therefore *larger* than $1/2\pi$; this is a familiar property of spaces of positive curvature. Note that for a radius larger than $\pi a/2$, the circumference *decreases* as the radius increases.

The surface of the sphere $r =$ [constant] surrounding the origin is

$$[\text{area}] = \int_0^{2\pi} \int_0^{\pi} r^2 \sin\theta \, d\theta d\phi = 4\pi r^2 \qquad [69]$$

Hence the ratio of the radius squared to the area is larger than $1/4\pi$. For a radius larger than $\pi a/2$, the area decreases as the radius increases.

The volume inside the sphere $r = [\text{constant}]$ is

$$[\text{volume}] = \int_0^{2\pi} \int_0^{\pi} \int_0^{r} \frac{r'^2}{\sqrt{1 - r'^2/a^2}} \, dr' \sin\theta \, d\theta d\phi$$

$$= 4\pi \left[\frac{a^3}{2} \sin^{-1} \frac{r}{a} - \frac{ra^2}{2} \sqrt{1 - r^2/a^2} \right] \qquad [70]$$

■ *Exercise 5.* Show that if $r \ll a$, then Eq. [70] gives a volume $4\pi r^3/3$, and explain why this makes sense. ■

To obtain the total volume of the three-sphere we must take $r = 0$ and $\sin^{-1} r/a = \pi$ ("bottom" of sphere). The total volume is then $2\pi^2 a^3$. Our three-sphere is a *closed space*; it has a finite volume even though it has no boundaries.

It is sometimes convenient to replace the radial coordinate r by a new "angular" coordinate χ such that

$$r = a \sin \chi, \qquad 0 < \chi < \pi \qquad [71]$$

The coordinate χ has an advantage over the coordinate r: it stands in unique correspondence to the points of space. In terms of χ, the distance inverval becomes

$$dl^2 = a^2[d\chi^2 + \sin^2\chi \, (d\theta^2 + \sin^2\theta \, d\phi^2)] \qquad [72]$$

In this expression, a plays the role of an overall scale factor that characterizes the distances in three-geometry. For instance, according to Eq. [67], the radial distance from the origin to the point χ is simply

$$l = [\text{radial distance}] = a\chi \qquad [73]$$

(ii) Negative Curvature. Next we want to find a homogeneous and isotropic geometry of negative curvature ($K < 0$). The spacetime interval for such a geometry can be obtained from Eq. [66] by simply replacing a^2 by $-a^2$:

$$dl^2 = (dx^1)^2 + (dx^2)^2 + (dx^3)^2 + \frac{(x^1 dx^1 + x^2 dx^2 + x^3 dx^3)^2}{-a^2 - (x^1)^2 - (x^2)^2 - (x^3)^2} \quad [74]$$

The geometrical arguments that precede Eq. [66] are, of course, not applicable when this replacement is made.

Since the metric tensor corresponding to Eq. [66] satisfies the condition [55] on the curvature tensor, the metric tensor corresponding to Eq. [74] will also satisfy this condition, the algebra being the same in both cases. The only difference is that the value of K is now negative,

$$K = -\frac{1}{a^2} \quad [75]$$

In "polar" coordinates, Eq. [74] becomes

$$dl^2 = \frac{dr^2}{1 + r^2/a^2} + r^2 d\theta^2 + r^2 \sin^2\theta \, d\phi^2 \quad [76]$$

The radius of a circle r = [constant] around the origin is

$$l = [\text{radius}] = a \sinh^{-1}\frac{r}{a} \quad [77]$$

This makes the ratio of radius to circumference smaller than $1/2\pi$. Likewise, for a sphere, the ratio of radius squared to area is smaller than $1/4\pi$. The volume of a sphere with r = [constant] is

$$[\text{volume}] = 4\pi\left(-\frac{a^3}{2}\sinh^{-1}\frac{r}{a} + \frac{ra^2}{2}\sqrt{1 + r^2/a^2}\right) \quad [78]$$

As $r \to \infty$, this volume diverges. The space of negative curvature is open and infinite.*

We can introduce a dimensionless coordinate χ such that

$$r = a \sinh \chi \quad [79]$$

In terms of this coordinate,

$$dl^2 = a^2[d\chi^2 + \sinh^2\chi \, (d\theta^2 + \sin^2\theta \, d\phi^2)] \quad [80]$$

This is the analog of Eq. [72].

---

* This is the "natural" topology for this space, suggested by the infinite range of values of r. We could try to make the space closed and finite by postulating that points are identical within periodically repeating intervals (toroidal topology); but this would be just as unnatural as imposing such a topology on flat spacetime.

(iii) Zero Curvature. Obviously, Eq. [55] also has the trivial solution $K = 0$ and $^{(3)}R_{mnks} = 0$. This corresponds to flat, Euclidean space. At first sight, such a flat geometry seems of no interest since we certainly expect that the matter distribution in the universe produces some curvature. However, we must remember that matter generates curvature in the four-geometry; it need not generate curvature in the three-geometry. As we will see, it is possible to construct solutions of Einstein's equations such that the three-geometry is flat at each instant of time, but the four-geometry is curved.*

The flat space is of course open and infinite.

Table 9.2 summarizes the properties of our homogeneous, isotropic three-geometries of constant curvature.

For these three-geometries we can now construct the four-geometry, according to $ds^2 = dt^2 - dl^2$. For the cases of positive and negative curvature this leads to, respectively,

$$ds^2 = dt^2 - \frac{dr^2}{1 - r^2/a^2} + r^2 d\theta^2 + r^2 \sin^2\theta \, d\phi^2 \qquad [81]$$

$$ds^2 = dt^2 - \frac{dr^2}{1 + r^2/a^2} + r^2 d\theta^2 + r^2 \sin^2\theta \, d\phi^2 \qquad [82]$$

TABLE 9.2 HOMOGENEOUS, ISOTROPIC THREE-GEOMETRIES

	Curvature	Distance interval*	Radial distance	Three-volume
(i) Positive curvature	$K = \dfrac{1}{a^2}$	$dl^2 = \dfrac{dr^2}{1 - r^2/a^2} + r^2 d\Omega^2$	$l = a\sin^{-1}r/a$	finite, closed
(ii) Negative curvature	$K = -\dfrac{1}{a^2}$	$dl^2 = \dfrac{dr^2}{1 + r^2/a^2} + r^2 d\Omega^2$	$l = a\sinh^{-1}r/a$	infinite, open
(iii) Zero curvature	$K = 0$	$dl^2 = dr^2 + r^2 d\Omega^2$	$l = r$	infinite open

* $d\Omega^2 = d\theta^2 + \sin^2\theta \, d\phi^2$

---

* Note that this depends on the choice of coordinates. If we introduce some new coordinates that mix the old space and time coordinates, then the three-geometry in the new coordinates will in general not be flat.

These spacetime intervals define the *Robertson-Walker* geometries of positive and of negative curvature. These spacetime intervals are often written in the form

$$ds^2 = dt^2 - a^2 \left[\frac{dr^2/a^2}{1 \mp r^2/a^2} + \frac{r^2}{a^2}d\theta^2 + \frac{r^2}{a^2} \sin^2\theta \; d\phi^2 \right] \qquad [83]$$

where the term in parentheses depends only on the dimensionless quantity r/a.*

Although the three-dimensional geometry, of positive or negative curvature, can be described equally well by the coordinates x^1, x^2, x^3, or r, θ, ϕ, or χ, θ, ϕ, these coordinates are not equally good for describing the expanding universe. Of these three sets of coordinates, only χ, θ, ϕ may be regarded as comoving with the galaxies. The reason is that isotropy and homogeneity of the expansion require that if $dl(t)$ is the distance between two galaxies at time t, then the distance at time $t + \Delta t$ must be proportional to the initial distance,

$$dl(t + \Delta t) = f(t)dl(t) \qquad [84]$$

where the proportionality factor $f(t)$ depends on time, but not on the position in space. Eq. [84] simply says that when the distance between a pair of galaxies increases by a factor of f, then the distance between any other pair also increases by the same factor.

In the expressions [62], [66], and [72] for the distance interval, the only quantity that can change in time is a; an expanding universe will have a time-dependent value of a. Only in the last of these expressions does a enter in such a way that Eq. [84] is satisfied. Thus, the coordinates χ, θ, ϕ may be regarded as comoving coordinates with respect to which the galaxies are at rest. Note that χ, θ, ϕ are not the only possible comoving coordinates. Any time-independent functions of χ, θ, ϕ can be regarded as comoving coordinates; for instance, r/a (= $\sin \chi$), θ, ϕ are comoving coordinates, and they are sometimes used instead of χ, θ, ϕ.

9.8 THE FRIEDMANN MODELS

The time dependence of a is determined by the Einstein equations. We will solve these equations for the separate cases of the Robertson-Walker geometries of positive, negative, and zero (spatial) curvature.

---

*   Many textbooks write the Robertson-Walker line element as $ds^2 = dt^2 - a^2[dr^2/(1 \mp r^2) + r^2d\theta^2 + r^2 \sin^2\theta \; d\phi^2]$. But the "$r$" in this way of writing the line element is actually the dimensionless coordinate r/a.

In order to write down the Einstein equations, we must know the energy-momentum tensor of the matter in the universe. We will assume that the energy-momentum tensor in the universe today is that of a uniform gas with zero pressure. The galaxies may be regarded as the "particles" out of which this gas is made, and since the velocities of the galaxies do not deviate much from uniform expansion, we can neglect the "pressure" of the gas of galaxies (in general, kinetic pressure is an indication of an rms deviation of the velocities from uniform velocity). However, in the very early, very hot universe, the pressure generated by radiation and by elementary particles must have been very important; we will explore this in the next chapter.

We will call the general homogeneous, isotropic models of the universe with mass density, pressure, and, possibly, a cosmological term the *Friedmann-Lemaître models*. The special models with zero pressure and zero cosmological constant are the Friedmann models, whereas models with a nonzero cosmological constant are the Lemaître models.* We begin with a study of the Friedmann models.

(i) Positive Curvature and $\Lambda = 0$*.* In the coordinates χ, θ, ϕ, the spacetime interval for the Robertson-Walker geometry of positive curvature is

$$ds^2 = dt^2 - a^2(t)[d\chi^2 + \sin^2\chi \, (d\theta^2 + \sin^2\theta \, d\phi^2)] \qquad [85]$$

The solution of the Einstein equations becomes simplest if we replace the time coordinate t by a time parameter η defined by

$$dt = ad\eta \qquad [86]$$

Then

$$ds^2 = a^2(\eta)[d\eta^2 - d\chi^2 - \sin^2\chi \, (d\theta^2 + \sin^2\theta \, d\phi^2)] \qquad [87]$$

and the metric tensor is

$$g_{\mu\nu} = \begin{pmatrix} a^2 & 0 & 0 & 0 \\ 0 & -a^2 & 0 & 0 \\ 0 & 0 & -a^2\sin^2\chi & 0 \\ 0 & 0 & 0 & -a^2\sin^2\chi \, \sin^2\theta \end{pmatrix} \qquad [88]$$

Some of the Christoffel symbols are easy to calculate:

---

* There is no general agreement on the naming of cosmological models; the terminology adopted here is as good as any.

$$\Gamma^0{}_{00} = \frac{\dot{a}}{a} \quad \Gamma^0{}_{kn} = -\frac{\dot{a}}{a^2} g_{kn} \quad \Gamma^k{}_{0n} = \frac{\dot{a}}{a} \delta^k_n$$

$$\Gamma^0{}_{k0} = 0 \quad \Gamma^k{}_{00} = 0 \qquad [89]$$

where

$$\dot{a} \equiv \frac{da}{d\eta} \qquad [90]$$

Here the dot stands for the derivative with respect to the time parameter η, contrary to the usage of Chapter 7 (see Eqs. [7.102]-[7.105]),
where the dot stood for the derivative with respect to proper time.
These derivatives are related as follows:

$$\frac{da}{d\tau} = \frac{da}{dt} = \frac{1}{a}\frac{da}{d\eta} = \frac{1}{a}\dot{a}$$

■ *Exercise 6.* Check the results given in Eq. [89]. ■

The R_{00} component of the Ricci tensor is entirely determined by
Eq. [89]:

$$R_{00} = \frac{3}{a^2}(a\ddot{a} - \dot{a}^2) \qquad [91]$$

The R_{kn} components can be easily evaluated by relating them to
$^{(3)}R_{kn}$; this saves some labor. We begin with the definition

$$R_{kn} = R^{\alpha}{}_{kn\alpha} = R^0{}_{kn0} + R^l{}_{knl} \qquad [92]$$

The term $R^0{}_{kn0}$ involves only Christoffel symbols of the type [89]. The
term $R^l{}_{knl}$ equals

$$R^l{}_{knl} = -\Gamma^l{}_{kn,l} + \Gamma^l{}_{kl,n} + (\Gamma^0{}_{kl}\Gamma^l{}_{0n} + \Gamma^s{}_{kl}\Gamma^l{}_{sn}) - (\Gamma^0{}_{kn}\Gamma^l{}_{0l} + \Gamma^s{}_{kn}\Gamma^l{}_{sl})$$
$$[93]$$

In this expression, the Christoffel symbols with one zero index are
known from Eq. [89]; the remaining symbols, with purely spatial indices, add up to the Ricci tensor of three-space:

$$R^l{}_{knl} = {}^{(3)}R^l{}_{knl} + \Gamma^0{}_{kl}\Gamma^l{}_{0n} - \Gamma^0{}_{kn}\Gamma^l{}_{0l} \qquad [94]$$

If we insert

$$^{(3)}R^l_{\ knl} = \frac{1}{a^2}\ ^{(3)}g^{lm}[^{(3)}g_{mn}\ ^{(3)}g_{kl} - {}^{(3)}g_{ml}\ ^{(3)}g_{kn}] \qquad [95]$$

according to Eq. [55], then everything on the left side of Eq. [94] is determined. The result is

$$R_{kn} = \frac{1}{a^4}(2\dot{a}^2 + \dot{a}^2 + a\ddot{a})g_{kn} \qquad [96]$$

The curvature scalar is then given by

$$R = R^0_{\ 0} + R^k_{\ k} = g^{00}R_{00} + g^{kn}R_{kn} = \frac{6}{a^3}(a + \ddot{a}) \qquad [97]$$

The Einstein equation, including the cosmological constant, is

$$R_\mu^{\ \nu} - \tfrac{1}{2}\delta_\mu^\nu R = -8\pi G T_\mu^{\ \nu} - \Lambda\delta_\mu^\nu \qquad [98]$$

We will concentrate on the $_{00}$ component of this equation,

$$R_0^{\ 0} - \tfrac{1}{2}R = -8\pi G T_0^{\ 0} - \Lambda \qquad [99]$$

With Eqs. [91] and [97] this becomes

$$-\frac{3}{a^4}(a^2 + \dot{a}^2) = -8\pi G T_0^{\ 0} - \Lambda \qquad [100]$$

Note that in this equation $\ddot{a}$ has canceled; we have to solve only a first-order differential equation. This equation suffices to determine $a(t)$ if $T_0^{\ 0}$ is given. The other components of the Einstein equation can be ignored because it turns out that they tell us nothing different from Eq. [100].

For a gas of galaxy "particles" without pressure, the energy-momentum tensor is

$$T_\mu^{\ \nu} = \rho u_\mu u^\nu \qquad [101]$$

where ρ is the proper mass density. In our comoving coordinates, the galaxies are at rest and hence

$$T_0^{\ 0} = \rho \qquad [102]$$

■ *Exercise 7.* Derive Eq. [102]. (Hint: Show that $u^0 = d\eta/d\tau = 1/a$ and $u_0 = a$.)
■

The volume of the universe varies in time as a^3; hence the density varies as a^{-3}, and we can write

$$\rho(t) = \frac{M}{2\pi^2 a^3} \qquad [103]$$

where M is a constant. Since $2\pi^2 a^3$ is the total volume of the universe, we find it convenient to say that M is the total "mass of the universe." Of course, from an operational point of view, it is meaningless to talk of the mass of the universe: in order to measure the mass of a system we would have to place ourselves outside of the system and measure how the system reacts to pushes and pulls; this we cannot do if the system is the whole universe. Furthermore, according to Eq. [103], M is really the sum of all the *proper* masses of the particles in the universe; we know from special relativity that such a sum will not give the total mass since it fails to take into account kinetic and potential energies.

Our general dynamical equation [100] now becomes

$$\frac{3}{a^4}(\dot{a}^2 + a^2) = 8\pi G\rho + \Lambda$$

$$= \frac{4GM}{\pi a^3} + \Lambda \qquad [104]$$

In the Friedmann models, it is assumed either that the cosmological constant is exactly zero, or else that it is so small that it can be neglected compared with ρ. Under this assumption, Eq. [104] reduces to

$$\frac{3}{a^4}(\dot{a}^2 + a^2) = \frac{4GM}{\pi a^3} \qquad [105]$$

This is the differential equation that describes the Friedmann model of positive curvature.

The equation of motion [105] can also be derived without making explicit use of Einstein's equations. In Section 9.3 we gave a simple argument which led to the following equation of motion for the distance between galaxies (see Eq. [27]):

$$\frac{d^2l}{dt^2} = \left(-\frac{Gm}{l^2} + \frac{\Lambda}{3}l \right) \qquad [106]$$

If we multiply this by dl/dt, we obtain

$$\frac{dl}{dt}\frac{d^2l}{dt^2} = \frac{dl}{dt}\left(-\frac{Gm}{l^2} + \frac{\Lambda}{3}l\right) \qquad [107]$$

that is,

$$\frac{1}{2}\frac{d}{dt}\left(\frac{dl}{dt}\right)^2 = \frac{d}{dt}\left(\frac{Gm}{l} + \frac{\Lambda}{6}l^2\right) \qquad [108]$$

The integral of Eq. [108] is

$$\frac{1}{2}\left(\frac{dl}{dt}\right)^2 + [\text{constant}] = \frac{4\pi}{3}G\rho l^2 + \frac{\Lambda}{6}l^2 \qquad [109]$$

where we have made the substitution $m = 4\pi l^3\rho/3$.

To compare this with Eq. [105], we note that distances in the universe vary as $a(t)$, that is, $l(t) \propto a(t)$. Furthermore, by Eqs. [86] and [90],

$$\frac{dl}{dt} \propto \frac{da}{dt} = \frac{da}{d\eta}\frac{d\eta}{dt} = \frac{\dot{a}}{a} \qquad [110]$$

which results in

$$\frac{1}{2}\left(\frac{\dot{a}}{a}\right)^2 + [\text{constant}] = \frac{4\pi G}{3}\rho a^2 + \frac{\Lambda}{6}a^2 \qquad [111]$$

This equation has the same form as Eq. [105], except that the value of the constant is left undetermined.*

The solution of Eq. [105] is

$$a = a_*(1 - \cos\eta) \qquad [112]$$

where

$$a_* = \frac{2GM}{3\pi} \qquad [113]$$

By integrating Eq. [86], we can then find the connection between the time t and the time parameter η,

---

* In the context of Newtonian physics we cannot determine the constant in Eq. [111], because we cannot measure $a(t)$ itself, but only the ratio of the values of $a(t)$ at two different times, that is, the factor by which distances expand in some time interval. In Einstein's theory we can measure $a(t)$ itself, by the curvature of space.

$$t = a_*(\eta - \sin \eta) \qquad [114]$$

Eqs. [112] and [114] may be regarded as parametric equations for $a(t)$. Fig. 9.13 shows a plot of a as a function of t. The curve given by this plot is a cycloid. At $t = 0$, $\pm 2\pi a_*$, $\pm 4\pi a_*,\ldots$, etc., $a(t)$ vanishes; that is, the universe contracts to a point. Since the density will become very large when this is about to happen, our approximate expression for the energy-momentum tensor will fail. For example, it will obviously be necessary to take into account the pressure exerted by matter and radiation in the dense, hot universe. We should also keep in mind the possibility that the classical Einstein equation becomes inapplicable at very high densities; quantum effects will be one obvious source of difficulties. It is therefore not clear exactly what happens at the singular points of Fig. 9.13, and we do not know whether the universe actually has the periodic behavior suggested by this figure.

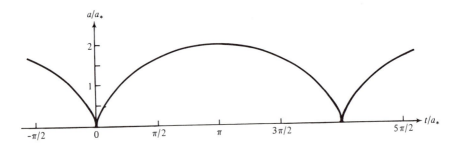

Fig. 9.13 Radius of curvature of the positive-curvature Friedmann universe as a function of time. The curve is a cycloid.

We will concentrate on one expansion-contraction cycle, with the Big Bang at $t = 0$, the moment of maximum expansion at $t = \pi a_*$, and total recollapse at $t = 2\pi a_*$.

(ii) Negative Curvature and $\Lambda = 0$. For the Robertson-Walker geometry of negative curvature, a calculation similar to that carried out in Eqs. [85]-[104] leads to an equation of motion

$$\frac{3}{a^4}(\dot{a}^2 - a^2) = 8\pi G\rho + \Lambda$$

$$= \frac{4GM}{\pi a^3} + \Lambda \qquad [115]$$

Here M is defined in exactly the same way as in Eq. [103],

$$\rho = \frac{M}{2\pi^2 a^3} \qquad [116]$$

Since the volume of the space is infinite, M cannot be interpreted as a total mass; but we may say that M is the sum of rest masses contained in a volume $2\pi^2 a^3$.

■ *Exercise 8.* Derive Eq. [115]. ■

Again, we introduce the assumption that Λ is zero, so

$$\frac{3}{a^4}(\dot{a}^2 - a^2) = \frac{4Gm}{\pi a^3} \qquad [117]$$

The model of the universe described by Eq. [117] is called the Friedmann model of negative curvature.

The solution of Eq. [117] is

$$a = a_*(\cosh \eta - 1) \qquad [118]$$

and it then follows that

$$t = a_*(\sinh \eta - \eta) \qquad [119]$$

Fig. 9.14 shows a as a function of t. The universe begins with a Big Bang and continues to expand forever. As $t \to \infty$, the universe gradually becomes flat. Again, the state near the singularity at $t = 0$ is not adequately described by our equations; some of the required modifications are understood, but the closer we approach $t = 0$, the less we understand.

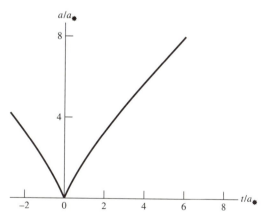

Fig. 9.14 Radius of curvature of the negative-curvature Friedmann universe as a function of time. The plot includes a continuation to negative values of t. Note that for small values of t, the behavior is similar to that shown in Fig. 9.13.

Fig. 9.15 Radius of curvature of the flat Friedmann universe as a function of time. The plot includes a continuation to negative values of t. The behavior at small values of t is, again, similar to that shown in Fig. 9.13.

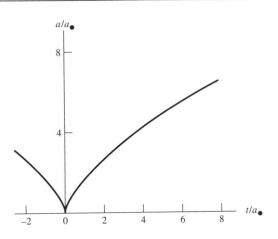

■ **Exercise 9.** Use Eqs. [118] and [119] to show that the universe becomes flat as $t \to \infty$. ■

(iii) Zero Curvature and $\Lambda = 0$. In the case of a universe with a flat three-geometry, it is convenient to introduce a dimensionless radial coordinate $\chi = r/a$. With this coordinate, we can write the spacetime interval in a form analogous to the other cases:

$$ds^2 = a^2(\eta)(d\eta^2 - d\chi^2 - \chi^2\, d\theta^2 - \chi^2 \sin^2\theta\, d\phi^2) \qquad [120]$$

so

$$g_{\mu\nu} = \begin{pmatrix} a^2 & 0 & 0 & 0 \\ 0 & -a^2 & 0 & 0 \\ 0 & 0 & -a\chi^2 & 0 \\ 0 & 0 & 0 & -a^2\chi^2\sin^2\theta \end{pmatrix}$$

Note that the spacetime interval [120] differs from flat spacetime only by the overall factor $a(\eta)$. A spacetime geometry that differs from the flat spacetime geometry only by such an overall factor is said to be *conformally flat*.

The equation of motion for $a(\eta)$ can then be derived as in Eqs. [88]-[105] with the result

$$\frac{3}{a^4}\, \dot{a}^2 = 8\pi G\rho + \Lambda \qquad [121]$$

where M is defined as before (see Eq. [103]).

In the Friedmann case, $\Lambda = 0$. With $\dot{a} = a\, da/dt$, Eq. [121] then reduces to

$$\left(\frac{da}{dt}\right)^2 = \frac{4GM}{3\pi} \frac{1}{a} \qquad [122]$$

which has the solution

$$a(t) = \left(\frac{3GM}{\pi}\right)^{1/3} t^{2/3} \qquad [123]$$

This function is plotted in Fig. 9.15. As $t \to \infty$, the four-geometry tends to become flat.

9.9 THE EMPTY LEMAÎTRE MODELS

The Lemaître models differ from the Friedmann models in that the cosmological constant is not zero. For the sake of simplicity, we will assume that the cosmological constant is much larger than $8\pi G T_0{}^0$. We can then neglect $T_0{}^0$ altogether, and we can regard the universe as empty. The behavior of the universe will then be controlled by the cosmological term; this term roughly has the same effect as a uniform mass density $\rho_{eff} = -\Lambda/4\pi G$, constant in space and time (see Eq. [7.51]). A positive value of Λ corresponds to a negative effective mass density (repulsion), and a negative value of Λ corresponds to a positive mass density (attraction). Hence, we expect that in a universe with a positive value of Λ, the expansion will tend to accelerate (monotonic behavior); whereas in a universe with a negative value of Λ, the expansion will slow down, stop, and reverse (oscillatory behavior). The calculations confirm this.

(i) Positive Curvature and $\Lambda \neq 0$. If we neglect $T_0{}^0$ in Eq. [100], we obtain

$$-\frac{3}{a^4}(a^2 + \dot{a}^2) = -\Lambda$$

With $\dot{a} = a\,da/dt$, this becomes

$$\left(\frac{da}{dt}\right)^2 = -1 + \frac{\Lambda a^2}{3} \qquad [124]$$

It is immediately obvious from Eq. [124] that $-1 + \Lambda a^2/3$ cannot be negative. This implies that $\Lambda > 0$, and that the value of a can never be less than $\sqrt{3/\Lambda}$. Thus, the empty Lemaître model with positive curvature requires a positive cosmological constant, and its radius of curvature is never zero (no Big Bang).

The integration of Eq. [124] is elementary and gives

$$a(t) = \sqrt{\frac{3}{\Lambda}} \cosh \sqrt{\frac{\Lambda}{3}} t \qquad [125]$$

where t has been set to zero at the time at which a has its minimum value. Fig. 9.16a gives a plot of $a(t)$. For $t > t_{min}$, the universe expands monotonically, and as $t \to \infty$, the universe becomes flat.

■ *Exercise 10.* Integrate [124] and obtain [125]. ■

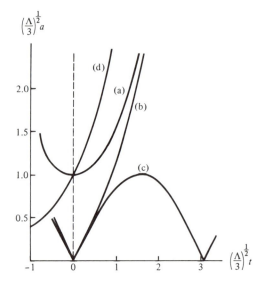

Fig. 9.16 Radius of curvature of the empty Lemaitre universes as a function of time. (a) Positive-curvature model, $\Lambda > 0$. (b) Negative-curvature model, $\Lambda > 0$. (c) Negative-curvature model, $\Lambda < 0$. (d) Flat model, $\Lambda > 0$.

(ii) Negative Curvature and $\Lambda \neq 0$. In this case, the equation analogous to Eq. [124] is

$$\left(\frac{da}{dt}\right)^2 = 1 + \frac{\Lambda a^2}{3}$$

which can be integrated to give

$$a(t) = \sqrt{\frac{3}{\Lambda}} \sinh \sqrt{\frac{\Lambda}{3}} t \qquad \text{for } \Lambda > 0$$

$$a(t) = \sqrt{\frac{3}{-\Lambda}} \sin \sqrt{\frac{-\Lambda}{3}} t \quad \text{for } \Lambda < 0 \qquad [126]$$

These functions are plotted in Figs. 9.16b and c, respectively. Note that both universes begin with a Big Bang at $t = 0$. The first of these expands monotonically, whereas the second oscillates with a period $2\pi\sqrt{3/(-\Lambda)}$. Of course, in our actual universe, the mass density near the singularity at $t = 0$ must have been very large, and hence the empty Lemaître models cannot be used to describe the behavior near this time.

(iii) Zero Curvature and $\Lambda \neq 0$. Our equation of motion is now

$$\left(\frac{da}{dt}\right)^2 = \frac{\Lambda a^2}{3} \qquad [127]$$

This equation makes sense only if $\Lambda > 0$, and it then has the solution

$$a(t) = a(0) \exp \sqrt{\frac{\Lambda}{3}} t \qquad [128]$$

This universe expands exponentially with a characteristic doubling time of $0.693\sqrt{3/\Lambda}$ (see Fig. 9.16d). The model described by Eq. [128] is usually called the *de Sitter model*. Note that Eq. [127] also has an exponentially decreasing solution, but this is of no relevance to our (expanding) universe.

Table 9.3 summarizes the characteristics of the Friedmann and empty Lemaître models. The differential equation for the general Lemaître model with both a matter density and a cosmological term cannot be integrated in terms of elementary functions. However, these general models can be roughly described as some combination of the cases listed in Table 9.3.

Consider a universe that begins with a Big Bang. In the early universe the mass density is very large, and we can neglect the cosmological term--that is, we have approximately a Friedmann universe. As the universe expands and the mass density decreases, the cosmological term will become more important. In the Friedmann cases of negative or zero curvature, the expansion and the decrease in mass density are monotonic (see Figs. 9.14 and 9.15) and hence the cosmological term, if any, will ultimately dominate the behavior of the universe. Thus, the universe gradually turns into an empty Lemaître universe of negative or zero curvature. Note that in the case of negative curvature with $\Lambda < 0$, the expansion of the Lemaître universe will stop at some later

TABLE 9.3 ISOTROPIC, HOMOGENEOUS MODELS OF THE
           UNIVERSE

	Three-volume	Time dependence
Friedmann, $\Lambda = 0$		
(i) Positive curvature	finite	oscillatory
(ii) Negative curvature	infinite	monotonic (for $t > 0$)
(iii) Zero curvature	infinite	monotonic (for $t > 0$)
Empty Lemaitre, $\Lambda \neq 0$		
(i) Positive curvature, $\Lambda > 0$	finite	monotonic (for $t > t_{min}$)
(ii) Negative curvature, $\Lambda > 0$	infinite	monotonic (for $t > 0$)
Negative curvature, $\Lambda < 0$	infinite	oscillatory
(iii) Zero curvature, $\Lambda > 0$	infinite	monotonic

time (see Fig. 9.16c); the universe reverses and we finally end up in a recontracting Friedmann universe of negative curvature.

In the case of a Friedmann universe of positive curvature, the mass density reaches a minimum when one-half the period has elapsed (see Fig. 9.13). Hence, the cosmological term will dominate the behavior of the universe only if it is sufficiently large compared with this minimum mass density. The critical value of Λ is (see Exercise 11)

$$\Lambda_E = \left(\frac{\pi}{2GM} \right)^2 \qquad [129]$$

If Λ is larger than Λ_E, then what began as a positive-curvature Friedmann universe (see Fig. 9.13) gradually turns into a positive-curvature, expanding Lemaître universe (see Fig. 9.16a). A typical example of the function $a(t)$ for $\Lambda > \Lambda_E$ is shown in Fig. 9.17. The transition from the nearly Friedmann to the nearly empty Lemaître universe can be fast or slow depending on the value of Λ. In the exceptional case $\Lambda = \Lambda_E$, the transition is never completed; the universe remains suspended at a constant value of a,

$$a = [\text{constant}] = \frac{1}{\sqrt{\Lambda_E}} \qquad [130]$$

The static universe with this constant value of a is known as the *Einstein universe*. However, the equilibrium at the value $a = 1/\sqrt{\Lambda_E}$ is unstable (see Problem 13). Any perturbation in a leads either to mon-

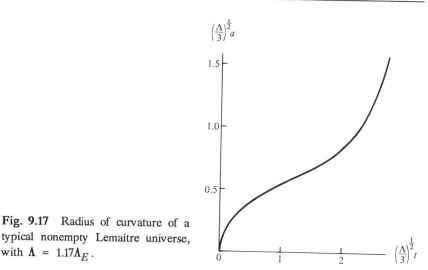

$$\left(\frac{\Lambda}{3}\right)^{\frac{1}{2}} a$$

Fig. 9.17 Radius of curvature of a typical nonempty Lemaître universe, with $\Lambda = 1.17\Lambda_E$.

otonic expansion (toward an expanding Lemaître universe) or to monotonic contraction (toward a contracting Friedmann universe).

■ *Exercise 11.* Show that [130] is a solution of [105] if $\Lambda = \Lambda_E$. ■

■ *Exercise 12.* Show that the equation of motion [105] can be written as

$$\left(\frac{da}{dt}\right)^2 + \left(1 - \frac{4GM}{3\pi a} - \frac{\Lambda a^2}{3}\right) = 0 \qquad [131]$$

The quantity $V(a) = (1 - 4GM/3\pi a - \Lambda a^2/3)$ can be regarded as an effective potential for the motion of the universe. Make a rough plot of $V(a)$ as a function of a. Show that $V(a)$ has a maximum at $a_m = (2GM/\pi\Lambda)^{1/3}$. If the value of this maximum is $V(a_m) < 0$, then da/dt is always positive, and the universe expands monotonically. If $V(a_m) > 0$, then da/dt vanishes at some value of a smaller than a_m; that is, a motion that begins at $a = 0$ reaches a turning point, and then the universe recontracts. Show that $V(a_m) < 0$ if $\Lambda > \Lambda_E$, and that $V(a_m) > 0$ if $\Lambda < \Lambda_E$. ■

Finally, note that Table 9.3 contains no empty Lemaître models that are monotonic and have $\Lambda < 0$. We can understand the absence of such models if we recall that a negative cosmological constant corresponds to a positive effective mass density of the vacuum; as the volume of the universe grows so does the total effective mass, and ultimately this stops the expansion and initiates a recontraction. We can draw an interesting conclusion: the general Lemaître model with a nonzero mass density and $\Lambda < 0$ must necessarily be of the oscillating type. If this universe were monotonic, it would gradually approach a monotonic empty model, which, as we have seen, does not exist.

9.10 PROPAGATION OF LIGHT; THE PARTICLE HORIZON

Light signals propagate along lightlike worldlines, that is, lines with $ds^2 = 0$. If we place the origin of coordinates at the position of the light source, then the light will move out radially ($d\theta = d\phi = 0$), and hence for all our models of the universe we obtain

$$0 = a(\eta)(d\eta^2 - d\chi^2)$$

The equation of the worldline of a light signal emitted at the initial time parameter η_i is then

$$\chi = \eta - \eta_i \qquad [132]$$

This means that in a χ-η spacetime diagram, the worldlines of light signals are at 45° (see Fig. 9.18).

Suppose that two light signals are sent out from the source at $\chi = 0$: the first when the time parameter has the value η_i, the second when it has the value $\eta_i + \Delta\eta_i$. The worldlines are, respectively,

$$\chi = \eta - \eta_i$$

$$\chi = \eta - (\eta_i + \Delta\eta_i) \qquad [133]$$

The two light signals therefore arrive at any given point χ with a difference

$$\Delta\eta(\chi) = \Delta\eta_i \qquad [134]$$

between their time parameters. This says that $\Delta\eta$ is a constant, independent of χ, and it implies that if time is measured by η, then there is no redshift. However, the atomic clocks in the galaxies do *not* measure η time, they measure τ time, or t time. Since $\Delta t = a\Delta\eta$, it follows from Eq. [134] that $\Delta t/a$ remains constant as the two signals propagate; the time interval Δt between the signals therefore changes in direct proportion to a. For a continuous wave train with a frequency ν, in which each peak may be regarded as a signal, the time interval between signals equals $\Delta t = 1/\nu$, and hence

$$a\nu = [\text{constant}] \qquad [135]$$

This gives us the redshift of light--in an expanding universe, a increases and hence ν must decrease. (Incidentally: We have already given an alternative derivation of Eq. [135] in terms of photons, by means of the simple argument following Eq. [33].)

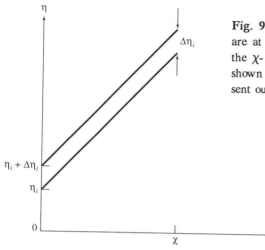

Fig. 9.18 Worldlines of light signals are at an angle of 45^0 with respect to the χ- and η-axes. The two worldlines shown here belong to two light signals sent out from $\chi = 0$ at different times.

According to Eq. [135], the frequency with which a wave arrives at the point χ is

$$\nu(\chi) = \nu(i)\ \frac{a(\eta_i)}{a(\eta)} = \nu(i)\ \frac{a(\eta - \chi)}{a(\eta)} \qquad [136]$$

where $\nu(i)$ is the emitted frequency.

If the distance between the emitter and receiver is small ($\chi \ll 1$), then we can approximate $a(\eta - \chi)$ by the first term in a power series,

$$\nu(\chi) \simeq \nu(i)\left[1 - \chi\ \frac{\dot{a}(\eta)}{a(\eta)} + \ldots\right] \qquad [137]$$

that is,

$$\frac{\nu(\chi) - \nu(i)}{\nu(i)} \simeq -\chi\ \frac{\dot{a}(\eta)}{a(\eta)} \qquad [138]$$

In terms of the wavelength $\lambda = c/\nu$, we can write this as

$$\frac{\lambda(\chi) - \lambda(0)}{\lambda(0)} \simeq +\chi\ \frac{\dot{a}(\eta)}{a(\eta)} \qquad [139]$$

The distance between emitter and receiver, at the instant η, is given by (see Eq. [73])

$$l = a(\eta)\chi \qquad [140]$$

and Eq. [139] therefore takes the form of Hubble's law (compare with Eqs. [8] and [9]):

$$\frac{\lambda(\chi) - \lambda(i)}{\lambda(i)} \simeq \frac{\dot{a}}{a^2} l \qquad [141]$$

From this we can identify the Hubble constant as

$$H = \frac{\dot{a}}{a^2} \qquad [142]$$

Since $\dot{a} = a\,da/dt$ (see Eq. [90]), we can also write this as

$$H = \frac{1}{a} \frac{da}{dt} \qquad [143]$$

In the above derivation of Hubble's law we assumed that the distance l is a metric distance, that is, a distance defined according to the metric of spacetime. In practice, astronomers calibrate distances by luminosities, and this means that the distance appearing in Hubble's law is actually defined as proportional to the square root of the area of a sphere, rather than the radius of a sphere. For small distances, where the effects of curvature of space are insignificant, the luminosity distance agrees with the metric distance. However, if we want to explore deviations from linearity in Hubble's law, we must take this distinction into account, and we must also take into account higher-order terms in the power series for $a(\eta - \chi)$ in Eq. [137].

Note that with the expression [143] for H, we can write the Einstein equation as

$$3\left[H^2 + \frac{K}{a^2}\right] = 8\pi G\rho + \Lambda \qquad [144]$$

where $K = +1$ or -1 for the cases of positive and of negative curvature, respectively. This leads to an expression for a in terms of observable parameters:

$$a^2 = \frac{K}{H^2(\Omega - 1 + \Lambda/3H^2)} \qquad [145]$$

We will find these expressions useful in the next section.

A curious feature of the Friedmann models and other models with an initial Big Bang is that the universe has a part that is visible to us and a part that is invisible. Consider a light signal sent out from the

origin $\chi = 0$ when the universe began, at $\eta_i = 0$. At the time characterized by η, this light signal will have reached a point $\chi = \eta$ (see Eq.
[132]). Conversely, a light signal emitted from this point χ when the
universe began will have reached the origin, but any light signal emitted from a more distant point will take a longer time to reach the
origin. Thus, for an observer at the origin, the surface

$$\chi = \eta \qquad [146]$$

represents the boundary of the visible universe. This boundary is
called the *particle horizon* or the *object horizon*, since it tells us where
we find the most distant particle or object we can see. In the spacetime diagram, this boundary is the intersection of the past light cone
with the initial surface $\eta_i = 0$ (see Fig. 9.19). Note that an observer in
the early universe, near the instant $\eta_i = 0$, can see only a small fraction of the universe. Even though the size of the universe at or near
the instant $\eta_i = 0$ is extremely small, light signals take a long time to
travel between two points separated by some radial interval χ, because
the distance between these points keeps on increasing while the signal
attempts to "catch up."

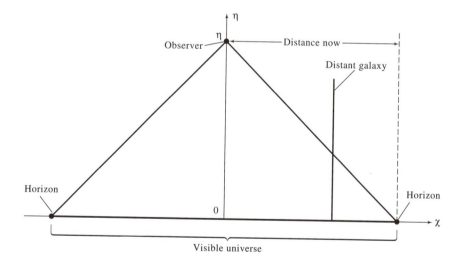

Fig. 9.19 The past light cone of an observer at a time characterized by η. The
observer can see only the portion of the universe that lies within this light cone.
The particle horizon for this observer is the boundary of the observable portion of
the universe at the initial instant of the Big Bang.

If we look at a faraway region of the universe, we see this region
as it was a long time ago, when the light reaching us now began its
journey. If we look at the horizon of our universe, we see the matter
as it was at the initial instant; we see the primordial fireball. As the

horizon expands, more and more of this fireball comes into our visible universe. Of course, at the horizon the redshift is infinite (see Eq. [136]), and hence the light will actually be too faint to be seen.

The actual distance measured at time η between the observer and the horizon point from which the light was emitted is $l = a(\eta)\chi$ (see Eq. [73]), or

$$l = a(\eta)\eta \qquad [147]$$

In this form, the equation is valid for all the models of the universe based on the Robertson-Walker geometry. To express this distance to the horizon as a function of t time we must evaluate a and η in terms of t. For the Friedmann models, the required equations for $a(\eta)$ and $\eta(t)$ were given in the preceding section. For these models, the horizon distance l is finite, that is, the horizon exists. For more general models, with some different functions $a(\eta)$ and $\eta(t)$, the existence of a horizon hinges on whether the factor η that appears on the right side of Eq. [147] is finite. Since $d\eta = dt/a$, we can express $\eta(t)$ as an integral over $1/a$,

$$\eta(t) = \int_0^t \frac{dt'}{a(t')} \qquad [148]$$

and the criterion for the existence of a horizon is that this integral be finite. Clearly, the integral [148] will be finite whenever $a(t)$ near $t = 0$ behaves like a fractional power of t, that is, $a(t) \to t^n$ with $0 < n < 1$.

While the horizon expands as a function of time, so does the distance $l = a(\eta)\eta$. The rate of increase of this distance to the horizon is

$$\frac{dl}{dt} = \frac{da}{dt}\eta + a\frac{d\eta}{dt} = Ha\eta + 1 = Hl + 1 \qquad [149]$$

It is interesting to compare this speed with the speed of recession of a comoving particle at the same distance. At time η, the distance to this particle is $l = a(\eta)\chi$ (where χ is fixed), and the speed is

$$\frac{dl}{dt} = \frac{da}{dt}\chi = \frac{da}{dt}\eta = Hl \qquad [150]$$

Thus, the speed of recession of the horizon exceeds the speed of recession of the particle by 1, that is, by the speed of light. Of course, this is in accord with our intuitive expectation that gradually more and more of the particles become included within the visible part of the universe.

For the positive-curvature model, the volume of the visible part of the universe is

$$2\pi^2 a^3(\eta - \sin\eta \cos\eta)/\pi \qquad [151]$$

■ *Exercise 13.* Derive Eq. [151]. ■

The factor $(\eta - \sin\eta \cos\eta)/\pi$ in Eq. [151] gives the fraction of the total volume visible to an observer at a fixed position. This fraction gradually increases. At the moment of maximum expansion ($\eta = \pi$), all of the volume is visible; at the final moment of collapse ($\eta = 2\pi$), the light completes a trip around the universe, and the entire universe is visible twice (that is, every point is visible from two opposite directions).

9.11 COMPARISON OF THEORY AND OBSERVATION

The observable cosmological parameters are the mass density, the Hubble constant, the deceleration parameter, and the age of the universe. Different cosmological models predict different relations among these parameters.

In this section we follow the convention that parameters with the subscript 0 (such as ρ_0, H_0, q_0, t_0) are evaluated at the present time; quantities without the subscript are evaluated at an arbitrary time.

According to Eq. [144],

$$3\left[H^2 + \frac{K}{a^2}\right] = 8\pi G\rho + \Lambda \qquad [152]$$

where $K = +1$, $K = 0$, or $K = -1$, for the cases of positive, zero, or negative curvature, respectively. It follows from this that the curvature of the universe is determined by H, ρ, and Λ, or, equivalently, H, Ω, and Λ:

$$\Omega > 1 - \frac{\Lambda}{3H^2} \quad \text{for positive curvature}$$

$$\Omega < 1 - \frac{\Lambda}{3H^2} \quad \text{for negative curvature}$$

$$\Omega = 1 - \frac{\Lambda}{3H^2} \quad \text{for zero curvature} \qquad [153]$$

Since Λ is not a directly observable parameter, it is best to eliminate it. According to Eq. [28],

$$q = \frac{\Omega}{2} - \frac{\Lambda}{3H^2} \qquad [154]$$

This relation among q, ρ, and Λ was derived in Section 9.3 without making explicit use of Einstein's equations. It is obvious from the arguments given in Section 9.8 (see Eqs. [106]-[111]) that Eq. [154] can be obtained by differentiating the Einstein equation once with respect to t.

■ *Exercise 14.* Multiply Eq. [105] by a^2 and differentiate with respect to t. Show that the result can be put in the form [154]. ■

By means of Eq. [154] we can eliminate Λ from Eq. [153] and obtain

$$\tfrac{3}{2}\Omega > 1 + q \quad \text{for positive curvature}$$

$$\tfrac{3}{2}\Omega < 1 + q \quad \text{for negative curvature}$$

$$\tfrac{3}{2}\Omega = 1 + q \quad \text{for zero curvature} \qquad [155]$$

From here on it is convenient to treat the Friedmann and Lemaître models separately. In the Friedmann models $\Lambda = 0$, and the situation is as follows:

(i) Positive Curvature and $\Lambda = 0$. According to Eq. [154], $q = \tfrac{1}{2}\Omega$, and hence we can express the condition [155] either in terms of q or in terms of Ω. This leads to

$$q > \tfrac{1}{2}, \quad \Omega > 1 \qquad [156]$$

We can also obtain an inequality between the true age of the universe and the Hubble age. The inequality is

$$0 < t < \tfrac{2}{3}H^{-1} \qquad [157]$$

■ *Exercise 15.* Show that

$$t = H^{-1} \frac{(\eta - \sin\eta)\sin\eta}{(1 - \cos\eta)^2} \qquad [158]$$

and show that if η is in the range $0 < \eta < \pi$, which corresponds to an expanding universe, then t is in the range given by Eq. [157]. ■

(ii) Negative Curvature and $\Lambda = 0$. The conditions on q and Ω are

$$q < \tfrac{1}{2}, \quad \Omega < 1 \tag{159}$$

and the age satisfies the inequalities

$$\tfrac{2}{3}H^{-1} < t < H^{-1} \tag{160}$$

■ *Exercise 16.* Derive Eq. [160]. ■

(iii) Zero Curvature and $\Lambda = 0$. In this last case

$$q = \tfrac{1}{2}, \quad \Omega = 1 \tag{161}$$

and

$$t = \tfrac{2}{3}H^{-1} \tag{162}$$

This takes care of the Friedmann models; we next consider the general Lemaître models with $\Lambda \neq 0$.

(i) Positive Curvature and $\Lambda \neq 0$. The condition on the density parameter and the deceleration parameter for this case is

$$\tfrac{3}{2}\Omega > 1 + q \tag{163}$$

According to the discussion at the end of Section 9.9, the universe will expand monotonically if $\Lambda > \Lambda_E$, and it will eventually recontract if $\Lambda < \Lambda_E$. Let us express the critical value Λ_E in terms of the observable parameters. By Eq. [145] we have

$$a^2 = \frac{1}{H^2(\Omega - 1 + \Lambda/3H^2)} \tag{164}$$

and hence

$$\Lambda_E = \left(\frac{\pi}{2GM}\right)^2 = \left[\frac{\pi}{2G(2\pi^2 a^3 \rho)}\right]^2$$

$$= \frac{1}{16\pi^2 G^2 \rho^2} H^6(\Omega - 1 + \Lambda/3H^2)^3 \tag{165}$$

The condition $\Lambda > \Lambda_E$ for monotonic behavior then reads

$$\tfrac{1}{2}\Omega - q > \frac{4}{\Omega^2}\left[\tfrac{1}{2}\Omega - \tfrac{1}{3}(q + 1)\right]^3 \quad \text{(monotonic)} \qquad [166]$$

where the expression $\Lambda = \tfrac{3}{2}\Omega H^2 - 3H^2 q$ has been used for Λ (see Eq. [154]). As this type of universe expands, it gradually approaches the empty Lemaître model (*i*) of Table 9.3.

The condition $\Lambda < \Lambda_E$ for oscillatory behavior is obtained by reversing the sign of the inequality in Eq. [166]:

$$\tfrac{1}{2}\Omega - q < \frac{4}{\Omega^2}\left[\tfrac{1}{2}\Omega - \tfrac{1}{3}(q + 1)\right]^3 \quad \text{(oscillatory)} \qquad [167]$$

Note that according to this inequality, a universe with a negative value of Λ, that is, a negative value of $\Omega - 2q$, must oscillate. The absence of monotonic models with $\Lambda < 0$ has already been explained in Section 9.9.

If both sides of [167] are equal, then the universe gradually approaches an Einstein universe; however, the instability of this configuration will eventually result in either expansion or recontraction.

(ii) Negative Curvature and $\Lambda \neq 0$. The condition for this case is

$$\tfrac{3}{2}\Omega < 1 + q \qquad [168]$$

The universe continues to expand if $\Lambda > 0$, and recontracts if $\Lambda < 0$. These two conditions may be expressed as

$$\Omega - 2q > 0 \quad \text{(monotonic)} \qquad [169]$$

$$\Omega - 2q < 0 \quad \text{(recontraction)} \qquad [170]$$

(iii) Zero Curvature and $\Lambda \neq 0$. In this case

$$\tfrac{3}{2}\Omega = 1 + q \qquad [171]$$

The conditions for continual expansion and for recontraction are as in case (*ii*):

$$\Omega - 2q > 0 \quad \text{(monotonic)} \qquad [172]$$

$$\Omega - 2q < 0 \quad \text{(recontraction)} \qquad [173]$$

Note that, in the Friedmann model, if any two of the four parameters Ω_0, q_0, H_0, and t_0 are determined from observation, then the model fixes the remaining two and the evolution of the universe. Thus, cosmology may be described, in Sandage's words, as "a search for two numbers" (Sandage, 1970). In the Lemaître models, there is an extra adjustable constant (Λ), and we must carry out a search for three numbers.

From the available observational data, we can set the following limits on Ω_0, q_0, H_0, and t_0 (see Sections 9.3, 9.4, 9.6):

$$0.1 \le \Omega_0 \le 2$$

$$|q_0| \le 1$$

$$40 \text{ km/s·Mpc} < H_0 < 100 \text{ km/s·Mpc}$$

$$10 \times 10^9 \text{ years} \le t_0 \le 20 \times 10^9 \text{ years}$$

Since these limits are not very tight, a large variety of models of the universe will fit the data. At present, we cannot decide whether the universe is Friedmann or Lemaître, or even whether it is open or closed, monotonic or recontracting (see Fig. 9.20).

Table 9.4 gives some examples of typical Friedmann models of our universe, with $\Omega_0 = 1$, 0.14, and 2. All these models are consistent with the observational data. As already mentioned, the value $\Omega_0 = 1$ is suggested by theoretical arguments concerning the early universe; however, this value of Ω_0 would require a large amount of nonbaryonic dark matter. The value $\Omega_0 = 0.14$ could permit most of the dark matter to be baryonic, and all of it would be in galaxies and their halos. The value $\Omega_0 = 2$ would require the largest amount of nonbaryonic dark matter, and has nothing to commend it.

Note that both the flat and the positive-curvature models have true ages close to the lower limit of 10×10^9 years. If improvements in the determination of the distance scale were to reveal a value of H_0 larger than about 65 km/s·Mpc, then both of these models would be ruled out by the data on the true age of the universe. However, Lemaître models (with $\Lambda > 0$) would not be ruled out by such a development.

Finally, it may be well to keep in mind that past experience in cosmology suggests that the observable cosmological parameters could be subject to substantial changes, in excess of the quoted uncertainties. For instance, the value of the Hubble constant has suffered several drastic revisions--according to recent determinations, the value of this constant is about six times smaller than the value proposed by Hubble. It will be a while before we can predict with confidence the future of our universe.

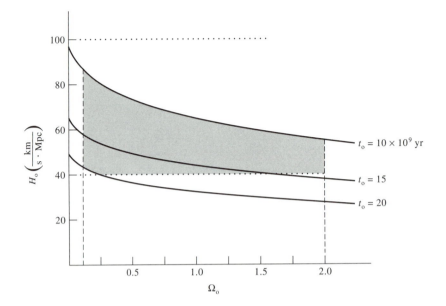

Fig. 9.20 Limits on the observational parameters displayed in a Ω_0-H_0 diagram for Friedmann models ($\Lambda = 0$). The screened region indicates the range of values consistent with the data. Positive-curvature, negative-curvature, and flat models fall in the consistent range.

TABLE 9.4 TYPICAL FRIEDMANN MODELS OF OUR UNIVERSE

	Flat	Negative curvature	Positive curvature
Density parameter Ω_0	1.0	0.14	2.0
Deceleration parameter q_0	0.5	0.07	1.0
Hubble constant H_0	60 km/s·Mpc	60 km/s·Mpc	55 km/s·Mpc
Mass density ρ_0	$6.8 \cdot 10^{-30}$g/cm^3	$9.5 \cdot 10^{-31}$g/cm^3	$1.1 \cdot 10^{-29}$g/cm^3
Cosmological constant Λ	0	0	0
Hubble age H_0^{-1}	$1.6 \cdot 10^{10}$ yr	$1.6 \cdot 10^{10}$ yr	$1.8 \cdot 10^{10}$ yr
True age t_0	$1.1 \cdot 10^{10}$ yr	$1.4 \cdot 10^{10}$ yr	$1.0 \cdot 10^{10}$ yr
Radius of curvature a_0	∞	$1.7 \cdot 10^{10}$ l-yr	$1.8 \cdot 10^{10}$ l-yr
Time for expansion-contraction cycle	N.A.	N.A.	$1.1 \cdot 10^{11}$ yr
Radius of curvature at maximum expansion	N.A.	N.A.	$3.6 \cdot 10^{10}$ l-yr

FURTHER READING

North, J. D., *The Measure of the Universe* (Clarendon Press, Oxford, 1965) traces the history and conceptual foundations of modern cosmology; it includes discussions of alternative theories of gravitation and of metaphysical issues and is rather wordy but very readable.

Both Misner, C. W., Thorne, K. S., and Wheeler, J. A., *Gravitation* (Freeman, San Francisco, 1973) and Peebles, P. J. E., *Physical Cosmology* (Princeton University Press, Princeton, 1971) include brief outlines of the history of modern cosmology. Hubble, E., *The Realm of the Nebulae* (Yale University Press, New Haven, 1936) is a classic which relates the discovery of the extragalactic systems.

Bertotti, B., Balbinot, R., Bergia, S., and Messina, A., *Modern Cosmology in Retrospect* (Cambridge University Press, Cambridge, 1990) is a collection of interesting papers dealing with the history of our modern ideas about the universe and the Big Bang. It includes an account by Wilson, R. M., "Discovery of the cosmic microwave background."

Peebles, P. J. E., *Principles of Physical Cosmology* (Princeton University Press, Princeton, 1993) gives an authoritative and excellent general introduction to cosmology, both in its observational and theoretical aspects.

Kolb, E. W., and Turner, M. S., *The Early Universe* (Addison-Wesley, Reading, 1990) and Börner, G., *The Early Universe* (Springer-Verlag, Berlin, 1988) include good concise summaries of the observed properties of the universe, and summaries of cosmological models. (Note: The numerical value of the critical density stated in Börner, p. 69, is wrong!)

Rubin, V. C., and Coyne, G. V., *Large-Scale Motions in the Universe* (Princeton University Press, Princeton, 1988) offers a collection of papers dealing with the large-scale structure of the mass distribution.

Jacoby et al. (1992) and Huchra (1992) give good reviews of the status of determinations of the cosmological distance scale and the search for an accurate value of H_0. De Vaucouleurs (1993) provides a long list of recent distance determinations of galaxies and clusters. Rowan-Robinson, M., in *The Cosmological Distance Ladder* (Freeman, New York, 1985) makes a gallant attempt at resolving the discrepancies between the distance scales advocated by Sandage and by de Vaucouleurs; he also provides a clear introduction to the astronomy and astrophysics on which the different methods of distance determination are based.

Carroll, Press, and Turner (1992) give a broad discussion of the physical basis for the cosmological constant and of the observational evidence for and against this constant.

The evidence for dark matter is discussed in a multitude of recent books and articles, such as:

Audouze, J., and Tran Than Van, J., eds., *Dark Matter* (Editions Frontieres, Gif-sur-Ivette, 1988).

Bahcall, J. N., Piran, T., and Weinberg, S., *Dark Matter in the Universe* (World Scientific Publishing, Singapore, 1987).

Blair, D. G., Buckingham, M. J., and Ruffini, R., *Proceedings of the Fifth Marcel Grossmann Meeting on General Relativity* (World Scientific Publishing, Singapore, 1989).

Galeotti, P., and Schramm, D. N., eds., *Dark Matter in the Universe* (Kluwer, Dordrecht, 1990).

Holt, S. S., Bennett, C. L., and Trimble, V., *After the First Three Minutes* (American Institute of Physics, New York, 1991).

Kormendy, J., and Knapp, G. R., *Dark Matter in the Universe* (Reidel, Dordrecht, 1987).

Lynden-Bell, D., and Gilmore, G., *Baryonic Dark Matter* (Kluwer, Dordrecht, 1990).

Rees, M. J., "Galaxy Formation and Dark Matter," in Hawking, S. W., and Israel, W., *Three hundred years of gravitation* (Cambridge University Press, Cambrige, 1987).

Tremaine, S., "Dynamical Evidence for Dark Matter," Physics Today, Feb. 1992.

Trimble, V., "Existence and Nature of Dark Matter in the Universe," Ann. Rev. Astron. Astrophys. 25, 425 (1987).

Fang, L. Z., Kiang, T., Cheng, F. H., and Hu, F. X., Q. J. R. Astron. Soc. 23, 363 (1982).

Cowan, Thielemann, and Truran (1991) review the determination of the age of the universe by radioactive dating of the elements (nucleocosmochronology), and make comparisons with ages obtained by other methods.

Bernstein, J., and Feinberg. G., eds., *Cosmological Constants* (Columbia University Press, 1986) contains a collection of reprints of "landmark" papers on cosmology and the early universe, from Einstein's first misguided attempt at constructing a static solution of his equations (1917) to Guth's inflationary model (1982).

The solution of Einstein's equations for an isotropic, homogeneous universe is of course given in many books on gravitational theory. The following contain a general discussion of open and closed models, with and without a cosmological constant: Misner, Thorne, and Wheeler, op. cit.; Robertson, H. P., and Noonan, T. W., *Relativity and Cosmology*; Tolman, R. C., *Relativity, Thermodynamics, and Cosmology* (Oxford University Press, Oxford, 1938); Weinberg, S., *Gravitation and Cosmology* (Wiley, New York, 1972).

Ryan, M. P., and Shepley, L. C., *Homogeneous Relativistic Cosmologies* (Princeton University Press, Princeton, 1975) presents a mathematical discussion of diverse cosmological models, including some rather weird ones which have little to do with our universe. This book includes neat reviews of differential forms and of the concept of "singularity."

Hawking S. W., and Ellis, G. F. R., *The Large Scale Structure of Space-Time* (Cambridge University Press, Cambridge, 1973) uses the general singularity theorems to prove rigorously that the observed properties of our universe *demand* an initial singularity, at least if classical physics is applicable.

Our knowledge and ignorance of what happens at the singularity, at the beginning of the universe, are summarized in the article Novikov, I. D., and Zeldovich, Ya. B.,"Physical Processes Near Cosmological Singularities, " Ann. Rev. Astron. Astrophys. 11, 387 (1973).

Felten, J. E., and Isaacman, R., Phys. Rev. 58, 689 (1986) gives plots of $a(\tau)$ for different values of Ω.

REFERENCES

Adams, W. S., Astrophys. J. **93**, 11 (1941).

Alpher, R. A., and Herman, R. C., Nature **162**, 774 (1948).

Alpher, R. A., Bethe, H. A., and Gamow, G., Phys. Rev. **73**, 803 (1948).

Bahcall, N. A., in Rubin, V. C., and Coyne, G. V., eds., *Large-Scale Motions in the Universe* (Princeton University Press, Princeton, 1988).

Bondi, H., and Gold, T., Mon. Not. Roy Astron. Soc. **108**, 252 (1948).

Carroll, S. M., Press, W. H., and Turner, E. L., Ann. Rev. Astron. Astrophys., **30**, 499 (1992).

Cowan, J. J., Thielemann, F.-K., and Truran, J. W., Ann. Rev. Astron. Astrophys. **29**, 447 (1991).

Davis, M., "Evidence for Dark Matter in Galactic Systems," in Kormendy, J., and Knapp, G. R., *Dark Matter in the Universe* (Reidel, Dordrecht, 1987).

Davis, M., and Peebles, P. J. E., Astrophys. J. **267**, 465 (1983).

de Vaucouleurs, G., Astrophys. J. **415**, 10 (1993).

Dicke, R. H., Peebles, P. J. E., Roll, G., and Wilkinson, D. T., Astrophys. J. **142**, 414 (1965).

Efstathiou, G., and Silk, J., Fund. Cosmic Phys. **9**, 1 (1983).

Efstathiou, G., Ellis, R. S., and Peterson, B. A., Mon. Not. Roy. Astron. Soc. **232**, 715 (1988).

Faber, S. M., and Gallagher, J. S., Ann. Rev. Astron. Astrophys. **17**, 135 (1979).

Faber, S. M., and Jackson, R. E., Astrophys. J. **204**, 668 (1976).

Fermi, E., and Turkevich, A. (1950), unpublished.

Gamow, G., Phys. Rev. **74**, 505 (1948).

Ganga, K., Cheng, E., Meyer, S., and Page, L., Astrophys. J. **410**, L57 (1993).

Geller, M. J., and Huchra, J. P., in Rubin and Coyne, 1988.

Gott, J. R. III, Gunn, J. E., Schramm, D. N., and Tinsley, B. M., Astrophys. J. **194**, 543 (1974).

Gunn, J. E., Ann. N.Y. Acad. Sci. **224**, 56 (1973).

Gunn, J. E., and Oke, J. B., Astrophys. J. **195**, 255 (1975).

Harrison, E. R., Ann. Rev. Astron. Astrophys. **11**, 155 (1973).

Hawking, S. W., and Israel, W., eds., *Three hundred years of gravitation* (Cambridge University Press, Cambridge, 1987).

Hewitt, J. N., in Akerlof, C. W., and Srednicki, M. A., eds, *Texas PASCOS 92: Relativistic Astrophysics & Particle Cosmology* (New York Academy of Sciences, New York, 1993).

Hills, R. E., and Bahcall, J. N., Ann. N.Y. Acad. Sci. **224**, 58 (1973).

Hoyle, F., Mon. Not. Roy Astron. Soc. **108**, 372 (1948).

Hubble, E., Astrophys. J. **64**, 321 (1926).

Hubble, E., *The Realm of the Nebulae* (Yale University Press, New Haven, 1936), p. 186.

Huchra, J. P., Science **256**, 321 (1992).

Jacoby, G. H., Branch, D., Ciardullo, R., Davies, R. L., Harris, W. E., Pierce, M. J., Pritchet, C. J., Tonry, J. L., and Welch, D. L., Pub. Astron. Soc. Pacific **104**, 599 (1992).

Kormendy, J., and Knapp, G. R., *Dark Matter in the Universe* (Reidel, Dordrecht,

1987).

Lauer, T. R., Astrophys. J., **292**, 104 (1985).

McKellar, A., Publ. Dominiom Astrophys. **7**, 251 (1941).

Maddox, S. J., Efstathiou, G., Sutherland, W. J., and Loveday, J., Mon. Not. Roy. Astron. Soc. **242**, 43P, 1990.

Mather, J. C., Cheng, E. S., Eplee, R. E., Isaacman, R. B., Meyer, S. S., Shafer, R. A., Weiss, R., Wright, E. L., Bennett, C. L., Boggess, N. W., Dwek, E., Gulkis, S., Hauser, M. G., Janssen, M., Kelsall, T., Lubin, P. M., Moseley, S. H., Murdock, T. L., Silverberg, R. F., Smoot, G. F., and Wilkinson, D. T., Astrophys. J. **354**, L37 (1990).

Mushotzky, R. F., "The Missing Mass in Clusters of Galaxies and Elliptical Galaxies," in Holt, S. S., Bennett, C. L., and Trimble, V., *After the First Three Minutes* (American Institute of Physics, New York, 1991).

Osmer, P. S., Astrophys. J. **253**, 28 (1982).

Peacock, J. A., and Gull, S. F., Mon. Not. Roy. Astron. Soc. **196**, 611 (1981).

Peebles, P. J. E., *Physical Cosmology* (Princeton University Press, Princeton, 1971).

Peebles, P. J. E, *The Large-Scale Structure of the Universe* (Princeton University Press, Princeton, 1980).

Peebles, P. J. E., *Principles of Physical Cosmology* (Princeton University Press, 1993).

Penzias, R. A., and Wilson, R. W., Astrophys. J. **142**, 419 (1965).

Reines, F., ed., *Cosmology, Fusion, and Other Matters* (Colorado Associated University Press, Boulder, 1972).

Rood, R. T., in Vangioni-Flam, E., Casse, M., Adouze, J., and Tran Than Van, T., eds., *Astrophysical Ages and Dating Methods* (Editions Frontieres, Gif sur Yvette, 1990).

Rowan-Robinson, M., *The Cosmological Distance Ladder* (Freeman, New York, 1985).

Rubin, V. C., and Coyne, G. V., eds., *Large-Scale Motions in the Universe* (Princeton University Press, Princeton, 1988).

Ryle, M., Ann. Rev. Astron. and Astrophys., **6**, 249 (1968).

Sancisi, R., and van Albada, T. S., in Kormendy, J., and Knapp, G. R., *Dark Matter in the Universe* (Reidel, Dordrecht, 1987).

Sandage, A., Astrophys. J. **152**, L149 (1968).

Sandage, A. R., Physics Today, February 1970, p. 34.

Sandage, A., and Cacciari, C., Astrophys. J., **350**, 645 (1990).

Sandage, A., and Tammann, G. A., Astrophys. J. **196**, 313 (1975a).

Sandage, A., and Tammann, G. A., Astrophys. J. **197**, 265 (1975b).

Sandage, A., and Tammann, G. A., Astrophys. J. **365**, 1 (1990).

Schmidt, M., Astrophys. J. **151**, 393 (1968).

Schmidt, M., Astrophys. J. **162**, 371 (1970); ibid., **176**, 273 (1972).

Schramm, D. N., Ann. Rev. Astron. Astrophys. **12**, 383 (1974).

Smoot, G. F., Bennett, C. L., Kogut, A., Wright, E. L., Aymon, J., Boggess, N. W., Cheng, E. S., De Amici, G., Gulkis, S., Hauser, M. G., Hinshaw, G., Jackson, P. D., Janssen, M., Kaita, E., Kelsall, T., Keegstra, P., Lineweaver, C., Loewenstein, K., Lubin, P., Mather, J., Meyer, S. S., Moseley, S. H., Murdock, T., Rokke, L., Silverberg, R. F., Tenorio, L., Weiss, R., and Wilkinson, D. T., Astrophys. J., **396**, L1

(1992).

Tammann, G. A., and Sandage, A., Astrophys. J. **294**, 81 (1985).

Tonry, J. L., in Akerlof, C. W., and Srednicki, M. A., eds., *Texas PASCOS 92: Relativistic Astrophysics & Particle Cosmology* (New York Academy of Sciences, New York, 1993).

Trimble, V., Ann. Rev. Astron. Astrophys. **25**, 425 (1987).

Tully, R. B., and Fisher, J. R., Astron. and Astrophys., **54**, 661 (1977).

VandenBerg, D. A., in Vangioni-Flam, E., Casse, M., Adouze, J., and Tran Than Van, T., eds., *Astrophysical Ages and Dating Methods* (Editions Frontieres, Gif sur Yvette, 1990).

Wall, J. V., Pearson, T. J., and Longair, M. S., Mon. Not. Roy Astron. Soc. **193**, 683 (1980); ibid., **196**, 597 (1981).

Wasserburg, G. J., Papanastassiou, D. A., Tera, F., and Huneke, J. C., Phil. Trans. A**285**, 7 (1977).

Wilkinson, D. T., in Kolb, E. W., et al., eds., *Inner Space/Outer Space* (University of Chicago Press, Chicago, 1986).

Winget, D. E., Hansen, C. J., Liebert, J., Van Horn, H. M., Fontaine, G., Nather, R. E., Kepler, S. O., and Lamb, D. Q., Astrophys. J., **315**, L77 (1987).

PROBLEMS

1. Consider a local geodesic reference frame whose origin is comoving with an expanding Friedmann universe. A particle of mass m is at a distance l from the origin. Use the equation of geodesic deviation (Eq. [6.120]) to show that the gravitational field of the universe exerts a (radial) tidal force

$$f = -mH_0^2 q_0 l$$

on this particle. Show that this force coincides with the Newtonian attractive force of the matter inside the sphere of radius l about the origin (see Eq. [18]).

2. Plot the value of the circumference of a circle as a function of the measured radius. Do this for the range $0 <$ [measured radius] $< \pi a$, for the uniformly curved three-geometries of both positive and negative curvature.

3. Plot the value of the square root of the area of a sphere as a function of the measured radius. Do this for the range $0 <$ [measured radius] $< \pi a$, for the uniformly curved three-geometries of both positive and negative curvature.

4. Expand the formulas for the volume of a sphere (Eqs. [70] and [78]) in powers of b/a; keep terms of order $(b/a)^2$. Consider a sphere with $b = 10^7$ light-years in a universe with $a = 10^{10}$ light-years. By what fraction does the volume deviate from $4\pi r^3/3$?

5. Some astrophysicists have proposed that the large redshifts of quasars are gravitational redshifts, rather than cosmological redshifts produced by the expansion of the universe. The quasar PKS 2000-330 exhibits a redshift of $z = 3.78$. Assume that this is a gravitational redshift, and assume that the quasar is a spherical mass of 10^6 $M_\odot$ that emits light from its surface. What must be the circumference of the sphere if the light is to suffer a redhshift of $z = 3.78$ as it escapes to large distance?

6. The space interval for the homogeneous isotropic geometry of positive curvature is (see Eq. [66])

$$dl^2 = \frac{dr^2}{1 - r^2/a^2} + r^2 d\theta^2 + r^2 \sin^2\theta \, d\phi^2$$

(a) What is the measured length of a path that starts at one point and proceeds once around the entire universe, on the straigthest possible line?

(b) Consider a circle $r = b$ around the origin. What is the surface area of this circle? Show that for $r \to 0$, your answer has the expected form.

(c) What is the maximum surface area that any circle in this universe can have?

7. Show that the deceleration parameter q can be expressed as follows as a function of η, for the Friedmann universes of positive and of negative curvatures, respectively:

$$q = \frac{1 - \cos\eta}{\sin^2\eta} \qquad q = \frac{\cosh\eta - 1}{\sinh^2\eta}$$

8. Show that the age of the universe can be expressed as follows in terms of H_0 and Ω_0, for Friedmann universes of positive and of negative curvatures, respectively,

$$t_0 = H_0^{-1} \frac{\Omega_0}{2(1 - \Omega_0)^{3/2}} \left[\cos^{-1}\left(\frac{2}{\Omega_0} - 1 \right) - \frac{2}{\Omega_0} (\Omega_0 - 1)^{1/2} \right]$$

$$t_0 = H_0^{-1} \frac{\Omega_0}{2(\Omega_0 - 1)^{3/2}} \left[\frac{2}{\Omega_0} (1 - \Omega_0)^{1/2} - \cosh^{-1}\left(\frac{2}{\Omega_0} - 1 \right) \right]$$

9. Find the true (measured) distance from the Earth to the quasar QSO(OH 471) which has a redshift of $z = 3.40$. Assume that our universe is a positive-curvature Friedmann universe with the parameters of Table 9.4.

10. Show that the motion of a particle in a Lemaitre universe is such that

$$ap = [\text{constant}]$$

Here p is the momentum of the particle as measured by an observer whose position coincides with that of the particle and whose velocity coincides with the local velocity of expansion of the universe (comoving observer). (Hint: Show that the $\mu = 1$

component of the geodesic equation is

$$\frac{d}{d\tau}\left[-a^2\,\frac{d\chi}{d\tau}\right] = 0$$

The distance measured by the observer is $dl = ad\chi$, and $p \propto dl/d\tau$.)

11. Derive the result of the preceding problem by noting that in quantum mechanics a particle of momentum p is represented by a wave of wavelength $\lambda = h/p$, and that this wavelength partakes of the expansion of the universe.

12. At present, the velocity of our Galaxy relative to the comoving coordinates of the universe is $\simeq 630$ km/s. Use the result of Problem 10 and the parameters for the negative-curvature Friedmann universe given in Table 9.4 to calculate the velocity of our Galaxy when the universe was $\frac{1}{10}$ of its present age. In this calculation, ignore the gravitational attraction between our Galaxy and any mass concentrations.

13. Prove that the Einstein universe is unstable, that is, prove that if a is slightly larger than $(\Lambda E)^{-1/2}$ then $\ddot{a} > 0$, and that if a is slightly smaller than $(\Lambda E)^{-1/2}$ then $\ddot{a} < 0$.

14. Find an expression, analogous to Eq. [151], for the volume of the visible universe in the case of the negative-curvature Friedmann model. Calculate the rate, in cubic light-years per year, at which the volume of the visible part of the universe is increasing. Also, calculate the rate at which the mass included in the visible universe is increasing. Use the values of the parameters given in Table 9.4 for the negative-curvature Friedmann universe. (Hint: The mass in the visible universe is directly proportional to the volume.)

15. The de Sitter universe has a spacetime interval

$$ds^2 = dt^2 - a(t)^2(d\chi^2 + \chi^2 d\theta^2 + \chi^2 \sin^2\theta\, d\phi^2)$$

where $\chi = r/a$ is the dimensionless radial coordinate, and

$$a(t) = a(0)e^{\sqrt{\Lambda/3}\,t}$$

(a) At time t, what is the true (measured) distance l from an observer located at the origin to the particle horizon?

(b) Verify that when the observer looks at the horizon, the light he receives has infinite redshift.

(c) What is the speed of recession dl/dt of the horizon?

16. A homogeneous, isotropic model of the universe with the spacetime interval

$$ds^2 = dt^2 - a(t)^2(d\chi^2 + \chi^2\,d\theta^2 + \chi^2\,\sin^2\phi\,d\phi^2)$$

has a radius of curvature that varies as follows with time:

$$a(t) = b\sqrt{t}$$

(a) What is the distance from our Galaxy to the particle horizon in this universe, that is, what is the distance to the most remote galaxy we can see at time t?

(b) Is there an event horizon in this universe, that is, is there some remote galaxy that we will never see?

17. Consider the de Sitter model with the spacetime interval given in Problem 13. Suppose that in this universe a light signal is sent out from the origin $\chi = 0$ at $t = 0$ and is received at $\chi > 0$ at time t.

(a) Find the redshift factor $\nu(\chi)/\nu(0)$ for this light signal.

(b) Show that for $\chi = 1/[a(0)\sqrt{\Lambda/3}]$, the redshift factor is zero, that is, the light signal is redshifted to zero frequency.

18. Suppose that our universe is a closed Friedmann universe with $a_* = 3.9 \times 10^{10}$ light-years and $a_0 = 2.5 \times 10^{10}$ light-years. Consider the quasar Q1208+10011, which has a redshift of $z = 4.80$.

(a) Find the coordinate interval χ from the Earth to this quasar.

(b) Find the true distance l (which would be measured with meter sticks) from the Earth to this quasar today.

(c) Find the true distance l between the Earth and this quasar at the time the light reaching us now was emitted.

19. Assume that our universe is a closed Friedmann model, with a mass density today of $\rho_0 = 8.5 \times 10^{-27}$ g/m^3 and a radius of curvature today of $a_0 = 2.5 \times 10^{10}$ light-years.

(a) What is the total volume of this universe today?

(b) What is the total mass in this volume? If all of this mass is in the form of baryons (protons and neutrons), what is the number of baryons in the universe?

(c) At what time will the universe reach maximum expansion?

(d) What will be the radius of curvature and the volume of the universe at maximum expansion?

20. Consider a closed Friedmann model of the universe; assume that $a_* = 3.8 \times 10^{10}$ light-years and that the radius of curvature today is $a_0 = 2.5 \times 10^{10}$ light-years.

(a) What is the distance (in light-years) from the origin to the particle horizon, that is, the distance to the most remote galaxy from which light has reached us?

(b) What is the volume (in cubic light-years) of the visible part of the universe?

(c) What is the speed of recession dl/dt of the particle horizon?

21. The positive-curvature Friedmann universe with the parameters given in Table 9.4 will begin to contract at an age of 1.2×10^{11} years. At this time redshifts of galaxies will begin to convert into blueshifts. Will all the redshifts suddenly turn to blueshifts? Will there be both redshifted galaxies and blueshifted galaxies? Is there any time at which we will see only blueshifted galaxies on the sky? For a universe aged $\simeq 18 \times 10^{11}$ years, make a rough (qualitative) plot of the observed frequency shift of galaxies as a function of distance from the Earth.

22. Consider a universe with the following values of the observational parameters:

$$\Omega_0 = 0, \quad q_0 = -0.5, \quad H_0 = 90 \text{ km s}^{-1} \text{ Mpc}^{-1}$$

Prepare a table listing the quantities that appear in Table 9.4.

23. Consider universes with the following values of the observational parameters:

(a) $\Omega_0 = 1.2, \quad q_0 = 0.6, \quad H_0 = 55 \text{ km s}^{-1} \text{ Mpc}^{-1}$
(b) $\Omega_0 = 1.0, \quad q_0 = 0.5, \quad H_0 = 55 \text{ km s}^{-1} \text{ Mpc}^{-1}$
(c) $\Omega_0 = 0.4, \quad q_0 = 0.2, \quad H_0 = 55 \text{ km s}^{-1} \text{ Mpc}^{-1}$

Prepare a table giving the quantities listed in Table 9.4 for each of these universes.

10. The Early Universe

LET THERE BE LIGHT.
Genesis, 1.3

Extrapolating the present motion of expansion of the universe backward in time, we conclude that the early universe must have been very dense. And extrapolating the (adiabatic) expansion of the cosmic blackbody radiation backward in time, we conclude that the early universe must have been very hot. Thus, at an early time, the universe must have been very different from what it is now. There were no stars and no galaxies, but only a uniform hot plasma, consisting of free electrons and free nuclei. The chemical composition of the early universe must also have been different. The heavy elements (elements other than hydrogen, deuterium, helium, and lithium) in our immediate environment were formed by nuclear reactions in the cores of stars, and these elements therefore did not exist in the early universe. At very early times, the violent thermal collisions would have prevented the existence of any kind of nucleus, and the matter in the universe must have been in the form of free electrons, protons, and neutrons. At the earliest times, even the protons and neutrons would have been disrupted, and the universe must have contained a mix of quarks and other elementary particles.

The observed expansion of the universe and the observed cosmic blackbody radiation provide the empirical basis for a Friedmann model of the universe with a Big Bang, sometimes called the *Standard*

Model. Further evidence supporting this model is provided by calculations of the synthesis of helium in the universe. Although stars make helium by the thermonuclear burning of hydrogen, most of the helium in the universe must be primordial, since it is found even in stars that have not yet burned long enough to accumulate a significant amount of helium. This primordial helium was formed by nuclear reactions in the early universe, and the abundance of this helium (relative to hydrogen) can be calculated by examining the thermal equilibrium attained by protons and neutrons in reactions in the early, hot universe. The numbers obtained in such calculations of the helium abundance are in excellent agreement with the observational data. The abundances of some other light elements formed in the early universe can be calculated similarly, again with excellent results.

At the earliest times, the universe must have contained a mixture of many kinds of elementary particles. Because of this, particle physics has come to play a central role in the study of cosmology. Grand Unified Theories (GUTs) predict that at energies of the order of 10^{14} GeV, the weak, electromagnetic, and strong interactions merge into a single interaction, endowed with a high symmetry. Energies of this order are far beyond anything attainable with accelerators. However, at a sufficiently early time, particles in the early universe could easily have had thermal energies of this order. Hence theories of elementary particles and their interactions at extreme energies help us understand the behavior of the early universe; conversely, the behavior of the early universe, as revealed by the presence (or the absence) of relic particles, helps us understand particle physics. Particle physicists have come to view the early universe as a testing ground for their theories.

10.1 THE TEMPERATURE OF THE EARLY UNIVERSE

At present, the universe is cold and it is matter dominated. At a temperature of only 2.7 K, the contribution of the cosmic blackbody radiation to the mass density of the universe is negligible compared with the contribution of massive particles. However, at an earlier time, the universe was much hotter and it was radiation dominated. To see this, let us compare the densities of matter and radiation. Since the volume of the universe is directly proportional to a^3, the density of matter is inversely proportional to a^3,

$$\rho_{matter} = \rho_0 \left(\frac{a_0}{a} \right)^3 \qquad [1]$$

According to the Stefan-Boltzmann law, the (mass) density of the radiation is proportional to the fourth power of the temperature,

$$\rho_{rad} = 4 \left(\frac{\pi^2 k^4}{60 \hbar^3} \right) T^4 \qquad [2]$$

where $\pi^2 k^4 / 60 \hbar^3 = \pi^2 k^4 / 60 \hbar^3 c^2 = 5.67 \times 10^{-5}$ g/s$^3 \cdot$K^4 is the Stefan-Botzmann constant (in cgs units, besides the factor of $1/c^2$ in the Stefan-Boltzmann constant, an extra factor of $1/c^3$ is required on the right side of Eq. [2]). Since $T \propto 1/a$, we obtain

$$\rho_{rad} = \rho_{rad,0} \left(\frac{a_0}{a} \right)^4 \qquad [3]$$

By comparing Eqs. [1] and [3], we see that in the early universe, when a is small, the density of radiation dominates.

The transition from a radiation-dominated universe to a matter-dominated universe occurs when $\rho_{matter} = \rho_{rad}$, that is, when

$$\rho_0 \left(\frac{a_0}{a} \right)^3 = \rho_{rad,0} \left(\frac{a_0}{a} \right)^4 \qquad [4]$$

The present density of the radiation is $\rho_{rad,0} = 4.5 \times 10^{-34}$ g/cm^3, which corresponds to $T = 2.7$ K in Eq. [2]. The present density of matter is somewhat uncertain. It is convenient to write this density as $\rho_0 = \Omega_0 \rho_c = \Omega_0 \times 1.88 \, h^2 \times 10^{-29}$ g/cm^3 (see Eq. [9.26]), so the uncertainties in the matter density are contained within Ω_0 and h. The condition [4] then leads to

$$\frac{a_0}{a} = \frac{\rho_0}{\rho_{rad,0}} = 2.3 \times 10^4 \, \Omega_0 h^2 \qquad [5]$$

This shows that matter and radiation had equal densities when the universe was about 10^4 times smaller than it is now. Correspondingly, the temperature of the cosmic blackbody radiation was about 10^4 times larger than it is now,

$$T = \frac{a_0}{a} T_0 = 2.3 \times 10^4 \, \Omega_0 h^2 \, T_0 = 6.3 \times 10^4 \text{ K} \times \Omega_0 h^2 \qquad [6]$$

To calculate the time at which the criterion [5] is satisfied, we need to specify the model of the universe and the values of Ω_0 and h. For a typical Friedmann model, such as contemplated in Section 9.11, the time is about 10^3 years. This time may be said to mark the beginning of the matter era and the end of the radiation era of the universe. However, the criterion of equal matter and radiation densities does not provide a sharp dividing line, since radiation will continue to contribute a significant fraction of the mass density and of the pressure for many thousand years after the instant of equal densities. Some cosmologists prefer to place the end of the radiation era at the time when the ionized hydrogen plasma in the universe changed into neutral hydrogen gas ("recombination" of hydrogen). At the prevailing density, this recombination occurs at a temperature of about 3000 K--at higher (earlier) temperatures, the hydrogen gas is almost totally ionized; at lower (later) temperatures, the hydrogen gas is almost entirely neutral. The time at which this transition occurs is about 10^5 years.

The recombination of hydrogen gas led to a decoupling of radiation and matter. Before recombination, there was a high density of free electrons, and the scattering of the photons by these free electrons maintained thermal equilibrium between matter and radiation. After recombination, the scattering of photons was rare--neutral hydrogen is almost completely transparent to radiation, and hardly ever scatters a photon. Thus, after recombination, the photons ceased to interact with matter, and their temperature became independent of the temperature of matter.

To formulate the Einstein equations for a radiation-filled universe, we need the energy-momentum tensor for the radiation. If we regard the radiation as a fluid with a mass density ρ and a pressure P, its energy-momentum tensor is (see Problem 2.28)

$$T_\mu{}^\nu = (P + \rho_{rad})u_\mu u^\nu - \delta_\mu{}^\nu P \qquad [7]$$

Here, u_μ is the flow velocity of the fluid. Since in our comoving coordinates, the fluid is at rest, $u_k = 0$ and $u_0 = a$, $u^0 = 1/a$ (see Exercise 9.7). Hence

$$T_0{}^0 = (\rho_{rad} + P)u_0 u^0 - P = (\rho_{rad} + P) - P = \rho_{rad} \qquad [8]$$

In a Friedmann universe of positive curvature, the Einstein equation [9.100] with $\Lambda = 0$ then becomes

$$\frac{3}{a^4}(a^2 + \dot{a}^2) = 8\pi G \rho_{rad} \qquad [9]$$

Note that according to this equation, the pressure of the radiation has no effect whatsoever on the expansion of the universe--the pressure cancels in Eq. [8] and does not appear in Eq. [9]. The reason why the pressure of a fluid cannot affect the dynamics of the universe is that the fluid is homogeneous. The pressure is the same at all points in the fluid, and there are no pressure gradients. Hence each element of fluid experiences exactly the same pressure forces from all sides, and there is no net force on the fluid that might accelerate or decelerate it.

Although the Big Bang is sometimes described as a primeval "explosion," it differs from ordinary explosions in a crucial way--its outward motion is the result of initial conditions, not the result of outward pressure forces.

The mass density of the radiation is proportional to $1/a^4$ (see Eq. [3]); hence we can write the Einstein equation as

$$\frac{3}{a^4}(a^2 + \dot{a}^2) = 8\pi G \rho_{rad,0} \frac{a_0^4}{a^4} \qquad [10]$$

In the early universe, a is small, and the term $\propto 1/a^4$ on the right side of this equation is much larger than the term $\propto 1/a^2$ on the left side. We can therefore neglect the latter term, and write the equation in the approximate form

$$\frac{3}{a^4}\dot{a}^2 = 8\pi G \rho_{rad,0} \frac{a_0^4}{a^4} \qquad [11]$$

or, in terms of the time derivative $da/dt = \dot{a}/a$,

$$\frac{3}{a^2}\left(\frac{da}{dt}\right)^2 = 8\pi G \rho_{rad,0} \frac{a_0^4}{a^4} \qquad [12]$$

The term we have neglected in Eqs. [11] and [12] is the "curvature" term, which distinguishes between the Friedmann universes of positive, negative, or zero curvature. Thus, the approximate equation [12] is equally valid for all these Friedmann models--near the Big Bang, the curvature of the space geometry makes no difference to the dynamics of the universe. We already noticed this insensitivity of the early universe to the curvature of space in Figs. 9.13-9.15, where we saw that near $t = 0$, the behavior of $a(t)$ is the same for all the Friedmann models.

The solution of Eq. [12] is trivial,

$$a(t) = \left[\frac{32\pi G}{3} a_0^4 \rho_{rad,0}\right]^{1/4} t^{1/2} \qquad [13]$$

Accordingly, the mass density of the radiation is

$$\rho_{rad}(t) = \rho_{rad,0} \frac{a_0^4}{a^4} = \frac{3}{32\pi G}\frac{1}{t^2} \qquad [14]$$

Note that the constant $\rho_{rad,0} a_0^4$ has canceled in this expression, and the dependence of the density on the time is uniquely fixed by Eq.

[14], which contains no reference to the conditions in the universe now. Likewise, the dependence of the temperature of the radiation on the time is

$$T = \left[\frac{\rho_{rad}}{4\pi^2k^4/60\hbar^3}\right]^{1/4} = \left[\frac{45\hbar^3}{32\pi^3k^4G}\right]^{1/4}\frac{1}{t^{1/2}} \qquad [15]$$

This shows that the temperature becomes arbitrarily large as $t \to 0$. Eq. [15] can be expressed concisely in terms of the Planck mass $m_{Pl} = \sqrt{\hbar/G} = \sqrt{\hbar c/G} = 2.18 \times 10^{-5}$ g $= 1.22 \times 10^{19}$ GeV:

$$k^2T^2 = \frac{1}{4\sqrt{2}}\sqrt{\frac{45}{\pi^3}}\frac{\hbar m_{Pl}}{t} = \frac{0.301}{\sqrt{2}}\frac{\hbar m_{Pl}}{t} \qquad [16]$$

It is easy to check that for the radiation-dominated universe, with $a \propto t^{1/2}$, the Hubble age is $H^{-1} = 2t$. As a function of the temperature this becomes

$$H^{-1} = 2t = \frac{1}{2\sqrt{2}}\sqrt{\frac{45}{\pi^3}}\frac{\hbar m_{Pl}}{k^2T^2} = \frac{0.602}{\sqrt{2}}\frac{\hbar m_{Pl}}{k^2T^2} \qquad [17]$$

■ **Exercise 1.** Show that for the radiation-dominated universe, the Hubble age equals $2t$, that is, $H^{-1} = a/(da/dt) = 2t$. ■

■ **Exercise 2.** Verify that the exact solution of Eq. [12] is

$$a(\eta) = b_*\sin\eta \qquad [18]$$

$$t = b_*(1 - \cos\eta) \qquad ■ \qquad [19]$$

At very early times, the universe was so extremely hot that the typical thermal energies of the blackbody photons were large enough to permit the creation of electron-positron pairs. The required temperature is given by $kT \simeq m_ec^2$, that is, $T \simeq 6 \times 10^9$ K. According to Eq. [15], this temperature corresponds to an age of $\simeq 10$ s.

At higher temperatures, the electrons, positrons, and also neutrinos form a relativistic gas of leptons. Such a relativistic gas of particles has properties that are very similar to those of the gas of photons making up the blackbody radiation. The energy density of the gas is proportional to the fourth power of the temperature, but because there are several kinds of leptons, the constant of proportionality is somewhat larger. Thus, the universe was lepton dominated at very early times.

The dynamical equation for the lepton-dominated universe is essentially the same as for the radiation-dominated universe. Only some of the constants of proportionality are somewhat altered; for rough estimates of the thermal history of the universe, we can ignore these alterations.

From Eq. [15] we find that at an age of 10^{-4} s, the universe had a temperature of 10^{12} K, which was sufficient to permit the creation of muons and pions by thermal fluctuations. At slightly earlier times even the creation of protons, neutrons, and other baryons and antibaryons as well as mesons and antimesons was possible, and the universe was dominated by strongly-interacting particles, or hadrons. The earliest time at which it is meaningful to speak of more or less well-defined spacetime geometry that can serve as background for the motion of particles is the *Planck time*, $\sqrt{\hbar G/c^5} \simeq 10^{-43}$ s. Before this time, the universe was dominated by quantum fluctuations in the geometry, about which we know nothing.

We can therefore distinguish four eras in the chronology of the universe (see Fig. 10.1):

(i) The Hadron Era ($t = 10^{-43}$ to 10^{-4} s; $T = 10^{31}$ to 10^{12} K; $kT = 10^{19}$ to 1 GeV)
During this era, the universe contained a highly relativistic plasma of elementary particles and antiparticles of all kinds--quarks, gluons, electrons, muons, taus, photons, W and Z bosons, neutrinos, and gravitons. The numbers of baryons and antibaryons in this primordial fireball were nearly equal, but there was a small excess of baryons (representing a small fraction of the whole). As the temperature dropped toward 10^{12} K, the baryons and antibaryons annihilated, leaving only the small excess of baryons which accounts for the number of baryons found in the present universe. Near the end of this era, at $t \simeq 10^{-6}$ s, the quarks combined into nucleons (such as protons and neutrons) and mesons (such as pions).

(ii) The Lepton Era ($t = 10^{-4}$ to 10 s; $T = 10^{12}$ to 4×10^9 K; $kT = 1$ GeV to 1 MeV)
In this era, only a much smaller number of particles survived. The remaining particles were electrons, muons, taus, neutrinos, and their corresponding antiparticles, and also photons, gravitons, protons, and neutrons. Even today the neutrinos and antineutrinos from this era should still be with us and fill our intergalactic space with a neutrino gas whose temperature is 71% of the cosmic blackbody radiation temperature. Since the interactions of neutrinos are extremely weak, there is not much hope for the direct detection of this cosmic neutrino gas.

(iii) The Radiation Era ($t = 10$ to 10^{12} s; $T = 4 \times 10^9$ to 10^4 K; $kT = 1$ MeV to 10 eV)
This era began when electrons annihilated, leaving only a relatively small residue which balanced the positive charge density of the protons. The density of the remaining electrons, and that of protons and neutrons, was negligible compared with the densities of photons and neutrinos.

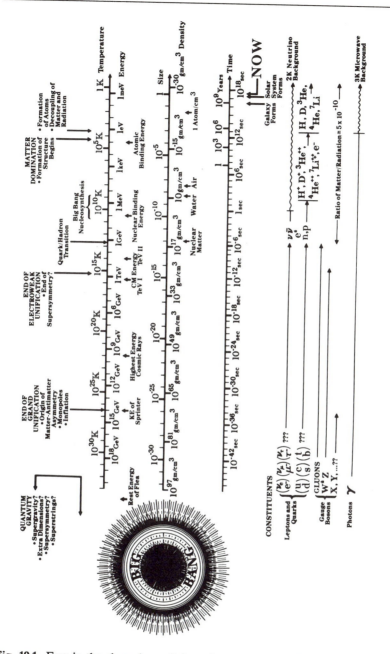

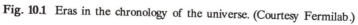

Fig. 10.1 Eras in the chronology of the universe. (Courtesy Fermilab.)

During this era, at a temperature of $\simeq 10^9$ K, the neutrons fused with the protons and formed ^{4}He nuclei. About 25% (by mass) of the matter was made into helium and some very small amounts were made into deuterium, lithium, and other light nuclei (see Section 10.4).

(iv) The Matter Era $(t > 10^{12}$ s; $T < 10^4$ K; $kT < 10$ eV$)$

The most recent era began when the density of radiation dropped below the density of matter. The exact birthdate of this era depends on the present mass density of the universe and is therefore somewhat uncertain. Near the beginning of this era, electrons combined with protons and formed neutral atoms of hydrogen and helium. This happened when the temperature dropped to $\simeq 3000$ K.

Once matter became neutral, the interaction between matter and radiation was drastically reduced, and there was insufficient exchange of energy to maintain thermal equilibrium. Matter and radiation expanded and cooled separately, each with its own temperature. The temperature of matter decreases as $T \propto 1/a^{2}$*; the temperature of radiation decreases as $T \propto 1/a$. Hence, the temperature of matter ultimately dropped below that of the radiation.

As long as matter and radiation were coupled, any small irregularities in the matter distribution were held in check by the diffusion of photons, which provides viscous damping. But once matter was on its own, a region with a slight excess of mass tended to grow by the gravitational attraction it exerted on its surroundings. By some such condensation process galaxies began to form at $t \simeq 10^{16}$ s. And within one of these galaxies the Solar System formed at $t \simeq 10^{17}$ s after the Big Bang.

10.2 PARTICLES IN THERMAL EQUILIBRIUM

We cannot say much about the behavior of the matter in the universe before the Planck time, 10^{-43} s. At these extremely early times, the universe was dominated by unknown quantum fluctuations in the geometry. But after $t = 10^{-43}$ s, the constituents of the matter in the universe must have been quarks, leptons, photons, and gauge bosons. In a static universe, the collisions and reactions among these particles would ensure thermal equilibrium between them. However, in the quickly expanding early universe, thermal equilibrium is not so easily attainable. The universe expands with a characteristic time of H^{-1}, and whether thermal equilibrium can be attained depends on how this characteristic expansion time compares with the characteristic reaction time of the particles. If the characteristic expansion time is long compared with the characteristic reaction time, then the reaction rate can keep up with the expansion rate, and the matter in the universe will

---

* This corresponds to the adiabatic expansion of a perfect gas.

attain a quasistatic thermal equilibrium, with a well-defined instantaneous temperature.

To compare these characteristic times, let us express them in terms of the temperature. Since all the particles in the early universe are extremely relativistic, H^{-1} is roughly given by $\hbar m_{Pl}/k^2T^2$, which is the result we found for photons (Eq. [17]). The characteristic reaction time for particles with a density n and a reaction cross section σ is $1/nv\sigma$, or $1/n\sigma$, since for the particles in the early universe the speed is $v \simeq 1$. The particle density is of the order of magnitude of the photon density, $n \simeq (kT)^3/\hbar^3$ (for a more accurate formula, see Eq. [36]). Cross sections for typical reactions mediated by massless bosons, such as the photon, are of the order of $\alpha^2\hbar^2/E^2 \simeq \alpha^2\hbar^2/k^2T^2$, where α is the fine-structure constant, $\alpha \simeq 1/137$. Hence the criterion $H^{-1} > 1/n\sigma$ for thermal equilibrium is $\hbar m_{Pl}/(kT)^2 > \hbar/\alpha^2kT$, or roughly

$$kT < \alpha^2 m_{Pl} \simeq 10^{16} \text{ GeV} \qquad [20]$$

Thus, in the very early stages of the hadron era, such reactions are not sufficient to maintain thermal equilibrium, but after about $t \simeq 10^{-38}$ s (corresponding to $T \simeq 10^{16}$ GeV), the reaction rate exceeds the expansion rate, and thermal equilibrium is attained. What happens before 10^{-38} s remains unclear; perhaps some other, unknown reactions maintain thermal equilibrium even at these very early times.

The equilibrium distribution of a gas of particles at a given temperature T is well known from statistical physics. For a gas of fermions (particles of half-integer spin), the number of particles in a given quantum state is

$$n(E) = \frac{1}{e^{(E - \mu)/kT} + 1} \qquad [21]$$

where μ is the chemical potential, or the Fermi energy. For a gas of bosons (particles of integer spin),

$$n(E) = \frac{1}{e^{(E - \mu)/kT} - 1} \qquad [22]$$

The number of states per unit volume in an energy interval dE is $g \times 4\pi p^2 dp/(2\pi\hbar)^3 = g/2\pi^2\hbar^3 \times \sqrt{E^2 - m^2}\,E\,dE$, where g is a statistical factor that represents the number of spin states (for massive particles, $g = 2s + 1$; for massless particles, such as photons, $g = 2$; for neutrinos, $g = 1$, since they are always left-handed). Accordingly, the number density of particles, the energy density, and the pressure are

$$n = \frac{g}{2\pi^2\hbar^3} \int_m^\infty \frac{\sqrt{E^2 - m^2}}{e^{(E-\mu)/kT} \pm 1} E \, dE \qquad [23]$$

$$\rho = \frac{g}{2\pi^2\hbar^3} \int_m^\infty \frac{\sqrt{E^2 - m^2}}{e^{(E-\mu)/kT} \pm 1} E^2 \, dE \qquad [24]$$

$$P = \frac{g}{6\pi^2\hbar^3} \int_m^\infty \frac{(E^2 - m^2)^{3/2}}{e^{(E-\mu)/kT} \pm 1} \, dE \qquad [25]$$

The chemical potentials of different species of particles are related by the reactions in which these particles participate. In any reaction, the sums of the chemical potentials before and after the reaction must be equal. Thus, by examining reactions, we can extract information about the chemical potentials. For example, since photons can be produced singly in reactions such as

$$e^- + p \rightarrow e^- + p + \gamma \qquad [26]$$

we immediately recognize that the chemical potential for photons must be zero. And since any particle can annihilate with its antiparticle and produce nothing but photons, for instance,

$$e^- + e^+ \rightarrow \gamma + \gamma \qquad [27]$$

we recognize that the chemical potential of the particle must necessarily equal the negative of the chemical potential of the antiparticle, for instance, $\mu_{e^-} = -\mu_{e^+}$.

The absolute conservation laws for the baryon number, the three "flavors" of lepton number (electron, mu, and tau), and electric charge restrict possible reactions among particles and therefore also restrict the information about the chemical potentials we can extract by examining such reactions. For instance, conservation of baryon number implies that the same number of baryons minus antibaryons must appear on each side of a reaction. Consequently, the reaction cannot reveal the value of a common additive constant in the chemical potentials of all the baryons. Similarly, reactions cannot reveal such common additive constants in the chemical potentials of the three flavors of leptons, and in the chemical potentials of charged particles. Furthermore, reactions cannot reveal an overall additive constant in *all* the chemical potentials. Altogether, there are then five numbers that must be determined by other means, or five chemical potentials that may be regarded as independent of the reactions. For example, we may take the five independent chemical potentials as μ_p, μ_{e^-}, μ_{ν_e}, $\mu_{\nu_{mu}}$, and $\mu_{\nu_{tau}}$. To evaluate these five chemical potentials, we need some other information, such as the baryon density, lepton density, and charge density in the early universe. Note that in the expanding universe, the chemical potentials are not necessarily constant; in general, they depend on the radius of the universe, that is, they depend on time. If the number of particles is conserved, this simply corresponds to the familiar dependence of the Fermi energy on the volume in which the gas is confined--when we increase the available volume, the Fermi energy decreases.

Unfortunately, information about the chemical potentials in the early universe is scarce. The chemical potentials of baryons are believed to be small. This belief is rooted in the observation that in the present universe the ratio of the baryon density and the photon density is very small, about 10^{-9}. The number of baryons, that is, the number of baryons minus the number of antibaryons, is exactly conserved during the expansion of the universe. The number of photons is approximately conserved (as we saw in Section 9.5, during the expansion, the wavelength of each photon changes, but no photons are created or destroyed). Hence the ratio of baryon density and photon density in the early universe must have been equally small,

$$\frac{n_B - n_{\bar{B}}}{n_\gamma} \simeq 10^{-9} \qquad [28]$$

The small value of the baryon density implies an equally small electron density. Such a match between the baryon density and the electron density is required for overall charge neutrality in the universe. At most, the positive electric charge density on the baryons is $e \times (n_B - n_{\bar{B}})$ (if almost all of the baryons have positive charge, like the proton). This positive charge must be matched by the negative charge density on the electrons, $-e \times (n_{e^-} - n_{e^+})$. Hence

$$\frac{n_{e^-} - n_{e^+}}{n_\gamma} < 10^{-9} \qquad [29]$$

We expect that during the lepton era, when the electrons were presumably the most abundant particles of negative charge, the electron density was close to the upper limit set by Eq. [29]. But if other negatively charged particles were present (such as muons), then the electron density might have been considerably below this upper limit.

From the limit on the electron density, we can obtain a limit on the electron chemical potential. According to Eq. [23], the net electron density is

$$n_{e^-} - n_{e^+} = n(\mu) - n(-\mu)$$

$$= \frac{g}{2\pi^2\hbar^3} \int_m^\infty E\sqrt{E^2 - m^2} \left[\frac{1}{e^{(E - \mu_{e^-})/kT} + 1} - \frac{1}{e^{(E + \mu_{e^-})/kT} + 1} \right] dE \quad [30]$$

With the approximation $m \ll kT$, the integral can be evaluated explicitly,

$$n_{e^-} - n_{e^+} = g \frac{(kT)^3}{6\pi^2\hbar^3} \left[\pi^2 \left(\frac{\mu_{e^-}}{kT} \right) + \left(\frac{\mu_{e^-}}{kT} \right)^3 \right] \quad [31]$$

■ *Exercise 3.* Evaluate the integral in Eq. [30]. ■

Since $n_\gamma \simeq (kT)^3/\hbar^3$, Eqs. [29] and [31] tell us that

$$\frac{\mu}{kT} < 10^{-9} \quad [32]$$

It is more difficult to place tight limits on the chemical potentials of baryons. The net baryon number $n_B - n_{\bar{B}}$ is the sum of the individual baryon numbers of all the distinct baryon species,

$$n_B - n_{\bar{B}} = \sum_i n(\mu_i) - n(-\mu_i) = \sum_i g_i \frac{(kT)^3}{6\pi^2\hbar^3} \left[\pi^2 \left(\frac{\mu_i}{kT} \right) + \left(\frac{\mu_i}{kT} \right)^3 \right] \quad [33]$$

where the second equality is valid only if all the baryons are relativistic (otherwise we must use the general expression [30]). The chemical potentials in Eq. [33] can be expressed as linear combinations of the five independent chemical potentials, μ_p, μ_{e^-}, μ_{ν_e}, $\mu_{\nu_{mu}}$, and $\mu_{\nu_{tau}}$. The chemical potentials of the leptons enter into these combinations because some of the baryon chemical potentials are related to lepton chemical potentials, for example, the beta-decay reaction

$$n \leftrightarrow p + e^- + \bar{\nu} \quad [34]$$

implies

$$\mu_n - \mu_p = \mu_{e^-} - \mu_\nu \quad [35]$$

Such relationships among baryon and lepton chemical potentials mean that if we want to use the limit [28] to establish that each of the baryon chemical potentials is small, we must first establish that the difference $\mu_{e^-} - \mu_\nu$ is small (and also that similar differences for the muon and tau lepton families are small). Unfortunately, we are unable to do this. The best available limit on $\mu_{e^-} - \mu_\nu$ comes from calculations of the nucleosynthesis of helium and calculations of the mass density contributed by a neutrino background in the universe. As we will see in Section 10.4, helium is produced in the early universe by the combination of neutrons with protons, and the resulting abundance of helium depends on the initial ratio of neutrons to protons. This ratio, in turn, is related to the difference $\mu_n - \mu_p$, or, alternatively, to the difference $\mu_{e^-} - \mu_\nu$ (see Eq. [35]). The calculations of the helium abundance set a limit of $|\mu_{e^-} - \mu_\nu| < \simeq kT$ at the time of nucleosynthesis (Bianconi, Lee, and Ruffini, 1990, 1991). Thus, the available information permits the chemical potentials to have values of the order of kT. The limits that calculations of the mass density impose on the chemical potentials of neutrinos will be discussed in Section 10.3; again these limits permit the chemical potentials to have values of the order of kT or even larger.

If some of the chemical potentials on the right side of Eq. [33] are as large as kT or larger, then any unknown chemical potential that we try to calculate from this equation (or from a similar equation that expresses the condition of zero electric charge density in the universe) may turn out to be of the order of magnitude of kT or larger. Thus, the small observed value of the net baryon number does not guarantee that μ_i/kT for each baryon species is small--the small value of the net baryon number may result from large cancellations between large values of μ_i/kT for distinct baryon species.

In view of the absence of conclusive information about the values of the chemical potentials in the early universe, we will assume that $|\mu| \ll kT$ for all the chemical potentials. Although this assumption is somewhat arbitrary, it has the virtue of simplicity.

Each species of particle (and each species of antiparticle) contributes a term of the form [24] or [25] to the total density or pressure in the universe. The dominant contributions to the total density or pressure arise from those particles that are relativistic at the given temperature, that is, particles such that $m \ll kT$. Going backward in time, we first encounter this dominance of relativistic particles when we reach the radiation era, where the density of (relativistic) photons exceeds the rest-mass density of matter, at $t \simeq 10^{12}$ s. As we go farther backward in time, the temperature increases, and electrons and muons become relativistic, and then protons and neutrons become relativistic, etc. Each relativistic species contributes a density and a pressure of about the same order of magnitude as the photon density and pressure. (The exceptions are particle species with a large negative value of the chemical potential, $\mu << - kT$; such particles have a strongly depressed density. However, for every such particle species, there is an antiparticle species with a large positive value of the chemical potential, $\mu >> kT$, and these antiparticles have a strongly enhanced density, and they contribute even more to the density and the pressure than a relativistic species of small chemical potential.)

In view of this dominance of relativistic particles, we can make the approximation $m << kT$ in Eqs. [23]-[25] when dealing with the early universe. Furthermore, as stated above, we will also make the approximation $|\mu| << kT$. These approximations mean that we can neglect m and μ in the integrands in Eqs. [23]-[25], and we can extend the range of integration to 0. The integrals can then be expressed in terms of Riemann zeta functions (Landau and Lifshitz, 1958). For fermions, this leads to the results

$$n = 1.202 \times \frac{3k^3}{4\pi^2\hbar^3} gT^3 \qquad [36]$$

$$\rho = \frac{7}{8} \frac{\pi^2 k^4}{30\hbar^3} gT^4 \qquad [37]$$

$$P = \frac{\rho}{3} \qquad [38]$$

and for bosons,

$$n = 1.202 \times \frac{k^3}{\pi^2\hbar^3} gT^3 \qquad [39]$$

$$\rho = \frac{\pi^2 k^4}{30\hbar^3} gT^4 \qquad [40]$$

$$P = \frac{\rho}{3} \qquad [41]$$

The total density and pressure are the sums of all the densities and pressures contributed by all the fermions and bosons. These totals can be expressed as

$$\rho = \frac{\pi^2 k^4}{30\hbar^3} g_* T^4 \qquad [42]$$

$$P = \frac{\rho}{3} = \frac{\pi^2 k^4}{90\hbar^3} g_* T^4 \qquad [43]$$

where g_* is a sum over the fermion and boson g factors,

$$g_* = \sum_{\text{bosons}} g_i + \frac{7}{8} \sum_{\text{fermions}} g_i \qquad [44]$$

Eq. [42] resembles Eq. [2] for photons. The only difference is that our new equation has an extra factor of g_*. Note that this overall statistical factor of our mix of fermions and bosons depends on temperature, and hence on the radius of the universe, since the particle species that contribute to the sum [44] are different in different eras. For instance, at the earliest times, when the particles are quarks, leptons, gauge bosons, gluons, and Higgs particles, $g_* \simeq 100$; whereas at the latest times, when the only relativistic particles that remain thermally coupled are photons, $g_* = 2$. This temperature dependence of g_* complicates the solution of the Einstein equations. But we can avoid this complication if we solve the Einstein equations separately for each era. Thus, for any time interval within which g_* is approximately constant, Eq. [14] for the time dependence of ρ remains approximately valid,

$$\rho = \frac{3}{32\pi G} \frac{1}{t^2} \qquad [45]$$

Consequently, we again obtain Eqs. [16] and [17]. The only change is that the factor of $\sqrt{2}$ must be replaced by $\sqrt{g_*}$ (that is, the statistical factor of 2 of the photon gas must be replaced by the statistical factor g_* of our mix of fermions and bosons),

$$k^2 T^2 = \frac{1}{4\sqrt{g_*}} \sqrt{\frac{45}{\pi^3}} \frac{\hbar m_{Pl}}{t} = \frac{0.301}{\sqrt{g_*}} \frac{\hbar m_{Pl}}{t} \qquad [46]$$

and

$$H^{-1} = \frac{1}{2\sqrt{g_*}} \sqrt{\frac{45}{\pi^3}} \frac{\hbar m_{Pl}}{k^2 T^2} = \frac{0.602}{\sqrt{g_*}} \frac{\hbar m_{Pl}}{k^2 T^2} \qquad [47]$$

Note that Eq. [46] is only an approximation. Since ρ_0 has a discontinuity whenever g_* has a discontinuity, whereas $a(t)$ must necessarily be continuous, we need to introduce an extra adjustable constant in the solution of the Einstein equation [13]. The general solution of this equation is then not $a \propto t^{1/2}$, but $a \propto (t + b)^{1/2}$, where b is a constant of integration. Whenever g_* has a discontinuity, we need to readjust the constant b so as to keep $a(t)$ continuous. This alters the relation [46] between the temperature and the time, but does not affect the relation [47] between H^{-1} and the temperature.

10.3 *NEUTRINO DECOUPLING*

When the temperature drops to $kT \simeq 1$ MeV, the weak interactions cannot continue to maintain thermal equilibrium of the neutrinos. At this point the neutrinos decouple from the rest of the matter in the universe, and the neutrino distribution evolves independently of the other particle distributions, which remain in thermal equilibrium. After decoupling, each neutrino behaves as a free particle. For any free particle moving in an expanding universe, the momentum decreases in inverse proportion to the radius of the universe: if the momentum is initially p, and the universe expands from a to a', then the momentum decreases to a new value p',

$$p' = \frac{a}{a'} p \qquad [48]$$

This redshift of the neutrino momentum can be explained in the same way as the redshift of the photon momentum and energy (after the photons have decoupled; see Section 9.5). Quantum mechanically, the neutrino can be regarded as a standing wave in a container of expanding size. If the universe expands by a factor a'/a, the wavelength expands by the same factor, and the momentum $p = h/\lambda$ decreases by the inverse factor, as in Eq. [48]. Alternatively, this result for the redshift of the neutrino momentum can be obtained by integration of the geodesic equation in the expanding universe (see Problem 9.10).

The initial, thermal distribution of neutrinos just before decoupling is given by the fermion distribution [21], with $E = \sqrt{p^2 + m^2}$. After decoupling, the number of neutrinos in each quantum state remains constant, and only the momentum of this state changes. Hence we obtain the redshifted distribution of neutrinos at any later time by replacing p by $(a'/a)p'$, which gives

$$n(p') = \frac{1}{\exp[(\sqrt{p^2 + m^2} - \mu)/kT] + 1} \qquad [49]$$

$$= \frac{1}{\exp\{[\sqrt{(a'/a)^2 p'^2 + m^2} - \mu]/kT\} - 1} \qquad [50]$$

If the neutrino mass is zero, this reduces to

$$n(p') = \frac{1}{\exp\{[(a'/a)p' - \mu]/kT\} + 1} \qquad [51]$$

which is simply a thermal fermion distribution with a reduced temperature $T' = (a/a')T$ and a reduced chemical potential $\mu' = (a/a')\mu$,

$$n(p') = \frac{1}{\exp[(p' - \mu')/kT'] + 1} \tag{52}$$

Thus, the neutrino distribution remains thermal, even though there are no interactions to thermalize it. This thermal behavior of zero-mass neutrinos is similar to the behavior of photons (after the photons decouple).

In spite of this similarity of zero-mass neutrinos and photons, the temperature of the neutrinos in the present universe is somewhat lower than the temperature of the photons. The reason is that the photons decouple later than the neutrinos, and in the meantime the electrons and positrons in the early universe annihilate, dumping their rest-mass energy into the universe in the form of heat. This reheating of the universe benefits the photons, but not the neutrinos, which are already decoupled, and therefore remain unaffected.

To calculate how much of an increment of the photon temperature results from the electron-positron annihilation, we note that before this annihilation, $g_* = 11/2$ (in Eq. [44], the photons contribute $g_\gamma = 2$, and the electrons and positrons contribute $g_{e^-} = 2$ and $g_{e^+} = 2$). After this annihilation, the only remaining relativistic thermal particles are the photons, with $g_* = 2$. Thus, g_* decreases from $11/2$ to 2. The effect of this decrease of g_* on the temperature T can be calculated by appealing to the conservation of entropy. The entropy is constant because the process of expansion is adiabatic--there is no external heat input into the universe. For a relativistic gas of bosons, such as photons, the entropy density is $(2\pi^2 k^4/45\hbar^3)T^3$, and for a mixture of several boson and fermion gases, the entropy density is $(2\pi^2 k^4/45\hbar^3)g_*T^3$. The condition that the entropy in a comoving, expanding volume remains constant can therefore be expressed as

$$g_* T^3 a^3 = [\text{constant}] \tag{53}$$

In general, g_* is a known function of T, and Eq. [53] tells us how the temperature varies with the radius of the universe. During the electron-positron annihilation, at roughly constant radius, Eq. [53] determines the increase of T resulting from the decrease of g_*. When g_* decreases by a factor of $(4/11)$, the temperature increases by a factor of $(11/4)^{1/3}$. Hence the photon temperature exceeds the (unchanged) neutrino temperature by this same factor. After this time, both the photon and the neutrino temperatures decrease jointly, in inverse proportion to the radius of the universe, and therefore the photon temperature remains permanently larger than the neutrino temperature. Given that the observed photon temperature now is 2.74 K, the neutrino temperature must be $2.74 \times (4/11)^{1/3} = 1.96$ K. However, there is no practical way to detect the neutrino gas and to measure its temperature.

The above calculation of the neutrino temperature assumed zero mass for the neutrinos. Some years ago, experiments on beta decay seemed to indicate the possibility of a nonzero mass of the electron neutrino, of the order of 30 eV. Neutrinos of such a mass could account for all the dark, missing mass in the universe. The most recent experiments (Robertson et al., 1991) indicate an upper limit of 9 eV for the neutrino mass, which is not quite enough to account for all the dark mass in the universe. However, the mu neutrino and the tau neutrino could be much more massive (the available experimental limits on their masses are 3 MeV and 35 MeV, respectively). For massive neutrinos, the redshifted distribution [50] is not a thermal distribution, and we cannot speak of a neutrino temperature. However, the energy density, the pressure, the mean kinetic energy, etc., of the neutrinos can be readily evaluated from Eq. [50]. By comparing the calculated energy density with the maximum conceivable energy density ($\Omega_0 \simeq 1$, or $\rho_0 = 1.9 \times 10^{-29} h_0{}^2$ g/cm^3), we can set limits on the mass of the neutrino, the chemical potential, or both. For instance, if the neutrinos have nonzero mass but zero chemical potential (Ruffini, Song, and Stella, 1983),

$$\sum_i m_i < 100 h_0{}^2 \text{ eV} \qquad [54]$$

where the sum extends over the three neutrino families. And if the neutrinos have zero mass but nonzero chemical potential (Weinberg, 1972),

$$\left(\sum_i \mu_i{}^4\right)^{1/4} < 50 h_0{}^{1/2} \text{ MeV} \qquad [55]$$

10.4 NUCLEOSYNTHESIS; THE ABUNDANCE OF PRIMORDIAL HELIUM

Helium is a ubiquitous component of stars. This is not surprising, since stars make helium by the thermonuclear burning of hydrogen. However, even quite young stars, which have not had time to make much helium, are found to contain a substantial amount of helium--in the youngest stars, the helium abundance is about 25%, or 1/4, by mass, and all the youngest stars have this same abundance. This uniform helium abundance in young stars indicates that their helium is primordial--the helium must be a relic from the Big Bang, and it must have been present in the gas out of which the stars formed.

Helium is synthesized in the early universe when the temperature drops sufficiently for neutrons to bind with protons and form deuterium nuclei. In turn, these bind with each other or with free protons or neutrons to form helium and other, heavier nuclei. The first calculations of such nucleosynthesis were performed by Gamow and his collaborators (Gamow, 1946, 1948; Alpher, Bethe, and Gamow, 1948; Alpher and Herman, 1948). Gamow recognized that the resulting abundances of diverse nuclei hinge on an interplay of the characteristic reaction time and the characteristic expansion time of the universe, and he gave a simple estimate of the temperature and density required for nucleosynthesis. Initially, in the early, hot universe, the neutrons and protons are free. When the temperature drops below 10^9 K, corresponding to the dissociation energy of deuterium, the neutrons and protons can form deuterium, via the radiative capture reaction

$$n + p \rightarrow d + \gamma$$

The characteristic reaction time is $1/nv\sigma$, where n is density of nucleons, σ the capture cross section, and v the mean thermal speed. If the resulting abundance of deuterium (and helium) is to be of the same order of magnitude as the abundance of hydrogen, the characteristic reaction time must be of the order of magnitude of the expansion time, or the Hubble time. Thus, $1/nv\sigma \simeq H^{-1}$, or

$$nv\sigma \simeq H \qquad\qquad [56]$$

According to Eq. [15], a temperature of 10^9 K corresponds to $t \simeq 200$ s, and $H \simeq 1/200$ s. Substituting these numbers and the known cross section into Eq. [56], Gamow estimated that the nucleon density at the time of deuterium formation must have been about $n \simeq 10^{18}$ /cm^3.

Alpher and Herman (1948) improved Gamow's calculations and obtained more accurate results by numerical integration of the equations for reaction rates. They also extrapolated the temperature of the universe to the present time, starting with the temperature and the density at an early time. This extrapolation is an immediate consequence of the dependence of T and n on the radius of the universe, $T \propto 1/a$, and $n \propto 1/a^3$, which implies $T^3/n = $ [constant] and $T^3/n = T_0^3/n_0$. Thus, the temperature T and the density n at the time of nucleosynthesis determine the present temperature T_0, if the present density n_0 is assumed known. Alpher and Herman estimated the present temperature at 5 K, but a more careful analysis (see Peebles, 1971) shows that the uncertainty in this prediction of the present temperature is rather large--any temperature in the range from 1 to 30 K is consistent with the data.

Gamow had hoped to explain the abundances of all kinds of elements by nucleosynthesis in the hot, early universe, but we now know that only the light nuclei are formed in the early universe, whereas heavier nuclei are formed in the cores of stars, by fusion of lighter nuclei.

In general, the calculations of element formation in the early universe require numerical integration of the reaction rates. But in the case of helium, we can make a simple estimate of the resulting abundance, as follows. Formation of helium is strongly favored over formation of other nuclei, and almost all the initially available free neutrons are ultimately bound into helium nuclei. Thus, the abundance of primordial helium hinges on the abundance of neutrons relative to protons. At early times, say, $t \leq 10^{-2}$ s, in the lepton era, neutrons and protons are in thermal equilibrium. Since the neutron is slightly more massive than the proton, thermal equilibrium favors protons over neutrons. The ratio of the neutron and proton abundances is given by the Boltzmann factor

$$\frac{[n]}{[p]} = e^{-Q/kT} \qquad [57]$$

where $Q = 1.29$ MeV is the difference between the neutron and proton rest-mass energies.

The thermal equilibrium between neutrons and protons is maintained by the following reactions, which involve the weak interaction:

$$n \leftrightarrow p + e^- + \bar{\nu} \qquad [58]$$

$$e^+ + n \leftrightarrow p + \bar{\nu} \qquad [59]$$

$$\nu + n \leftrightarrow p + e^- \qquad [60]$$

The first of these reactions is by far the slowest, and it does not contribute much to the equilibrium. (Proceeding in the forward direction, this reaction is neutron decay, with a half-life of about 10 min, which is much longer than the typical time scale, or Hubble age, of the early universe. Proceeding in the backward direction, this reaction requires a simultaneous collision among three particles, which is much less likely than a collision between two particles.) Hence, the important equilibrium reactions are [59] and [60]. The cross sections for these reactions can be calculated from the theory of the weak interactions, and these cross sections are found to be of the order of magnitude of $\sigma \simeq G_F{}^2 \hbar^{-4} E^2$, where $G_F = 1.4 \times 10^{-49}$ erg·cm^3 is the Fermi constant, and E is the lepton energy. The characteristic reaction time for a neutron is then $1/n\sigma v$, where n is the density of leptons and v their speed. The lepton density is $n \simeq (kT/\hbar)^3$ and the speed is $\simeq 1$, since the leptons are relativistic at the high temperatures of interest; furthermore, the lepton energy is $\simeq kT$. Combining these factors, we find a characteristic time

$$\frac{1}{n\sigma} = \frac{\hbar^7}{G_F{}^2(kT)^5} \qquad [61]$$

The reactions will be able to maintain thermal equilibrium provided the characteristic reaction time is smaller than the characteristic expansion time H^{-1} of the universe. For the lepton era, H^{-1} is given by Eq. [47], and, if we ignore g_* and other numerical factors, the condition for thermal equilibrium is

$$\frac{\hbar m_{Pl}}{(kT)^2} > \frac{\hbar^7}{G_F{}^2(kT)^5}$$

or

$$(kT)^3 > (0.8 \text{ MeV})^3 \qquad [62]$$

From this we see that in the early universe, at high temperature, the characteristic reaction time is small compared with the Hubble time, which confirms that the reactions [59] and [60] are able to maintain thermal equilibrium between neutrons and protons. While thermal equilibrium lasts, the ratio of neutrons to protons is given by the Boltzmann factor, as in Eq. [57].

But when the universe cools to a lower temperature, the characteristic reaction time becomes long compared with the Hubble time, and thermal equilibrium fails. For a rough estimate of the residual abundance of neutrons, we can assume that the neutron–proton ratio becomes frozen at the fixed value corresponding to the instant when the Hubble time equals the reaction time. According to Eq. [62], this happens at a temperature of $kT \simeq 0.8$ MeV (at a time $t \simeq 1$ s). The neutron–proton ratio at this critical time is

$$\frac{[n]}{[p]} \simeq e^{-1.29/0.8} \simeq \frac{1}{5} \qquad [63]$$

The abundance of neutrons, expressed as a fraction of the total amount of baryonic matter, is therefore

$$\frac{[n]}{[n] + [p]} \simeq \frac{1}{6} \qquad [64]$$

After the temperature drops to about 0.1 MeV (at $t \simeq 1$ min), the neutrons quickly bind with protons to make deuterium nuclei, and the deuterium nuclei fuse to make ^{4}He nuclei. Within 2 or 3 minutes, almost all the neutrons are incorporated into helium nuclei. The only neutrons to avoid this fate are a few that decay during the time interval between 1 s and 3 min (since the half-life of the neutron is 10 min, there is not time for a large amount of decay), a few that remain as deuterium, and a few that become incorporated into lithium. If we ignore these small losses, we see that the fraction of ^{4}He of the total amount of baryonic matter must be $\simeq 1/3$ (of which $1/6$ is due to the neutrons and $1/6$ is due to the equal number of protons in ^{4}He). The result of this rough estimate is in surprisingly good agreement with the observed helium abundance of $\simeq 1/4$. Note that we were able to obtain this result without specifying initial conditions for the universe--thermal equilibrium implicitly specifies all that is needed by way of initial conditions. Also note that our result is independent of the mass density in the universe.

Although our rough calculation gives us a clear picture of the basic physics involved in the synthesis of primordial helium, in an accurate calculation we need to take into account that the cross sections for the reactions [59] and [60] are somewhat suppressed because some of the final states are already occupied by the leptons of the fermion gas, and these states are then forbidden to the leptons emerging from the reactions [59] and [60]. Furthermore, the thermal equilibrium does not cease at one instant, but it fails gradually over a time interval around $t \simeq 1$ s. This means that to evaluate the final neutron abundance we need to integrate the rates for the reactions [59] and [60] numerically. And to evaluate the helium abundance, we need to integrate the rates of all the reactions involved in helium synthesis, as well as the competing reactions involved in the synthesis of light elements other than helium. Fig. 10.2 shows the results of such a numerical calculation of the synthesis of ^{4}He and other light elements. The final abundance of each element is plotted as a function of the density of baryon matter.

Note that in the case of ^{4}He, the result is nearly independent of the baryon density--the abundance is between 0.20 and 0.25 over a wide range of values of ρ_0. This insensitivity to the baryon density was already suggested by our crude calculation. The slight decrease of the abundance at small values of ρ_0 revealed in Fig. 10.2 is a consequence of neutron decay; a universe of small ρ_0 takes a longer time to cool to the temperature at which neutrons can bind with protons, and this delay permits more neutrons to decay. For other light nuclei, the abundance is quite sensitive to the baryon density. For instance, Fig. 10.2 shows that if the baryon density is high, the abundance of deuterium will be low. The reason for this inverse relationship is that although the rate of formation of deuterium increases when there are more neutrons and protons, the rate of destruction of the deuterium increases even more. The abundances of ^{3}He and ^{7}Li are also sensitive to the baryon density.

Heavy elements cannot be formed in any significant amounts in the early universe. The lack of stable nuclei at mass numbers 5 and 8 constitutes a "barrier" that blocks the further growth of a nucleus by stepwise capture of neutrons or protons. This blockage of nuclear growth

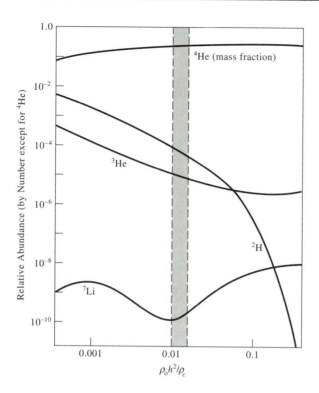

Fig. 10.2 Calculated abundances of light elements as a function of the baryon density ρ_0. The band indicates the range of baryon densities permitted by the observational data on the abundances. (From Schramm, 1991.)

was first demonstrated in a calculation by Fermi and Turkevich (unpublished).

Observational data on the abundances of light elements come from a large variety of sources. Intensities of spectral lines of ^{4}He can be measured in stars, planetary nebulas, and H II regions of galaxies. Spectral lines of ^{7}Li have been measured in metal-poor stars (a low abundance of heavy elements indicates that the star was formed from primordial, uncontaminated gas). Deuterium abundances have been measured in the Solar System and in molecular gas in the interstellar medium. And ^{3}He abundances have been measured in the Solar System and in H II regions. With the exception of ^{4}He, all these abundances are afflicted with substantial uncertainties. Table 10.1 summarizes the observational data on the abundances of light elements.

The observed abundances are consistent with the calculations displayed in Fig. 10.2, provided that the baryon density ρ_0 lies between 1.9×10^{-31} and 3.0×10^{-31} g/cm^3, which means that Ω_0 lies between $0.010h^{-2}$ and $0.016h^{-2}$ (Walker et al., 1991). This consistency of the theoretical calculations of nucleosynthesis and observation must be reckoned as a triumph of the Standard Model of the Big Bang. In essence, in Fig. 10.2 we are attempting to fit four distinct pieces of data by adjusting one single parameter (ρ_0). The Standard Model of the Big Bang meets this challenge with spectacular success. Furthermore, the required value of ρ_0 is consistent with observational data on the (baryon) mass density of the universe.

TABLE 10.1 ABUNDANCES OF LIGHT ELEMENTS

Element	Abundance (by mass, as fraction of baryon abundance)
^{4}He	0.24
^{3}He	2×10^{-5} to 3×10^{-4}
^{2}D	1×10^{-5} to 3×10^{-5}
^{7}Li	1×10^{-10} to 8×10^{-10}

Another test of the Standard Model is provided by an examination of the number of neutrino families, characterized by distinct "flavors." Although only the electron type of neutrino participates in the reactions [59] and [60], the presence of other kinds of neutrinos affects the abundance of ^{4}He because it changes the temperature at which the neutron-proton ratio freezes out of equilibrium. The presence of other neutrinos increases $g*$, and therefore decreases H^{-1} (see Eq. [47]); hence the freeze-out occurs at a higher temperature, with a higher abundance of neutrons, which leads to a higher abundance of helium. For consistency with the observed helium abundance, the Standard Model demands that the number of neutrino families be no more than 4. This is in excellent agreement with data from recent experiments at the LEP accelerator, according to which the most likely number of neutrino families is 3.

10.5 DENSITY PERTURBATIONS; THE JEANS MASS

Although the universe is fairly smooth on a large scale, it has prominent clumps on a small scale, where we find galaxies and clusters of galaxies whose mass density is much in excess of the average density of the universe. The development of such clumps in the universe is due to a gravitational instability: if there is some small initial enhancement of the density in some volume, the gravitational attraction tends to contract this volume, which leads to a further enhancement of the density. Thus, density perturbations are gravitationally unstable, and they tend to grow. However, the gravitational contraction is opposed by pressure forces, which tend to maintain the density uniform. We will have to investigate under what conditions a density perturbation is unstable, so it will continue to grow.

For the sake of simplicity we begin with an analysis of the density perturbations in a uniform, static fluid governed by Newtonian physics. This means we ignore the expansion of the universe and we ignore the curvature of the space and time. Our result is therefore only an approximation, applicable only to a small region of our expanding and curved universe. However, as we will see, the qualitative features of our result are of wider applicability.

The Newtonian equations that determine the motion of the fluid under the influence of pressure forces and gravity are

$$\frac{\partial \rho}{\partial t} + \nabla \cdot (\rho \mathbf{v}) = 0 \tag{65}$$

$$\frac{\partial \mathbf{v}}{\partial t} + (\mathbf{v} \cdot \nabla)\mathbf{v} + \frac{1}{\rho}\nabla P + \nabla \Phi = 0 \tag{66}$$

$$\nabla^2 \Phi = 4\pi G \rho \tag{67}$$

The first of these equations is simply the continuity equation; the second is Newton's second law (the first two terms are the "total" derivative of the velocity, and the other two terms are the negatives of the pressure force and the gravitational force per unit mass); and the third is Poisson's equation for the gravitational potential. Suppose that $\bar{\rho}$, $\bar{P}$, $\bar{\mathbf{v}}$, and $\bar{\Phi}$ are the quantities appropriate for the static, uniform fluid. If the fluid suffers a small perturbation, these quantities are changed to $\bar{\rho} + \delta\rho$, $\bar{P} + \delta P$, $\bar{\mathbf{v}} + \delta\mathbf{v}$, and $\bar{\Phi} + \delta\Phi$. To first order in small quantities, Eqs. [65]–[67] become

$$\frac{\partial \delta\rho}{\partial t} + \bar{\rho}\nabla \cdot (\delta\mathbf{v}) = 0 \tag{68}$$

$$\frac{\partial \delta\mathbf{v}}{\partial t} + \frac{v_s^2}{\bar{\rho}}\nabla\delta\rho + \nabla\delta\Phi = 0 \tag{69}$$

$$\nabla^2\delta\Phi = 4\pi G\delta\rho \tag{70}$$

where

$$v_s = \sqrt{\frac{\delta P}{\delta\rho}} \tag{71}$$

is the speed of sound in the fluid.

If we take the divergence of Eq. [69] and use the other two equations, we obtain a single differential equation for $\delta\rho$:

$$\frac{\partial^2 \delta\rho}{\partial t^2} - v_s{}^2 \nabla^2 \delta\rho = 4\pi G \bar\rho \, \delta\rho \qquad [72]$$

To investigate the character of the solutions of Eq. [72], consider a plane-wave solution of the form

$$\delta\rho(\mathbf{x}, t) = A \, e^{-i\mathbf{k}\cdot\mathbf{x} - \omega t} \qquad [73]$$

Upon substituting this into Eq. [72] we find that ω and $\mathbf{k}$ must satisfy the dispersion relation

$$\omega^2 - v_s{}^2 k^2 = -4\pi G \bar\rho \qquad [74]$$

This tells us that if k is small, ω will be real, and the perturbation will oscillate harmonically. But if k is large, ω will be imaginary, and the perturbation will grow exponentially. The critical value of k above which the perturbation will grow is $k_J = \sqrt{4\pi G \bar\rho / v_s{}^2}$, and the critical wavelength is

$$\lambda_J = \frac{2\pi}{k_J} = \sqrt{\frac{\pi v_s{}^2}{G \bar\rho}} \qquad [75]$$

This is called the *Jeans wavelength*.

If we want to construct a localized contracting perturbation of finite extent, we must form a superposition of such plane waves. Since the minimum wavelength for a contracting perturbation is $\simeq \lambda_J$, the extent of the superposition cannot be smaller than $\simeq (\lambda_J)^3$, and the amount of mass in this volume is

$$M_J = (\lambda_J)^3 \bar\rho = \pi^{3/2} \frac{v_s{}^3}{\sqrt{G^3 \bar\rho}} \qquad [76]$$

This is called the *Jeans mass*; it is the minimum mass that leads to growth of a density perturbation.*

We can understand the growth of a density perturbation by a simple physical argument. Suppose that the initial density perturbation occurs in a spherical volume of radius λ within which the density is slightly in excess of the normal density. The mass M in this volume will then begin to fall toward the center under the influence of its

---

* The Jeans mass is sometimes defined without the factor $\pi^{3/2}$ in Eq. [76], and sometimes with a different numerical factor.

own gravitational attraction. The typical speed of fall is $\sqrt{GM/\lambda}$, and the characteristic "free-fall" time for the process of contraction is the radius λ divided by the typical speed, $\simeq \lambda/\sqrt{GM/\lambda} \simeq 1/\sqrt{G\bar{\rho}}$. The pressure forces that oppose the contraction are propagated in the form of sound waves. Hence the characteristic response time for the pressure forces is $\simeq \lambda/v_s$. If the latter time is longer than the former, the pressure forces will respond too slowly to prevent the contraction. Hence, the condition for instability and for growth of the density perturbation is $\lambda/v_s > 1/\sqrt{G\bar{\rho}}$, which says that λ must exceed the Jeans wavelength and the mass must exceed the Jeans mass, in agreement with our perturbation analysis based on Eq. [72].

Although in the above analysis we took as starting point a static background, without expansion, we can perform a similar analysis for a homogeneously expanding fluid. The Newtonian equations describing such a fluid again show that the Jeans mass [76] is the minimum mass that leads to growth of density perturbations (Weinberg, 1972).

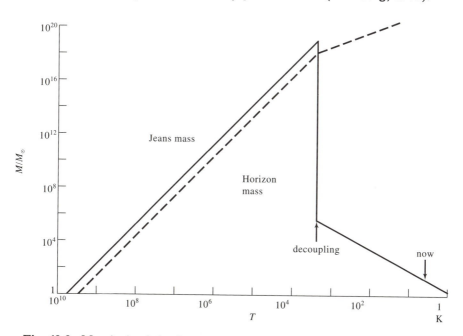

Fig. 10.3 Magnitude of the Jeans mass as a function of the temperature T for a universe containing baryons, without "dark" matter. The plot assumes $\Omega_0 h^2 = 0.05$. The temperature T decreases monotonically with time; hence this plot indicates the time dependence of the Jeans mass. The dashed line indicates the horizon mass.

Fig. 10.3 is a plot of the magnitude of the Jeans mass vs. temperature, for a universe dominated by baryons at late times. During the radiation era, the pressure in the universe is supplied by the radiation, which is strongly coupled to the baryonic plasma. The pressure of the radiation is high (compared with the mass density), and this leads to a high and constant sound velocity, nearly equal to the speed of light. According to Eq. [76], the Jeans mass therefore increases as the density of the universe decreases. The Jeans mass reaches a maximum value of $10^{19} M_\odot$ at the time of recombination. But upon recombination, the radiation decouples from the baryonic matter, and the pressure affecting the baryons is then their own, relatively low, kinetic pressure. This leads to a drastic drop in the sound velocity and a corresponding drop in the Jeans mass. Immediately after recombination, the Jeans mass is only $10^6 M_\odot$, and it continues to decrease thereafter.

One defect of the simple Newtonian perturbation analysis is that the Jeans mass in the radiation era is larger than the horizon mass, that is, the mass included within the distance to the particle horizon. This horizon mass is indicated in Fig. 10.3 by the dashed line. Since perturbations in the gravitational potential cannot propagate beyond this horizon, the Newtonian equation [72] for small perturbations of the gravitational field becomes invalid. However, a relativistic analysis confirms that the Jeans mass [76] still serves as the criterion for distinguishing between growing and oscillatory perturbations (Weinberg, 1972).

Let us consider the evolution of a perturbation of mass $10^{12} M_\odot$, which is approximately the mass of a typical galaxy. During the early radiation era, this mass is in excess of the Jeans mass, and the perturbation can therefore grow, that is, the density of the clump of mass can begin to increase relative to the density of the background (since the background density of the expanding universe is decreasing, a growth in the density pertubation merely means that the density of the perturbation does not decrease as fast as the density of the background). At a time of about 10^8 s, the Jeans mass reaches 10^{12} $M_\odot$, and the character of the perturbation changes from growing to oscillatory. The perturbation then oscillates until the time of recombination, where the Jeans mass drops, and the character of the perturbation changes back from oscillatory to growing. The perturbation then continues to grow, and finally it evolves into a galaxy.

As a next step, we need to examine the time dependence of growing perturbations in an expanding universe. Although it is possible to perform such an analysis with the Newtonian equations for an expanding fluid (compare the Newtonian treatment of the expanding universe in Section 9.3), this approach is of limited validity since it cannot deal with growing perturbations that extend beyond the horizon distance, such as the growing perturbations in the radiation era. The full relativistic analysis of the growth of perturbations is somewhat complicated, and instead of treating the general case, let us deal only with the simple and special case of a uniform density perturbation in a Friedmann universe of zero curvature. The results for this case are representative of the general case. In a Friedmann universe of zero curvature, the unperturbed Einstein equation is (see Eq. [9.144])

$$3H^2 = 8\pi G\rho \qquad [77]$$

If we increase the density slightly, to a new value ρ', the universe changes to a Friedmann universe of positive curvature with an Einstein equation

$$3\left[H^2 + \frac{1}{a^2}\right] = 8\pi G\rho' \qquad [78]$$

It is customary to describe the perturbation by the fractional difference δ between the densities, called the density contrast:

$$\delta = \frac{\rho' - \rho}{\rho} \qquad [79]$$

From Eqs. [77] and [78], with equal values of the Hubble constant, we find that the density contrast obeys the equation

$$\delta = \frac{\rho' - \rho}{\rho} = \frac{3}{8\pi G\rho a^2} \qquad [80]$$

The time dependence of a and ρ therefore determines the time dependence of δ. In the radiation-dominated era, $a \propto t^{1/2}$ and $\rho \propto t^{-2}$ (see Eqs. [13] and [14]); hence

$$\delta \propto t \quad \text{(radiation dominated)} \qquad [81]$$

In the matter-dominated era, $a \propto t^{2/3}$ and $\rho \propto t^{-2}$ (see Eqs. [9.123] and [9.103]); hence

$$\delta \propto t^{2/3} \quad \text{(matter dominated)} \qquad [82]$$

Although we have obtained these results for the time dependence of the growing perturbations from a special model, the general analysis leads to the same results: the density contrast for a perturbation involving a mass much larger than the Jeans mass grows in proportion to t in the radiation era, and in proportion to $t^{2/3}$ in the matter era.

It is of interest to compare these results for the growth of perturbations with observational data on density contrasts in the early universe. As mentioned in Section 9.5, on a small angular scale, the cosmic blackbody radiation is uniform to better than 1 part in 10^4, that is, at most $\Delta T/T \simeq 10^{-4}$. Such a temperature contrast $\Delta T/T$ would presumably indicate a local (adiabatic) compression of the photon gas, with $\Delta \rho_{rad}/\rho_{rad} = 3\Delta T/T$. Since the baryonic matter is strongly coupled to the photons up to recombination, it would have the same density distribution and the same density contrast,

$$\frac{\Delta \rho}{\rho} = 3 \frac{\Delta T}{T} \qquad [83]$$

Thus, the upper limit on the temperature contrast implies a roughly equal limit on the density contrast, $\Delta \rho/\rho \simeq 10^{-4}$.

In the matter-dominated universe after recombination, the density contrast of the clumps of matter grows in proportion to $t^{2/3}$. Hence, between the time of recombination ($t = 10^{13}$ s) and now ($t = 10^{17}$ s), the density contrast increases by a factor of $(10^{17}/10^{13})^{2/3} = 500$, and the density contrast now should be

$$500 \times 10^{-4} = 5 \times 10^{-2} \qquad [84]$$

This is much smaller than the density contrast observed in the matter distribution in the universe now. The density contrast in the largest clusters in the universe now is roughly 10^2 or 10^3. Such a high density contrast is the result of perturbative growth ($\propto t^{2/3}$) until $\delta \rho/\rho$ reaches a value near 1, followed by nonlinear (nonperturbative) growth to the final value now. But according to Eq. [84], the perturbative growth never reaches anywhere near 1, and nonlinear growth never begins!

The conflict between the observed density contrast at the time of recombination and the density contrast now can be resolved by taking into account the dark matter in the universe. As we saw in Section 9.6, most of the mass in the universe is dark matter. If this dark matter consists of WIMPs (*w*eakly *i*nteracting *m*assive *p*articles) that couple only weakly to the baryonic matter and to the blackbody radiation, then the dark matter can decouple from the radiation at a time much earlier than the recombination time, and the density contrast can start to grow from this earlier time onward, according to the power law [81] or [82]. Thus, the density contrasts in the dark matter are not constrained by the observational limits on the temperature contrast in the cosmic blackbody radiation, and they can be much larger than these limits. These large density contrasts can easily grow to $\simeq 1$ within the available time (and then continue to grow nonlinearly). Although the baryonic matter is initially (at $t = 10^{13}$ s) coupled to the radiation, with its lower density contrast, it becomes gravitationally coupled to the dark matter after recombination, and quickly acquires the higher density contrast of this dark matter. This means that the growth of the perturbations in the baryonic, visible matter is not governed by the power law [82], but by the gravitational attraction that the clumps of dark matter exert on the visible matter.

Accordingly, the large-scale structure of the universe is governed by the properties of the dark matter. The hypothesized forms of dark matter fall into two categories: hot dark matter and cold dark matter. Hot dark matter consists of particles of fairly small mass (say, $m \leq 100$ eV), such as neutrinos of small but nonzero mass, which would form a relativistic gas at the temperatures prevailing toward the end of the radiation era. Cold dark matter consists of particles of large mass (say, $m \geq 10$ keV), which would form a nonrelativistic gas at these temperatures. Thus, the words *hot* and *cold* do not really describe the actual temperature, but whether the particles experience this temperature as a high temperature (particle of low mass, $m \leq kT$) or as a low temperature (particle of high mass, $m >> kT$).

Table 10.2 lists some popular dark-matter candidates.

TABLE 10.2 DARK-MATTER CANDIDATES (WEAKLY INTERACTING MASSIVE PARTICLES)*

WIMP	Mass	Origin (t, kT)	Abundance, now (cm^{-3})
HOT DARK MATTER			
Invisible axion	10^{-5} eV	10^{-30} s, 10^{12} GeV	10^9
Light neutrino	30 eV	1 s, 1 MeV	10^2
COLD DARK MATTER			
Photino, gravitino	$\simeq$keV	10^{-4} s, 10^2 MeV	10
Photino, Sneutrino, neutralino, axino, heavy neutrino	$\simeq$GeV	10^{-3} s, 10 MeV	10^{-5}
Magnetic monopoles	10^{16} GeV	10^{-34} s, 10^{14} GeV	10^{-21}
Pyrgons, Maximons, Newtorites	10^{19} GeV	10^{-43} s, 10^{19} GeV	10^{-24}
Quark nuggets	$\simeq 10^{15}$ g	10^{-5} s, 300 MeV	10^{-44}
Primordial black holes	$>10^{15}$ g	$>10^{-12}$ s, $<10^3$ GeV	$<10^{-44}$

* From Kolb and Turner, 1990.

The perturbations that ultimately grow into galaxies and clusters of galaxies begin as primeval fluctuations in the very early universe. For instance, in the inflationary model (see Section 10.6), such fluctuations arise from quantum processes. But the evolution of such perturbations into galaxies and clusters of galaxies proceeds in different ways in a universe filled with hot dark matter and a universe filled with cold dark matter. In the hot-matter-dominated universe, the large structures form first, and the small structures form later, by fragmentation of the larger objects (such an evolution of structure is called "from the top down"). In contrast, in the cold-matter-dominated universe, the small structures form first, and the larger structures form later, by the merging of small objects (such an evolution of structure is called "from the bottom up").

This difference in the evolution of structures arises from a difference in the behavior of the small-scale density perturbations in these two kinds of universes. In the hot-matter-dominated universe, any small-scale density perturbations tend to be wiped out by the free-streaming motion of the particles. The hot-matter particles are relativistic ($E \simeq kT \geq m$), and their high-speed motion tends to disperse any initial density perturbation. The only density perturbations that survive this free-streaming motion are those of a diameter so large that the particles do not have enough time to stream out. In the absence of small-scale density perturbations, the large-scale perturbations will be the first to contract gravitationally, and they will form large-scale structures.

In the cold-matter-dominated universe, the particles move slowly, and small-scale density perturbations do not tend to disperse. According to the inflationary model of the early universe, the small-scale density perturbations predominate, and when they contract gravitationally, they will form small-scale structures.

The evolution of structure in a universe filled either with hot matter or with cold matter has been explored by means of numerical simulations that track the motion of a large number of mass points interacting gravitationally. In the hot-matter-dominated universe, the initial gravitational contraction of each of the large objects is strongly anisotropic, usually with one preferential direction of collapse. The result is that the initial lumps squash into quite thin "pancakes" (as originally proposed by Zeldovich, 1970), which subsequently fragment into smaller objects. Large voids are left between the "pancakes." Although some such voids have been observed in the actual universe, the simulations indicate that in the hot-matter-dominated universe the large-scale structures tend to overdevelop, with excessive contrasts between the voids and the thin pancakes.

The cold-matter-dominated universe suffers from the opposite problem: large-scale structures tend to underdevelop. For example, Figs. 10.4 a-d show the result of a numerical simulation of a universe filled with cold dark matter. If we compare the slices a-d with a slice through the actual universe (see Fig. 9.3), we see that the mass distribution in the simulation is excessively uniform. One scheme that has been proposed to explain this discrepancy is that maybe the visible galaxies do not trace the true mass distribution in the universe, that is, light does not trace mass. The voids we observe might be filled with a more or less uniform distribution of faint galaxies, which do not show up in the surveys on which Fig. 9.3 is based. Thus, the mass distribution in our universe might be much more uniform than suggested by Fig. 9.3, and the pronounced structure we see in Fig. 9.3 might represent only the contribution from bright galaxies that have formed preferentially at the location of the strongest clumps in the mass dis-

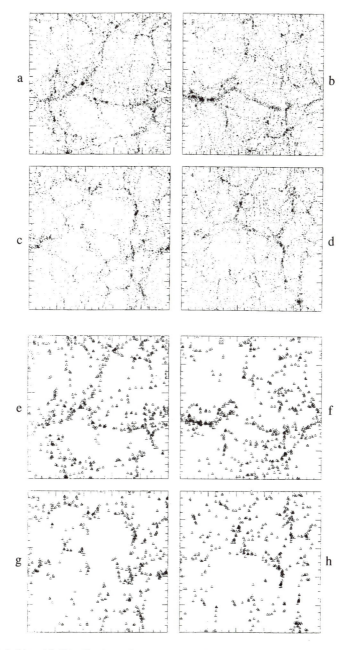

Fig. 10.4 (a) - (d) Distribution of mass points obtained from a numerical simulation of the evolution of a universe with cold dark matter. The diagrams show several adjacent slices through the cubical volume used in the calculations. (e) - (h) Distribution of the largest concentrations of points found in the preceding diagrams. Note that this distribution of the largest concentrations displays much more contrast than the complete distribution. (From Davis and Efstathiou, 1988.)

tribution. This scheme for the preferential formation of bright galaxies within the strong clumps of the mass distribution is called "biased" galaxy formation; but the astrophysical mechanism that might lead to biased galaxy formation is not known. Figs. 10.4 e-h indicate that biasing does indeed produce enhanced density contrasts. In these figures, only the points located in strong clumps of the mass distribution have been plotted; all other points have been deleted. Comparing Figs. 10.4 e-h with Fig. 9.3, we see that the density contrasts and the overall appearance of the biased matter distribution produced by the numerical simulations are in remarkably good agreement with what we see in the actual universe. Because of the good agreement between the numerical simulations and the actual universe, cosmologists have come to prefer cold dark matter over hot dark matter.

10.6 THE INFLATIONARY MODEL

The Standard Model of the early universe, described in the previous sections, is quite successful in dealing with the development of the universe from an age of about 10^{-5} s onward. Thus, the model gives us an account of the history of our universe over a range of ages spanning 23 powers of ten. Its most impressive success lies in the calculations of the abundances of helium and other light elements.

However, when we seek to extrapolate the Standard Model to earlier times, the model presents us with several puzzles for which it offers no answers:

(i) The Flatness Puzzle. As we saw in the preceding chapter, the density parameter of the universe now is $\Omega_0 \simeq 0.1$ to 2 (the higher value would include a large amount of nonbaryonic dark matter). In any case, Ω_0 is of the order of magnitude of 1, that is to say, it is of the order of magnitude of the value required for a flat Friedmann universe.

What makes this puzzling is that an old universe, such as ours, can achieve a value $\Omega_0 \simeq 1$ only by an extremely delicate tuning of the value of Ω at an early time. At an early time, the value of Ω was necessarily closer to 1 than it is now. We can see this from Eq. [9.145], with $\Lambda = 0$:

$$\Omega = 1 + \frac{K}{H^2 a^2} \qquad [85]$$

In the early universe, $a \propto t^{1/2}$ and $H \propto 1/t$ (see Eqs. [13] and [17]). Hence the second term on the right side of Eq. [85] is proportional to t, and therefore $\Omega \rightarrow 1$ as $t \rightarrow 0$. Now suppose that at some short time after the primeval singularity, say at $t = 10^{-43}$ s, the value of Ω is somewhat different from 1. The difference will then grow drastically, and the universe will quickly attain either a very large value of Ω (if the initial value of Ω is larger than 1) or else a very small value (if the initial value of Ω is smaller than 1). This means the universe either recontracts very quickly, or else it expands very quickly and disperses. Only if the initial value of Ω is *extremely* close to 1 will it remain of the order of magnitude of 1 for a long time.

Thus, our universe is a very exceptional universe: it must have started with an initial value of Ω almost exactly equal to 1 in order to avoid recontraction or dispersion by now. Expressed another way, the initial universe must have been almost exactly a flat universe in order to permit our present universe to be not too far from a flat universe. A simple calculation shows that at $t = 10^{-43}$ s, the value of Ω must have differed from 1 by no more than $\pm 10^{-60}$!

Although such a fine tuning of the initial value of Ω seems puzzling, it may not have any special significance. *A priori*, any value of Ω is as likely as any other value, and maybe at $t = 10^{-43}$ s the universe had an initial value of, say, $\Omega = 1 - 10^{-61}$ by coincidence. This, of course, is a very lucky coincidence for us, since if Ω had been outside the range 1 ± 10^{-60}, then the universe would not have permitted us to exist. But the coincidence $\Omega = 1 - 10^{-61}$ would require an explanation only if we had several universes before our eyes, and we found that in all of them Ω has this value. In observing only a single universe, and in finding some value of Ω, we are not faced with a pattern, and we require no explanation.

Furthermore, the puzzle disappears completely if we assume that our universe is a flat universe. In such a universe, Ω is not a free parameter, but is permanently fixed at $\Omega = 1$. Maybe the above arguments indicating that, in the early universe, Ω is extremely close to 1 should be taken as a strong hint that our universe is indeed flat.

Since the flatness puzzle is not much of a puzzle, it does not indicate much of a deficiency in the Standard Model. However, if we can invent an alternative model of the universe that gives us a value of Ω of the order of 1 at the present time without invoking any special conditions, then we might judge such an alternative model somewhat superior to the Standard Model.

(ii) The Smoothness Puzzle. On a large scale, the present universe is fairly homogeneous, or smooth. We have relied on this smoothness of the universe in our treatment of the Friedmann models in the preceding chapter. The best observational evidence for this smoothness is the isotropy of the cosmic blackbody radiation--the temperature of this radiation is uniform over the sky to 1 part in 10^5.

This uniformity poses a true puzzle, because it cannot be attributed to thermal contact and thermal equilibrium in the early universe. Different regions of the early universe are out of contact with each other; they cannot communicate by thermal signals or even by light signals, because not enough time has elapsed to permit signals to reach from one region to the other. Consider, for instance, two opposite patches in the sky, separated by 180°. The blackbody photons reaching the Earth now were emitted by the hot hydrogen plasma, just before recombination. Since the photons from one of these patches are reaching the Earth only now, they have not yet reached the opposite patch--the two patches are out of contact.

The maximum distance that a signal can reach from one patch is limited by the available travel time; as we saw in Section 9.10, this maximum distance defines the particle horizon. Since the smoothness problem hinges on the existence of the particle horizon, it is also called the *horizon puzzle*. To appreciate how much the opposite patches in the sky are out of contact, let us express the distance between them as a multiple of the horizon distance. The coordinate distance χ of each patch from the Earth can be calculated by examining the worldline of a light signal from the patch to the Earth,

$$\chi = \eta_0 - \eta_{rec} \qquad [86]$$

where η_0 is the time parameter now and η_{rec} is the time parameter at recombination. The coordinate distance between the two patches is therefore

$$\Delta\chi = 2(\eta_0 - \eta_{rec}) \qquad [87]$$

The horizon coordinate distance associated with each patch at the recombination time equals η_{rec}. (This is the distance reached by a light signal that has left the patch at $t = 0$. Actually, the earliest time at which it is meaningful to speak of the travel of light signals is the Planck time, 10^{-43} s; but since the recombination time is much larger than the Planck time, it hardly matters whether we take η_{rec} or $\eta_{rec} - \eta_{Pl}$ as our horizon distance.) Hence the number of horizon distances lying between the two patches is

$$[\text{number of horizon distances}] = \frac{2(\eta_0 - \eta_{rec})}{\eta_{rec}} \qquad [88]$$

The values of η_0 and η_{rec} that appear in this ratio can be obtained from the appropriate expressions for $\eta(t)$. For the matter-dominated universe, the dependence of η on t is given (implicitly) by Eq. [9.114] or [9.119]. For the radiation-dominated universe, $\eta(t)$ is given by the integral $\eta = \int dt/a(t)$, with $a(t) \propto t^{1/2}$, as in Eq. [13]. The result for the number of horizon distances lying between the two patches is about 120. This means that at recombination, there were about 60 causally disconnected regions, each of diameter equal to two horizon distances, between the two patches. The number of such causally disconnected regions around the circumference surrounding us is approximately $\pi \times 60 \simeq 180$, and the angular diameter of each is therefore $\simeq 2^0$. Accordingly, when we look at the sky, we would expect the temperatures in patches separated by 2^0 or more to be completely independent--yet the observational data show these temperatures to be almost exactly the same!

■ *Exercise 4.* Show that for the radiation-dominated universe,

$$\eta(t) = \int \frac{dt}{a(t)} = \frac{2}{(32\pi G a_0{}^4 \rho_{rad,0}/3)^{1/4}} \, t^{1/2} \quad ■$$

An accidental coincidence of the temperatures in all such patches of the sky is beyond belief. There ought to be some other explanation of the smoothness of the temperature distribution of the universe, and, likewise, of the smoothness in the mass distribution, the smoothness of the expansion rate, and so on. One possible explanation is to suppose that the present expanding universe was preceded by a contracting universe, which passed through a moment of greatest contraction. Such a behavior of $a(t)$ is illustrated in Fig. 9.13. However, to make the continuation from one universe to the next meaningful, $a(t)$ must remain finite at the moment of maximum contraction, that is, there must be no singularity. The absence of a singularity at the first moment of the Big Bang would seem to be in contradiction with general singularity theorems that have been proved for the solutions of Einstein's equations. But these theorems deal only with the classical regime; they can be circumvented in the very early universe, where matter is in the form of quantum fields, and where even the geometry is quantized. Furthermore, if the solution of the Einstein equations develops a singularity, this merely indicates a breakdown in the equations, not necessarily an actual singularity in the real world. If the Einstein equations break down, it is natural to seek some modification of the equations that avoids the singularity, and attempts in this direction have been made (Brandenberger, 1985, 1992). Thus, the question of whether there is an actual singularity at the first moment of the Big Bang still remains open.

A contracting and reexpanding universe, without a singularity, has no horizon, and no horizon problem. This explanation of the horizon problem has not found much favor among cosmologists, because we do not (yet) know how to calculate the evolution of the universe from the contracting stage to the expanding stage. The inflationary explanation (see below) is popular because some of its details can be calculated.

(iii) The Monopole Puzzle. The generally accepted theory of fundamental interactions between particles (the "Standard Model" of electromagnetic, weak, and strong interactions) predicts that at high energies,

the interactions become unified. Such a unification means that different interactions become merely different aspects of a single underlying interaction. Thus, at an energy of about 300 GeV, the electromagnetic and weak interactions become unified into an electroweak interaction, and at an energy of about 10^{14} GeV, the strong interaction should become unified with the electroweak interaction (Grand Unified Theory, or GUT). In the unified phase, the particles and the vacuum are endowed with special symmetries. At the highest energies, the particle states have the maximum symmetry; at lower energies, the symmetry decreases ("spontaneous breakdown" of the symmetry), and, concurrently, the unification of the interactions is destroyed.

The extreme temperatures in the very early universe supply more than enough energy to bring about the complete unification of the strong and electroweak interactions. This grand unification lasts until the universe cools to 10^{26} K ($\simeq 10^{14}$ GeV), where there is a first spontaneous breakdown of symmetry. The unification of the electromagnetic and weak interactions lasts until the universe cools to 10^{15} K ($\simeq 300$ GeV), where there is a further spontaneous breakdown of the remaining symmetry. Fig. 10.1 summarizes the unification eras of the universe.

Unfortunately, because of the presence of horizons, the breakdown of the symmetry does not proceed smoothly. Regions separated by a horizon distance are causally disconnected, and hence the directions of symmetry breaking are expected to be different in such regions (in this context, the "direction" is not a direction in ordinary space, but a direction in the internal-symmetry space). Hence, there will be discontinuities, or defects, in the directions of symmetry breakdown ("topological defects"). Among such defects that form during the breakdown of grand unification are pointlike defects, which have the character of magnetic monopoles, that is, sources or sinks of magnetic field lines. The number of defects is expected to be at least one per horizon volume. Since the horizon volume is quite small and the mass of each monopole is large, about 10^{16} GeV, the mass density of magnetic monopoles is enormous; their mass density would be 10^{14} times as large as the baryonic mass density! This catastrophic prediction provided the original motivation for the inflationary model of the early universe.

The inflationary model was first proposed by Guth (1981), and it was later modified and improved by others (Linde, 1982). The essential assumption of the inflationary model is that, at about the time of the grand unification symmetry breakdown, the universe goes through an enormous and quick expansion, an expansion so large that a single horizon volume of the pre-expansion universe encompasses the entire observable universe that surrounds us now. Such an inflation solves the smoothness puzzle, since before inflation, the matter in the entire observable universe was in causal and thermal contact, and it could attain a uniform temperature. The inflation also solves the monopole puzzle--one monopole per horizon volume implies at most one monopole within our observable universe. And inflation also solves the flatness problem. During inflation, the radius of curvature $a(t)$ increases by an enormous factor, perhaps a factor of 10^{43}. This makes the initially curved geometry almost exactly flat, and it makes the final value of Ω almost exactly 1, regardless of what the initial value of Ω was.

The inflation is caused by a large energy density of the initial vacuum state. At the highest temperatures, in the grand unified phase of the universe, the vacuum is stable--it is the state of lowest energy. But when the temperature reaches the critical value of 10^{14} GeV, the symmetric vacuum becomes unstable. At and below this temperature, the symmetric vacuum has a higher energy than the symmetry-broken vacuum. Thus, the universe makes a phase transition from the higher-energy phase to the lower-energy phase. This phase transition is believed to occur gradually, and until it proceeds to completion, the vacuum still retains a large amount of energy. As we know from Section 7.3, a vacuum energy shows up in the Einstein equations in the form of a cosmological term, with nonzero cosmological constant. A positive energy density corresponds to a positive cosmological constant. The energy density associated with ordinary particles is negligible compared with the energy density of the vacuum; consequently, the Einstein equation for the inflationary development of the universe is the de Sitter equation [9.127], and the solution for $a(t)$ is Eq. [9.128],

$$a(t) = a(t_I) \exp \sqrt{\frac{\Lambda}{3}} (t - t_I) \qquad [89]$$

where t_I is the time at the beginning of inflation. Thus, the inflation of the radius of curvature is exponential.

Of course, the enormous expansion also leads to an enormous decrease of the temperature of the universe, to a temperature near zero. However, the energy released in the transition of the vacuum is changed into thermal energy, and this thermal energy is transferred to the particles in the universe, reheating them to about the same temperature they would have had without inflation.

Fig. 10.5 is a sketch of the radius of curvature and the temperature in the inflationary model as a function of time.

The inflationary model makes the definite assertion that Ω_0 is now almost exactly 1. This assertion can serve as one observational test of the model.

Another observational test of the model emerges from the study of the primordial density fluctuations that serve as seeds for the formation of the clumps of matter that ultimately grow into galaxies and clusters of galaxies. In the inflationary model, such density fluctuations in the post-inflationary era can be traced to fluctuations of whatever quantum fields existed in the de Sitter spacetime. Since the initial spectrum of these fluctuations is known from quantum theory, the post-inflationary spectrum can be calculated. The prediction of the inflationary model is that the resulting spectrum of primordial fluctuations in the mass distribution is uniform, so $\delta M/M$ is a constant, independent of the length scale λ of the fluctuation δM. This kind of

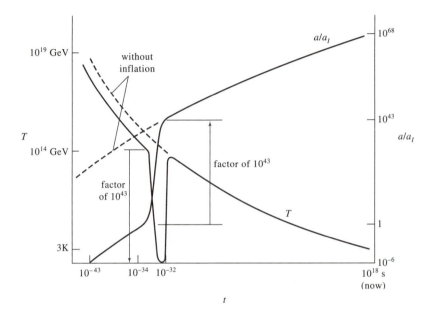

Fig. 10.5 $a(t)$ and $T(t)$ in the inflationary model.

scale-invariant spectrum is called the Harrison–Zeldovich spectrum. The primordial fluctuations in the mass distribution cause observable fluctuations in the temperature distribution of the cosmic blackbody radiation. For primordial fluctuations of short wavelength (and a small angular scale), the relationship between δM and δT is complicated, since such mass fluctuations enter the horizon at an early time, and they evolve in various ways before the time of decoupling. But for fluctuations of long wavelength, and an angular scale of more than 1^0 or 2^0, the relationship between δM and δT is simple, since such fluctuations enter their horizons at or after the time of decoupling. They then evolve only gravitationally, and they can affect the cosmic blackbody radiation only by the perturbation they produce in the gravitational potential--wherever the gravitational potential is deeper or shallower, the blackbody photons suffer a relative redshift or a blueshift, and, correspondingly, δT will be negative or positive (Sachs–Wolfe effect). Thus, the inflationary model makes an unambiguous prediction for the spectrum of the temperature fluctuations in the cosmic blackbody radiation on angular scales of 1^0 or more.

Comparison of the predicted spectrum of temperature fluctuations and the measured spectrum serves as an observational test of the inflationary model. The fluctuations found by Smoot et al. (see Section 9.5) are in good agreement with the prediction of the inflationary model.

FURTHER READING

Comprehensive and well-organized expositions of all aspects of the physics of the early universe are givne by Börner, G., *The Early Universe* (Springer Verlag, Berlin, 1988) and by Kolb, E. W., and Turner, M. S., *The Early Universe* (Addison-Wesley, Reading, 1990).

A wealth of information about a variety of topics will be found in the following references:

Kolb, E. W., and Turner, M. S., *The Early Universe: Reprints* (Addison-Wesley, Reading, 1990). (A companion volume to Kolb and Turner, op. cit., containing reprints of original papers.)

Peacock, J. A., Heavens, A. F., and Davies, A. T., *Physics of the Early Universe* (Edinburgh University Press, Edinburgh, 1990).

Unruh, W. G., and Semenoff, G. W. *The Early Universe* (Reidel, Dordrecht, 1988).

Gibbons, G. W., Hawking, S. W., and Siklos, S. T. C., eds. *The Very Early Universe* (Cambridge University Press, Cambridge, 1983).

Big-Bang nucleosynthesis is reviewed in Boesgard, A. M., and Steigman, G., Ann. Rev. Astron. Astrophys. **23**, 319 (1985); Peebles, P. J. E, Schramm, D. N., Turner, E. L., and Kron, R. G., Nature, **352**, 769 (1991); and Sato, K., and Adouze, J., *Primordial Nucleosynthesis and Evolution of Early Universe* (Kluwer, Dordrecht, 1991). The simple, rough estimate of the neutron and helium abundances in Section 10.4 is based on Peebles, P. J. E., *Physical Cosmology* (Princeton University Press, Princeton, 1971).

A historical account of the first attempts at calculations of nucleosynthesis and the first prediction of the cosmic blackbody radiation temperature is given by Alpher, R. A., and Herman, R., "Early Work on 'big-bang' cosmology and the cosmic blackbody radiation," in Bertotti, B., Balbinot, R., Bergia, S., and Messina, A., *Modern Cosmology in Retrospect* (Cambridge University Press, Cambridge, 1990). These first attempts mistakenly assumed that all elements were synthesized in the early universe; it was later recognized by Burbidge, Burbidge, Fowler, and Hoyle that heavy elements are, instead, synthesized in stars. This evolution of our ideas about nucleosynthesis is described in Wagoner, R. V., "Deciphering the nuclear ashes of the early universe: a personal perpective," in Bertotti et al., op. cit. A code for numerical calculations of nucleosynthesis was developed by Wagoner, and, with some revisions, this has become the standard method of calculation.

Galaxy formation is discussed in Peebles, P. J. E., *Principles of Physical Cosmology* (Princeton University Press, Princeton, 1993) and in Rees, M. J., "The emergence of structure in the universe: galaxy formation and dark matter," in Hawking, S. W., and Israel, W., eds., *Three Hundred Years of Gravitation* (Cambridge

University Press, Cambridge, 1987).

A clear and vivid elementary description of the evolution of the early universe intended for a "general" audience is given by Weinberg, S., *The First Three Minutes* (Basic Books, New York, 1977).

A good introduction to the particle physics required for the investigation of the early universe is provided by Collins, P. D. B., Martin, A. D., and Squires, E. J., *Particle Physics and Cosmology* (Wiley, New York, 1989). For a brief summary of the experimental and theoretical basis for the unification of forces, see Dimopoulos, S., Raby, S. A., and Wilczek, F., "Unification of Couplings," Physics Today, October, 1991.

The inflationary scenario is clearly and concisely laid out in the article "Inflation for Astronomers" by Narlikar, J. V., and Padmanabhan, T., Ann. Rev. Astron. Astrophys. **29**, 325 (1991). Other good references are:

Olive, K. A., Physics Reports, **190**, 307 (1990).

Blau, S. K., and Guth, A. H., "Inflationary cosmology," in Hawking and Israel, op. cit.

Linde, A. D., "Inflation and quantum cosmology," in Hawking and Israel, op. cit.

Linde, A. D., *Inflation and Quantum Cosmology* (Academic Press, Boston, 1990).

Linde, A. D., *Particle Physics and Inflationary Cosmology* (Harwood Academic Publishers, New York, 1990).

A radical alternative answer to the flatness and the smoothness puzzles is offered by the "anthropic principle," according to which the universe has to be more or less the way it is, since otherwise intelligent life could not arise in it, and we could not contemplate the universe and ask questions about it. Details will be found in Breuer, R., *The Anthropic Principle* (Springer-Verlag, Berlin, 1991) and in Barrow, J. D., and Tipler, F. J., *The Anthropic Cosmological Principle* (Clarendon Press, Oxford, 1986). According to Weinberg, S., Phys. Rev. Lett., **59**, 2607 (1987), the anthropic principle also sets an upper bound on the magnitude of the cosmological constant.

REFERENCES

Alpher, R. A., Bethe, H., and Gamow, G., Phys. Rev. **73**, 803 (1948).

Alpher, R. A., and Herman, R., Nature **162**, 774 (1948).

Bianconi, A., Lee, H. W., and Ruffini, R., Nuovo Cimento **105**, 1055 (1990); see also Astron. Astrophys. **241**, 343 (1991).

Brandenberger, R. H., Rev. Mod. Phys. **57**, 1 (1985).

Brandenberger, R. H., in Sanchez, N., and Zichichi, A., eds., *Proceedings of the International School of Physics "D. Chalonge"* (World Scientific, Singapore, 1992).

Davis, M., and Efstathiou, G., "N-Body Simulations of a Universe Dominated by Cold Dark Matter," in Rubin, V. C., and Coyne, G. V., eds., *Large-Scale Motions in the Universe* (Princeton University Press, Princeton, 1988).

Gamow, G., Phys. Rev. **70**, 572 (1946).

Gamow, G., Nature **162**, 680 (1948).

Guth, A. H., Phys. Rev. D **23**, 347 (1981).

Kolb, E. W., and Turner, M. S., *The Early Universe* (Addison-Wesley, Reading,

1990).

Landau, L. D., and Lifshitz, E. M., *Statistical Physics* (Addison-Wesley, Reading, 1958), p. 164.

Linde, A. D., Phys. Lett. **108B**, 389 (1982).

Peebles, P. J. E., *Physical Cosmology* (Princeton University Press, Princeton, 1971), p. 128.

Peebles, P. J. E., Schramm, D. N., Turner, E. L., and Kron, R. G., Nature **352**, 769 (1991).

Robertson, R. G. H., et al., Phys. Rev. Lett. **67**, 957 (1991). For a general review of the neutrino mass limits, see Phys. Rev. D **45**, S1 (1992).

Ruffini, R., and Song, D. J., Astron. Astrophys. **179**, 3 (1987).

Ruffini, R., Song, D. J., and Stella, L., Astron. Astrophys. **125**, 265 (1983).

Schramm, D. N., Physica Scripta T **36**, 22 (1991).

Walker, T. P., Steigman, G., Schramm, D. N., Olive, K. A., and Kang, H.-S., Astrophys. J. **376**, 51 (1991).

Weinberg, S., *Gravitation and Cosmology* (Wiley, New York, 1972).

Zeldovich, Ya. B., Astron. Astrophys. **5**, 84 (1970).

PROBLEMS

1. Assume that our universe is described by a flat Friedmann model, with $\Omega_0 = 1$, $h = 1$, $t_0 = 10^{11}$ years, and a radiation temperature $T = 2.7$ K. Under these assumptions, calculate the time at which the densities of matter and radiation were equal, and calculate the time at which the radiation attained a temperature of 3000 K (recombination).

2. Show that for the flat Friedmann universe, the horizon distance is $3t$ if the universe is matter dominated and $2t$ if the universe is radiation dominated. In our universe, what is the horizon distance now? What was it at the recombination time? What was it at the Planck time? Assume that the values of the cosmological parameters are as given in Problem 1. How do these answers change if inflation is taken into account?

3. Show that Eq. [12] has a general solution $a \propto (t + b)^{1/2}$, where b is a constant. If $b \simeq 0$ before neutrino decoupling, estimate the value of b after decoupling.

4. Derive the expression [25] for the pressure of a relativistic fermion or boson gas. (Hint: Use the familiar arguments of elementary kinetic theory to show that the pressure P is proportional to the mean value of the product of momentum, velocity, and number of particles per unit volume, $P = \frac{1}{3}\langle npv\rangle$.)

5. Derive Eqs. [36] and [39] for the densities of relativistic fermion or boson particles in a gas at temperature T.

6. If our universe has $\Omega_0 = 1$, $h = 1$, and all of the dark matter is baryons, what is the average number of baryons per unit volume? For a radiation temperature of

2.7 K, what is the average number of photons per unit volume? What is the ratio of the number of baryons to the number of photons?

7. Assume that our universe is flat, with $\Omega_0 = 1$, $h = 1$, $t_0 = 10^{11}$ years, and a radiation temperature $T = 2.7$ K. If we go back in time, to the radiation era, the electrons become relativistic and, at an even earlier time, the muons and the baryons become relativistic. Calculate the times at which the electrons, muons, and baryons become relativistic, respectively.

8. Before recombination, the plasma filling the universe is closely coupled to the radiation, and any irregularities in the mass distribution (and in the photon distribution coupled to it) tend to be smoothed out by diffusion of photons from high-density regions into low-density regions. The mean free path of a photon is $\lambda \simeq 1/n_e \sigma_T$, where $\sigma_T = 2e^2/3m_e c^2 = 6.65 \times 10^{-25}$ cm^2 is the Thomson cross section and n_e the electron density, $n_e \simeq \rho/m_p$.
(a) Show that in a time t, a photon typically diffuses over a distance $\sqrt{\lambda ct}$ (this is a random-walk problem).
(b) The matter within a sphere of radius, say, 10 times smaller than the diffusion distance will then become very smoothly distributed. Show that the mass MS within such a sphere is

$$MS \simeq \frac{10^{-3}}{\sqrt{\rho}} \times \left[\frac{m_p ct}{\sigma_T} \right]^{3/2}$$

To find the maximum mass that will be smoothed out by this diffusion process, we must set t large as possible, which means we must set $t = 10^{13}$ s, which is the decoupling time. What is the resulting mass, called the *Silk mass*?

9. (a) From thermodynamics, derive the expression $(2\pi^2 k^4/45\hbar^3)g_* T^3$ for the entropy density of a (fully) relativistic gas of bosons or fermions. Show that except for a numerical factor of the order of 1, the entropy density is k times the particle density.
(b) The entropy of our universe is mostly due to the photons and neutrinos. For $T = 2.7$ K, what is the entropy density?

10. What is the speed of pressure waves ("sound" waves) in a photon gas at temperature T?

11. Estimate the number of neutrons that decay during nucleosynthesis in the early universe before they have a chance to become trapped in deuterium and ^{4}He nuclei.

12. Estimate roughly by what factor the helium abundance would be different if instead of 4 neutrino families there were, say, 10.

13. From Fig. 10.5, estimate the lowest temperature that the universe attains during inflation, before reheating.

14. The distance to the edge of the observable universe now is about 10^{10} light-years. What was the size of this portion of the universe at the Planck time, before inflation?

Appendix

THE VARIATIONAL PRINCIPLE IN FIELD THEORY AND THE CANONICAL ENERGY-MOMENTUM TENSOR

A.1 THE LAGRANGIAN EQUATIONS FOR A SYSTEM OF PARTICLES

The Lagrangian formalism for a system of fields is a generalization of the formalism for a system of particles. We therefore begin with a short review of the latter.

Suppose that we have a system of particles which can be described by a finite set of generalized coordinates. We write these coordinates as $q_i(t)$, where i = 1, 2, 3,..., N. The Lagrangian formalism rests on the assumption that the dynamical equations of motion can be derived from Hamilton's variational principle as follows: We define the *action* I as the integral

$$I = \int_{t_1}^{t_2} L(q_i(t), \dot{q}_i(t))\, dt \qquad [1]$$

where the Lagrangian $L(q_i, \dot{q}_i)$ is some given function of the coordinates q_i and the velocities $\dot{q}_i = dq_i/dt$. The equations of motion are then obtained from the requirement that the action remain stationary for infinitesimal variations of the functions $q_i(t)$.* These variations are arbitrary except for the constraint that they vanish at the times t_1 and t_2.

---

* We will not consider the possibility that L has an *explicit* time dependence [that is, an extra time dependence in addition to the time dependence implied by $q_i(t)$ and $\dot{q}_i(t)$]. An explicit time dependence is possible only if external forces act on the system.

If the variation of the coordinate q_i at the time t is $\delta q_i(t)$, then the velocities suffer a corresponding change $\delta \dot{q}_i(t) = (d/dt)\delta q_i(t)$, and consequently the action changes by

$$\delta I = \int \sum_{i=1}^{N} \left[\frac{\partial L}{\partial q_i} \delta q_i + \frac{\partial L}{\partial \dot{q}_i} \delta \dot{q}_i \right] dt$$

$$= \int \sum_{i=1}^{N} \left[\frac{\partial L}{\partial q_i} \delta q_i + \frac{\partial L}{\partial \dot{q}_i} \frac{d}{dt} \delta q_i \right] dt \qquad [2]$$

The terms containing the time derivatives $(d/dt)\delta q_i(t)$ can be integrated by parts. Since the variations δq_i vanish at $t = t_1$ and at $t = t_2$, we obtain

$$\delta I = \int \sum_{i=1}^{N} \left[\frac{\partial L}{\partial q_i} + \frac{d}{dt} \frac{\partial L}{\partial \dot{q}_i} \right] \delta q_i \, dt \qquad [3]$$

The action will be stationary for arbitrary choices of the functions $\delta q_i(t)$ if and only if all these functions have zero coefficients, that is, if

$$\frac{d}{dt} \frac{\partial L}{\partial \dot{q}_i} - \frac{\partial L}{\partial q_i} = 0 \qquad [4]$$

These are the Euler-Lagrange equations for the system. Note that the number of these differential equations is N, the same as the number of degrees of freedom of the system.

The equations [4] can also be written as

$$\frac{d}{dt} \pi_i - \frac{\partial L}{\partial q_i} = 0 \qquad [5]$$

where

$$\pi_i = \frac{\partial L}{\partial \dot{q}_i} \qquad [6]$$

is the canonical momentum conjugate to q_i.

Note that in the integral [1] and in the Lagrange equations [4] the time t plays the role of independent variable. In Section 3.4 we derived Lagrangian equations using the proper time τ as the independent variable. Although the use of τ is more elegant, we are now concerned with the development of the Lagrangian formalism for fields, and for that purpose the use of t is more suitable.

The equations [4] have the consequence that the *Hamiltonian*,

$$H = \sum_{i=1}^{N} \left(\dot{q}_i \frac{\partial L}{\partial \dot{q}_i} \right) - L \qquad [7]$$

is a constant of the motion. The proof is trivial:

$$\frac{dH}{dt} = \sum_{i=1}^{N} \left[\ddot{q}_i \frac{\partial L}{\partial \dot{q}_i} + \dot{q}_i \frac{d}{dt} \frac{\partial L}{\partial \dot{q}_i} \right] - \sum_{i=1}^{N} \left[\frac{\partial L}{\partial q_i} \dot{q}_i + \frac{\partial L}{\partial \dot{q}_i} \ddot{q}_i \right]$$

$$= \sum_{i=1}^{N} \dot{q}_i \left[\frac{d}{dt} \frac{\partial L}{\partial \dot{q}_i} - \frac{\partial L}{\partial q_i} \right] = 0 \qquad [8]$$

By way of illustration, let us apply the above results to the very simple case of a system consisting of just one free particle. In this case it is easy to verify that the Lagrangian

$$L = - m\sqrt{1 - (\dot{x}^2 + \dot{y}^2 + \dot{z}^2)} \qquad [9]$$

gives the correct equation of motion and the correct (relativistic) momenta p_x, p_y, p_z. Furthermore, the corresponding Hamiltonian H equals the total energy (rest-mass energy plus kinetic energy) of the particle:

$$H = \frac{m}{\sqrt{1 - (\dot{x}^2 + \dot{y}^2 + \dot{z}^2)}} \qquad \cdot[10]$$

■ *Exercise 1.* Evaluate Eqs. [4] and [6] for the Lagrangian L and also derive Eq. [10]. ■

Note that the equality between H and energy hinges on making the correct choice for L. Obviously, there are many Lagrangians that will give the same equation of motion for a free particle. What distinguishes the Lagrangian [9] is that the canonical momentum coincides with the relativistic momentum.

■ *Exercise 2.* Consider the Lagrangian

$$L = \tfrac{1}{2}m(\dot{x}^2 + \dot{y}^2 + \dot{z}^2)$$

Evaluate Eqs. [4] and [6] for this L and compare with the results of Exercise 1. ∎

We will always make the choice [9] for the Lagrangian of a free particle. Furthermore, we will insist that for a system of interacting particles the Lagrangian have the form [9] in the limiting case of large separation between the particles (no interaction). We can then make the following generalization: the total Hamiltonian of any system of interacting particles and fields equals the total energy of the particles and fields. But before we justify this statement we must develop the Lagrangian formalism for fields.

The conservation law for the Hamiltonian is directly related to invariance under time translation, that is, $dH/dt = 0$ is a direct consequence of the absence of any explicit time dependence in the Lagrangian (L depends on time only implicitly, through its dependence on q_i and $\dot{q}_i$). Furthermore, momentum conservation is a direct consequence of invariance under space translations, that is, $d\pi_i/dt = 0$ if L is independent of q_i; see Eq. [5]. Thus, the conservation of energy and momentum is intimately connected with symmetry under time and space translations. The general connection between a (continuous) symmetry group of the Lagrangian and conservation laws is known as *Noether's theorem*.

A.2 THE LAGRANGIAN EQUATIONS FOR FIELDS

Suppose now that we have a "system" which consists of a field. For the sake of simplicity we will first deal with the case of a single one-component field (scalar field). Of course we are ultimately interested in the electromagnetic field (vector field) and in the gravitational field (tensor field). But the equations for these many-component fields will not be hard to discover if we first reach an understanding of the simple case of a one-component field.

A one-component field can be described by a function $\psi(\mathbf{x}, t)$ of space and time. At a given time, the function $\psi(\mathbf{x}, t)$ gives the amplitude of the field at the point $\mathbf{x}$. This amplitude plays a role analogous to the set of generalized coordinates q_i: to describe the field system we must specify $\psi(\mathbf{x}, t)$ for all $\mathbf{x}$; to describe the particle system we must specify q_i for all i. In fact, this correspondence goes beyond a mere analogy because we can describe the field approximately by giving its values at a discrete set of points in space (a procedure often used in the numerical integration of field equations). Suppose that we divide all of space into cubical cells of volume ΔV and consider only the amplitudes of the field at the center of each cube. If the coordinates of these centers are $\mathbf{x}_1$, $\mathbf{x}_2$, $\mathbf{x}_3$, ..., then we can replace the fields by a set of generalized coordinates

$$q_i = \psi(\mathbf{x}_i) \qquad i = 1, 2, 3, \ldots \tag{11}$$

In this approximation the field equations are replaced by a set of Euler-Langrange equations of the type [4]. The exact field may be regarded as the limiting case, with $\Delta V \to 0$, of such a discrete set of generalized coordinates. Note that the approximation [11] involves a discrete infinity of degrees of freedom while the exact field has a continuous infinity of degrees of freedom.

The action integral for a field can now be written down by analogy with the particle case (see Table A.1). The Lagrangian appearing in [1] of course contains a *summation* over *i*; the Lagrangian for fields will therefore contain an *integration* over *x*. We will indicate this integration explicitly by writing

$$L = \int \mathcal{L}(\psi, \partial\psi/\partial t, \partial\psi/\partial x^k) \, d^3x \tag{12}$$

where $\mathcal{L}(\psi, \partial\psi/\partial t, \partial\psi/\partial x^k)$ is the *Lagrangian density*. It is assumed that $\mathcal{L}$ is some expression constructed out of ψ and the *first* derivatives of ψ. The presence of ψ and of the time derivative $\partial\psi/\partial t$ is obvious from the analogy with the Lagrangian for particles. The presence of the space derivative $\partial\psi/\partial x^k$ is less obvious but can be understood in the context of the approximation [11]: Ignoring *y* and *z* we have

$$\frac{\partial\psi}{\partial x} \simeq \frac{q_{i+1} - q_i}{x_{i+1} - x_i} \propto q_{i+1} - q_i \tag{13}$$

where x_i is the *x* coordinate of the center of one cube and x_{i+1} that of the center of the adjacent cube. Hence the presence of $\partial\psi/\partial x$ simply means that the Lagrangian contains some "potential" energy associated with a difference in the values of adjacent generalized coordinates.

The action integral is then

$$I = \int_{t_1}^{t_2} \int \mathcal{L}(\psi, \partial\psi/\partial t, \partial\psi/\partial x^k) \, d^3x \, dt \tag{14}$$

A variation $\delta\psi(\mathbf{x}, t)$ in the field produces corresponding variations

$$\delta\left(\frac{\partial\psi}{\partial t}\right) = \frac{\partial}{\partial t}\,\delta\psi \tag{15}$$

and

$$\delta\left(\frac{\partial\psi}{\partial x^k}\right) = \frac{\partial}{\partial x^k}\,\delta\psi \qquad\qquad [16]$$

in the derivatives. This leads to the following change in the action:

$$\delta I = \int_{t_1}^{t_2}\int \left[\frac{\partial\mathcal{L}}{\partial\psi}\,\delta\psi + \frac{\partial\mathcal{L}}{\partial(\partial\psi/\partial t)}\,\delta\left(\frac{\partial\psi}{\partial t}\right) + \frac{\partial\mathcal{L}}{\partial(\partial\psi/\partial x^k)}\,\delta\left(\frac{\partial\psi}{\partial x^k}\right) \right] d^3x\,dt$$

$$= \int_{t_1}^{t_2}\int \left[\frac{\partial\mathcal{L}}{\partial\psi}\,\delta\psi + \frac{\partial\mathcal{L}}{\partial(\partial\psi/\partial t)}\,\frac{\partial}{\partial t}\delta\psi + \frac{\partial\mathcal{L}}{\partial(\partial\psi/\partial x^k)}\,\frac{\partial}{\partial x^k}\delta\psi \right] d^3x\,dt \qquad [17]$$

We integrate the term containing $(\partial/\partial t)\delta\psi$ by parts and impose the usual constraint that $\delta\psi = 0$ at $t = t_1$ and at $t = t_2$. We also integrate the terms containing $(\partial/\partial x^k)\delta\psi$ by parts and make the assumption that $\delta\psi = 0$ at the limits of the x, y, and z integrations. The result is

$$\delta I = \int_{t_1}^{t_2}\int \left[\frac{\partial\mathcal{L}}{\partial\psi} - \frac{\partial}{\partial t}\frac{\partial\mathcal{L}}{\partial(\partial\psi/\partial t)} - \frac{\partial}{\partial x^k}\frac{\partial\mathcal{L}}{\partial(\partial\psi/\partial x^k)} \right]\delta\psi\, d^3x\,dt \qquad [18]$$

Since $\delta\psi$ is an arbitrary function, the action will be stationary if and only if the term in brackets is zero. With the Einstein summation convention and the comma notation for derivatives this can be written

$$\frac{\partial}{\partial x^\mu}\frac{\partial\mathcal{L}}{\partial\psi_{,\mu}} - \frac{\partial\mathcal{L}}{\partial\psi} = 0 \qquad\qquad [19]$$

This partial differential equation is the field equation for the given Lagrangian. Note that it resembles Eq. [4], but contains both time and space derivatives (see Table A.1).

The Hamiltonian for a system of fields resembles Eq. [7], but the summation over i must be replaced by an integration over x,

$$H = \int \psi_{,0}\frac{\partial\mathcal{L}}{\partial\psi_{,0}}\,d^3x - L$$

$$= \int \left(\psi_{,0} \frac{\partial \mathscr{L}}{\partial \psi_{,0}} - L \right) d^3x \qquad [20]$$

The field equations can be used to show that $dH/dt = 0$; we will give the proof of this in the next section. Taking the conservation of H for granted, let us first inquire into the physical interpretation of this quantity. This physical interpretation has already been mentioned in the preceding section: the Hamiltonian of a closed system* is the total energy of the system.

TABLE A.1   CORRESPONDENCE BETWEEN A PARTICLE SYSTEM AND A FIELD SYSTEM

	Particle system	Field system
State of system described by	$q_i(t)$	$\psi(\mathbf{x}, t)$
Independent variables	i, t	$\mathbf{x}, t$
Lagrangian	$L = L(q_i, \dot{q}_i)$	$L = \int \mathscr{L}(\psi, \partial\psi/\partial t, \partial\psi/\partial x^k) d^3x$
Equation of motion	$\dfrac{d}{dt}\dfrac{\partial L}{\partial \dot{q}_i} - \dfrac{\partial L}{\partial q_i} = 0$	$\dfrac{\partial}{\partial x^\mu}\dfrac{\partial \mathscr{L}}{\partial \psi_{,\mu}} - \dfrac{\partial \mathscr{L}}{\partial \psi} = 0$
Hamiltonian (energy)	$H = \Sigma \dot{q}_i \dfrac{\partial L}{\partial \dot{q}_i} - L$	$H = \int \psi_{,0} \dfrac{\partial \mathscr{L}}{\partial \psi_{,0}} d^3x - L$

To understand why this is so, consider a system of interacting particles and fields. The detailed mathematical treatment of interactions is beyond the scope of our discussion. Suffice it to say that the total Lagrangian, and hence also the total Hamiltonian, of an interacting system of particles and fields must contain some terms in which the particles and field variables appear together, in the form of a product. The total Hamiltonian will then consist of a sum of terms of the types [7], and [20], and more, and this total Hamiltonian will be conserved. Suppose that initially the particles are very widely separated and the fields absent. At a later time the particles come together, collide, produce strong fields, emit radiation, etc. (We may even suppose that an equal number of particles and antiparticles were present initially and that in the end they all annihilate, leaving nothing but fields!) Since the system initially consists of free particles, we know that the value

---

* A closed system is a system not subject to any external forces.

of the Hamiltonian is initially equal to the energy (see Eq. [10]). We also know that both the Hamiltonian and the energy are conserved as the particles come together and generate fields. Hence we can conclude that the Hamiltonian and the energy are equal at all times. Thus, the identification of the Hamiltonian and the energy is justified by the conservation law.*

A.3 THE ENERGY-MOMENTUM TENSOR

We are now ready to construct the energy-momentum tensor. Since H is the energy, the integrand appearing in Eq. [20] can be regarded as the energy density,

$$t_0{}^0 = \psi_{,0} \frac{\partial \mathscr{L}}{\partial \psi_{,0}} - \mathscr{L} \qquad [21]$$

The complete energy-momentum tensor must then be such that its $t_0{}^0$ component agrees with Eq. [21]. Obviously, the quantity

$$t_\mu{}^\nu = \psi_{,\mu} \frac{\partial \mathscr{L}}{\partial \psi_{,\nu}} - \mathscr{L} \qquad [22]$$

satisfies this condition. In fact, arguments of the type given in Section 2.5 (see Exercise 2.21) show that $t_\mu{}^\nu$ is the *only* tensor that satisfies this condition. This tensor is called the *canonical energy-momentum tensor*. Next, we must prove that the conservation law

$$\frac{\partial}{\partial x^\nu} t_\mu{}^\nu = 0 \qquad [23]$$

is satisfied. The calculation is quite similar to that given in Eq. [8]:

$$\frac{\partial}{\partial x^\nu} t_\mu{}^\nu = \psi_{,\mu,\nu} \frac{\partial \mathscr{L}}{\psi_{,\nu}} + \psi_{,\mu} \frac{\partial}{\partial x^\nu} \frac{\partial \mathscr{L}}{\partial \psi_{,\nu}} - \frac{\partial \mathscr{L}}{\partial \psi} \psi_{,\mu} - \frac{\partial \mathscr{L}}{\partial \psi_{,\alpha}} \psi_{,\alpha,\mu} \qquad [24]$$

The first and last terms cancel, and the other two terms add to zero by

---

* Texts on classical mechanics often delight in pointing out examples in which the Hamiltonian differs from the energy. The reason for this apparent contradiction is that in mechanics one considers only the *mechanical energy* (energy that can be more or less directly associated with a particle), rather than the total energy. There is not always a simple connection between the mechanical Hamiltonian and the mechanical energy. Furthermore, the Hamiltonians used in mechanics often have an explicit time dependence.

virtue of the field equation [19].

The differential conservation law [23] implies the conservation of the energy (or Hamiltonian). We have

$$\frac{dH}{dt} = \frac{d}{dt} \int t_0{}^0 \, d^3x = \int \frac{\partial}{\partial t} t_0{}^0 \, d^3x$$

$$= - \int \frac{\partial}{\partial x^k} t_0{}^k \, d^3x \qquad [25]$$

Using Gauss' theorem, we can change the volume integral of the "divergence" $(\partial/\partial x^k) t_0{}^k$ into a surface integral. The surface must include all volume, that is, the surface must be at infinity. Under the assumption that $t_0{}^k$ is exactly zero beyond some large distance or at least tends to zero faster than $1/r^2$, the surface integral vanishes and therefore H is constant.

By a similar argument we can show that the total momentum

$$P_k = \int t_k{}^0 \, d^3x \qquad [26]$$

is also constant.

■ *Exercise 3.* Prove this. ■

Before we consider some examples, we must issue a warning concerning the canonical energy-momentum tensor. It can happen that $t^{\mu\nu}$ is not symmetric in μ, ν (see, for example, Eq. [36]). Since the gravitational field equations [3.48] are based on a symmetric energy-momentum tensor, this is a very serious defect.* But this defect can always be corrected by adding to the tensor $t^{\mu\nu}$ of Eq. [22] an extra term $\bar{t}^{\mu\nu}$ chosen so that the final result is symmetric. This alteration of the energy-momentum tensor is permissible provided it leads to no change in the total energy [20] and momentum [26] and to no change in the conservation law [23]. Mathematically this amounts to the requirements that $\int \bar{t}_0{}^0 \, d^3x = \int \bar{t}_k{}^0 = 0$ and $\partial_\nu \bar{t}_\mu{}^\nu$. Thus, the extra term redistributes the energy (and momentum) in space, but does not change its total value. The freedom to add extra terms to $t^{\mu\nu}$ implies an ambiguity in the energy-momentum tensor. This ambiguity is a nuisance and it can be eliminated only by introducing some extra assumptions. For example, we might introduce the assumption that the

---

* A lack of symmetry in the energy-momentum tensor also leads to troubles with the conservation of angular momentum.

energy-momentum tensor should be gauge invariant (see the case of electromagnetism discussed below).

As a simple example of our results, let us consider the case of the one-component field ψ with a Lagrangian density

$$\mathcal{L} = \tfrac{1}{2}(\psi_{,\alpha}\psi^{,\alpha} - m^2\psi^2)$$

$$= \tfrac{1}{2}(\eta^{\alpha\beta}\psi_{,\alpha}\psi_{,\beta} - m^2\psi^2) \qquad [27]$$

where m is a constant. The Lagrangian field equation is

$$\frac{\partial}{\partial x^\alpha}\eta^{\mu\alpha}\psi_{,\alpha} + m^2\psi = 0$$

that is,

$$\psi_{,\mu}{}^{,\mu} + m^2\psi = 0 \qquad [28]$$

This is a famous equation of quantum mechanics--it is the Klein-Gordon equation for a free scalar field (of course we are now regarding ψ as a purely classical field rather than a quantum field).

The corresponding canonical energy-momentum tensor is

$$t_\mu{}^\nu = \psi_{,\mu}\eta^{\nu\alpha}\psi_{,\alpha} - \delta_\mu{}^\nu\mathcal{L}$$

$$= \psi_{,\mu}\psi^{,\nu} - \tfrac{1}{2}\delta_\mu{}^\nu(\psi_{,\alpha}\psi^{,\alpha} - m^2\psi^2) \qquad [29]$$

■ *Exercise 4.* Show that the energy density is

$$t_0{}^0 = \tfrac{1}{2}(\psi_{,0})^2 + \tfrac{1}{2}(\nabla\psi)^2 + \tfrac{1}{2}m^2\psi^2 \qquad ■ \qquad [30]$$

Now that we have explored the case of the one-component field, we are ready to write down the equations for a multi-component field. It is of course obvious that if instead of the one-component field ψ we have a field with several components (such as A_μ or $h_{\mu\nu}$), then when we perform the variation of the action, each component produces a term of the type shown in the brackets of Eq. [18]. Since the different components have independent variations, each of these terms must vanish separately and we obtain as many field equations as there are components in the field. Thus, the equations for a four-component (vector) field are

$$\frac{\partial}{\partial x^\mu}\frac{\partial \mathscr{L}}{\partial A^\nu{}_{,\mu}} - \frac{\partial \mathscr{L}}{\partial A^\nu} = 0 \qquad \nu = 0, 1, 2, 3 \qquad [31]$$

and those for a sixteen-component (tensor) field are

$$\frac{\partial}{\partial x^\mu}\frac{\partial \mathscr{L}}{\partial h^{\alpha\beta}{}_{,\mu}} - \frac{\partial \mathscr{L}}{\partial h^{\alpha\beta}} = 0 \qquad \alpha, \beta = 0, 1, 2, 3 \qquad [32]$$

Equation [31] yields the field equations for a free electromagnetic field provided we take

$$\mathscr{L}_{(em)} = -\frac{1}{16\pi}(A_{\mu,\nu} - A_{\nu,\mu})(A^{\mu,\nu} - A^{\nu,\mu}) \qquad [33]$$

Equation [32] yields the linear field equations for a free gravitational field provided we take*

$$\mathscr{L}_{(1)} = \frac{1}{4}(h_{\mu\nu,\lambda}h^{\mu\nu,\lambda} - 2h_{\mu\nu}{}^{,\mu}h^{\nu\lambda}{}_{,\lambda} + 2h_{\mu\nu}{}^{,\mu}h^{,\nu} - h_{,\nu}h^{,\nu}) \qquad [34]$$

■ **Exercise 5.** Show that $\mathscr{L}_{(em)}$ leads to the field equation [3.14] with $j^\nu = 0$.
■

■ **Exercise 6.** Show that $\mathscr{L}_{(1)}$ leads to the field equation [3.48] with $T^{\mu\nu} = 0$. (Hint: When working out the partial derivatives of $\mathscr{L}_{(1)}$ with respect to $h^{\alpha\beta}{}_{,\mu}$ you may treat $h_{\alpha\beta}$ as independent of $h_{\beta\alpha}$, that is, you need not worry about the symmetry of the tensor field. The terms in the Lagrangian [34] are already arranged in such a way that they directly lead to a symmetric field equation.)   ■

The canonical energy-momentum tensors can be calculated from the appropriate generalization of Eq. [20]:

$$t_{(em)\mu}{}^\nu = A^\alpha{}_{,\mu}\frac{\partial \mathscr{L}_{(em)}}{\partial A^\alpha{}_{,\nu}} - \delta_\mu{}^\nu \mathscr{L}_{(em)} \qquad [35]$$

$$t_{(1)\mu}{}^\nu = h^{\alpha\beta}{}_{,\mu}\frac{\partial \mathscr{L}_{(1)}}{\partial h^{\alpha\beta}{}_{,\nu}} - \delta_\mu{}^\nu \mathscr{L}_{(1)} \qquad [36]$$

■ **Exercise 7.** Use [33] and [35] to show that

---

* The subscript $_{(1)}$ on $\mathscr{L}$ indicates that we are dealing with the linear approximation.

$$t_{(\text{em})\mu}{}^{\nu} = -\frac{1}{4\pi} \left[-A^{\alpha}{}_{,\mu} F^{\nu}{}_{\alpha} - \frac{1}{4} \delta_{\mu}{}^{\nu} F^{\alpha\beta} F_{\alpha\beta} \right] \qquad [37]$$

where

$$F^{\alpha\beta} = A^{\alpha,\beta} - A^{\beta,\alpha} \quad \blacksquare$$

■ **Exercise 8.** Use [34] and [36] to find an expression for $t_{(1)\mu}{}^{\nu}$. Show that if the gauge condition $h^{\mu\nu}{}_{,\nu} = \frac{1}{2} h^{,\nu}$ is inserted in your expression, then the result is

$$t_{(1)\mu}{}^{\nu} = \frac{1}{4} [2h^{\alpha\beta}{}_{,\mu} h_{\alpha\beta}{}^{,\nu} - h_{,\mu} h^{,\nu} - \delta_{\mu}{}^{\nu} (h^{\alpha\beta}{}_{,\lambda} h_{\alpha\beta}{}^{,\lambda} - \frac{1}{2} h_{,\lambda} h^{,\lambda})] \quad \blacksquare \qquad [38]$$

Eq. [38] resembles Eq. [3.62]. To see that these equations are actually identical, we need only express Eq. [38] in terms of $\phi^{\alpha\beta}$ by means of $h^{\alpha\beta} = \phi^{\alpha\beta} - \frac{1}{2}\eta^{\alpha\beta}\phi$.

Eq. [37] resembles Eq. [2.127], but is *not identical* to it. In fact, the canonical energy-momentum tensor [37] suffers from the defect mentioned previously: it is not symmetric. This defect is easily repaired by adding an extra term

$$-\frac{1}{4\pi} \frac{\partial}{\partial x^{\alpha}} (A_{\mu} F^{\nu\alpha}) \qquad [39]$$

to $t_{(\text{em})\mu}{}^{\nu}$. This extra term has a divergence which is identically zero,

$$\frac{\partial}{\partial x^{\nu}} \frac{\partial}{\partial x^{\alpha}} (A_{\mu} F^{\nu\alpha}) = 0 \qquad [40]$$

and hence has no effect on the conservation law.

■ **Exercise 9.** Prove Eq. [40]. (Hint: Note that the differential operator $\partial/\partial x^{\nu} \partial/\partial x^{\alpha}$ is symmetric in ν, α, whereas $F^{\nu\alpha}$ is antisymmetric.) ■

Furthermore, if the $\nu = 0$ components of the extra term are integrated over all volume, the result is zero,

$$\int \frac{\partial}{\partial x^{\alpha}} (A_{\mu} F^{0\alpha}) \, d^3x = 0 \qquad [41]$$

and hence the values of the energy and momentum in Eqs. [20] and [26] are not altered.

■ *Exercise 10.*   Prove Eq. [41], assuming that the fields vanish at infinity. (Hint: Use $F^{00} = 0$ and use Gauss' theorem.) ■

Since $(\partial/\partial x^\alpha)F^{\alpha\nu} = 0$ for a free electromagnetic field, we can also write the expression [39] as

$$- \frac{1}{4\pi} A_{\mu,\alpha} F^{\nu\alpha} \qquad [42]$$

which when added to the right side of Eq. [37] gives

$$\frac{1}{4\pi} \left(-A^\alpha_{,\mu} F^\nu_\alpha - \tfrac{1}{4}\delta_\mu^\nu F^{\alpha\beta} F_{\alpha\beta} \right) - \frac{1}{4\pi} A_{\mu,\alpha} F^{\nu\alpha} \qquad [43]$$

This is easily seen to be identical to [2.127].

One might ask whether the choice of the extra term in Eq. [39] is unique. Could we have used a different extra term to symmetrize the energy-momentum tensor? The choice of the extra term is determined by gauge invariance. If the energy-momentum tensor is to serve as an unambiguous source of gravitational fields, it must obviously be left invariant by the electromagnetic gauge transformation of Eq. [3.19]. The tensor [43] has this property; any other choice does not.

Incidentally: the gravitational energy-momentum tensor [38] is *not* invariant under the gravitational gauge transformation of Eq. [3.49]. But this does no harm. When we use this term as a source of gravitation in Eq. [3.64], both the right and left sides of this equation have gauge-dependent terms, but the complete equation does not--the gauge-dependent terms cancel and there is no ambiguity in the final result.

A.4   THE VARIATIONAL PRINCIPLE FOR EINSTEIN'S EQUATIONS

We have seen in the preceding section that the approximate linear field equations for gravitation can be obtained from the variational principle with the Lagrangian $\mathcal{L}_{(1)}$ of Eq. [34]. The exact nonlinear Einstein equations can also be obtained from a variational principle. The corresponding Lagrangian may be taken as

$$\mathcal{L} = \frac{1}{\kappa^2} R\sqrt{-g} \qquad [44]$$

where R is the curvature scalar and $\kappa^2 = 16\pi G/c^4$. This Lagrangian contains not only first derivatives of the fields $g_{\mu\nu}$, but also second derivatives. In general, the presence of second derivatives leads to an

Euler–Lagrange equation somewhat more complicated than [19]. However, it turns out (see below) that in the special case of the Lagrangian [44], the terms contributed by the second derivatives cancel identically. Hence, the Lagrangian [44] effectively depends only on first derivatives.

To see how the second derivatives are eliminated, we begin with the identity (see Problem 6.25)

$$R\sqrt{-g} = \frac{\partial}{\partial x^\alpha}[g^{\mu\nu}\sqrt{-g}(-\Gamma^\alpha_{\mu\nu} + \delta_\mu{}^\alpha\Gamma^\beta_{\nu\beta})]$$
$$- g^{\mu\nu}\sqrt{-g}(\Gamma^\beta_{\mu\alpha}\Gamma^\alpha_{\nu\beta} - \Gamma^\alpha_{\mu\nu}\Gamma^\beta_{\alpha\beta}) \qquad [45]$$

Second derivatives of $g_{\mu\nu}$ appear only in the first term (with square brackets) in Eq. [45]; note that this term is a divergence. The contribution of this term to the action

$$I = \frac{1}{\kappa^2}\int R\sqrt{-g}\, d^3x dt \qquad [46]$$

is of the form

$$\int \frac{\partial}{\partial x^\alpha}[\quad]\, d^3x dt \qquad [47]$$

This is the (four-dimensional) volume integral of a divergence. Hence, by integration by parts, [47] can be converted into a surface integral over the boundary of the volume. Since the variational principle assumes that the variation $\delta g_{\mu\nu}$ vanishes on the boundary (see Eqs. [17], [18]), it follows that [47] gives no contribution to the variation of the action. We can, therefore, ignore the first term on the right side of Eq. [45] and regard

$$\mathcal{L} = \frac{1}{\kappa^2} g^{\mu\nu}\sqrt{-g}(\Gamma^\beta_{\mu\alpha}\Gamma^\alpha_{\nu\beta} - \Gamma^\alpha_{\mu\nu}\Gamma^\beta_{\alpha\beta}) \qquad [48]$$

as the effective Lagrangian. This Lagrangian contains first derivatives only.

It is a straightforward, but tedious, exercise to check that the Einstein equations do indeed follow from [48]. We can save ourselves some labor by taking advantage of the linear approximation. For this purpose, we begin with the observation that the Euler–Lagrange equation will necessarily be a tensor equation. This can best be seen by writing the variation of the action in the form

$$\delta \int \frac{1}{\kappa^2} R \sqrt{-g} \, d^3x dt$$

$$= \int \left[\frac{1}{\sqrt{-g}} \left(\frac{\partial \mathscr{L}}{\partial g_{\mu\nu}} - \frac{\partial}{\partial x^\alpha} \frac{\partial \mathscr{L}}{\partial g_{\mu\nu,\alpha}} \right) \delta g_{\mu\nu} \right] \sqrt{-g} \, d^3x dt \qquad [49]$$

Since R is a scalar, the quantity in brackets on the right side must also be a scalar. But this quantity consists of the product of an (arbitrary) tensor $\delta g_{\mu\nu}$ with

$$\frac{1}{\sqrt{-g}} \left(\frac{\partial \mathscr{L}}{\partial g_{\mu\nu}} - \frac{\partial}{\partial x^\alpha} \frac{\partial \mathscr{L}}{\partial g_{\mu\nu,\alpha}} \right) \qquad [50]$$

Hence the quantity [50] must be a tensor and, therefore, the differential equation obtained by setting [50] equal to zero is a tensor equation.

We now want to check that the Einstein equations coincide with the Euler-Lagrange equations derived from Eq. [48]. Since we are dealing with *tensor* equations, we need only to check this in some special, convenient coordinates. The tensor character of the equations guarantees that if agreement obtains in some special coordinates, then it also obtains in general coordinates. As our special coordinates, we take local geodesic coordinates (at one point). In these coordinates, the field equation will then contain only second-order derivatives linearly. To single out these terms we write

$$g_{\mu\nu} = \eta_{\mu\nu} + \kappa h_{\mu\nu} \qquad [51]$$

and

$$\Gamma^\alpha{}_{\mu\nu} = \frac{\kappa}{2} \eta^{\alpha\beta} (h_{\nu\beta,\mu} + h_{\beta\mu,\nu} - h_{\mu\nu,\beta}) \qquad [52]$$

and express the Lagrangian [48] as

$$\mathscr{L} = -\frac{1}{4} \eta^{\mu\nu} [(h_\alpha{}^\beta{}_{,\mu} + h^\beta{}_{\mu,\alpha} - h_{\mu\alpha}{}^{,\beta})(h_\beta{}^\alpha{}_{,\nu} + h^\alpha{}_{\nu,\beta} - h_{\nu\beta}{}^{,\alpha})$$
$$- (h_\nu{}^\alpha{}_{,\mu} + h^\alpha{}_{\mu,\nu} - h_{\mu\nu}{}^{,\alpha})h^\beta{}_{\beta,\alpha}] + \dots \qquad [53]$$

where the dots stand for extra terms which are of no interest to us. Multiplying out the terms in the parentheses, and using the symmetry of $h_{\mu\nu}$ in μ, ν, we readily obtain

$$\mathcal{L} = \tfrac{1}{4}(h_{\alpha\beta,\mu}\,h^{\alpha\beta,\mu} - 2h_{\alpha\beta,\mu}\,h^{\mu\alpha,\beta} + 2h_{\mu\alpha}{}^{,\mu}\,h_{,\alpha} - h_{,\alpha}h^{,\alpha}) + \dots \quad [54]$$

This Lagrangian agrees with $\mathcal{L}_{(1)}$ (see Eq. [34]) except that the term $-\tfrac{1}{2}h_{\mu\nu}{}^{,\mu}\,h^{\nu\lambda}{}_{,\lambda}$ which appears in the latter has been replaced by a term $-\tfrac{1}{2}h_{\alpha\beta,\mu}\,h^{\mu\alpha,\beta}$. But it is easy to check that this replacement does not make any difference at all for the differential equations generated by these Lagrangians.

■ *Exercise 11.* Show that both the above-mentioned terms give exactly the same contribution to the Euler-Lagrange equations. ■

We have, therefore, established that in local geodesic coordinates the Euler-Lagrange equations obtained from [48] agree with the usual equations of the linear approximation. A coordinate transformation from geodesic to general coordinates then tells us that the Euler-Lagrange equations always coincide with the Einstein equations *in vacuo*. The Einstein equations in the presence of matter can of course also be obtained from a variational principle, but we will not deal with this generalization here.

A rather more elegant derivation of the Einstein equations from the Lagrangian [44] is provided by the Palatini method, according to which the metric tensor $g_{\mu\nu}$ and the Christoffel symbols $\Gamma^{\lambda}{}_{\mu\nu}$ are treated as independent variables. In terms of these variables, the Lagrangian [44] is

$$\mathcal{L} = \frac{1}{\kappa^2}\,\sqrt{-g}\,g^{\beta\alpha}\,R_{\beta\alpha}$$

$$= \frac{1}{\kappa^2}\,\sqrt{-g}\,g^{\beta\alpha}(-\Gamma^{\iota}{}_{\beta\alpha,\iota} + \Gamma^{\iota}{}_{\beta\iota,\alpha} + \Gamma^{\sigma}{}_{\beta\iota}\,\Gamma^{\iota}{}_{\sigma\beta} - \Gamma^{\sigma}{}_{\beta\alpha}\Gamma^{\iota}{}_{\sigma\iota}) \quad [55]$$

Note that this Lagrangian contains the Christoffel symbols $\Gamma^{\lambda}{}_{\mu\nu}$ and their derivatives, but it contains the metric tensor without its derivatives. The variation of the action $I = \int \mathcal{L}\,d^3x dt$ is therefore

$$\delta I = \frac{1}{\kappa^2}\int R_{\beta\alpha}\left[\frac{\partial(g^{\beta\alpha}\sqrt{-g})}{\partial g^{\mu\nu}}\right]\delta g^{\mu\nu}\,d^3x dt$$

$$+ \frac{1}{\kappa^2}\int\left[\sqrt{-g}\,g^{\beta\alpha}\frac{\partial R_{\beta\alpha}}{\partial \Gamma^{\lambda}{}_{\mu\nu}} - \frac{\partial}{\partial x^{\rho}}\left(\sqrt{-g}\,g_{\alpha\beta}\frac{\partial R_{\beta\alpha}}{\partial \Gamma^{\lambda}{}_{\mu\nu,\rho}}\right)\right]\delta\Gamma^{\lambda}{}_{\mu\nu}\,d^3x dt$$

$$[56]$$

Since the derivative of the determinant g is simply $\partial g/\partial g^{\mu\nu} = -gg_{\mu\nu}$ (see Problem 6.16), the first of these integrals reduces to

$$\frac{1}{\kappa^2} \int (R_{\mu\nu} - \tfrac{1}{2}g_{\mu\nu}R)\sqrt{-g}\; \delta g^{\mu\nu}\; d^3xdt \qquad [57]$$

and the condition that this vanish for an arbitrary variation $\delta g^{\mu\nu}$ immediately yields the Einstein equations $R_{\mu\nu} - \tfrac{1}{2}g_{\mu\nu}R = 0$.

Although we have thereby achieved our goal of deriving the Einstein equations, we must still examine the second of the integrals in Eq. [56] and verify that the equations resulting from an arbitrary (and independent) variation $\delta\Gamma^\lambda{}_{\mu\nu}$ do not lead to an inconsistency. With $R_{\beta\alpha}$ $= -\Gamma^\iota{}_{\beta\alpha,\iota} + \Gamma^\iota{}_{\beta\iota,\alpha} + \Gamma^\sigma{}_{\beta\iota}\Gamma^\iota{}_{\sigma\beta} - \Gamma^\sigma{}_{\beta\alpha}\Gamma^\iota{}_{\sigma\iota}$ it is straightforward to evaluate the term in brackets in the second of the integrals in Eq. [56] and to obtain

$$\frac{1}{\kappa^2} \int \Big[\sqrt{-g}(\Gamma^\nu{}_{\lambda\alpha}g^{\mu\alpha} + \Gamma^\mu{}_{\beta\lambda}g^{\beta\nu} - \Gamma^\alpha{}_{\lambda\alpha}g^{\mu\nu} - \delta_\lambda{}^\nu\Gamma^\mu{}_{\beta\alpha}g^{\beta\alpha})$$

$$+ \frac{\partial}{\partial x_\lambda}(g^{\mu\nu}\sqrt{-g}) - \frac{\partial}{\partial x^\rho}(g^{\mu\rho}\sqrt{-g}\delta_\lambda{}^\nu) \Big]\; \delta\Gamma^\lambda{}_{\mu\nu}\; d^3xdt \qquad [58]$$

If this is to vanish for an arbitrary variation $\delta\Gamma^\lambda{}_{\mu\nu}$, the term in brackets must be zero,

$$\sqrt{-g}(\Gamma^\nu{}_{\lambda\alpha}g^{\mu\alpha} + \Gamma^\mu{}_{\beta\lambda}g^{\beta\nu} - \Gamma^\alpha{}_{\lambda\alpha}g^{\mu\nu} - \delta_\lambda{}^\nu\Gamma^\mu{}_{\beta\alpha}g^{\beta\alpha})$$

$$+ \frac{\partial}{\partial x_\lambda}(g^{\mu\nu}\sqrt{-g}) - \frac{\partial}{\partial x^\rho}(g^{\mu\rho}\sqrt{-g}\delta_\lambda{}^\nu) = 0 \qquad [59]$$

Taking into account that $\partial\sqrt{-g}/\partial x^\lambda = \sqrt{-g}\Gamma^\iota{}_{\lambda\iota}$ (see Problem 6.18), we find that Eq. [59] reduces to an equation for the covariant derivative of the metric tensor,

$$g^{\mu\nu}{}_{;\lambda} + \delta_\lambda{}^\nu g^{\mu\rho}{}_{;\rho} = 0 \qquad [60]$$

This condition merely says that the covariant derivative of the metric tensor is zero. As we know from Section 6.4, this condition can be used to find the expression for $\Gamma^\lambda{}_{\mu\nu}$ in terms of the derivatives of the metric tensor. Hence the condition [60] merely tells us how the Christoffel symbols are related to the metric, that is, the condition [60] merely restores the dependence between the Christoffel symbols and the metric which we abandoned when we treated these quantities as independent for the purposes of the Palatini method.*

Incidentally: According to Eq. [46] we can say that the Einstein equations represent the condition for an extremum in the (four-dimensional) volume integral of the curvature. This circumstance justifies Whittaker's well-known remark: "Gravitation simply represents a continual effort of the universe to straighten itself out."

FURTHER READING

The last chapter of Goldstein, H., *Classical Mechanics* (Addison-Wesley, Reading, 1959), gives a nice introduction to the Lagrangian and Hamiltonian equations for fields with emphasis on the analogy with the corresponding equations for particles.

Wentzel, G., *Quantum Theory of Fields* (Interscience, New York, 1949), contains an excellent discussion of the Lagrangian methods in field theory. Although this book deals with quantum fields, any results that do not involve commutation relations are of course equally valid in the classical case. The canonical energy-momentum tensor is derived by Wentzel, and also by Landau, L. D., and Lifshitz, E. M., *The Classical Theory of Fields* (Addison-Wesley, Reading, 1962).

Two other good references on classical field theory are Soper, D. E., *Classical Field Theory* (Wiley, New York, 1976) and Davis, W. R., *Classical Fields, Particles, and the Theory of Relativity* (Gordon and Breach, New York, 1970).

---

* Note that if we had attempted to apply the Palatini method to the Lagrangian [48] instead of [44], we would have obtained a wrong and inconsistent set of equations. Although these Lagrangians are effectively equivalent when we treat $\Gamma^{\lambda}_{\mu\nu}$ as dependent on $g_{\mu\nu}$, they cease to be equivalent when we treat $\Gamma^{\lambda}_{\mu\nu}$ as independent of $g_{\mu\nu}$. The success of the Palatini method hinges on the right choice of Lagrangian.

ANSWERS TO EVEN-NUMBERED PROBLEMS

When a problem asks for the components of a tensor, the answer usually lists only those components that are nonzero.

Chapter 1

4. $r_0 = \hbar^2/Gm_e^2M = 2.0 \times 10^4$ cm; $E = - (GMm_e)^2 m_e/2\hbar^2 = - 9.5 \times 10^{-25}$ eV

6. 1.5×10^2 cm/yr; 6.3×10^4 s/yr

8. 1.7×10^{-3}

10. $Q^{11} = - s^2M/2$, $Q^{22} = s^2M$, $Q^{33} = - s^2M/2$

12. 1.7×10^{-3} rad; 1.7×10^{-12} rad

14. (a) 1.2×10^9 cm

16. 7.7×10^{-10} N

18. $\simeq 1 \times 10^{-10}$

20. 3×10^5 kW

22. 10^{-6} dyne·cm (according to Renner); 10^{-32}/cm², or 10^{-11}/s²

Chapter 2

2. (a) 5.9×10^{-2} cm/s; (b) $\simeq 1$ cm/s

6. $- 6.4$ cm

12. no

14. $2(v_0/a_0)\sqrt{1 - v_0^2}$, $(v_0/a_0)\sqrt{1 - 4v_0^2} + (1/2a_0) \sin^{-1}2v_0$; the first

20. $p_0 = \hbar\omega$, $\sqrt{p_x^2 + p_y^2 + p_z^2} = \hbar\omega/c$

22. (a) $T'^{\mu\nu} = [(1 - v)/(1 + v)] T^{\mu\nu}$; (b) $E = \sqrt{(1 - v)/(1 + v)}\, E$

24. (a)

$$F^{\mu\nu} = \begin{pmatrix} 0 & 0 & 0 & -2\lambda/z \\ 0 & 0 & 0 & 0 \\ 0 & 0 & 0 & 0 \\ 2\lambda/z & 0 & 0 & 0 \end{pmatrix}$$

(b)

$$F'^{\mu\nu} = \frac{1}{\sqrt{1 - v^2}} \begin{pmatrix} 0 & 0 & 0 & -2\lambda/z \\ 0 & 0 & 0 & 2\lambda v/z \\ 0 & 0 & 0 & 0 \\ 2\lambda/z & -2\lambda v/z & 0 & 0 \end{pmatrix}$$

(c) $B'_y = \dfrac{2\lambda v}{z\sqrt{1 - v^2}}$

(d) yes, $B'_y = \dfrac{2I}{z}$

26.

$$T^{\mu\nu} = \frac{1}{4\pi} \begin{pmatrix} E^2 & 0 & 0 & E^2 \\ 0 & 0 & 0 & 0 \\ 0 & 0 & 0 & 0 \\ E^2 & 0 & 0 & E^2 \end{pmatrix}$$

32. $A_{00} = -6$, $A_{01} = -2$, $A_{02} = 8$, $A_{10} = -9$, $A_{11} = -3$, $A_{12} = 12$

34. $F = \omega \cos \omega(t - z)\, dx^0 \wedge dx^1 + \omega \cos \omega(t - z)\, dx^1 \wedge dx^3$
  $*F = \omega \cos \omega(t - z)\, (dx^0 \wedge dx^2 + dx^2 \wedge dx^3)$

Chapter 3

2. $h^{\mu\nu} \rightarrow h^{\mu\nu} + \partial^\mu \Lambda^\nu + \partial^\nu \Lambda^\mu + \dfrac{1 - a}{2a - 1}\, \eta^{\mu\nu} \partial_\alpha \Lambda^\alpha$

Chapter 4

2. (a) $h_{00} = h_{11} = h_{22} = h_{33} = 2\Phi/\kappa$, with $\Phi = -GM/r$ for $r > R$ and $\Phi = GMr^2/2R^3 - 3GM/2R$ for $r < R$
(b) $-2\pi GM/3 = -0.93$ cm

4. $h_{00} = h_{11} = h_{22} = h_{33} = 2\Phi/\kappa$, with $\Phi = G(M_1 + M_2)/r$ for $r > R_2$ and $\Phi = -GM_1/r - GM_2/R_2$ for $R_1 < r < R_2$

6. (a) $h_{00} = \gamma^2(w^2 + 1)(2\Phi/\kappa)$, $h_{11} = \gamma^2(w^2 + 1)(2\Phi/\kappa)$, $h_{22} = 2\Phi/\kappa$, $h_{33} = 2\Phi/\kappa$, $h_{10} = -2w\gamma^2(2\Phi/\kappa)$ with $\Phi = 2G\lambda \ln r + [\text{constant}]$ and $\gamma = 1/\sqrt{1 - w^2}$

(b) $\dfrac{du_x}{d\tau} = 8w\gamma^2\, \dfrac{G\lambda}{r}\, \dfrac{v_y}{1 - v_y{}^2}$

8. (b) $du_x/d\tau = 16\pi G\sigma\gamma^2 wu^0 u_y$, $du_y/d\tau = -16\pi G\sigma\gamma^2 wu^0 u_x$, $du_z/d\tau = 0$
(c) helix, similar to orbit of charged particle in magnetic field

10. $dr_1/dr_2 = \sqrt{(1 - 2GM/r_1)/(1 - 2GM/r_2)}$; 1.1×10^{-14}

14. $\omega \simeq \omega_0(1 + GM/R - 3GM/2r + \Omega^2 R^2/2)$

16. $\Delta\nu/\nu = \pm 1.7 \times 10^{-6}$
18. 1.1×10^{-4} s
20. $\Delta t = 8GM/3 = 6.6 \times 10^{-5}$ s
22. $2\sqrt{2}GM/R = 6.0 \times 10^{-6}$ rad
24. $2GM/R = 1.4 \times 10^{-9}$ rad
26. 1.5 l-yr
28. (a) 3.6×10^{10} l-yr, 5.9×10^{-6} rad; (b) 4.2×10^{-6} rad

Chapter 5

4. (a)

$$t^{\mu\nu}_{(1)} = A^2\omega^2 \sin^2\omega(t - z) \begin{pmatrix} 1 & 0 & 0 & 1 \\ 0 & 0 & 0 & 0 \\ 0 & 0 & 0 & 0 \\ 1 & 0 & 0 & 1 \end{pmatrix}$$

(b)

$$t'^{\mu\nu}_{(1)} = A^2\omega^2 \sin^2\omega(t - z) \begin{pmatrix} \gamma^2 & -\gamma v & 0 & \gamma \\ -\gamma v & \gamma^2 v^2 & 0 & -\gamma v \\ 0 & 0 & 0 & 0 \\ \gamma & -\gamma v & 0 & 1 \end{pmatrix}$$

with $\gamma = 1/\sqrt{1 - v^2}$
(c) $\theta = \tan^{-1} v/\sqrt{1 - v^2}$
6. $h = $ [constant] $- (\kappa/4g)\ A\omega^2(1 - 2\ \cos^2\phi)R^2\sin^2\theta\ \cos\omega t$ (for a wave incident along the z-axis)
8. $\simeq 10^{-17}$ erg/s
10. 2.0×10^9 erg/s; $- 1.1 \times 10^{-18}$ cm/s
12. 2.4×10^{-31} erg/s

14. (a) $\dfrac{32G}{5c^5}m^2r^4\omega^6$; (b) $\dfrac{2e^2}{3c^3}r^2\omega^4$, $mv = \sqrt{\dfrac{5}{48}\dfrac{ec}{\sqrt{G}}} = 1.8 \times 10^4$ g·cm/s;

(c) 1×10^{83} s
16. (a)

$$Q^{kl} = \frac{ml^2}{2} \begin{pmatrix} 2\cos^2\omega t - \sin^2 \omega t & 3\cos\omega t \sin\omega t & 0 \\ 3\cos\omega t \sin\omega t & 2\sin^2\omega t - \cos^2\omega t & 0 \\ 0 & 0 & -1 \end{pmatrix}$$

(b) $\dfrac{72G}{45c^5}(ml^2\omega^3)^2$; (c) $\simeq 10^{-23}$ erg/s
18. type $\oplus$; 2×10^{-19} erg; no
20. $\kappa h^{11} = - \kappa h^{22} = - 3Gmv^2/r$; 1×10^{-39}
22. $a' = 4a/\pi^2$; $a' = 4a/9\pi^2$; the second

Chapter 6

2. $\Gamma^\mu_{\alpha\beta} = 0$, $R^\alpha_{\beta\mu\nu} = 0$, $R_{\beta\mu} = 0$, $R = 0$

4. $\delta B^1 = 0$, $\delta B^2 = 0$; differs by an angle of $\pi/2$

6. $A^0_{;1} = N'/2$

8. (a) $g_{11} = 1 + 4(x^1)^2$, $g_{22} = (x^1)^2$

(b) $\Gamma^1_{11} = \dfrac{4x^1}{1 + 4(x^1)^2}$, $\Gamma^1_{22} = \dfrac{-x^1}{1 + 4(x^1)^2}$, $\Gamma^2_{22} = \dfrac{1}{x^1}$

(c) $R^1_{\ 212} = \dfrac{4(x^1)^2}{1 + 4(x^1)^2}$, $R^2_{\ 112} = -4$

12. (a) $(A^0, A^1, A^2, A^3) = \left(\tfrac{1}{2}, 2, \tfrac{1}{2}, \tfrac{1}{2}\right)$; (b) yes, different worldline

14. (a) $\Gamma^1_{33} = -\sin\chi \cos\chi \sin^2\theta$; $\Gamma^3_{13} = \cot \chi$
(b) $(0, \cos(2\pi \cos\chi), 0, -\sin\chi \sin(2\pi \cos\chi))$

28. $R^1_{\ 010} = -\dfrac{\kappa}{2} A_\oplus \omega^2 \cos\omega t$, $R^2_{\ 020} = \dfrac{\kappa}{2} A_\oplus \omega^2 \cos\omega t$

Chapter 7

8. $\dfrac{4\pi}{3} \sqrt{1 - \dfrac{2GM}{r}} \left[r^3 + \dfrac{5GMr^2}{2} + \dfrac{15G^2M^2r}{2}\right] + 5G^3M^3 \tanh^{-1}\left(1 - \dfrac{2GM}{r}\right) \Bigg|_{r_1}^{r_2}$;

1.4×10^{28} cm

10. (a) $\dfrac{dr}{dt} = 1 - \dfrac{2GM}{r}$; (b) $2(r_1 - r_2) + 4GM \log \dfrac{r_1 - 2GM}{r_2 - 2GM}$

(c) $\sqrt{1 - \dfrac{2GM}{r_1}} \left[2(r_1 - r_2) + 4GM \log \dfrac{r_1 - 2GM}{r_2 - 2GM}\right]$

20. $t = 6\pi GM/\sqrt{3}$; $\tau = 6\pi GM$

24. polar: $\Omega_{LT} = \dfrac{1}{2} \dfrac{G}{r_0^3} S_E = 0.041$ arcsec/yr

equatorial: $\Omega_{LT} = -\dfrac{G}{r_0^3} S_E = -0.082$ arcsec/yr

26. 1×10^{21} rev/s

28. (a) Schwarzschild with $C = 2Gm_2$ for $r > R_2$, $C = 2Gm_1$ for $R_1 < r < R_2$

Chapter 8

2. $ds^2 = (1 - \rho^2\omega^2)dt^2 - d\rho^2 - \rho^2 d\phi^2 - dz^2 - 2\omega\rho^2 d\phi dt$

4. $48(GM)^2/r^6$

6. (b) $8r_S$

8. $ds^2 = (1 - r_S/r)d\tilde{u}^2 + 2d\tilde{u}dr - r^2d\theta^2 - r^2\sin^2\theta\,d\phi^2$

12. (b) $r = Q^2/(M - \sqrt{M^2 - Q^2/G})$

Chapter 9

4. $\dfrac{4\pi}{3}b^3\left[1 \pm \dfrac{1}{30}\dfrac{b^2}{a^2}\right]$; $\pm3.3 \times 10^{-8}$

6. (a) $2\pi a$; (b) $2\pi a^2(1 - \sqrt{1 - r^2/a^2})$; (c) $4\pi a^2$

12. 1680 km/s

14. $V = 2\pi a^3(-\eta + \sinh\eta\,\cosh\eta)$, with $\eta = 2.18$; $dV/dt = 1.9 \times 10^{23}$ l-yr^3/yr; $dm/dt = 1.5 \times 10^{47}$ g/yr

16. (a) $\dfrac{1}{\sqrt{\Lambda/3}}(e^{\sqrt{\Lambda/3}t} - 1)$; (b) no

18. (a) $\chi = 0.729$; (b) $a_0\chi = 1.82 \times 10^{10}$ l-yr; (c) $a\chi = 0.314 \times 10^{10}$ l-yr

20. (a) 3.1×10^{10} l-yr; (b) 8.8×10^{31} l-yr^3; (c) 2.7

22. $\rho_0 = 0$, $\Lambda = 1.3 \times 10^{-35}$/s^2; $H_0^{-1} = 1.1 \times 10^{10}$ yr; $t_0 = 1.4 \times 10^{10}$ yr; $a_0 = 1.5 \times 10^{10}$ l-yr

Chapter 10

2. 6.5×10^9 l-yr; $\simeq 4 \times 10^5$ l-yr; 3.2×10^{-33} cm; no change

6. 1.13×10^{-5}/cm^3; 403/cm^3; 3.6×10^7

8. $2 \times 10^{12}\ M_\odot$

10. $1/\sqrt{3}$

12. 36% instead of 33%

14. 3×10^{-47} cm

Subject Index

For alphabetical lists of authors quoted, see the reference sections that appear at the ends of the individual chapters.

FUNDAMENTAL CONSTANTS[a]

Speed of light	$c = 3.00 \times 10^{10}$ cm/sec
Planck's constant	$\hbar = 1.05 \times 10^{-27}$ erg sec
	$= 6.58 \times 10^{-22}$ MeV sec
Gravitational constant	$G = 6.67 \times 10^{-8}$ cm^3 g^{-1} sec^{-2}
	$\kappa = (16\pi G/c^4)^{1/2} = 2.04 \times 10^{-24}$ sec (cm g)$^{-1/2}$
Proton charge	$e = 4.80 \times 10^{-10}$ esu
Fine structure constant	$\alpha = e^2/\hbar c = 1/137.0$
Electron mass	$m_e = 0.911 \times 10^{-27}$ g
	$= 0.511$ MeV/c^2
Proton mass	$M_p = 1.67 \times 10^{-24}$ g
	$= 938$ MeV/c^2
Neutron mass	$M_n = M_p + 2.31 \times 10^{-27}$ g
	$= M_p + 1.29$ MeV/c^2
Compton wavelength	$\lambdabar_C = \hbar/m_e c = 3.86 \times 10^{-11}$ cm
Bohr radius	$a_0 = \hbar^2/m_e e^2 = 0.529 \times 10^{-8}$ cm
Rydberg constant	$hcR_\infty = \frac{1}{2} m_e c^2 \alpha^2 = 13.6$ eV
Boltzmann constant	$k = 1.38 \times 10^{-16}$ erg/°K
	$= 8.62 \times 10^{-5}$ eV/°K
Stefan-Boltzmann constant	$\sigma = \pi^2 k^4/60\hbar^3 c^2 = 5.67 \times 10^{-5}$ g sec^{-3} °K^{-4}

CONVERSION CONSTANTS

1 year (y)	$= 3.16 \times 10^7$ sec
1 astronomical unit (A.U.)	$= 1.50 \times 10^{13}$ cm
1 light year (ly)	$= 0.946 \times 10^{18}$ cm
1 parsec (pc)	$= 3.26$ light years
	$= 3.09 \times 10^{18}$ cm
1 second of arc (″)	$= 4.85 \times 10^{-6}$ radians
1 electron volt (eV)	$= 1.60 \times 10^{-12}$ erg

[a] The values of all constants have been rounded off to three significant figures; this is convenient and (usually) sufficient.